Lecture Notes in Physics

Volume 951

The Lecture Notes in Physics

The series Lecture Notes in Physics (LNP), founded in 1969, reports new developments in physics research and teaching-quickly and informally, but with a high quality and the explicit aim to summarize and communicate current knowledge in an accessible way. Books published in this series are conceived as bridging material between advanced graduate textbooks and the forefront of research and to serve three purposes:

- to be a compact and modern up-to-date source of reference on a well-defined topic
- to serve as an accessible introduction to the field to postgraduate students and nonspecialist researchers from related areas
- to be a source of advanced teaching material for specialized seminars, courses and schools

Both monographs and multi-author volumes will be considered for publication. Edited volumes should, however, consist of a very limited number of contributions only. Proceedings will not be considered for LNP.

Volumes published in LNP are disseminated both in print and in electronic formats, the electronic archive being available at springerlink.com. The series content is indexed, abstracted and referenced by many abstracting and information services, bibliographic networks, subscription agencies, library networks, and consortia.

Proposals should be sent to a member of the Editorial Board, or directly to the managing editor at Springer:

Christian Caron
Springer Heidelberg
Physics Editorial Department I
Tiergartenstrasse 17
69121 Heidelberg/Germany
christian.caron@springer.com

More information about this series at http://www.springer.com/series/5304

Ilarion V. Melnikov

An Introduction
to Two-Dimensional
Quantum Field Theory
with (0,2) Supersymmetry

 Springer

Ilarion V. Melnikov
Department of Physics and Astronomy
James Madison University
Harrisonburg, VA, USA

ISSN 0075-8450 ISSN 1616-6361 (electronic)
Lecture Notes in Physics
ISBN 978-3-030-05083-2 ISBN 978-3-030-05085-6 (eBook)
https://doi.org/10.1007/978-3-030-05085-6

Library of Congress Control Number: 2018964707

This Springer imprint is published by the registered company Springer Nature Switzerland AG
The registered company address is: Gewerbestrasse 11, 6330 Cham, Switzerland

To Victor Melnikov

Preface

Complex geometry has played a key role in most developments of the last 30 years in quantum field theory and string theory. This has come about not only via the analytic S-matrix but more generally through the beautiful interrelations between supersymmetry and complex geometry. Both structures introduce a great deal of rigidity compared to the more general categories of non-supersymmetric theories and real differential geometry, and this rigidity allows for general conceptual results and detailed quantitative predictions. Among the highlights in these developments we might recall the web of dualities between ten-dimensional string theories, the Seiberg-Witten solution of the low energy dynamics of $N = 2$ supersymmetric gauge theories, mirror symmetry and Seiberg duality, and, more recently, the construction and investigation of a large class of non-Lagrangian field theories associated with the (2,0) superconformal theory in six dimensions and its compactifications.

On closer examination, we find that most of these remarkable results have used relations between theories with four or more supercharges and Kähler geometry. While Kähler complex manifolds constitute the most familiar class of complex manifolds, a generic complex manifold does not admit a Kähler metric but still has a great deal more rigidity compared to a generic real manifold.

While great progress has been made in understanding theories with four or more supercharges, it is important to extend these successes as far as possible to theories with less supersymmetry. In that sense, two-dimensional quantum field theories with (0,2) supersymmetry are the "ultimate frontier" where we can still use tools from complex geometry to constrain the kinematics and dynamics of a non-trivial theory. Explorations of this frontier have been playing a growing role in modern mathematical physics.

The (0,2) theories were introduced relatively early on in string theory. Perturbative heterotic string theory involves the least supersymmetric theory of them
all: a (0,1) two-dimensional supergravity theory which describes the propagation
of a string in a ten-dimensional background. The resulting spacetime theories
are chiral and propagate non-abelian gauge fields and have been studied at great
length with the goal of providing unified models for elementary particle physics. A
critical heterotic string background with minimal four-dimensional super-Poincaré
invariance requires the internal theory to be a (0,2) superconformal field theory.
This realization led to an intensive study of such theories. A number of general
techniques were constructed for obtaining a large class of models of this sort.
The relation through compactification to supersymmetric four-dimensional theories
leads to powerful constraints on properties of the two-dimensional theories, and
insights gained from this point of view have proven useful in more general (0,2)
quantum field theories.

More recently, (0,2) theories have come to prominence as descriptions of surface
defects and low energy dynamics of solitonic strings in four-dimensional supersymmetric theories, where they provide some of the probes of a four-dimensional
theory beyond perturbation theory. In addition, such theories naturally arise in the
context of holography, as well as compactifications of the (2,0) six-dimensional
superconformal theories on four-manifolds.

There is another conceptual reason for interest in (0,2) quantum field theories:
they may be considered as models for $N = 1$ field theories in four dimensions:
some (0,2) theories exhibit confinement and supersymmetry breaking, while others
have a rich IR dynamics controlled by superconformal theories with chiral symmetries, marginal deformations, and accidental symmetries. So, one can develop
useful analogies with four-dimensional dynamics in the context of simpler two-
dimensional theories. The analogy becomes a concrete relation through compactification of $N = 1$ $d = 4$ theories to two dimensions on an appropriate background.

The purpose of these lecture notes is to introduce the reader to these fascinating
theories. The audience is assumed to have some basic background in conformal
theory, quantum field theory, and general relativity/differential geometry at the level
of Volume I of Polchinski's string theory text and some exposure to supersymmetry.
A major theme will be to point out and utilize the relations between structures
from complex geometry and field theory. To that end, we will need to introduce
a number of mathematical concepts. Our treatment of these will not be complete,
but we will strive to explain the essential results and ideas, as well as to provide
references for further study. Some of these are given in the appendix, while other
notions are developed in the main body of the text. Throughout the text the reader
will find a number of exercises. Rather than being afterthoughts tacked on to the
end of a section, the exercises are an integral part of the text and develop results or
point out subtleties that are used in subsequent developments. There has been no
deliberate attempt at obfuscation but also no attempt to have a grand unified theory
of notation: the reader should not be dismayed if the same symbol is sometimes
used for different purposes.

We will begin with a thorough examination of the basic structures of (0,2) quantum field theory and conformal field theory. While setting down the fundamentals, this will also help us to establish a set of conventions and notation that we will use in what follows. Next, we will turn to a simple class of Lagrangian theories—the (0,2) Landau-Ginzburg models—and discuss the resulting renormalization group flows, dynamics, and symmetries. We will also make contact with the more familiar (2,2) theories and compare and contrast the (0,2) and (2,2) theories. Having gotten some experience with this simplest class of models, we will examine (0,2) non-linear sigma models. These theories have a rich geometric structure and yield an important generalization of familiar Kähler geometry. They are also more delicate and exhibit anomalies that break global symmetries or even invalidate a particular theory but are particularly fascinating because of a direct connection with compactification of the heterotic string. After developing these structures, we will be in a position to appreciate the many simplifications offered by the (0,2) linear sigma model approach, which provides a unified framework for treating non-linear sigma models and Landau-Ginzburg theories. Here we will touch on the rich subject of mirror symmetry, mainly developed in the context of (2,2) theories and only recently generalized to classes of (0,2) models.

There are several glaring omissions in these notes and the following three deserve special mention. First, this is by no means a complete catalogue of every (0,2) application or result. Although we will meet many concrete examples, each illustrating either a general feature or a particular subtlety, there are many more to be found in the literature. Second, our point of view will be very much two-dimensional, so we will not discuss the many ways to obtain (0,2) theories from higher dimensional constructions. Finally, we do not present the modern localization tools that have been and can be applied with great success to these theories.

Acknowledgements

I would like to first thank my collaborators who taught me most of what I know of this beautiful subject: it has been fun, and I look forward to more of it in your brilliant company! I am especially grateful to R. Minasian, R. Plesser, S. Sethi, and S. Theisen for their continued encouragement to undertake and finish this book. Thanks also to my family who heard more (0,2) complaints than they deserve. This work was completed with support from the College of Science and Mathematics at James Madison University, the 4-VA Initiative grant "Frontiers in string geometry," the KITP Scholar program and the National Science Foundation under Grant No. NSF PHY-1748958, and the Max Planck Institute for Gravitational Physics.

Harrisonburg, VA, USA   Ilarion V. Melnikov

Contents

Chapter 1
(0,2) Fundamentals

Abstract In this chapter we introduce a number of notational conventions, describe our primary object of study—the (0,2) supersymmetry algebra, and give a Lagrangian field realization of this structure.

1.1 The Lorentz Group and Light-Cone Coordinates

Consider Minkowski space $\mathbb{R}^{1,1}$ with coordinates $x^\mu = (x^0, x^1)$ and line element

$$ds^2 = \eta_{\mu\nu}dx^\mu dx^\nu = -(dx^0)^2 + (dx^1)^2 . \qquad (1.1.1)$$

We can use this coordinate basis to write 1-forms and vector fields as

$$\omega = \omega_0 dx^0 + \omega_1 dx^1 , \qquad\qquad v = v^0\partial_0 + v^1\partial_1 , \qquad (1.1.2)$$

where we use the abbreviation $\partial_\mu = \frac{\partial}{\partial x^\mu}$.

The proper orthochronous two-dimensional Lorentz group $SO(1, 1)/\mathbb{Z}_2$ is generated by boosts, and the corresponding transformations act on vector field components as

$$\Lambda(\xi) \cdot \begin{pmatrix} v^0 \\ v^1 \end{pmatrix} = \begin{pmatrix} \cosh\xi & \sinh\xi \\ \sinh\xi & \cosh\xi \end{pmatrix} \begin{pmatrix} v^0 \\ v^1 \end{pmatrix} , \qquad (1.1.3)$$

where $\xi \in \mathbb{R}$ is the boost parameter.

© Springer Nature Switzerland AG 2019
I. V. Melnikov, *An Introduction to Two-Dimensional Quantum Field Theory
with (0,2) Supersymmetry*, Lecture Notes in Physics 951,
https://doi.org/10.1007/978-3-030-05085-6_1

1.1.1 Light-Cone Conventions

We will use light-cone coordinates $x^{\pm\pm} = x^0 \pm x^1$ and vector fields

$$\partial_{++} = \tfrac{1}{2}(\partial_0 + \partial_1) , \qquad\qquad \partial_{--} = \tfrac{1}{2}(\partial_0 - \partial_1) . \qquad (1.1.4)$$

A vector then takes the form

$$v = v^{++}\partial_{++} + v^{--}\partial_{--} , \qquad\qquad v^{\pm\pm} = (v^0 \pm v^1) . \qquad (1.1.5)$$

The dual basis to $\partial_{\pm\pm}$ is given by $dx^{\pm\pm} = (dx^0 \pm dx^1)$, and 1-forms are written as

$$\omega = \omega_{++}dx^{++} + \omega_{--}dx^{--} , \qquad\qquad \omega_{\pm\pm} = \tfrac{1}{2}(\omega_0 \pm \omega_1) . \qquad (1.1.6)$$

The line element is

$$ds^2 = -dx^{++}dx^{--} , \qquad \Longleftrightarrow \qquad v^\mu w^\nu \eta_{\mu\nu} = -\tfrac{1}{2}(v^{++}w^{--} + v^{--}w^{++}) . \qquad (1.1.7)$$

In other words, the components of the metric are $\eta_{++--} = \eta_{--++} = -1/2$. We use the metric to raise and lower indices; for example, starting with the vector v as above, we obtain a 1-form with components $v_{\mp\mp} = -\tfrac{1}{2}v^{\pm\pm}$.

1.1.2 Spinors in Two Dimensions

Since $SO(1, 1)/\mathbb{Z}_2$ is abelian, its irreducible representations are one-dimensional; this is evident for the vectors and one-forms, where the $\pm\pm$ components yield the eigenvectors:

$$\Lambda(\xi) \cdot \begin{pmatrix} v^{++} \\ v^{--} \end{pmatrix} = \begin{pmatrix} e^\xi v^{++} \\ e^{-\xi} v^{--} \end{pmatrix} , \qquad \Lambda(\xi) \cdot \left(\omega_{++}\ \ \omega_{--} \right) = \left(e^{-\xi}\omega_{++}\ \ e^\xi \omega_{--} \right) . \qquad (1.1.8)$$

There are two inequivalent irreducible real spinor representations, with components denoted by $\psi_\pm$, and transformations

$$\Lambda(\xi) \cdot \psi_+ = e^{-\xi/2}\psi_+ , \qquad\qquad \Lambda(\xi) \cdot \psi_- = e^{\xi/2}\psi_- . \qquad (1.1.9)$$

These are Majorana–Weyl spinors, and the $\psi_\pm$ transform as "square roots" of the $\omega_{\pm\pm}$. We will come back to this later, but at this point the observation is sufficient for explaining the $dx^{\pm\pm}$ notation. A little bit more generally we will say that an operator that transforms as $\Lambda(\xi) \cdot A = e^{s\xi}$, has spin s. We will only consider theories where $s \in \frac{1}{2}\mathbb{Z}$.

Exercise 1.1 The established notation allows us to write down the action for a single Majorana–Weyl fermion, an anti-commuting spinor field ψ_+:

$$S = \frac{i}{2\pi} \int d^2x \ \psi_+ \partial_{--} \psi_+ \ .$$

Verify that S is Lorentz-invariant, i.e. invariant under the transformation

$$\psi_+(x) \mapsto \Lambda(\xi) \cdot \psi_+(\Lambda^{-1}(\xi) \cdot x) = e^{-\xi/2} \psi_+(\Lambda^{-1}(\xi) \cdot x) \ .$$

Moreover, show that S is real if products of Majorana–Weyl fermions are conjugated according to

$$\overline{\chi^1 \chi^2 \cdots \chi^k} = \chi^k \chi^{k-1} \cdots \chi^1 \ .$$

□

The equation of motion for the Majorana–Weyl fermion is $\partial_{--}\psi_{++} = 0$. We will call such fields "left-moving," and those that are annihilated by ∂_{++} "right-moving."

1.2 The (0,2) Supersymmetry Algebra

Consider a Poincaré-invariant two-dimensional quantum field theory (QFT) defined on Minkowski space $\mathbb{R}^{1,1}$. By assumption the theory is endowed with a conserved 4-momentum operator P^μ, the components of which define the Hamiltonian $H = P^0$, as well as the spatial momentum $P = P^1$, and our theory possesses a (0,2) supersymmetry if there exists a conserved complex fermionic *supercharge* operator $Q_+ = Q_{1+} + i Q_{2+}$, with Q_{1+} and Q_{2+} real and independent, and its Hermitian conjugate $\overline{Q}_+ = Q_{1+} - i Q_{2+}$ satisfying

$$\{Q_+, \overline{Q}_+\} = 2(H - P) = -4P_{++} \ . \tag{1.2.1}$$

We will restrict attention to theories in which these conserved charges arise from conserved currents:

$$P^\mu = \int dx^1 T^{0\mu} \ , \qquad Q_+ = \int dx^1 S_+^0 \ , \qquad \overline{Q}_+ = \int dx^1 \overline{S}_+^0 \ , \tag{1.2.2}$$

where $T^{\mu\nu}$ is the energy-momentum tensor, and S_+^μ and its complex conjugate $\overline{S}_+^\mu$ are supercurrents. To explore these currents further, it will be convenient to introduce a superspace structure.

1.3   Minkowski (0,2) Superspace

The commuting operators H and P generate time and spatial translations. This means, essentially by definition, that for any field ϕ we have

$$[H, \phi] = i\,\partial_0\phi \, , \qquad\qquad [P, \phi] = -i\,\partial_1\phi \tag{1.3.1}$$

To similarly realize the action of the supercharges in terms of differential operators we introduce the (0,2) superspace, where the usual coordinates (x^0, x^1) are amended by a complex Grassmann coordinate θ^+, i.e. the full superspace has coordinates $(x^0, x^1; \theta^+, \overline{\theta}^+)$.[1] The fields are organized into superfields, and the most general superfield takes the form

$$\Phi(x; \theta) = \phi(x) + \theta^+ \psi_+(x) + \overline{\theta}^+ \overline{\chi}_+(x) + \theta^+ \overline{\theta}^+ \rho_{++}(x) \, ; \tag{1.3.2}$$

the fields ϕ and ρ_{++} have the same statistics, which are opposite to those of ψ_+ and $\overline{\chi}_+$. We call ϕ the lowest and ρ_{++} the highest component of the superfield Φ. All of our theories will have a fermion number operator $(-1)^F$: $F = 0 \mod 2$ for bosonic fields and $F = 1 \mod 2$ for fermionic fields. Equivalently, for any bosonic field Φ and any fermionic field Ψ

$$[(-1)^F, \Phi] = 0 \, , \qquad\qquad \{(-1)^F, \Psi\} = 0 \, . \tag{1.3.3}$$

In particular, $(-1)^F$ anticommutes with supercharges. Our superfields will always have definite statistics, and we will refer to them as bosonic or fermionic; occasionally, we will also speak of Grassmann even or Grassmann odd quantities.

We now realize the action of supersymmetry as follows. Let ξ^+ be an anticommuting supersymmetry parameter. Then the supersymmetry action is generated by the Hermitian operator

$$X_\xi = \xi^+ Q_+ - \overline{\xi}^+ \overline{Q}_+ \, , \tag{1.3.4}$$

i.e. the transformation of any field is given by $\delta_\xi \Phi = i[X_\xi, \Phi]$. On the other hand, the supersymmetry action on a superfield is realized via differential operators:

$$\delta_\xi \Phi(x; \theta) = \left(\xi^+ \mathcal{Q}_+ - \overline{\xi}^+ \overline{\mathcal{Q}}_+ \right) \Phi(x; \theta) \, , \tag{1.3.5}$$

---

[1]The Grassmann coordinates $\theta^+, \overline{\theta}^+$ transform under global Lorentz transformations as duals to the ψ_+ spinors introduced above. The geometry of supermanifolds governs the extension of this structure to local Lorentz invariance [391]. These ideas play an important role in covariant string perturbation theory.

where

$$\mathcal{Q}_+ = \frac{\partial}{\partial\theta^+} + 2i\overline{\theta}^+\partial_{++}\,, \qquad \overline{\mathcal{Q}}_+ = -\frac{\partial}{\partial\overline{\theta}^+} - 2i\theta^+\partial_{++} \qquad (1.3.6)$$

realize the supersymmetry action on superspace:

$$\mathcal{Q}_+^2 = 0\,, \qquad \overline{\mathcal{Q}}_+^2 = 0\,, \qquad \{\mathcal{Q}_+, \overline{\mathcal{Q}}_+\} = -4i\,\partial_{++}\,. \qquad (1.3.7)$$

Exercise 1.2 In this exercise we make some checks of the various definitions and gain a little practice with manipulating supercharges. First, show that our definitions are consistent with the Jacobi identity. That is, show that the following holds

$$\delta_{\xi_2}\left[\delta_{\xi_1}\Phi\right] - \delta_{\xi_1}\left[\delta_{\xi_2}\Phi\right] = [X_{\xi_1},[X_{\xi_2},\Phi]] - [X_{\xi_2},[X_{\xi_1},\Phi]] = [[X_{\xi_1}, X_{\xi_2}],\Phi]$$

by using the differential representation on the left and (1.2.1) on the right.

Second, show that if a superfield Φ has definite statistics, i.e. $(-1)^F\Phi = \pm 1$, then

$$\overline{\mathcal{Q}_+\Phi} = (-1)^{F\Phi}\overline{\mathcal{Q}}_+\overline{\Phi} \qquad (1.3.8)$$

if we use the Grassmann-reversing conjugation conventions as above.   □

1.4 Superspace Derivatives and Multiplets

There is another set of differential operators on superspace that anti-commute with $\mathcal{Q}_+$ and $\overline{\mathcal{Q}}_+$:

$$\mathcal{D}_+ = \frac{\partial}{\partial\theta^+} - 2i\overline{\theta}^+\partial_{++}\,, \qquad \overline{\mathcal{D}}_+ = -\frac{\partial}{\partial\overline{\theta}^+} + 2i\theta^+\partial_{++}\,. \qquad (1.4.1)$$

They satisfy $\mathcal{D}_+^2 = \overline{\mathcal{D}}_+^2 = 0$ and $\{\mathcal{D}_+, \overline{\mathcal{D}}_+\} = 4i\,\partial_{++}$. These supercovariant derivatives allow us to introduce a wide class of constrained superfields and hence supersymmetry multiplets. The simplest are the *chiral* superfields, which satisfy

$$\overline{\mathcal{D}}_+\Phi = 0 \qquad \Longleftrightarrow \qquad \Phi = \phi - 2i\theta^+\psi_+ - 2i\theta^+\overline{\theta}^+\partial_{++}\phi\,, \qquad (1.4.2)$$

as well as anti-chiral superfields

$$\mathcal{D}_+\overline{\Phi} = 0 \qquad \Longleftrightarrow \qquad \overline{\Phi} = \overline{\phi} - 2i\overline{\theta}^+\overline{\psi}_+ + 2i\theta^+\overline{\theta}^+\partial_{++}\overline{\phi}\,. \qquad (1.4.3)$$

Here the ψ_+ and $\overline{\psi}_+$ are components of a right-moving Weyl fermion, and the bar reflects the naive conjugation. That is, if we write ψ_+ in terms of Majorana–Weyl components as $\psi_+ = \psi_+^1 + i\psi_+^2$, then $\overline{\psi}_+ = \psi_+^1 - i\psi_+^2$.

Exercise 1.3 Derive the supersymmetry transformations for the components of chiral and anti-chiral superfields:

$$\delta_\xi \phi = -2i\xi^+ \psi_+ \,, \qquad \delta_\xi \psi_+ = 2\overline{\xi}^+ \partial_{++}\phi \,; \qquad \delta_\xi \overline{\phi} = -2i\overline{\xi}^+ \overline{\psi}_+ \,,$$

$$\delta_\xi \overline{\psi}_+ = 2\xi^+ \partial_{++}\overline{\phi} \,.$$

$\square$

We can generalize the chirality condition by setting the $\overline{\mathcal{D}}_+$ derivative of a superfield to be a chiral superfield E. Such *E-chiral* superfields will play an important role in describing left-moving degrees of freedom. Another important class of superfields is obtained by imposing a reality condition $V = \overline{V}$.

A superfield is *irreducible* if and only if it does not contain a proper sub-multiplet that is closed under supersymmetry. A reducible superfield is *decomposable* if and only if it can be written as a sum of two non-trivial irreducible superfields. Thus, a general chiral superfield is irreducible; an E-chiral superfield is reducible (because E is a non-trivial sub-multiplet) but indecomposable (because its complement is not a closed sub-multiplet); a superfield given by a sum a chiral superfield and an anti-chiral superfield is reducible and decomposable.

There is more that can be said about the structure of superspace for two-dimensional theories; we refer the reader to [376] for a wealth of details and background. The (0,2) superspace was introduced in [136].

1.5   Supersymmetric Actions and Fermi Multiplets

The superspace formalism is particularly suited to constructing supersymmetric Lagrangians. As in four dimensions, the top component of any superfield transforms as a total derivative, as does the middle component of a chiral or an antichiral superfield. Up to boundary terms a real supersymmetric action is

$$S = \frac{1}{\pi} \int d^2x \left\{ \mathcal{D}_+ \overline{\mathcal{D}}_+ \, K_{--} - \frac{i}{2}\mathcal{D}_+ W_- - \frac{i}{2}\overline{\mathcal{D}}_+ \overline{W}_- \right\}\Big|_{\theta^+ = \overline{\theta}^+ = 0}, \qquad (1.5.1)$$

where K_{--} is a bose real superfield, W_- is a fermi chiral superfield, and $\overline{W}_-$ is its conjugate. We will have occasion to write many actions of similar form in what follows, and we will often leave the evaluation of the derivatives at $\theta = \overline{\theta} = 0$ as understood.

The first term is a D-term and will contain kinetic terms and their supersymmetric completion, while the last two terms are F-term superpotential terms; they will enjoy non-renormalization theorems that follow from holomorphy, symmetries, and selection rules.

To write a supersymmetric potential interaction we introduce left-moving fermions. The simplest way to do this is with a *chiral fermi* superfield

$$\Gamma_- = \gamma_- - 2i\theta^+ G_{-+} - 2i\theta^+\overline{\theta}^+ \partial_{++}\gamma_- \tag{1.5.2}$$

and has a conjugate antichiral superfield

$$\overline{\Gamma}_- = \overline{\gamma}_- + 2i\overline{\theta}^+ \overline{G}_{-+} + 2i\theta^+\overline{\theta}^+ \partial_{++}\overline{\gamma}_- . \tag{1.5.3}$$

We can now write a superpotential of the form $W_- = \Gamma_- J(\Phi)$. To generalize, set the Γ_- to be E-chiral fermi superfields satisfying

$$\overline{\mathcal{D}}_+ \Gamma_- = -2i E(\Phi) . \tag{1.5.4}$$

This leads to the component expansion

$$\Gamma_- = \gamma_- - 2i\theta^+ G_{-+} - 2i\theta^+\overline{\theta}^+ \partial_{++}\gamma_- + 2i\overline{\theta}^+ E(\Phi) ,$$
$$\overline{\Gamma}_- = \overline{\gamma}_- + 2i\overline{\theta}^+ \overline{G}_{-+} + 2i\theta^+\overline{\theta}^+ \partial_{++}\overline{\gamma}_- - 2i\theta^+\overline{E}(\overline{\Phi}) . \tag{1.5.5}$$

Exercise 1.4 Compute the supersymmetry variations for the fermi multiplets:

$$\delta_\xi \gamma_- = -2i\xi^+ G_{-+} + 2i\overline{\xi}^+ E(\phi) ,$$
$$\delta_\xi \overline{\gamma}_- = 2i\overline{\xi}^+ \overline{G}_{-+} - 2i\xi^+ \overline{E} ,$$
$$\delta_\xi G_{-+} = -2i\overline{\xi}^+ \partial_{++}\gamma_- + 2\overline{\xi}^+ E'(\phi)\psi_+ ,$$
$$\delta_\xi \overline{G}_{-+} = -2i\xi^+ \partial_{++}\overline{\gamma}_- - 2i\xi^+ \overline{E}'(\overline{\phi})\overline{\psi}_+ .$$

□

Suppose we have M E-chiral multiplets Γ_-^A satisfying $\overline{\mathcal{D}}_+\Gamma_-^A = -2i E^A(\Phi)$, and we set the superpotential to be

$$W_- = \sum_{A=1}^{M} \Gamma_-^A J_A(\Phi) . \tag{1.5.6}$$

The action (1.5.1) will be supersymmetric if and only if W_- is an off-shell chiral superfield; equivalently, the E and J satisfy the supersymmetry constraint

$$\sum_{A=1}^{M} E^A(\Phi) J_A(\Phi) = 0 . \tag{1.5.7}$$

We will refer to this as the $E \cdot J$ *supersymmetry constraint* in what follows. There are many solutions to the constraint; a particularly simple one is $E^A = 0$ *or* $J_A = 0$ for all A.

1.6 (0,2) Yukawa Models

We now have all the ingredients to write down the most general classical 2-derivative $(0, 2)$-supersymmetric action where all fields have spin $s \leq 1/2$. In this case the bosons and their superpartners are arranged in bose chiral multiplets, while the left-moving fermions are lowest components of E-chiral fermi multiplets just discussed.

A particularly important class of theories are those with free kinetic terms for bose chiral multiplets, E-chiral fermi multiplets and their conjugates, which we will call the *(0,2) Yukawa models*. The kinetic terms are quadratic in the fields and take the form

$$K_{--} = -\frac{i}{8} \left[\overline{\Phi} \partial_{--} \Phi - \Phi \partial_{--} \overline{\Phi} \right] - \frac{1}{4} \overline{\Gamma}_- \Gamma_- . \tag{1.6.1}$$

Exercise 1.5 Compute the component expansion and verify that it leads to the Lagrangian

$$\pi \mathcal{L} = \frac{1}{2}(\partial_{++}\overline{\phi}\partial_{--}\phi + \partial_{++}\phi\partial_{--}\overline{\phi}) + i\overline{\psi}_+\partial_{--}\psi_+ + i\overline{\gamma}_-\partial_{++}\gamma_- + \overline{G}_{-+}G_{-+}$$
$$+ \overline{\psi}_+\overline{E}'\gamma_- + \overline{\gamma}_-E'\psi_+ - \overline{E}E ,$$

where $E'(\phi) = \partial E/\partial \phi$.   □

As we see from the exercise, when we set $E = 0$ we get a free massless action. More generally, we obtain an interacting theory, and we can spruce up the interactions further by introducing the potential J as above. A short computation shows that this leads to additional terms in the action

$$\pi \mathcal{L}_J = -G_{-+}J(\phi) - \overline{G}_{-+}\overline{J}(\phi) + \gamma_-J'(\phi)\psi_+ + \overline{\psi}_+\overline{J}'(\overline{\phi})\overline{\gamma}_- . \tag{1.6.2}$$

Hence, after eliminating the auxiliary G_{-+} fields by their equations of motion, we obtain a sum of three terms:

$$\pi \mathcal{L}_{\text{kin}} = \frac{1}{2}(\partial_{++}\overline{\phi}\partial_{--}\phi + \partial_{++}\phi\partial_{--}\overline{\phi}) + i\overline{\psi}_{+}\partial_{--}\psi_{+} + i\overline{\gamma}_{-}\partial_{++}\gamma_{-},$$

$$\pi \mathcal{L}_{\text{E}} = +\overline{\gamma}_{-}E'\psi_{+} + \overline{\psi}_{+}\overline{E}'\gamma_{-} - \overline{E}E,$$

$$\pi \mathcal{L}_{\text{J}} = +\gamma_{-}J'\psi_{+} + \overline{\psi}_{+}\overline{J}'\overline{\gamma}_{-} - \overline{J}J. \qquad (1.6.3)$$

We need to solve the $E \cdot J$ constraint in order for the action to be supersymmetric. With just one fermi multiplet our choices are meager: we must set either $E = 0$ or $J = 0$. The two classes of theories so obtained are equivalent by the simple relabeling $\gamma_{-} \to \overline{\gamma}_{-}, \overline{\gamma}_{-} \to \gamma_{-}$.

We presented the action for just one bose and one fermi multiplet, but it is easy to generalize it to many multiplets. The E and J couplings now carry indices as above, and field redefinitions allow us to bring the quadratic kinetic term to a canonical form (written with the summation convention for repeated indices)

$$K_{--} = -\frac{i}{8}\delta_{a\overline{b}}\left[\overline{\Phi}^{\overline{b}}\partial_{--}\Phi^{a} - \Phi^{a}\partial_{--}\overline{\Phi}^{\overline{b}}\right] - \frac{1}{4}\delta_{A\overline{B}}\overline{\Gamma}^{\overline{B}}_{-}\Gamma^{A}_{-}, \qquad (1.6.4)$$

and it is easy to add the additional index structure to the component action; for example, we have

$$\gamma_{-}J'\psi_{+} = \sum_{A,b}\gamma^{A}_{-}\frac{\partial J_{A}}{\partial \phi^{b}}\psi^{b}_{+}. \qquad (1.6.5)$$

We will keep the indices as light as possible and only write out expressions in full glory when absolutely necessary. One useful simplification we will use repeatedly is

$$J_{A,b} = \frac{\partial J_{A}}{\partial \phi^{b}}. \qquad (1.6.6)$$

Exercise 1.6 At first sight we might be tempted to consider an exciting generalization for the quadratic action:

$$\Delta K_{--} = \frac{1}{8}\tau_{AB}\Gamma^{A}_{-}\Gamma^{B}_{-} - \frac{1}{8}\tau^{*}_{AB}\overline{\Gamma}^{A}_{-}\overline{\Gamma}^{B}_{-}$$

for some constant anti-symmetric matrix τ_{AB}. Show that this is a mirage: this coupling can be absorbed into a shift $J^{\text{new}}_{B} = J^{\text{old}}_{B} + E^{A}\tau_{AB}$; the shift is consistent with the $E \cdot J$ constraint. □

This is the basic structure of a renormalizable Lagrangian (0,2) field theory. There are two important generalizations that we will discuss in detail below:

1. we will consider a non-linear kinetic term and write a (0,2) non-linear sigma model—this will lead to many important connections with heterotic geometry;
2. we can also introduce vector fields by gauging some global symmetry of our theory—this is the domain of (0,2) gauged linear sigma models.

One can also discuss the most general gauging of a symmetry on some curved target-space geometry, but that is usually too much of a good thing, and we will stick to these simpler generalizations.

1.6.1   Superspace Equations of Motion

In this section we will illustrate a convenient superspace trick for deriving the equations of motion. The trick relies on three points:

1. if $\mathcal{D}_+ \overline{\mathcal{D}}_+ [AX]_{\theta^+, \bar{\theta}^+ = 0} = 0$ for all (0,2) superfields X, then $A = 0$;
2. the variation of a bose anti-chiral field can be written as $\delta \overline{\Phi} = \mathcal{D}_+ \mathcal{X}_-$ for some general fermi (0,2) superfield $\mathcal{X}_-$;
3. the variation of a fermi E-anti-chiral field can be written as $\delta \overline{\Gamma}_- = \mathcal{D}_+ Y_{--} - 2i \overline{E}' \mathcal{X}_-$, where the second term imposes the constraint $\mathcal{D}_+ \delta \overline{\Gamma}_- = -2i \delta \overline{E}(\overline{\Phi})$.

Plugging these variations into the action and using the $E \cdot J$ constraint leads to following equations of motion for the (0,2) theory with K_{--} as in (1.6.4) and potential W_- as in (1.5.6).

$$\mathcal{D}_+ \Gamma_- + 2i \overline{J} = 0 , \qquad \partial_{--} \mathcal{D}_+ \Phi - 2\Gamma_- \overline{E}' - 2\overline{\Gamma}_- \overline{J}' = 0 . \qquad (1.6.7)$$

The lowest components of these are

$$G_{-+} - \overline{J} = 0 , \qquad i \partial_{--} \psi_+ + \overline{J}' \overline{\gamma}_- + \overline{E}' \gamma_- = 0 , \qquad (1.6.8)$$

and this matches the equations of motion from the component action.

There is another perspective on the E-couplings that involves a fermionic gauge symmetry motivated by (2,2) super-gauge invariance described in (0,2) multiplets [143]; we will not use it here.

1.7   The Supercurrent Algebra via Superspace

An important yet accessible structure in a QFT is the current algebra—the equal time commutation relations of conserved currents associated to continuous global symmetries. In the context of (0,2) QFT there are some natural questions: what is the structure of the (0,2) supercurrent algebra? what superspace multiplets encode this structure? Although aspects of this structure were discussed long ago [242], a complete study was carried out much more recently in [158], and we will now summarize the assumptions and results of that work.

1.7.1 Assumptions on the Current Algebra

The main result of [158] is that the supercurrent algebra and the supercurrent multiplet are fixed by several basic assumptions:

1. the multiplet includes the real, conserved, and symmetric energy-momentum tensor $T_{\mu\nu}$;
2. the multiplet includes the supercurrents $S_{+\mu}$ and $\overline{S}_{+\mu}$;
3. the multiplet does not have any operators with absolute value of spin greater than 2;
4. the multiplet is not decomposable;
5. the multiplet is consistent with (1.2.1).

This is worked out in detail for theories with ≤ 4 supercharges and dimensions, but we will just present the results for two-dimensional (0,2) theories.

While the first two assumptions are perhaps "obvious," it is the third that is most constraining on the form of the relevant superfields. To start, consider the energy-momentum tensor

$$T_{\mu\nu}dx^\mu dx^\nu = T_{++++}dx^{++}dx^{++} + 2T_{++--}dx^{++}dx^{--} + T_{----}dx^{--}dx^{--},$$
$$(1.7.1)$$

with

$$T_{++--} = T_{--++}, \qquad \partial_{++}T_{\pm\pm--} + \partial_{--}T_{\pm\pm++} = 0. \qquad (1.7.2)$$

The T_{----} and T_{++++} must appear as, respectively, the lowest and highest components of two distinct bose superfields. Furthermore, the lowest component of the superfield containing T_{++++} must be a spin -1 object j_{++}.
     Since

$$P_{++} = -\int dx^1\,(T_{++++} + T_{++--}), \qquad Q_+ = -\int dx^1\,(S_{+++} + S_{+--}),$$
$$(1.7.3)$$

in order to obtain (1.2.1), the current algebra must have the terms

$$\{\overline{Q}_+, S_{+++}\} = -4T_{++++} + \dots, \qquad \{\overline{Q}_+, S_{+--}\} = -4T_{++--} + \dots,$$
$$(1.7.4)$$

where the ... refer to Schwinger terms that integrate to zero without additional operator insertions. Demanding that these relations arise from the superfields as above leads to an intricate set of algebraic conditions. These conditions determine the general supercurrent multiplet [158].

1.7.2 The $\mathcal{S}$ and $\mathcal{R}$ Multiplets

The most general supercurrent multiplet, *the $\mathcal{S}$-multiplet*, consists of two real bose superfields $\mathcal{S}_{++}$ and $\mathcal{T}_{----}$, as well as a complex fermi superfield $\mathcal{W}_-$ subject to relations[2]

$$4\partial_{--}\mathcal{S}_{++} = \mathcal{D}_+\mathcal{W}_- + \overline{\mathcal{D}_+\mathcal{W}_-}\,, \quad \overline{\mathcal{D}}_+\mathcal{T}_{----} = \partial_{--}\mathcal{W}_-\,, \quad \overline{\mathcal{D}}_+\mathcal{W}_- = C\,.$$

$$(1.7.5)$$

These have the component expansions

$$\mathcal{S}_{++} = j_{++} - i\theta^+ S_{+++} - i\overline{\theta}^+ \overline{S}_{+++} - 4\theta^+\overline{\theta}^+ T_{++++}\,,$$

$$\mathcal{W}_- = -\overline{S}_{+--} - i\theta^+(4T_{++--} + 2i\partial_{--}j_{++}) - \overline{\theta}^+ C + 2i\theta^+\overline{\theta}^+ \overline{S}_{+--,++}\,,$$

$$\mathcal{T}_{----} = 2T_{----} - \theta^+\partial_{--}S_{+--} + \overline{\theta}^+\partial_{--}\overline{S}_{+--} + 2\theta^+\overline{\theta}^+\partial^2_{--}j_{++}\,,$$

$$(1.7.6)$$

Exercise 1.7 Show that the relations imply that the energy-momentum tensor and supercurrents are conserved, while C is a constant.   □

Exercise 1.8 Check that the $\mathcal{S}$-multiplet implies the following supercurrent algebra:

$$\{Q_+, \overline{S}_{+++}\} = -4T_{++++} + 2i\partial_{++}j_{++}\,,$$

$$\{Q_+, \overline{S}_{+--}\} = -4T_{++--} - 2i\partial_{--}j_{++}\,,$$

$$\{Q_+, S_{+++}\} = 0\,,$$

$$\{Q_+, S_{+--}\} = i\overline{C}\,.$$

$$(1.7.7)$$

If and only if $C = 0$, the $\mathcal{S}$-supercurrent algebra leads to

$$Q_+^2 = \overline{Q}_+^2 = 0\,, \qquad\qquad \{Q_+, \overline{Q}_+\} = -4P_{++}\,. \qquad\qquad (1.7.8)$$

□

---

[2]We have slightly different conventions for the light-cone and conversions of bispinors from [158] that account for the different factors here and in the algebra below.

While any (0,2) QFT must admit an $\mathcal{S}$-multiplet, some admit a more restricted structure. An important simplification occurs when $\mathcal{W}_- = i\overline{\mathcal{D}}_+\mathcal{R}_{--}$ for some real operator $\mathcal{R}_{--}$. This requires $C = 0$ and, after relabeling $\mathcal{S}_{++} \to \mathcal{R}_{++}$, yields the $\mathcal{R}$-multiplet:

$$\partial_{--}\mathcal{R}_{++} + \partial_{++}\mathcal{R}_{--} = 0, \qquad \overline{\mathcal{D}}_+(\mathcal{T}_{----} - i\partial_{--}\mathcal{R}_{--}) = 0, \qquad (1.7.9)$$

with component expansions

$$\mathcal{R}_{\pm\pm} = j_{\pm\pm} - i\theta^+ S_{+\pm\pm} - i\overline{\theta}^+ \overline{S}_{+\pm\pm} - 4\theta^+\overline{\theta}^+ T_{++\pm\pm}. \qquad (1.7.10)$$

An $\mathcal{R}$-multiplet requires that the theory has a conserved R-current with components $j_{\pm\pm}$. The resulting supercurrent algebra is

$$\{Q_+, \overline{S}_{+\pm\pm}\} = -4T_{++\pm\pm} + 2i\partial_{++}j_{\pm\pm}, \quad \{Q_+, S_{+\pm\pm}\} = 0,$$

$$[Q_+, j_{\pm\pm}] = S_{+\pm\pm}, \qquad\qquad\qquad [\overline{Q}_+, j_{\pm\pm}] = -\overline{S}_{+\pm\pm}. \qquad (1.7.11)$$

The conserved R-charge is

$$R = -\int dx^1 \, (j_{++} + j_{--}), \qquad (1.7.12)$$

the $\mathcal{R}$-supercurrent algebra leads to

$$\{Q_+, \overline{Q}_+\} = -4P_{++}, \quad [R, Q_+] = -Q_+, \quad [R, \overline{Q}_+] = +\overline{Q}_+. \qquad (1.7.13)$$

An $\mathcal{R}$ multiplet exists in most of the (0,2) theories we will discuss.

1.7.3   A Superconformal Theory

A larger symmetry structure emerges when $\mathcal{W}_- = \partial_{--}\overline{\mathcal{D}}_+U$ for some real operator U.

Exercise 1.9 Show that given a solution to (1.7.5) and a real operator U there exists an improvement

$$\mathcal{W}_-^{\text{new}} = \mathcal{W}_- - \partial_{--}\overline{\mathcal{D}}_+U, \qquad \mathcal{S}_{++}^{\text{new}} = \mathcal{S}_{++} - \tfrac{1}{4}[\mathcal{D}_+, \overline{\mathcal{D}}_+]U,$$

$$\mathcal{T}_{----}^{\text{new}} = \mathcal{T}_{----} - \partial_{--}^2 U$$

such that the "new" operators also satisfy (1.7.5). Thus, whenever $\mathcal{W}_- = -\partial_{--}\overline{\mathcal{D}}_+U$, we can obtain a new multiplet with $\mathcal{W}_-^{\text{new}} = 0$.   □

When we are able to set $\mathcal{W}_- = 0$ by an improvement transformation the theory is in fact superconformal. That is, $T_{++--} = 0$, so that the energy-momentum tensor is traceless, T_{----} is invariant under (0,2) supersymmetry, and $\mathcal{S}_{++}$ is a right-moving multiplet: $\partial_{--}\mathcal{S}_{++} = 0$.

1.7.4 A Few Comments on the Supermultiplets

The structure of the supercurrent multiplets can be used to make many exact statements about renormalization group (RG) flow of a theory. A number of these are discussed in [158], and it is pointed out that one must keep a few basic facts in mind. Consider an RG flow obtained by deforming a conformal theory by a relevant operator. In this case, the extreme UV theory has a superconformal multiplet, and the relevant deformation causes it to mix with other operators, leading to a larger multiplet. However, once we determine the multiplet structure at a large but finite cut-off, it must persist at all non-zero energy scales. It is possible that in the extreme IR we find another superconformal multiplet: that is an example of what we will call an "accident." When the RG flow is accident free, we can learn about the extreme IR theory from the possibly simpler UV theory.

The form of the supercurrent algebra should not to be confused with properties of the vacuum. Consider, for instance, a theory with a UV supercurrent $\mathcal{S}$-multiplet with the constant $C = 0$. In this case it is sensible to ask whether (0,2) supersymmetry is spontaneously broken by the vacuum, i.e. whether the ground state is annihilated by $\boldsymbol{P}_{++}$. As pointed out in [379] in a more general context, the supersymmetry algebra implies that either all or none of the supercharges will be spontaneously broken.

On the other hand, if we find that the UV theory has an $\mathcal{S}$-multiplet with $C \neq 0$, then we say that the supercurrent algebra is *deformed* from its standard form. It is a simple exercise to show that if we set

$$\Sigma_{+\pm\pm} = e^{i\alpha}\mathcal{S}_{+\pm\pm} + e^{-i\alpha}\overline{\mathcal{S}}_{+\pm\pm} \,, \tag{1.7.14}$$

and denote by $\boldsymbol{Q}'_+ = e^{i\alpha}\boldsymbol{Q}_+ + e^{-i\alpha}\overline{\boldsymbol{Q}}_+$, then we can choose a phase α so that the real supercurrent Σ belongs to a standard (0,1) current algebra

$$\{\boldsymbol{Q}'_+, \Sigma_{+\pm\pm}\} = -4T_{++\pm\pm} \,. \tag{1.7.15}$$

It is then sensible to ask whether $\boldsymbol{Q}'_+$ is spontaneously broken—a difficult question since we no longer have the nice tools of holomorphy at our disposal.[3]

---

[3]The case when $\boldsymbol{Q}'_+$ is not spontaneously broken is what is sometimes referred to as "partial supersymmetry breaking." We will not use this terminology.

1.7.5 The Supercurrent Algebra of the Yukawa Models

The Yukawa model constructed above provides a simple realization of these multiplets. Let

$$\mathcal{S}_{++} = \frac{1}{4}\overline{\mathcal{D}}_+\overline{\Phi}\mathcal{D}_+\Phi \, ,$$

$$\mathcal{W}_- = -2(E\overline{\Gamma}_- + \Gamma_- J) \, ,$$

$$\mathcal{T}_{----} = -2\partial_{--}\Phi\partial_{--}\overline{\Phi} - i\left(\overline{\Gamma}_-\partial_{--}\Gamma_- + \Gamma_-\partial_{--}\overline{\Gamma}_-\right) \, . \qquad (1.7.16)$$

Exercise 1.10 Use the $E \cdot J$ constraint and the superspace equations of motion from (1.6.7) to show that these components satisfy the $\mathcal{S}$-multiplet relations of (1.7.5). □

If we set $E = 0$, then up to the equations of motion we have

$$\mathcal{W}_- = i\overline{\mathcal{D}}_+\mathcal{R}_{--} \, , \qquad\qquad \mathcal{R}_{--} = \Gamma_-\overline{\Gamma}_- \, . \qquad (1.7.17)$$

This is sensible since the theory has an R-symmetry that assigns charge $+1$ to θ^+ and Γ^- and leaves the bose chiral superfields invariant. This determines an $\mathcal{R}$-multiplet.

Finally, if we set $E = J = 0$, we obtain a free superconformal field theory with a superconformal multiplet.

1.8 Euclidean Worldsheet

We will most often find it convenient to work on a Euclidean world-sheet Σ ; typically this will be $\mathbb{R}^2 \simeq \mathbb{C}$, but for some applications we will find it convenient to compactify this to a sphere $S^2 \simeq \mathbb{P}^1$, and sometimes we will discuss the theory on the cylinder $\mathbb{R} \times S^1$ or the torus T^2.

Starting with the Minkowski theory, we define the Euclidean theory on $\mathbb{R}^2$ with coordinates (y^1, y^2) by analytic continuation $x^0 = -iy^1$ and a convenient relabeling $y^2 = -x^1$. We define a holomorphic coordinate $z = y^1 + iy^2$, so that the analytic continuation leads to

$$x^{++} = -iy^1 - y^2 = -i\overline{z} \, , \qquad\qquad x^{--} = -iy^1 + y^2 = -iz \, ,$$

$$\partial_{++} = i\frac{\partial}{\partial\overline{z}} \, , \qquad\qquad \partial_{--} = i\frac{\partial}{\partial z} \, . \qquad (1.8.1)$$

We will often abbreviate the derivatives $\partial/\partial\overline{z}$ and $\partial/\partial z$ as $\overline{\partial}$ and ∂. We will also use, as in [328], the conventions that the integration measure on $\mathbb{R}^2$ is $d^2z = idz \wedge d\overline{z} =$

$2dy^1dy^2$, while the Dirac delta function is defined as $\delta^2(z, \bar{z}) = \frac{1}{2}\delta(y^1)\delta(y^2)$. This implies the useful identity

$$\partial_z \bar{z}^{-1} = 2\pi\delta^2(z, \bar{z}) \, . \tag{1.8.2}$$

With these conventions, given a theory defined a Minkowski action $S_M = \int dx^0 dx^1 \mathcal{L}_M$, the corresponding Euclidean action $S_E = \int dy^1 dy^2 \mathcal{L}_E$ is obtained via

$$\mathcal{L}_E = -\mathcal{L}_M[\partial_{++} = i\bar{\partial}; \ \partial_{--} = i\partial] \, . \tag{1.8.3}$$

1.8.1 Euclidean Fermions

The Euclidean Lorentz group is SO(2) = U(1), and its action on various representations is obtained by replacing $\xi \to i\xi$ in (1.1.8) and similar expressions above. In particular, a 1-form $\omega = \omega_z dz + \omega_{\bar{z}} d\bar{z}$ has components that transform as

$$\Lambda(\xi) \cdot \left(\omega_{\bar{z}} \ \omega_z\right) = \left(e^{-i\xi}\omega_{\bar{z}} \ e^{i\xi}\omega_z\right) \, . \tag{1.8.4}$$

Similarly, irreducible spinor representations are the two 1-dimensional Weyl representations,

$$\Lambda(\xi) \cdot \psi_+ = e^{-i\xi/2}\psi_+ \, , \qquad\qquad \Lambda(\xi) \cdot \psi_- = e^{i\xi/2}\psi_- \, . \tag{1.8.5}$$

Clearly, unlike in two-dimensional Minkowski space, it is not possible to define a Majorana–Weyl spinor. This sometimes leads to some unnecessary confusion, so let us be clear: it is perfectly sensible to discuss the Euclidean version of a Minkowski theory of a Majorana–Weyl fermion. Starting with the action in Exercise 1.1, and performing the continuation we obtain the Euclidean action

$$S_E = \frac{1}{2\pi} \int d^2z \ \psi_+ \partial_z \psi_+ \, . \tag{1.8.6}$$

S_E does not have any obvious reality properties, but this is no cause for alarm, precisely because it is a Euclidean continuation of a unitary Minkowski theory. A related point is that when we continue a Weyl fermion with components $(\psi, \bar{\psi})$, then in the Euclidean theory we must treat the two components as independent degrees of freedom.

1.8.2 Superspace

Finally, we give our conventions for Euclidean superspace. Now θ^+ and $\bar\theta^+$ should be treated as independent Grassmann variables, and we define

$$\theta^+ = \frac{i}{\sqrt{2}}\theta \,, \qquad\qquad \bar\theta^+ = -\frac{i}{\sqrt{2}}\bar\theta \,, \qquad (1.8.7)$$

so that

$$\mathcal{D}_+ = -i\sqrt{2}\mathcal{D} \,, \quad \overline{\mathcal{D}}_+ = -i\sqrt{2}\overline{\mathcal{D}} \,, \quad \mathcal{Q}_+ = -i\sqrt{2}\mathcal{Q} \,, \quad \overline{\mathcal{Q}}_+ = -i\sqrt{2}\overline{\mathcal{Q}}$$
$$(1.8.8)$$

with

$$\mathcal{D} = \partial_\theta + \bar\theta\bar\partial \,, \qquad\qquad \mathcal{Q} = \partial_\theta - \bar\theta\bar\partial \,,$$

$$\overline{\mathcal{D}} = \partial_{\bar\theta} + \theta\bar\partial \,, \qquad\qquad \overline{\mathcal{Q}} = \partial_{\bar\theta} - \theta\bar\partial \,. \qquad (1.8.9)$$

The non-trivial anti-commutators for these are $\{\mathcal{D}, \overline{\mathcal{D}}\} = 2\bar\partial$ and $\{\mathcal{Q}, \overline{\mathcal{Q}}\} = -2\bar\partial$.

Moving on to the superfields, we will make the $\theta, \bar\theta$ substitutions as above, and we will also drop the $\pm$ on the fermions to stream-line notation. This leads to the Euclidean superfields

$$\Phi = \phi + \sqrt{2}\theta\psi + \theta\bar\theta\bar\partial\phi \,, \quad \Gamma = \gamma + \sqrt{2}\theta G + \theta\bar\theta\bar\partial\gamma + \sqrt{2}\bar\theta E(\Phi) \,,$$

$$\overline{\Phi} = \bar\phi - \sqrt{2}\bar\theta\bar\psi - \theta\bar\theta\bar\partial\bar\phi \,, \quad \overline{\Gamma} = \bar\gamma + \sqrt{2}\bar\theta\overline{G} - \theta\bar\theta\bar\partial\bar\gamma + \sqrt{2}\theta\overline{E}(\overline{\Phi}) \,. \qquad (1.8.10)$$

These satisfy $\overline{\mathcal{D}}\Phi = 0$ and $\overline{\mathcal{D}}\Gamma = \sqrt{2}E$, as well as $\mathcal{D}\overline{\Phi} = 0$ and $\mathcal{D}\overline{\Gamma} = \sqrt{2}\overline{E}$. As we mentioned above, the Euclidean fermions ψ and $\bar\psi$, as well as γ and $\bar\gamma$, are now to be treated as independent variables.[4] However, the continued theory still has a charge conjugation operator $\mathcal{C}$ which acts as follows: it complex-conjugates the Grassmann-even fields, reverses the order of the Grassmann-odd fields in a product, and sends

$$\mathcal{C}: \quad \theta \mapsto -i\bar\theta \,, \quad \bar\theta \mapsto -i\theta \,, \quad \psi \mapsto +i\bar\psi \,, \quad \bar\psi \mapsto +i\psi \,, \quad \gamma \mapsto -i\bar\gamma \,,$$
$$\bar\gamma \mapsto -i\gamma \,. \qquad (1.8.11)$$

Clearly $\mathcal{C}^2 = 1$, $\mathcal{C}(\Phi) = \overline{\Phi}$, and $\mathcal{C}(\Gamma) = -i\overline{\Gamma}$.

---

[4]The bars now do double-duty: they distinguish the components of the Weyl fermions, and they distinguish the world-sheet complex coordinates.

Exercise 1.11 Show that the following conjugation relations hold:

$$\mathcal{C}\overline{\mathcal{D}}\mathcal{C} = -i\mathcal{D}(-1)^F \,, \qquad \mathcal{C}\mathcal{D}\mathcal{C} = -i\overline{\mathcal{D}}(-1)^F \,, \qquad \mathcal{C}\mathcal{D}\overline{\mathcal{D}} = \overline{\mathcal{D}}\mathcal{D}\mathcal{C} \,.$$

Show that these also hold with $\mathcal{D}, \overline{\mathcal{D}} \to \mathcal{Q}, \overline{\mathcal{Q}}$.   □

1.8.3   Euclidean Yukawa Theory

We can now easily write down the Euclidean continuation of the Yukawa theory. With our substitutions as above, we set

$$K_z = -K_{--} = \frac{1}{4}\left[\overline{\Phi}\partial\Phi - \Phi\partial\overline{\Phi}\right] - \frac{1}{2}\overline{\Gamma}\Gamma \tag{1.8.12}$$

and obtain

$$S_{\text{Euc}} = \frac{1}{2\pi}\int d^2z \left\{ \mathcal{D}\overline{\mathcal{D}}[K_z] + \frac{1}{\sqrt{2}}\mathcal{D}(\Gamma J) + \frac{1}{\sqrt{2}}\overline{\mathcal{D}}(\overline{\Gamma J}) \right\}\Bigg|_{\theta=\overline{\theta}=0} \,. \tag{1.8.13}$$

Using $\mathcal{C}(K_z) = -K_z$ and the properties of $\mathcal{C}$ derived above, it is easy to see that S_{Euc} is $\mathcal{C}$-invariant. The action has the component expansion

$$S_{\text{Euc}} = \frac{1}{2\pi}\int d^2z \left\{ \overline{\partial}\overline{\phi}\partial\phi + \overline{\psi}\partial\psi + \overline{\gamma}\overline{\partial}\gamma \right.$$

$$\left. -\gamma J'\psi - \overline{\psi}\overline{J}'\overline{\gamma} - \overline{\gamma}E'\psi - \overline{\psi}\overline{E}'\gamma + E\overline{E} + J\overline{J} \right\} \,. \tag{1.8.14}$$

Exercise 1.12 Derive the superspace equations of motion that follow from S_{Euc}:

$$\mathcal{D}\Gamma - \sqrt{2}\overline{J} = 0 \,, \qquad \partial\mathcal{D}\Phi - \sqrt{2}\overline{J}'\overline{\Gamma} - \sqrt{2}\overline{E}'\Gamma = 0 \,.$$

Next, evaluate the supersymmetry variations. Define

$$\delta_\xi = \frac{1}{\sqrt{2}}\left[\xi\mathcal{Q} - \overline{\xi}\,\overline{\mathcal{Q}}\right] \,, \tag{1.8.15}$$

and compute

$$\delta_\xi\phi = \xi\psi \,, \qquad\qquad\qquad \delta_\xi\psi = -\overline{\xi}\,\overline{\partial}\phi \,,$$

$$\delta_\xi\overline{\phi} = \overline{\xi}\,\overline{\psi} \,, \qquad\qquad\qquad \delta_\xi\overline{\psi} = -\xi\partial\overline{\phi} \,,$$

$$\delta_\xi\gamma = \xi G - \overline{\xi}E(\phi) \,, \qquad\qquad \delta_\xi G = \overline{\xi}E'(\phi)\psi - \overline{\xi}\partial\gamma \,,$$

$$\delta_\xi\overline{\gamma} = -\overline{\xi}\,\overline{G} + \xi\overline{E}(\overline{\phi}) \,, \qquad\qquad \delta_\xi\overline{G} = \xi\overline{E}'(\overline{\phi})\overline{\psi} + \xi\partial\overline{\gamma} \,. \tag{1.8.16}$$

Although the supersymmetry variations depend on the holomorphic function E, the algebra closes without the use of equations of motion. Finally, using $\mathcal{C}\mathcal{Q}\mathcal{C} = -i\,\overline{\mathcal{Q}}(-1)^F$ show that $[\mathcal{C}, \delta_\xi] = 0$ provided we set $\mathcal{C}(\xi) = i\bar{\xi}$ and $\mathcal{C}(\bar{\xi}) = i\xi$. $\qquad\square$

Let Q and $\overline{Q}$ be the supercharge operators defined the action on any field X via

$$[\bar{\xi}\,\overline{Q} - \xi\,Q, X] = \delta_\xi X \,. \tag{1.8.17}$$

We introduce a useful shorthand: $Q \cdot X = [Q, X]$ for a bosonic operators and $Q \cdot X = \{Q, X\}$ for a fermionic operator (and similarly for the action of $\overline{Q}$). Using the results of the exercise we obtain a useful table:

$$
\begin{aligned}
&\overline{Q} \cdot \phi = 0\,, && Q \cdot \phi = -\psi\,, && \overline{Q} \cdot \gamma = -E(\phi)\,, && Q \cdot \gamma = -G\,,\\
&\overline{Q} \cdot \bar{\phi} = \bar{\psi}\,, && Q \cdot \bar{\phi} = 0\,, && \overline{Q} \cdot \bar{\gamma} = -\overline{G}\,, && Q \cdot \bar{\gamma} = -\overline{E}(\bar{\phi})\,,\\
&\overline{Q} \cdot \psi = -\bar{\partial}\phi\,, && Q \cdot \psi = 0\,, && \overline{Q} \cdot G = E'(\phi)\psi - \partial\gamma\,, && Q \cdot G = 0\,,\\
&\overline{Q} \cdot \bar{\psi} = 0\,, && Q \cdot \bar{\psi} = \partial\bar{\phi}\,, && \overline{Q} \cdot \overline{G} = 0\,, && Q \cdot \overline{G} = -\overline{E}'(\bar{\phi})\bar{\psi}\\
& && && && \qquad\quad -\bar{\partial}\bar{\gamma}\,.
\end{aligned}
$$

$$\tag{1.8.18}$$

It is easy to see explicitly that $Q^2 = 0$, $\overline{Q}^2 = 0$, and $\{Q, \overline{Q}\} = \bar{\partial}$.

Chapter 2
Conformalities

Abstract In this chapter we give an overview of two-dimensional conformal field theories and properties of the N=2 superconformal algebra and discuss its representations. We also discuss additional global symmetries and constraints from unitarity and compactness. These are probably familiar to many readers, but we introduce them here as a reminder and for later reference; the emphasis is on the results and perspective most relevant for (0,2) exploration. The author's favorite introduction to the subject is reference Ginsparg (Applied conformal field theory. http://arxiv.org/abs/hep-th/9108028). We also give an elementary discussion of conformal perturbation theory. This notion is at the heart of much of what we discuss in the rest of the book.

2.1 The Basics

2.1.1 The Conformal Group in Two Dimensions

The global conformal group in two dimensions is $SO(3, 1) = PSL(2, \mathbb{C})$. Its algebra is $\mathfrak{sl}_2\mathbb{C}$, and it is generated by the vector fields[1]

$$v_{-1} = -\frac{\partial}{\partial z}, \qquad v_0 = -z\frac{\partial}{\partial z}, \qquad v_1 = -z^2\frac{\partial}{\partial z}. \qquad (2.1.1)$$

These infinitesimal transformations are realized on the operators and states of the QFT by the operators L_{-1}, L_0, and L_1, which are, respectively, the infinitesimal generators of translations, dilatations, and special conformal transformations that satisfy the algebra

$$[L_1, L_{-1}] = 2L_0, \qquad [L_0, L_{\pm 1}] = \mp L_{\pm 1}. \qquad (2.1.2)$$

---

[1] This is the action of the generators on holomorphic functions; more generally, the vector fields are $v = a_{-1}v_{-1} + a_0v_0 + a_1v_1 + \text{c.c.}$ for $a_{0, \pm 1} \in \mathbb{C}$.

© Springer Nature Switzerland AG 2019

I. V. Melnikov, *An Introduction to Two-Dimensional Quantum Field Theory with (0,2) Supersymmetry*, Lecture Notes in Physics 951, https://doi.org/10.1007/978-3-030-05085-6_2

There are also the corresponding anti-holomorphic generators $\overline{L}_{0,\pm1}$ that commute with the $L_{0,\pm1}$. This split allows for an important simplification special to two dimensions: we can formally regard z and $\overline{z}$ as independent variables and only impose $\overline{z} = z^*$ when we need to extract a physical observable such as a correlation function.

In contrast with conformal symmetry in $d > 2$ dimensions, there are local conformal transformations $z \to z - \epsilon(z)$ and $\overline{z} \to \overline{z} - \overline{\epsilon}(\overline{z})$ for any meromorphic functions $\epsilon(z)$ and $\overline{\epsilon}(\overline{z})$. We can therefore organize the local operators according to their transformations under these, treating z and $\overline{z}$ as independent variables. A local operator $\Phi(z, \overline{z})$ is primary if under $z \to f(z)$ and $\overline{z} \to \overline{f}(\overline{z})$ it transforms as

$$\Phi(z, \overline{z}) \to \left(\frac{\partial f}{\partial z}\right)^{h_\Phi} \left(\frac{\partial \overline{f}}{\partial \overline{z}}\right)^{\overline{h}_\Phi} \Phi(f(z), \overline{f}(\overline{z})) . \qquad (2.1.3)$$

The pair $(h_\Phi, \overline{h}_\Phi)$ are the left- and right-moving conformal weights of the field Φ; the sum $\Delta_\Phi = h_\Phi + \overline{h}_\Phi$ is the scaling dimension (i.e. the eigenvalue of the dilatation generator), while the difference $s_\Phi = h_\Phi - \overline{h}_\Phi$ is the field's spin. An operator is quasi-primary if the transformation law above holds for $f(z), \overline{f}(\overline{z})$ restricted to global conformal transformations, i.e.

$$f(z) = \frac{az + b}{cz + d} , \qquad (2.1.4)$$

with $ad - bc = 1$ and $(a, b, c, d) \sim (-a, -b, -c, -d)$.

2.1.2 Unitary Compact CFTs

We will mostly discuss theories satisfying some "self-evident" assumptions that define the class of unitary and compact CFTs.

1. The global conformal symmetry and its associated algebra $\mathfrak{sl}_2\mathbb{C}$ is represented by the current algebra constructed from components of a conserved traceless energy-momentum tensor.
2. Local fields transform in representations of $\mathfrak{sl}_2\mathbb{C}$, and each highest weight field is labeled by its spin s and scaling dimension Δ. These are the quasi-primary fields. Any other field can be constructed as a linear combination of quasi-primary fields and their derivatives.
3. For any $\Delta_* \in \mathbb{R}$ there is a finite number of quasi-primary fields with $\Delta < \Delta_*$. A CFT with this property is called compact.
4. The theory is unitary. When defined in Minkowski space the theory will have a Hilbert space of states, a unique $\mathfrak{sl}_2\mathbb{C}$-invariant ground state $|0\rangle$, and a Hermitian Hamiltonian operator. This leads to a notion of a Hilbert space and Hermiticity in the Euclidean formulation.

5. The set of operators is complete, in the sense that every state can be generated by some local operator acting on the vacuum, and the operator product expansion for local fields defines an associative algebra. This last assumption is quite constraining and is fundamental to the bootstrap program in either its ancient [65] or modern [325] incarnation.

A few comments are in order. First, there is a more general class of scale-invariant QFTs, where $T_\mu^\mu = \partial_\mu V^\nu$ for some virial current V; however, a unitary two-dimensional QFT with a discrete spectrum is scale invariant if and only if it is conformally invariant [326].[2]

Second, the assumption of compactness excludes a number of theories. Some are "trivially non-compact," where the source of non-compactness is traced to the presence of a free scalar field; more generally, there are also interacting theories, such as Liouville theory or a conformal non-linear sigma model on an ALE space. In these cases conformal invariance, and in particular conformal Ward identities for correlation functions of quasi-primary local operators, can still be used to extract non-trivial information about the theory; see, e.g. [361] for a discussion in the context of Liouville theory. Some of the results that we will routinely use will not apply in this wider setting.

Finally, non-unitary theories are not without interest: there are also "trivial" examples of these, such as the free bc-ghost system of string theory, but there are also interacting theories, for instance various pure spinor formulations of the general type II string world-sheet [76], as well as statistical mechanics systems without reflection positivity.

2.1.3 The Energy-Momentum Tensor

We decompose the energy-momentum tensor of any Euclidean QFT with respect to the $z, \bar{z}$ coordinates as

$$\Theta = T_{z\bar{z}} = \tfrac{1}{4}(T_{11} + T_{22}) \, ,$$

$$T = T_{zz} = \tfrac{1}{4}(T_{11} - T_{22}) - \tfrac{i}{2}T_{12} \, ,$$

$$\overline{T} = T_{\bar{z}\bar{z}} = \tfrac{1}{4}(T_{11} - T_{22}) + \tfrac{i}{2}T_{12} \, . \tag{2.1.5}$$

The conservation equation $\partial^\mu T_{\mu\nu} = 0$ is equivalent to

$$\bar{\partial}T + \partial\Theta = 0 \, , \qquad\qquad \bar{\partial}\Theta + \partial\overline{T} = 0 \, . \tag{2.1.6}$$

---

[2]As discussed in [327], unitarity and a discrete spectrum are both necessary for the implication to hold.

A CFT has a traceless energy-momentum tensor and therefore operator equations

$$\Theta(z, \bar{z}) = 0 , \qquad \bar{\partial} T(z, \bar{z}) = 0 , \qquad \partial \bar{T}(z, \bar{z}) = 0 \qquad (2.1.7)$$

that hold in correlation functions when the point $(z, \bar{z})$ is not coincident with any other insertions. T and $\bar{T}$ are therefore separately conserved and are, respectively, holomorphic and anti-holomorphic. T and $\bar{T}$ enjoy a number of universal properties that we will now review. We will focus on T, keeping in mind that similar properties hold for $\bar{T}$.

Many CFT properties are clarified in the framework of radial quantization, where the worldsheet is taken to be $S^1 \times \mathbb{R}$.[3] A conformal theory on $S^1 \times \mathbb{R}$ can be mapped to the complex plane via $z = e^{i\theta + \tau}$, so that $z = 0$ corresponds to $\tau = -\infty$, while equal time slices are mapped to circles with constant $|z|$. The radial Hamiltonian is then given by

$$H = \oint_{C(0)} \frac{dz}{2\pi i} z T(z) + \oint_{C(0)} \frac{d\bar{z}}{2\pi i} \bar{z} \bar{T}(\bar{z}) , \qquad (2.1.8)$$

where $C(0)$ is a contour enclosing the origin and both the dz and $d\bar{z}$ integrations are oriented counter-clockwise.

$T(z)$ is holomorphic and has conformal weight $(2, 0)$, so we can expand it in modes L_n with conformal weight $(-n, 0)$ according to radial quantization:

$$T(z) = \sum_{n \in \mathbb{Z}} L_n z^{-n-2} . \qquad (2.1.9)$$

We can invert the mode expansion via

$$L_n = \oint_{C(0)} \frac{dz}{2\pi i} z^{n+1} T(z) , \qquad (2.1.10)$$

where $C(0)$ is a contour around $z = 0$. The L_n are the generators of the infinitesimal conformal transformations, with $L_{0, \pm 1}$ giving the global conformal transformations as above. Evidently the radial Hamiltonian is $H = L_0 + \bar{L}_0$. The action of the L_n on any local operator Φ is determined by the T–Φ operator product expansion (OPE):

$$[L_n, \Phi(w, \bar{w})] = \oint_{C(w)} \frac{dz}{2\pi i} z^{n+1} T(z) \Phi(w, \bar{w}) . \qquad (2.1.11)$$

---

[3]Radial quantization can also be used in d-dimensional theories by working on $S^{d-1} \times \mathbb{R}$.

In particular, for a quasi-primary field, which satisfies

$$[L_1, \Phi(0,0)] = 0 , \quad [L_0, \Phi(0,0)] = h\Phi(0,0) , \quad [L_{-1}, \Phi(0,0)] = \partial\Phi(0,0) ,$$
(2.1.12)

the OPE must take the form

$$T(z)\Phi(w,\overline{w}) \sim \cdots + \frac{A(w,\overline{w})}{(z-w)^4} + \frac{h\Phi(w,\overline{w})}{(z-w)^2} + \frac{\partial\Phi(w,\overline{w})}{(z-w)} ,$$
(2.1.13)

where the $\cdots$ denote more singular terms. Φ is a primary field if and only if

$$T(z)\Phi(w,\overline{w}) \sim \frac{h\Phi(w,\overline{w})}{(z-w)^2} + \frac{\partial\Phi(w,\overline{w})}{(z-w)} ,$$
(2.1.14)

with the corresponding commutators

$$[L_n, \Phi(w,\overline{w})] = h(n+1)w^n \Phi(w,\overline{w}) + w^{n+1}\partial\Phi(w,\overline{w}) .$$
(2.1.15)

The global conformal algebra requires the T–T OPE to take a standard form

$$T(z)T(w) \sim \frac{c/2}{(z-w)^4} + \frac{2T(w)}{(z-w)^2} + \frac{\partial T(w)}{z-w} ,$$
(2.1.16)

where the constant c is the left (or holomorphic) central charge of the CFT, and it follows from the mode expansion that the modes L_n satisfy the Virasoro algebra

$$[L_m, L_n] = (m-n)L_{m+n} + \frac{c}{12}m(m-1)(m+1)\delta_{m,-n} .$$
(2.1.17)

Setting $c = 0$ leads to the Witt algebra—the algebra of meromorphic vector fields on the Riemann sphere. Up to a change of basis every central extension of the Witt algebra over $\mathbb{C}$ is isomorphic to the Virasoro algebra for some choice of c [339].

2.1.4 Ward Identities for T

The OPE (2.1.14), together with the conservation law $\overline{\partial}T = 0$ lead to a Ward identify for a correlation function of primary operators Φ_i:

$$\langle T(z)\Phi_1(w_1)\cdots\Phi_n(w_n)\rangle = \sum_{i=1}^n \left[\frac{h_i}{(z-w_i)^2} + \frac{\partial_{w_i}}{z-w_i} \right] \langle \Phi_1(w_1)\cdots\Phi_n(w_n)\rangle .$$
(2.1.18)

Expanding this for $|z| \gg |w_i|$, we obtain

$$\langle T(z)\Phi_1(w_1) \cdots \Phi_n(w_n) \rangle \tag{2.1.19}$$

$$= \left[-z^{-1}\mathcal{L}_{-1} - z^{-2}\mathcal{L}_0 - z^{-3}\mathcal{L}_1 + O(z^{-4}) \right] \langle \Phi_1(w_1) \cdots \Phi_n(w_n) \rangle ,$$

where

$$\mathcal{L}_{-1} = -\sum_i \partial_{w_i} , \quad \mathcal{L}_0 = -\sum_i (h_i + w_i \partial_{w_i}) , \quad \mathcal{L}_1 = -\sum_i (2h_i w_i + w_i^2 \partial_{w_i}) \tag{2.1.20}$$

form a representation of the $\mathfrak{sl}_2\mathbb{C}$ algebra. In view of (2.1.13) and (2.1.19) holds for any quasi-primary insertions, since the more singular terms in (2.1.13) simply modify the $O(z^{-4})$ terms on the right-hand side of (2.1.19).

Exercise 2.1 Use the Ward identity above and the TT OPE to reduce the correlator $\langle T(z_1)T(z_2)\Phi_1(w_1) \cdots \Phi_n(w_n) \rangle$ to a differential operator acting on $\langle \Phi_1(w_1) \cdots \Phi_n(w_n) \rangle$. Assume the Φ_i are primary fields.   □

The Euclidean correlation functions have a natural interpretation in radial quantization. Once we radially order the operators $|w_1| > |w_2| > \cdots > |w_n|$, we have ,

$$\langle \Phi(w_1) \cdots \Phi(w_n) \rangle = \langle 0 | \Phi(w_1) \cdots \Phi(w_n) | 0 \rangle , \tag{2.1.21}$$

where $|0\rangle$ is the unique $\mathfrak{sl}_2\mathbb{C}$-invariant vacuum state in the Hilbert space $\mathcal{H}$, while $\langle 0 |$ denotes the state in the dual space $\mathcal{H}^*$ canonically isomorphic to $|0\rangle$. Consider (2.1.19) from this point of view. On one hand, the limit $\lim_{z \to 0} T(z)|0\rangle$ should be a state in the Hilbert space and in particular should be regular. The mode expansion implies

$$\lim_{z \to 0} T(z)|0\rangle = L_{-2}|0\rangle \tag{2.1.22}$$

if and only if

$$L_{n \geq -1}|0\rangle = 0 . \tag{2.1.23}$$

For $n = 0, \pm 1$ this is the familiar requirement that the vacuum should be $\mathfrak{sl}_2\mathbb{C}$-invariant, but regularity implies the stronger condition.

On the other hand, in CFT a special conformal transformation can map $\lim_{z \to 0} T(z)|0\rangle$ to a dual "out" state

$$\lim_{z \to \infty} z^4 \langle 0 | T(z) . \tag{2.1.24}$$

Demanding that this limit is regular leads to the requirement $\langle 0|L_n = 0$ for $n \leq 1$ and shows that the correlator $\langle T(z)\Phi_1(w_1)\cdots\Phi_n(w_n)\rangle$ must decay as $O(z^{-4})$ for large $|z|$. This, together with (2.1.19), implies

$$\mathcal{L}_{0,\pm 1}\langle \Phi_1(w_1)\cdots\Phi_n(w_n)\rangle = 0 . \tag{2.1.25}$$

These global conformal identities for correlators of quasi-primary fields determine the position dependence of two- and three-point functions. The two-point function of operators Φ and Ψ is zero unless they have equal weights $(h_\Phi, \overline{h}_\Phi) = (h_\Psi, \overline{h}_\Psi) = (h, \overline{h})$, in which case

$$\langle \Phi(z, \overline{z})\Psi(0)\rangle = \frac{G}{z^{2h}\overline{z}^{2\overline{h}}} \tag{2.1.26}$$

for a constant G. The three-point function of quasi-primary fields is determined up to the "structure constants" C_{ABC}[4]:

$$\langle \Phi_{A1}(z_1)\Phi_{B2}(z_2)\Phi_{C3}(z_3)\rangle = \frac{C_{ABC}}{z_{12}^{h_1+h_2-h_3} z_{23}^{h_2+h_3-h_1} z_{31}^{h_1+h_3-h_2}} \times \text{``c.c.''} . \tag{2.1.27}$$

By allowing one of the operators to be the identity operator, we see that C_{ABC} also include the coefficients G of the two-point function.

Exercise 2.2 Suppose that either all three operators in the three-point function are bosonic with integer spins or two operators are fermionic with half-integral spin and the third is bosonic with integral spin. Prove that the constants C_{ABC} defined by the correlation function above obey

$$C_{BAC} = (-1)^{F_A F_B}(-1)^{s_A+s_B+s_C} C_{ABC} . \tag{2.1.28}$$

$\square$

The higher-point correlation functions are not directly fixed by conformal invariance, but they can in principle be obtained from the three-point functions by repeated application of the OPE. A modern discussion of this procedure can be found in [322].

---

[4]The suppressed $\overline{z}$ dependence of the denominator, denoted by "c.c.", can be recovered via the substitution $z \to \overline{z}$ and $h \to \overline{h}$.

2.1.5 Operator-State Correspondence

The construction of the state $L_{-2}|0\rangle = \lim_{z \to 0} T(z)|0\rangle$ generalizes. For any local field $\Phi(z, \bar{z})$ there is a corresponding state

$$|\Phi\rangle = \lim_{z,\bar{z} \to 0} \Phi(z, \bar{z})|0\rangle , \qquad (2.1.29)$$

which creates a state by a local operator insertion in the infinite past with respect to radial quantization. On the other hand, any state $|\Phi\rangle$ on the cylinder defines a corresponding local operator $\Phi(0, 0)$, and since we assume the states form a Hilbert space, there is a 1:1 map between states and operators, so that the set of local operators is also endowed with a Hilbert space structure. As we will see below, the inner product on this Hilbert space is determined by the two-point correlation functions of primary fields.

Since $L_{n \geq -1}|0\rangle = 0$, the notion of a primary field is isomorphic to that of a primary state: $|\Phi\rangle$ is primary if and only if $L_{n>1}|\Phi\rangle = 0$; similarly $|\Phi\rangle$ is quasi-primary if and only if $L_1|\Phi\rangle = 0$. Equivalently, a primary state is a highest weight state of the Virasoro algebra, while a quasi-primary state is a highest weight with respect to the global conformal algebra.

Let $\mathcal{H}_{\text{qp}}$ denote the Hilbert space of quasi-primary states at fixed weights:

$$\mathcal{H}_{\text{qp}} = \{\lim_{z \to 0} \Phi(z, \bar{z})|0\rangle \mid \Phi \text{ is a quasi-primary with weights } (h, \bar{h})\} . \qquad (2.1.30)$$

Denote by $\Phi^\dagger(z, \bar{z})$ the Euclidean continuation of the standard Hermitian conjugate of the Minkowski continuation of Φ. Using this, we construct a state in the dual space via

$$\langle\Phi| = \lim_{z,\bar{z} \to \infty} z^{2h}\bar{z}^{2\bar{h}}\langle 0|\Phi^\dagger(z, \bar{z}) \qquad (2.1.31)$$

The Hermitian inner product on $\mathcal{H}_{\text{qp}}$ is then determined by

$$(\Phi, \Psi) = \lim_{z,\bar{z} \to \infty} \lim_{w,\bar{w} \to 0} z^{2h}\bar{z}^{2\bar{h}}\langle 0|\Phi^\dagger(z, \bar{z})\Psi(w, \bar{w})|0\rangle \qquad (2.1.32)$$

$$= \lim_{z,\bar{z} \to \infty} z^{2h}\bar{z}^{2\bar{h}}\langle\Phi^\dagger(z, \bar{z})\Psi(0, 0)\rangle .$$

Unitarity requires the two-point function on the right-hand side, known as the Zamolodchikov metric, to be positive-definite and Hermitian, so we obtain the desired inner product on $\mathcal{H}_{\text{qp}}$.

As usual, the inner product induces a notion of an adjoint operator, which we will call the conformal adjoint. This is given by

$$[\Phi(z, \bar{z})]^\circ = \bar{z}^{-2h}z^{-2\bar{h}}\Phi^\dagger(\bar{z}^{-1}, z^{-1}) . \qquad (2.1.33)$$

The adjoint action on the coordinates is perhaps a little unfamiliar, but in fact it is a standard notion in Euclidean QFT with a Minkowski antecedent: the definition of Euclidean time via $\tau = ix^0$ requires that we compensate the conjugation on the extra i factors with the $z \to \bar{z}^{-1}$ map. At any rate, as the following exercise shows, this conformal adjoint leads to the expected property $(\Phi, \Psi) = (\Psi, \Phi)^*$.

Exercise 2.3 Use the definition of the conformal adjoint to show that

$$(\Psi, \Phi)^* = \lim_{z,\bar{z}\to\infty} \langle \Phi^\dagger(z, \bar{z})\Psi(0, 0)\rangle = (\Phi, \Psi) . \tag{2.1.34}$$

□

Once an operator is expanded in modes, the compatibility of the mode expansion with the conformal adjoint yields the Hermitian properties of the modes. For instance, using the mode expansion for T we have

$$[T(z)]^\circ = \left[\sum_{m\in\mathbb{Z}} L_m z^{-m-2}\right]^\circ = \sum_m L_m^\dagger (z^{-m-2})^* , \tag{2.1.35}$$

while from the direct definition and the fact that T is Hermitian in Minkowski space, we have

$$[T(z)]^\circ = \bar{z}^{-4} T(\bar{z}^{-1}) = \sum_m L_m \bar{z}^{m-2} . \tag{2.1.36}$$

Compatibility of these two forms fixes $L_m^\dagger = L_{-m}$.

2.1.6 Some Key Properties

With the basic preliminaries in hand, we can now summarize key features of unitary compact CFTs. The elementary proofs of these statements are collected in Appendix A.

1. *The vacuum state $|0\rangle$ is primary and has $(h, \bar{h}) = (0, 0)$. It corresponds to the identity operator.*
2. *The central charge c is positive.*
3. *L_0 has a non-negative spectrum.*
4. *Every state is a sum of primary states and their Virasoro descendants.* For any primary state $|\Phi\rangle$ and an ordered partition of K $[K] = (k_1, k_2, \cdots, k_p)$, with $k_i \geq k_{i+1}$, we can define the descendant

$$L_{[K]} = L_{-k_1} L_{-k_2} \cdots L_{-k_p} |\Phi\rangle .$$

The statement is that every state in the Hilbert space is a linear combination of the primaries and their descendants.

5. *A local operator is anti-holomorphic if and only if it has weight $h = 0$.*

   If $h = \overline{h} = 0$ then the operator must be position-independent, and therefore in any local quantum field theory a constant multiple of the identity; the corresponding state is then a constant multiple of the vacuum. More generally, a theory can have a non-trivial topological sector, where the operators with $h = \overline{h} = 0$ constitute a finite-dimensional vector space and have position-independent correlation functions.

2.1.7 Minimal Models

Given any $|\Phi\rangle$ primary state with weight h, we can construct the full tower of descendants by acting with the Virasoro raising operators, which is known as a Verma module. We can organize the resulting states by their level, i.e. the integer K in the raising operators L_K defined above. For each K the result is clearly a finite-dimensional vector space, but its dimension may be smaller than the naive $P(K)$—the number of ordered partitions of K. Furthermore, the inner product on these states is completely determined by the Virasoro algebra in terms of h, the central charge c, and the norm of $|\Phi\rangle$. When $c < 1$, this inner product fails to be unitary unless

$$c = 1 - \frac{6}{m(m+1)}\,, \qquad m = 3, 4, 5, \ldots\,, \qquad (2.1.37)$$

and the weight h takes one of the $\binom{m}{2}$ values determined by integers p and q:

$$h = \frac{[(m+1)p - mq]^2 - 1}{4m(m+1)}\,, \qquad 1 \le q \le p \le m - 1\,. \qquad (2.1.38)$$

The GKO coset construction shows that a unitary CFT realizes these values of central charge and corresponding spectrum of operators [198]—these are the Virasoro minimal models.

These minimal models provide the simplest examples of solvable conformal field theories, where the principles discussed so far, together with OPE associativity and algebraic methods lead to a solution for the three-point functions of primary fields, and, more generally, for any correlation function of local operators.

A generalization of this sort of classification approach is in principle available in the class of theories known as rational CFTs. Such theories are characterized by the presence of a chiral algebra of integer spin holomorphic currents. We will not discuss these structures and refer the reader to [309] for more details. Additional very readable discussion of rational CFTs and the intimately related W-algebras is given in the excellent review [96].

2.1.8 Decomposable CFTs

Given two CFTs with energy momentum tensors T_1 and T_2, we can always form a direct product theory with energy-momentum tensor $T_{\text{tot}} = T_1 + T_2$. The "total" theory will then have a second linearly independent conserved current with spin 2 in addition to T_{tot}. Such constructions and their deformations provide many examples of CFTs composed of simpler building blocks.

It is amusing to ponder a converse to the construction. In other words, suppose that we are given a CFT with a number of quasi-primary spin 2 holomorphic fields T_α, such that $T_{\text{tot}} = \sum_\alpha T_\alpha$. Can we decompose the Virasoro algebra into components corresponding to the T_α? As the following example shows, there are non-trivial conditions for such a decomposition to hold.

Consider the simplest case, where $\alpha = 1, 2$, and $T_{\text{tot}} = T_1 + T_2$. In order to get the desired decomposition it is necessary that the OPE $T_\alpha(z_1)T_\beta(z_2)$ has the form

$$T_\alpha(z_1)T_\beta(z_2) \sim \frac{c_\alpha \delta_{\alpha\beta}}{2z_{12}^4} + \frac{2}{z_{12}^2}\delta_{\alpha\beta}T_\beta(z_2) + \frac{1}{z_{12}}\delta_{\alpha\beta}\partial T_\beta(z_2) \,. \tag{2.1.39}$$

On the other hand, if we simply demand that $T_\alpha T_\beta$ OPE closes on these operators and their descendants, $T_{\text{tot}} = T_1 + T_2$, and that the modes of T_α satisfy the Jacobi identity (i.e. the four-point function has crossing symmetry), we find that the OPE must take the form

$$T_\alpha(z_1)T_\beta(z_2) \sim \frac{G_{\alpha\beta}}{z_{12}^4} + \frac{1}{z_{12}^2}\sum_\gamma C_{\alpha\beta}^\gamma T_\gamma(z_2) + \frac{1}{2z_{12}}\sum_\gamma C_{\alpha\beta}^\gamma \partial T_\gamma(z_2) \,,$$

and the constants $G_{\alpha\beta}$ and $C_{\alpha\beta}^\gamma$ are symmetric in α, β and satisfy

$$\sum_{\alpha,\beta} G_{\alpha\beta} = \frac{c_{\text{tot}}}{2} \,, \quad \sum_\alpha C_{\alpha\beta}^\gamma = 2\delta_\beta^\gamma \,, \quad \sum_\gamma C_{\alpha\beta}^\gamma C_{\gamma\delta}^\rho = \sum_\gamma C_{\alpha\delta}^\gamma C_{\gamma\beta}^\rho \,. \tag{2.1.40}$$

There are solutions to these conditions that are not compatible with decomposition. For instance, there is nothing to be done if $G_{12} \neq 0$. Even if $G_{\alpha\beta} = c_\alpha \delta_{\alpha\beta}/2$ as desired, then for any real x a solution for the $C_{\alpha\beta\gamma}$ is given by

$$C_{111} = c_1 - x \,, \quad C_{112} = x \,, \quad C_{122} = -x \,, \quad C_{222} = c_2 + x \,. \tag{2.1.41}$$

Exercise 2.4 Verify the assertions in this section by studying the OPE of $T_\alpha T_\beta$ as well as the four-point function. □

It should be clear that having a decomposable energy-momentum tensor leads to stronger constraints on the theory. For instance, unitarity now requires that for any primary operator the total weight $h_{\text{tot}} = \sum_\alpha h_\alpha$, and the weight h_α with respect to T_α is non-negative.

2.2 Free Fields

2.2.1 A Majorana-Weyl Fermion

The simplest example of a unitary and compact CFT is provided by a single free
Majorana-Weyl fermion λ with action as in (1.8.6):

$$S = \frac{1}{4\pi} \int d^2z \, \lambda \bar{\partial} \lambda \, . \tag{2.2.1}$$

The equation of motion $\bar{\partial}\lambda = 0$ means that $\lambda(z)$ is purely holomorphic, and a
standard path-integral manipulation easily produces the OPE:

$$\int_\lambda \frac{\delta}{\delta\lambda(z,\bar{z})} \left[e^{-S}\lambda(w,\overline{w}) \right] = \int_\lambda e^{-S} \left[-\frac{1}{2\pi}\bar{\partial}_{\bar{z}}\lambda(z,\bar{z})\lambda(w,\overline{w}) + \delta^2(z-w) \right] \, . \tag{2.2.2}$$

The left-hand side is the integral of a total derivative; we assume that this vanishes
even if we multiply $\lambda(w,\overline{w})$ by any product of local operators inserted away from
$(w,\overline{w})$, and this leads to the OPE

$$\lambda(z)\lambda(w) \sim \frac{1}{z-w} \, . \tag{2.2.3}$$

From the action we see that λ has weight $(1/2, 0)$, and, using the energy-momentum
tensor[5]

$$T(z) = \lim_{w \to z} \left[-\frac{1}{2}\lambda(w)\partial_z\lambda(z) - \frac{1}{2(z-w)^2} \right] = -\frac{1}{2} : \lambda\partial\lambda :(z) \, , \tag{2.2.4}$$

we see that

$$T(z)\lambda(w) \sim \frac{\frac{1}{2}\lambda(w)}{(z-w)^2} + \frac{\partial\lambda(w)}{z-w} \, . \tag{2.2.5}$$

We used "conformal normal ordering" to produce a well-defined local energy-
momentum tensor $T(z)$. More generally, we can use the OPE to define products
of local operators by subtracting off the singularities in the operator product.[6] In

---

[5]The classical energy-momentum tensor can be obtained by the Noether procedure from the action;
this can be normal-ordered to yield the operator in the quantum theory, and the normalization is
fixed by the fact that T is quasi-primary of dimension 2.

[6]This is discussed systematically in [131].

this free theory, this is just Wick's theorem:

$$\lambda(z)\lambda(w) = \frac{1}{z-w} + :\lambda(z)\lambda(w): \,, \qquad (2.2.6)$$

where the second term on the right-hand side is regular in the limit $z \to w$.

Exercise 2.5 Use Wick's theorem to compute the $T-T$ OPE and verify that the central charge is given by $c = 1/2$. □

From what we said so far we have a CFT with $c = 1/2$ and $\bar{c} = 0$, and a single primary field λ of weight $(h, \bar{h}) = (1/2, 0)$. There is, however, more data that is to be determined. The fermionic field λ is a spinor with spin is $s = h - \bar{h} = 1/2$ and need not be single-valued under $z \to e^{2\pi i}z$. On the plane, or equivalently on the cylinder, the allowed monodromies are valued in $\mathbb{Z}_2$ and are known as

Neveu-Shwarz : $\lambda(ze^{2\pi i}) = \lambda(z)$ and   Ramond : $\lambda(ze^{2\pi i}) = -\lambda(z)$
$$(2.2.7)$$

boundary conditions. We can treat these democratically by setting

$$\lambda(ze^{2\pi i}) = -e^{2\pi i \nu}\lambda \,, \qquad (2.2.8)$$

where $\nu = 0$ for the R sector and $\nu = 1/2$ for the NS sector. This leads to a mode expansion

$$\lambda(z) = \sum_{r \in \mathbb{Z} - \nu} \lambda_r z^{-r-1/2} \,, \qquad (2.2.9)$$

and inserting this into the OPE yields the anti-commutation relations

$$\{\lambda_r, \lambda_s\} = \delta_{r,-s} \,. \qquad (2.2.10)$$

From the $T-\lambda$ OPE we also learn that these modes satisfy $[L_m, \psi_r] = (m/2 - r)\psi_{r+m}$. We also have a fermion number operator $(-1)^{F_\lambda}$, which satisfies $\{\lambda(z), (-1)^{F_\lambda}\} = 0$.

Equations such as our ν-dependent boundary condition at first appear a little bit strange: what data in the quantum field theory determines which of these boundary conditions we use? The answer is that the choice of monodromy of a local field around $z = 0$ is encoded in a choice of an "in" state.[7] So, more appropriately, we

---

[7]This is perhaps the simplest example of the more general concept of a defect in a quantum field theory; 't Hooft operators and their extensions are more sophisticated examples that occur in four-dimensional gauge theories.

consider a sub-space of the Hilbert space

$$\mathcal{H}_\nu = \{|\Psi\rangle \in \mathcal{H} \mid \lambda(ze^{2\pi i})|\Psi\rangle = -e^{2\pi i \nu}\lambda(z)|\Psi\rangle\}. \tag{2.2.11}$$

Without loss of generality we can decompose $\mathcal{H}_\nu$ into eigenspaces of L_0.

In our free theory the ground state $|0\rangle$ is defined by demanding that $\lim_{z\to 0}\lambda(z)|0\rangle$ is regular, which means $|0\rangle \in \mathcal{H}_{NS}$. So, we have $\lambda_r|0\rangle = 0$ for $r > 0$, and $\lim_{z\to 0}\lambda(z)|0\rangle = \lambda_{-1/2}|0\rangle$. Clearly any descendant of $|0\rangle$ of the form $\lambda_{-r_1}\cdots\lambda_{-r_k}|0\rangle$ will belong to $\mathcal{H}_{NS}$; in fact, all states in $\mathcal{H}_{NS}$ are constructed in this fashion. For this reason it is typical to speak of $|0\rangle$ as the NS-vacuum. We define $(-1)^F|0\rangle = |0\rangle$, and we can decompose

$$\mathcal{H}_{NS} = \mathcal{H}_{NS}^+ \oplus \mathcal{H}_{NS}^- \tag{2.2.12}$$

based on the eigenvalue of $(-1)^{F_\lambda}$.

Evidently, the states that lead to the Ramond boundary conditions cannot be obtained from the $\mathfrak{sl}_2\mathbb{C}$-invariant vacuum by acting with any operators that are polynomial in λ and derivatives. To describe these mystery states, we first observe that they must furnish a representation of $(-1)^{F_\lambda}$ as well as the zero mode λ_0. Since $\{(-1)^{F_\lambda}, \lambda_0\} = 0$, the smallest such representation is two-dimensional, so that we must have at least a pair of states $|\Sigma_\pm\rangle$ with $L_0|\Sigma_\pm\rangle = h_\Sigma|\Sigma_\pm\rangle$. Furthermore, we will demand that $(-1)^{F_\lambda}$ is diagonalizable (otherwise it is challenging to make sense of any spin-statistics relation), so that the two-dimensional representation takes the form

$$(-1)^{F_\lambda} = \begin{pmatrix} \alpha & 0 \\ 0 & -\alpha \end{pmatrix}, \qquad \psi_0 = \frac{1}{\sqrt{2}}\begin{pmatrix} 0 & 1 \\ 1 & 0 \end{pmatrix}. \tag{2.2.13}$$

Once we have these ground states in hand, we can obtain the remaining states in $\mathcal{H}_R$ by acting by various raising operators $\lambda_{-r_1}\cdots\lambda_{-r_k}|\Sigma_\pm\rangle$.

The operators corresponding to the states $|\Sigma_\pm\rangle$ are known as spin fields or disorder operators, and they are characterized by a non-local OPE with the λ[8]:

$$\lambda(z)\Sigma_+(w) \sim \frac{e^{i\pi/4}}{\sqrt{2}}\frac{1}{(z-w)^{1/2}}\Sigma_-(w). \tag{2.2.14}$$

The branch cut indicates a pathology in the CFT. Its cure involves a GSO-type projection. This can either combine the states into left-right symmetric combinations (this leads to the so-called diagonal invariant), or it can involve a more elaborate left-right asymmetric projection. The CFT constructions that we will study in detail will admit such projections once they are tensored with additional free fields.

---

[8]For a derivation of the phase see [197].

Leaving the issue of mutual locality of the OPE aside, we can investigate these fields further. A useful quantity to consider are the expectation values

$$\langle 0|\lambda(z)\lambda(w)|0\rangle \qquad\qquad \langle \Sigma_+|\lambda(z)\lambda(w)|\Sigma_+\rangle . \qquad (2.2.15)$$

The first one is easily evaluated from the OPE:

$$\langle 0|\lambda(z)\lambda(w)|0\rangle = \frac{1}{z-w} , \qquad (2.2.16)$$

but we can also compute it by an explicit mode expansion. We have the Hermiticity condition $\lambda_r^\dagger = \lambda_{-r}$, as well as the relation $\lambda_r|0\rangle = 0$ for $r > 0$, and this leads to (for $|z| > |w|$)

$$\langle 0|\lambda(z)\lambda(w)|0\rangle = \sum_{r>0,s<0} z^{-r-1/2} w^{-s-1/2} \langle 0|\{\lambda_r, \lambda_s\}|0\rangle \qquad (2.2.17)$$

$$= \sum_{m=0}^{\infty} z^{-m-1} w^m = \frac{1}{z-w} .$$

Exercise 2.6 Use the mode expansion in the Ramond sector to show

$$\langle \Sigma_+|\lambda(z)\lambda(w)|\Sigma_+\rangle = \frac{\sqrt{\frac{z}{w}} + \sqrt{\frac{w}{z}}}{z-w} .$$

Next, evaluate the weight of Σ_+ by the following trick. On one hand $T(z)|\Sigma_+\rangle = hz^{-2}|\Sigma_+\rangle$; on the other hand, we have the "point-split" definition of $T(z)$ in (2.2.4). Apply the latter to the expectation value just computed and compare with the former to obtain $h_\Sigma = 1/16$. $\qquad\qquad\square$

2.2.2 The Infamous Scalar

The first conformal field theory that is typically encountered by a young string theorist is that of a free scalar with action

$$S = \frac{1}{4\pi} \int d^2z\, \partial\phi\bar{\partial}\phi . \qquad (2.2.18)$$

The same sort of path integral manipulation as we carried out above for the fermion leads to the equations of motion and OPE

$$\partial\bar{\partial}\phi = 0 , \qquad \phi(z,\bar{z})\phi(w) \sim -\log|z-w|^2 . \qquad (2.2.19)$$

In this case the energy-momentum tensor has non-trivial components

$$T = -\frac{1}{2}\partial\phi\partial\phi \,, \qquad\qquad \overline{T} = -\frac{1}{2}\bar{\partial}\phi\bar{\partial}\phi \,, \qquad\qquad (2.2.20)$$

leading to $(c, \bar{c}) = (1, 1)$. The equation of motion allows us to separate the field into holomorphic and anti-holomorphic modes: $\phi = \phi_L(z) + \phi_R(\bar{z})$, with

$$\phi_L(z)\phi_L(w) \sim -\log(z - w) \,. \qquad\qquad (2.2.21)$$

A single "chiral" scalar ϕ_L will be a useful formal device in our investigations, even though it does not have a sensible Euclidean action.

This is a free field theory, but it is not a unitary compact CFT. Clearly ϕ itself is not a quasi-primary field, and the two-point function

$$\langle\phi(z, \bar{z})\phi(0, 0)\rangle = -\log |z|^2 \qquad\qquad (2.2.22)$$

is neither dilatation covariant or positive. The primary operators include

$$J = i\partial\phi(z) \,, \qquad \overline{J} = i\bar{\partial}\phi(\bar{z}) \,, \qquad V_k =: e^{ik\phi} : (z, \bar{z}) \,. \qquad (2.2.23)$$

These have, respectively, conformal dimensions $(1, 0)$, $(0, 1)$, and $(k^2/2, k^2/2)$. The $\mathfrak{sl}_2\mathbb{C}$-invariant vacuum belongs to a continuum of states labeled by the momentum $k \in \mathbb{R}$.

The diseases of this theory can be traced to the non-compact zero mode and a corresponding divergence of the Euclidean path integral. This can be cured by imposing a periodicity condition $\phi \sim \phi + 2\pi\rho$. Now ϕ is no longer a well-defined operator, and more generally, the continuum V_k is replaced by a discrete set of operators. We will return to this nice example a number of times.

2.3 Global Symmetries

A CFT may have global symmetries beyond the SL(2, $\mathbb{C}$) of the global conformal group. Suppose a continuous symmetry is realized by a set of conserved spin 1 currents. That is, we have conserved charges

$$Q = \oint \frac{dz}{2\pi} J(z, \bar{z}) + \oint \frac{d\bar{z}}{2\pi} \overline{J}(z, \bar{z}) \qquad\qquad (2.3.1)$$

that transform as scalars under the global conformal transformations and in the adjoint of a Lie algebra $\mathfrak{g}$ of the symmetry group. This requires the components

J and $\overline{J}$ to have weights, respectively, $(1, 0)$ and $(0, 1)$ and to obey the conservation law

$$\overline{\partial} J + \partial \overline{J} = 0 .$$

In a unitary and compact CFT this implies $\overline{\partial} J = 0$ and $\partial \overline{J} = 0$ separately. This has a number of implications. Most immediately, we have a doubling of the global symmetry to $\mathfrak{g}_L \oplus \mathfrak{g}_R$ for the separately conserved currents unless $J = 0$ or $\overline{J} = 0$. Let us focus for the moment on the holomorphic side corresponding to $\overline{\partial} J = 0$. The resulting structure is that of a Kac-Moody (KM), or affine Lie algebra, symmetry. Its implications for conformal field theory are discussed in the classic work [269]. We will present the simplest aspects after reviewing useful properties of Lie algebras and stating our conventions.

2.3.1 Some Lie Algebra Conventions

Suppose $\mathfrak{g}$ is the Lie algebra associated to a simple compact Lie group. The structure of $\mathfrak{g}$ is fixed up to the normalization of the Cartan-Killing metric—a single constant. We will work in the normalization conventions where the longest roots have length-squared 2, and we can choose a basis of generators for the adjoint representation, which we can express in terms of explicit structure constants as

$$(t^a_{\text{adj}})^b_c = i f^{ab}{}_c , \tag{2.3.2}$$

where $a, b, c = 1, \ldots, \dim \mathfrak{g}$, so that

$$\text{tr}_{\text{adj}} \, t^a_{\text{adj}} t^b_{\text{adj}} = \sum_{c,d} f^{ac}{}_d f^{bd}{}_c = 2h(\mathfrak{g}) \delta^{ab} , \tag{2.3.3}$$

where $h(\mathfrak{g})$ is the dual Coxeter number of $\mathfrak{g}$. This choice fixes the normalizations for the generators in every finite-dimensional unitary representation r of $\mathfrak{g}$. That is, for any representation r with Hermitian generators t^a_r, which by definition satisfy

$$[t^a_r, t^b_r] = i \sum_c f^{ab}{}_c t^c_r , \tag{2.3.4}$$

we have

$$\text{tr}_r \, t^a_r t^b_r = \ell(r) \delta^{ab} , \tag{2.3.5}$$

where $\ell(r)$ is the Dynkin index of representation r. We give some values of these in the following table.

$\mathfrak{g}$	$\mathfrak{su}(n)$	$\mathfrak{so}(n)$	$\mathfrak{sp}(n)$	$\mathfrak{e}_6$	$\mathfrak{e}_7$	$\mathfrak{e}_8$	$\mathfrak{f}_4$	$\mathfrak{g}_2$
dim $\mathfrak{g}$	$n^2 - 1$	$n(n-1)/2$	$n(2n+1)$	78	133	248	52	14
$h(\mathfrak{g}) = \frac{1}{2}\ell(\text{adj})$	n	$n-2$	$n+1$	12	18	30	9	4
$\ell(\text{fund})$	1	2	1	6	12	60	6	2
dim(fund)	n	n	$2n$	27	56	248	26	7

$$(2.3.6)$$

Here "fund" denotes the fundamental basic irreducible representation; with the exception of the spinor representations, all others basic representations are constructed from tensor products of the fundamental.[9] In our conventions $\mathfrak{sp}(1) = \mathfrak{su}(2)$. For $\mathfrak{so}(n)$ we also consider the spinor representations. For $\mathfrak{so}(2k+1) = B_k$ the spinor representation W has dimension 2^k and $\ell(W) = 2^{k-2}$; for $\mathfrak{so}(2k) = D_k$, each spinor representation W has dimension 2^{k-1} and $\ell(W) = 2^{k-3}$. The indices of the remaining basic irreducible representations are easily computed from these by using the identities

$$\ell(r_1 \oplus r_2) = \ell(r_1) + \ell(r_2), \qquad \ell(r_1 \otimes r_2) = \dim r_1 \ell(r_2) + \dim r_2 \ell(r_1).$$
$$(2.3.7)$$

It is also useful to recall that the Dynkin index is preserved for any regular sub-algebra $\mathfrak{h} \subset \mathfrak{g}$, i.e. a sub-algebra obtained by striking nodes from the extended Dynkin diagram of $\mathfrak{g}$. In this case, for the decomposition

$$\mathfrak{g} \supset \mathfrak{h}$$
$$R = r_1 \oplus r_2 \oplus \cdots \oplus r_n \qquad (2.3.8)$$

we have $\ell_{\mathfrak{g}}(R) = \ell_{\mathfrak{h}}(r_1) + \cdots + \ell_{\mathfrak{h}}(r_n)$.[10] Finally, the quadratic Casimir associated to an irreducible representation r is obtained as

$$\mathbf{C}_2(r) = \sum_a t_r^a t_r^a = C_2(r) \mathbb{1}_r, \qquad (2.3.9)$$

with

$$C_2(r) = \frac{\dim \mathfrak{g}}{\dim r} \ell(r). \qquad (2.3.10)$$

---

[9] A final unraveling of notation: recall that a basic irreducible representation has Dynkin labels $[0, \ldots, 1, \ldots, 0]$.

[10] For an irregular sub-algebra the two sides are related by another integer—the index of the embedding.

2.3.2 The Kac-Moody Current Algebra

We now return to our CFT with a global symmetry algebra of a simple compact Lie algebra $\mathfrak{g}$. The holomorphic Kac-Moody (KM) current algebra is generated by weight $(1,0)$ currents $J^a(z)$, $a = 1, \ldots, \dim \mathfrak{g}$. These currents have a mode expansion

$$J^a(z) = \sum_{n \in \mathbb{Z}} J_n^a z^{-n-1} , \qquad J_n^a = \oint \frac{dz}{2\pi i} z^n J^a(z) , \qquad (2.3.11)$$

and the J_0^a are the global symmetry charges. In order for these global charges to be well-defined with respect to the $\mathfrak{sl}_2\mathbb{C}$ transformations, the $J^a(z)$ must be quasi-primary fields; unitarity then ensures that they are Virasoro primary, so that the modes satisfy

$$[L_m, J_n^a] = -n J_{m+n}^a . \qquad (2.3.12)$$

The global charges obey the commutation relations of $\mathfrak{g}$ $[J_0^a, J_0^b] = i \sum_c f^{ab}_{\ c} J_0^c$, which fixes the OPE up to a c-number k, known as the KM level:

$$J^a(z) J^b(0) \sim \frac{k \delta^{ab}}{z^2} + \frac{i f^{ab}_{\ c}}{z} J^c(0) . \qquad (2.3.13)$$

The mode expansion $J(z) = \sum_{n \in \mathbb{Z}} J_n z^{-n-1}$ and the KM OPE lead to the KM algebra $\widehat{\mathfrak{g}}_k$

$$[J_m^a, J_n^b] = i f^{ab}_{\ c} J_{m+n}^c + k m \delta_{m,-n} \delta^{ab} . \qquad (2.3.14)$$

The Hermitian adjoint for the modes is given by $(J_n^a)^\dagger = J_{-n}^a$.

The global symmetry algebra allows us to organize the states according to unitary representations of $\mathfrak{g}$, with $|\Phi_r\rangle$ satisfying

$$J_0^a |\Phi_r\rangle = t_r^a |\Phi_r\rangle . \qquad (2.3.15)$$

A state is Kac-Moody primary if and only if it is a highest weight state with respect to the global algebra generated by the J_0^a, and it is annihilated by J_n^a for all $n > 0$ and all a. The notions of Virasoro primary and KM primary states are distinct: a state can be one without being the other; we will say a state is KMV primary if it is primary with respect to both. The corresponding operators are characterized by the OPE

$$J^a(z) \Phi_r(w) \sim \frac{t_r^a \cdot \Phi_r}{z - w} , \qquad (2.3.16)$$

and satisfy the KM Ward identities

$$\langle J^a(z)\Phi_{r_1}(w_1)\cdots\Phi_{r_n}(w_n)\rangle = \sum_{i=1}^{n} \frac{1}{z-w_i}\langle\Phi_{r_1}(w_1)\cdots\left[t_{r_i}^a\cdot\Phi_{r_i}\right]\cdots\Phi_{r_n}(w_n)\rangle .$$

(2.3.17)

Regularity of $\lim_{z\to 0} J(z)|0\rangle$ and $\lim_{z\to\infty}\langle 0|J(z)z^2$, which is equivalent to $J_n^a|0\rangle = 0$ for $n \geq 0$, implies that the left-hand side must vanish as $O(z^{-2})$ for $|z| \gg |w_i|$, so that we obtain the expected charge conservation Ward identity

$$0 = \sum_{i=1}^{n}\langle\Phi_{r_1}(w_1)\cdots\left[t_{r_i}^a\cdot\Phi_{r_i}\right]\cdots\Phi_{r_n}(w_n)\rangle .$$

(2.3.18)

We now enumerate some familiar properties of KM symmetries in unitary compact CFTs (the proofs of these can be found in the references);

1. The level k is a positive integer.
2. There is a Sugawara decomposition. That is, we can write $T(z)$ as a sum of two commuting operators $T(z) = T_{KM}(z) + T'(z)$ such that $T'(z)$ has a regular OPE with the currents $J^a(z)$. The central charge of T_{KM} is

$$c_{KM} = \frac{k \dim \mathfrak{g}}{k + h(\mathfrak{g})} .$$

(2.3.19)

It follows that the total central charge can be written as $c = c_{KM} + c'$, and unitarity requires $c' \geq 0$. Furthermore, the KM and Virasoro primary states take the form $|\Phi_r\rangle \otimes |\Psi'\rangle$, where $|\Phi_r\rangle$ is primary with respect to T_{KM}, and $|\Psi'\rangle$ is primary with respect to T'. We will call such operators KMV primaries.
3. Every state is a linear combination of Virasoro and KM primary states, as well as their descendants.
4. Unitarity constrains the representations r and their conformal weights for any fixed level k. Restricting attention to T_{KM}, we have

$$L_0^{KM} = \frac{1}{2(k + h(\mathfrak{g}))}\left[\sum_a J_0^a J_0^a + \sum_{n>0}\sum_a J_{-n}^a J_n^a\right] ,$$

so that on a KM-primary state we have

$$L_0^{KM}|\Phi_r\rangle = \frac{C_2(r)}{2(k + h(\mathfrak{g}))}|\Phi_r\rangle = \frac{c_{KM}}{2k}\frac{\ell(r)}{\dim r}|\Phi_r\rangle .$$

Furthermore, if λ denotes the highest weight of r and ψ the highest root of $\mathfrak{g}$, then unitarity requires $\lambda \cdot \psi \leq k$.[11] This means that in a unitary theory there is a finite number of KM primary representations at any fixed level k.

2.3.3 The $\widehat{\mathfrak{u}(1)}_r$ Current Algebra and Twisting

The simplest of all KM algebras corresponds to a $\mathfrak{u}(1)$ symmetry algebra. In this case, we have just one current $J(z)$, along with KM primaries $\Phi_q(z)$ labeled by $q \in \mathbb{R}$, with the OPEs

$$J(z)J(w) \sim \frac{r}{(z-w)^2}\,, \qquad J(z)\Phi_q(w) \sim \frac{q\Phi_q(w)}{z-w}\,. \qquad (2.3.20)$$

Unitarity requires $r > 0$ as long as J is not identically zero. Unlike the case of a simple Lie algebra, there is no intrinsic integrality condition on the level r: by taking $J' = sJ$, we obtain $r' = s^2 r$ and $q' = sq$ for the $J'J'$ and $J'\Phi_q$ OPEs. However, suppose that the charge of every state is an integer multiple of some minimal charge. In that case there exists a unique s such that $q' \in \mathbb{Z}$ and $\gcd(q'_1, q'_2, \ldots) = 1$. When this holds, we will fix the normalization of the current accordingly and speak of a $\widehat{\mathfrak{u}(1)}_r$ KM algebra.

The Sugawara decomposition takes a particularly simple form for a $\widehat{\mathfrak{u}(1)}_r$ KM algebra. We represent the current as $J = i\sqrt{r}\partial\phi$, where ϕ is a free chiral boson with OPE

$$\phi(z)\phi(w) \sim -\log(z-w)\,. \qquad (2.3.21)$$

A KMV primary operator Φ_q with weights $(h, \bar{h})$ can be decomposed as

$$\Phi_q(z) = e^{iq\phi/\sqrt{r}}(z)\widetilde{\Phi}_{h-q^2/2r, \bar{h}}(z)\,, \qquad (2.3.22)$$

where $\widetilde{\Phi}$ is a $\mathfrak{u}(1)$-neutral KMV primary operator. The Sugawara energy-momentum tensor takes the form

$$T(z) = \frac{1}{2r} : JJ : (z) + T' = -\frac{1}{2} : \partial\phi\partial\phi : + T'\,, \qquad (2.3.23)$$

where T' has central charge $c - 1$.

Let $\mathcal{H}^0$ denote the Hilbert space of states in a CFT with a $\widehat{\mathfrak{u}(1)}_r$ algebra. The KMV primary states in $\mathcal{H}^0$ satisfy $L_{m>0}|\Phi\rangle = 0$, $\bar{L}_{m>0}|\Phi\rangle = 0$, and $J_{m>0}|\Phi\rangle = 0$

---

[11]The inner product $\lambda \cdot \psi$ is computed with the Cartan-Killing metric normalized as above: $\psi \cdot \psi = 2$.

and can be organized by the weights and charges $(h, q; \bar{h})$. Starting from $\mathcal{H}^0$, we can construct an isomorphic η-twisted Hilbert space $\mathcal{H}^\eta$ by introducing a background charge for J_0. We can think of this as defining a new ground state

$$|\eta\rangle = \lim_{z \to 0} e^{-i\eta\sqrt{r}\phi}(z)|0\rangle . \tag{2.3.24}$$

Evidently $J_0|\eta\rangle = -\eta r|\eta\rangle$ and $L_0|\eta\rangle = \frac{\eta^2 r}{2}|\eta\rangle$.

We call this a twist because the original fields now have a branch cut at $z = 0$:

$$\Phi_q(z)e^{-i\eta\sqrt{r}\phi}(w) \sim (z - w)^{-\eta q} : e^{iq\phi/\sqrt{r}}(z)e^{-i\eta\sqrt{r}\phi}(w) : \widetilde{\Phi}_{h-q^2/2r,\bar{h}} ,$$
$$\tag{2.3.25}$$

so that

$$\Phi_q(z^{2\pi i})|\eta\rangle = e^{-2\pi i \eta q}\Phi_q(z)|\eta\rangle . \tag{2.3.26}$$

If $\eta q \in \mathbb{Z}$ for all q in the spectrum, then we have returned to the original untwisted boundary conditions. With our assumptions this holds if and only if $\eta \in \mathbb{Z}$. More generally, we now have a family of isomorphic Hilbert spaces $\mathcal{H}^\eta$ for $\eta \in [0, 1)$, where the isomorphism $\mathcal{H}^\eta \simeq \mathcal{H}^0$ identifies states with $h^\eta = h - \eta q + \eta^2 r/2$ and $q^\eta = q - \eta r$. As the following exercise shows, the isomorphism between $\mathcal{H}^\eta$ and $\mathcal{H}$ respects the KM and Virasoro algebras.

Exercise 2.7 Let

$$L_n^\eta = L_n - \eta J_n + \frac{\eta^2 r}{2}\delta_{n,0} , \qquad J_n^\eta = J_n - \eta r \delta_{n,0} . \tag{2.3.27}$$

Show that the L_n^η and J_n^η realize an isomorphic algebra for every value of η. □

Let us stress several features of a CFT with a $\widehat{u(1)}_r$ KM symmetry uncovered by these twisted considerations.

1. The action of the Virasoro and KM generators on a state in the η-twisted sector is isomorphic to the action of L_n^η and J_n^η on the corresponding state in the untwisted sector with $\eta = 0$.
2. Under the twist the vacuum state $|0\rangle$ is mapped to $|\eta\rangle$ with $(h, q, ; \bar{h}) = (\eta^2 r/2, -\eta r, 0)$; since $\mathcal{H}^1 = \mathcal{H}^0$, it follows that $\mathcal{H}^0$ must contain a holomorphic operator with $(h, q; \bar{h}) = (r/2, -r; 0)$, and unitarity requires the conjugate holomorphic operator with same weight and opposite charge to be present in $\mathcal{H}^0$ as well.
3. The level r must be an integer: otherwise, once we twist by $\eta = 1$ we reach a contradiction with the assumption that $q \in \mathbb{Z}$. Unitarity also requires $r > 0$.
4. $\mathcal{H}$ must contain an infinite number of KMV primary fields with different q, and in particular an infinite number of holomorphic operators with $h = rk^2/2$ and $q = \pm rk$.

When $\widehat{u(1)}_r$ is contained in a simple $\widehat{g}_k$ KM symmetry the latter organizes the infinite number of KMV primary representations of the former into a finite set of KMV primary representations.

Exercise 2.8 In this exercise we explore the twisting in the context of a compact scalar CFT. A compact scalar is obtained by imposing a periodicity condition $\phi \sim \phi + 2\pi\rho$. This has two effects: first, it allows us to introduce twisted sectors: $\phi(ze^{2\pi i}, \bar{z}e^{-2\pi i}) = \phi(z, \bar{z}) + 2\pi\rho$, and it quantizes the mode expansion in each sector. The primary operators are[12]

$$ J = i\partial\phi, \qquad \bar{J} = i\bar{\partial}\phi, \qquad V_k = e^{ik_L\phi_L + ik_R\phi_R}, \qquad k_{L,R} = \frac{n}{\rho} \pm \frac{m\rho}{2}, $$

where n and m label, respectively, the momentum and winding modes, and $J, \bar{J}$ are conserved currents associated to the translation symmetry $\phi \rightarrow \phi + \text{const}$ of the scalar theory.

Show that there is a $\widehat{u(1)}_r$ KM algebra present if and only if $\rho = \sqrt{2a/b}$ for co-prime integers a, b, with $r = 2ab$. Verify that in that case the spectrum contains the holomorphic operators promised above.

Set $\rho = \sqrt{2}$: this is the self-dual radius $SU(2)_L \times SU(2)_R$ point, where $\sqrt{2}J$ and $\sqrt{2}\bar{J}$ generate the Cartan generators for $SU(2)_L \times SU(2)_R$ with the proper normalization. Find the remaining 4 currents and check that the current algebras have level 1.

Finally, check that away from $\rho = \sqrt{2}$ the operators have integral charges with respect to $\frac{\rho}{2}(J + \bar{J})$ and $\frac{1}{\rho}(J - \bar{J})$. This shows that, as expected, the $SU(2)_L \times SU(2)_R$ symmetry ensures that the charges are integral with respect to an unbroken $U(1) \times U(1)$ subalgebra, but it is only at the rational values $\rho = \sqrt{2a/b}$ that we have holomorphic and anti-holomorphic $\widehat{u(1)}_r$ algebras. □

Exercise 2.9 Consider the CFT of r free Weyl fermions γ^i and their conjugates $\bar{\gamma}_i$, with the only non-trivial OPE

$$ \gamma^i(z_1)\bar{\gamma}_j(z_2) \sim \frac{\delta^i_j}{z_{12}}. $$

Show that the currents $\gamma^i\gamma^j$, $\bar{\gamma}_i\bar{\gamma}_j$, and $:\gamma^i\bar{\gamma}_j:$ generate a level 1 $\mathfrak{so}(2r)$ KM algebra, with the terms corresponding to the decomposition

$$ \mathfrak{so}(2r) \supset \mathfrak{su}(r) \oplus \mathfrak{u}(1), $$

$$ \mathbf{2r} = \mathbf{r}_{+1} \oplus \bar{\mathbf{r}}_{-1}, $$

$$ \wedge^2\mathbf{r} = \text{adj}(\mathfrak{su}(r))_0 \oplus \mathbf{1}_0 \oplus (\wedge^2\mathbf{r})_{-2} \oplus (\wedge^2\bar{\mathbf{r}})_{+2}, \tag{2.3.28} $$

---

[12]In presenting these we are neglecting co-cycles for the V_k; these are discussed in [328].

where the $u(1)$ current is the diagonal combination

$$J =: \gamma^1 \bar{\gamma}_1 : + : \gamma^2 \bar{\gamma}_2 : + \cdots + : \gamma^r \bar{\gamma}_r : . \tag{2.3.29}$$

This system is an important example of the twisting construction. Consider the Cartan sub-algebra of $u(r)$, generated by the currents $J_i =: \gamma^i \bar{\gamma}_i$. For each Weyl fermion we can define its fermion number by $(-1)^{F_i} = e^{i\pi J_{i0}}$. This definition allows for a generalization of the treatment we gave for a single Majorana-Weyl fermion above.

We define twisting parameters $0 \leq \nu_i \leq 1$, so that

$$\gamma^i(ze^{2\pi i}) = -e^{2\pi i \nu_i} \gamma^i(z), \qquad \bar{\gamma}_i(ze^{2\pi i}) = -e^{-2\pi i \nu_i} \bar{\gamma}_i(z), \tag{2.3.30}$$

and the mode expansions are

$$\gamma^i(z) = \sum_{s \in \mathbb{Z} - \nu_i} \gamma_s^i z^{-s-1/2}, \qquad \bar{\gamma}_i(z) = \sum_{s \in \mathbb{Z} + \nu_i} \bar{\gamma}_{is} z^{-s-1/2}. \tag{2.3.31}$$

The twisted ground state $|\nu\rangle$ is the twist of the NS vacuum, and it is the state annihilated by all positive modes γ_s^i and $\bar{\gamma}_{is}$ for $s > 0$. Show that its charges with respect to J_i and its weights with respect to

$$T_i = -\gamma^i \partial \bar{\gamma}_i - \bar{\gamma}_i \partial \gamma^i \tag{2.3.32}$$

are

$$q_i = \nu_i - \frac{1}{2}, \qquad\qquad h_i = \frac{(1 - 2\nu_i)^2}{8}. \tag{2.3.33}$$

This can be done directly by evaluating the 1-point functions for J_i and T_i in the twisted state through the mode expansion and canonical anti-commutation relations—the only non-zero one is $\{\gamma_s^i, \bar{\gamma}_{jt}\} = \delta_j^i \delta_{s,-t}$. Compare this to the general twisting discussion given above.

The NS and R sectors correspond to, respectively, $\nu_i = 1/2$ and $\nu_i = 0$ for all i. The $\nu_i \to 0$ limit is particularly interesting, since the fermions then have zero modes. With our conventions we have $\bar{\gamma}_{i0}|0\rangle = 0$, while the γ_0^i act as raising operators.

The ground state carries $|\nu\rangle$ carries a (typically fractional) fermion number with respect to $(-1)^{F_\gamma} = e^{i\pi J_0} = \prod_i (-1)^{F_i}$. Apply this to the Ramond sector with $\nu_i = 0$, for all i and verify

$$J_0|0\rangle = -\frac{r}{2}|0\rangle, \qquad L_0|0\rangle = \frac{r}{8}|0\rangle, \qquad (-1)^{F_\gamma}|0\rangle = e^{-i\pi r/2}|0\rangle. \tag{2.3.34}$$

The full set of ground states decomposes into spinor representations of $\mathfrak{so}(2r)$:

$$2^{r-1} = |0\rangle_{-r/2} \oplus_{i>j} \gamma^i\gamma^j|0\rangle_{-r/2+2} \oplus \text{terms with even number of } \gamma s$$

$$(2^{r-1})' = \oplus_i \gamma^i|0\rangle_{-r/2+1} \oplus_{i>j>k} \gamma^i\gamma^j\gamma^k|0\rangle_{-r/2+3}$$

$$\oplus \text{ terms with odd number of } \gamma s\,. \tag{2.3.35}$$

These are distinguished by the action of the chirality matrix, which, up to a phase, acts as $(-1)^{F_\gamma}$.[13]

It is sometimes convenient to work with a different parametrization for the twist parameter, for instance one where $\gamma(e^{2\pi i}z) = -e^{2\pi i\tilde{\nu}}\gamma(z)$ and $-1 < \tilde{\nu} \le 0$, with $\tilde{\nu} = -1/2$ corresponding to the untwisted theory. Work out the details of the twisting for a single Weyl fermion by the current $J = q : \gamma\overline{\gamma} :$ which assigns charge q to γ. You should find

$$J_0|\tilde{\nu}\rangle = q\left(\tilde{\nu} + \tfrac{1}{2}\right)|\tilde{\nu}\rangle\,, \qquad L_0|\tilde{\nu}\rangle = \tfrac{1}{8} + \tfrac{1}{2}\tilde{\nu}\,(\tilde{\nu}+1)\,|\tilde{\nu}\rangle\,, \tag{2.3.36}$$

and that for $\tilde{\nu} = 0$ the γ_0 annihilates the ground state, while $\overline{\gamma}_0$ acts as a raising operator.   □

2.4 The Superconformal Algebra

Having gone through a brief review of the basic structures, we now turn our attention to the N=2 superconformal algebra.

2.4.1 The Global Superconformal Algebra

We are already familiar with its most important component—the N=2 supersymmetry algebra—from Chap. 1. Modifying the notation slightly to suit our applications here, we have the supercharges $G^\pm_{-1/2}$, as well as the translation generator L_{-1} and the R-charge J_0, which satisfy[14]

$$\{G^\pm_{-1/2}, G^\pm_{-1/2}\} = 0\,, \quad \{G^+_{-1/2}, G^-_{-1/2}\} = 2L_{-1}\,, \quad [J_0, G^\pm_{-1/2}] = \pm G^\pm_{-1/2}\,. \tag{2.4.1}$$

---

[13]Recall that when r is odd the spinor representations of $\mathfrak{so}(2r)$ are complex and conjugate; when $r \in 4\mathbb{Z}$, the spinor representations are real; when $r \in 2 + 4\mathbb{Z}$, the spinor representations are pseudoreal.

[14]There is a delightful clash of two reasonable uses of $\pm$ in the literature: the first denotes world-sheet chirality (we used this in the Minkowski conventions of Chap. 1), and the second denotes the U(1) charges. The clash is particularly painful when working with (2,2) theories. Since we mostly work with a Euclidean world-sheet and just one copy of the N=2 algebra, we typically use the second notation.

In a conformal theory this is enlarged to a global superconformal algebra that includes additional charges L_0, L_1, and $G^{\pm}_{1/2}$. These are, respectively, the generators of dilatations, special conformal, and superconformal transformations. The non-vanishing commutators are

$$[L_m, L_n] = (m - n)L_{m+n} , \qquad [L_m, G^{\pm}_r] = (\tfrac{m}{2} - r)G^{\pm}_{m+r} , \qquad (2.4.2)$$
$$[J_0, G^{\pm}_r] = \pm G^{\pm}_r , \qquad \{G^+_r, G^-_s\} = 2L_{r+s} + (r - s)J_{r+s} ,$$

with $m = 0, \pm 1$ and $r, s = \pm 1/2$.

2.4.2 Unitary Representations

The states transform in representations of this algebra, and we organize them as follows. First, L_0 and J_0 commute with all operators and are Hermitian with respect to the inner product defined above. We will assume that these can be diagonalized (a sensible assumption in a unitary compact theory), and we will organize operators into subspaces with fixed L_0 and J_0 eigenvalues.

A state $|\phi\rangle$ is N=2 quasi-primary if and only if

$$L_0|\phi\rangle = h_\phi|\phi\rangle , \qquad J_0|\phi\rangle = q_\phi|\phi\rangle , \qquad G^{\pm}_{1/2}|\phi\rangle = 0 . \qquad (2.4.3)$$

The state $|\phi\rangle$ is also a conformal quasi-primary since $L_1 = \frac{1}{2}\{G^+_{1/2}, G^-_{1/2}\}$. Among all the states we distinguish two important classes, the chiral and anti-chiral states, defined by

$$\text{chiral} : \quad G^+_{-1/2}|\phi\rangle = 0 , \quad \text{anti-chiral} : \quad G^-_{-1/2}|\phi\rangle = 0 . \qquad (2.4.4)$$

The vacuum is the unique state annihilated by all of the symmetry generators: it is quasi-primary, chiral, and anti-chiral.

Unitary superconformal representations satisfy some important constraints. Following the logic below Exercise 2.3, the Hermitian conjugates of the superconformal generators are

$$L^{\dagger}_n = L_{-n} , \qquad J^{\dagger}_n = J_{-n} , \qquad (G^{\pm}_r)^{\dagger} = G^{\mp}_{-r} . \qquad (2.4.5)$$

Hence, for any state $|\phi\rangle$ of weight and charge (h, q), we have

$$\|G^{\mp}_{1/2}|\phi\rangle\|^2 + \|G^{\pm}_{-1/2}|\phi\rangle\|^2 = (2h \mp q)\||\phi\rangle\|^2 . \qquad (2.4.6)$$

It follows that $2h \geq \pm q$ and $h \geq 0$ for all states. The bound $2h \geq \pm q$ is saturated by two distinguished types of states:

1. chiral quasi-primary states, which satisfy $G^{\mp}_{\pm 1/2}|\phi\rangle = 0$ and $h_\phi = q_\phi/2$;
2. anti-chiral quasi-primary states, which satisfy $G^{\pm}_{\mp 1/2}|\phi\rangle = 0$ and $h_\phi = -q_\phi/2$.

Given any N=2 quasi-primary state $|\phi\rangle$, we can construct an N=2 multiplet by first acting with the raising operators $G^{\pm}_{-1/2}$ to obtain

$$|\psi^{\pm}\rangle = G^{\pm}_{-1/2}|\phi\rangle \ . \tag{2.4.7}$$

These states, if non-trivial, are conformal quasi-primary and have $(h, q) = (h_\phi + 1/2, q_\phi \pm 1)$. To complete the multiplet let

$$|X\rangle = G^{+}_{-1/2}G^{-}_{-1/2}|\phi\rangle - (1 + \tfrac{q_\phi}{2h_\phi})L_{-1}|\phi\rangle \ . \tag{2.4.8}$$

This state is constructed to satisfy $L_1|X\rangle = 0$, and the N=2 superconformal algebra closes on the "long multiplet" with conformal quasi-primary states $\{|\phi\rangle, |\psi^+\rangle, |\psi^-\rangle, |X\rangle\}$.

Exercise 2.10 Verify the previous assertion by showing

$$G^{+}_{-1/2}|\psi^+\rangle = G^{-}_{-1/2}|\psi^-\rangle = 0,$$

$$G^{+}_{-1/2}|\psi^-\rangle = |X\rangle + (1 + \tfrac{q}{2h})L_{-1}|\phi\rangle, \quad G^{-}_{-1/2}|\psi^+\rangle = -|X\rangle + (1 - \tfrac{q}{2h})L_{-1}|\phi\rangle,$$

$$G^{+}_{1/2}|X\rangle = -\frac{(2h+q)(2h+1)}{2h}|\psi^+\rangle, \quad G^{-}_{1/2}|X\rangle = \frac{(2h-q)(2h+1)}{2h}|\psi^-\rangle,$$

$$G^{+}_{-1/2}|X\rangle = -(1 + \tfrac{q}{2h})L_{-1}|\psi^+\rangle, \quad G^{-}_{-1/2}|X\rangle = (1 - \tfrac{q}{2h})L_{-1}|\psi^-\rangle. \quad \square$$

Chiral N=2 quasi-primary states reside in shortened multiplets because the $|\psi^+\rangle$ descendant is null, so that the multiplet consists of $\{|\phi\rangle, |\psi^-\rangle\}$, with

$$G^{\pm}_{1/2}|\phi\rangle = 0 \ , \quad G^{+}_{-1/2}|\phi\rangle = 0 \ , \qquad G^{-}_{-1/2}|\phi\rangle = |\psi^-\rangle \ ,$$

$$G^{-}_{\pm 1/2}|\psi^-\rangle = 0 \ , \quad G^{+}_{1/2}|\psi^-\rangle = 2q|\phi\rangle \ , \quad G^{+}_{-1/2}|\psi^-\rangle = 2L_{-1}|\phi\rangle \ . \tag{2.4.9}$$

There is an analogous story for anti-chiral N=2 quasi-primary states and corresponding short multiplets $\{|\phi\rangle, |\psi^+\rangle\}$.

2.4.3 A Little More Superspace

The superconformal algebra organizes local operators into supermultiplets. The N=2 quasi-primary operators reside in long or short multiplets. For example, the operators in a chiral multiplet obey

$$[G^+_{-1/2}, \phi] = 0\,, \qquad\qquad \{G^+_{-1/2}, \psi^-\} = 2[L_{-1}, \phi] = 2\partial\phi\,,$$

$$[G^-_{-1/2}, \phi] = \psi\,, \qquad\qquad [G^-_{-1/2}, \psi] = 0\,. \qquad\qquad (2.4.10)$$

It is useful to arrange the quasi-primary fields in superfields. To do so, we return to the Euclidean superspace discussed in the previous chapter but now, following [132, 376] apply it to our operators rather than free fields. Since we have also been discussing a holomorphic N=2 algebra, we also send $\bar\partial \to \partial$ to obtain the super-derivatives and supercharges

$$\mathcal{D} = \partial_\theta + \bar\theta\partial\,, \qquad\qquad \mathcal{Q} = \mathcal{G}^-_{-1/2} = \partial_\theta - \bar\theta\partial\,,$$

$$\overline{\mathcal{D}} = \partial_{\bar\theta} + \theta\partial\,, \qquad\qquad \overline{\mathcal{Q}} = \mathcal{G}^+_{-1/2} = \partial_{\bar\theta} - \theta\partial\,. \qquad\qquad (2.4.11)$$

The components of a long multiplet can now be arranged as

$$\Phi = \phi + \theta\psi^- + \bar\theta\psi^+ + \theta\bar\theta(X + \tfrac{q}{2h}\partial\phi)\,, \qquad\qquad (2.4.12)$$

and the action of the supercharges is encoded by

$$[\xi G^-_{-1/2} - \bar\xi G^+_{-1/2}, \Phi] = (\xi\mathcal{G}^-_{-1/2} - \bar\xi\mathcal{G}^+_{-1/2})\Phi\,. \qquad\qquad (2.4.13)$$

The two definitions of chirality coincide: on one hand a chiral multiplet has $2h = q$ and $\psi^+ = X = 0$; on the other hand, $\overline{\mathcal{D}}\Phi = 0$ holds if and only if $X = (1 - q/2h)\partial\phi$ and $\psi^+ = 0$, in which case the superfield takes the familiar form

$$\Phi = \phi + \theta\psi^- + \theta\bar\theta\partial\phi\,. \qquad\qquad (2.4.14)$$

Exercise 2.11 In this exercise we develop a differential representation of the global superconformal algebra. To start, we recall the operators

$$\mathcal{L}^{(z)}_{-1} = -\partial\,, \qquad\qquad \mathcal{L}^{(z)}_0 = -z\partial\,, \qquad\qquad \mathcal{L}^{(z)}_1 = -z^2\partial$$

that represent the $\mathfrak{sl}_2\mathbb{C}$ algebra on the plane. It is easy to extend this to a representation of the global superconformal algebra on the superspace $(z, \theta, \bar\theta)$ in a few steps. First, we already know that $\mathcal{L}_{-1} = -\partial$ and $\mathcal{G}^\pm_{-1/2}$ yield a representation of the supersymmetry algebra. Second, the R-symmetry assigns charge $+1$ to θ and -1 to $\bar{\theta}$ and leaves z invariant, so that

$$\mathcal{J}_0 = \theta \partial_\theta - \bar{\theta} \partial_{\bar{\theta}}$$

satisfies $[\mathcal{J}_0, \mathcal{G}^\pm_{-1/2}] = \pm \mathcal{G}^\pm_{-1/2}$. We also know that z and $\theta, \bar{\theta}$ have dilatation weights, respectively -1 and $-1/2$, so that

$$\mathcal{L}_0 = -z\partial - \tfrac{1}{2}(\theta \partial_\theta + \bar{\theta} \partial_{\bar{\theta}}).$$

Show that to close $\mathfrak{sl}_2\mathbb{C}$ we then need

$$\mathcal{L}_1 = -z^2 \partial - z(\theta \partial_\theta + \bar{\theta} \partial_{\bar{\theta}}).$$

Use this to derive the remaining generators

$$\mathcal{G}^\pm_{1/2} = [\mathcal{L}_1, \mathcal{G}^\pm_{-1/2}] = (z \mp \theta\bar{\theta})\mathcal{G}^\pm_{-1/2}. \qquad \square$$

A motivation for the superfield construction is that it gives a succinct presentation of the correlation functions for the multiplets, and the global Ward identities that follow from the superconformal invariance are realized as differential operators that annihilate the correlation functions. That is, analogously to (2.1.19), a correlation function

$$\langle \Phi_1(z_1)\Phi_2(z_2)\cdots\Phi_n(z_n)\rangle \tag{2.4.15}$$

that depends on the $z_i = (z_i, \theta_i, \bar{\theta}_i; \bar{z}_i)$ must be annihilated by the operators

$$\mathcal{J}_0 = \sum_i \left[\theta_i \partial_{\theta_i} - \bar{\theta}_i \partial_{\bar{\theta}_i} - q_i\right], \quad \mathcal{L}_0 = \sum_i \left[-h_i - z_i \partial_{z_i} - \tfrac{1}{2}(\theta_i \partial_{\theta_i} + \bar{\theta}_i \partial_{\bar{\theta}_i})\right],$$

$$\mathcal{L}_{-1} = \sum_i \left[-\partial_{z_i}\right], \quad \mathcal{L}_1 = \sum_i \left[-q\theta_i\bar{\theta}_i - 2z_i h_i - z_i^2 \partial_{z_i}\right.$$

$$\left. -z_i(\theta_i \partial_{\theta_i} + \bar{\theta}_i \partial_{\bar{\theta}_i})\right],$$

$$\mathcal{G}^+_{-1/2} = \sum_i \overline{\mathcal{Q}}_i, \quad \mathcal{G}^+_{1/2} = \sum_i \left[(z_i - \theta_i\bar{\theta}_i)\overline{\mathcal{Q}}_i - (2h_i + q_i)\theta_i\right]$$

$$\mathcal{G}^-_{-1/2} = \sum_i \mathcal{Q}_i, \quad \mathcal{G}^-_{1/2} = \sum_i \left[(z_i + \theta_i\bar{\theta}_i)\mathcal{Q}_i - (2h_i - q_i)\bar{\theta}_i\right].$$

$$\tag{2.4.16}$$

The holomorphic dependence of the two-point function is determined by these identities: $\langle\Phi_1(z_1)\Phi_2(z_2)\rangle$ is zero unless $h_1 = h_2 = h$, $\bar{h}_1 = \bar{h}_2 = \bar{h}$, and $q_1 = -q_2 = q$, in which case it takes the form

$$\langle\Phi_1(z_1)\Phi_2(z_2)\rangle = \frac{C_{12}}{\zeta_{12}^{2h}\,\bar{z}_{12}^{2\bar{h}}}\left(1 - \frac{q\theta_{12}\bar{\theta}_{12}}{\zeta_{12}}\right), \tag{2.4.17}$$

where

$$\zeta_{12} = z_{12} - \theta_1\bar{\theta}_2 - \bar{\theta}_1\theta_2\,, \qquad \theta_{12} = \theta_1 - \theta_2\,, \qquad \bar{\theta}_{12} = \bar{\theta}_1 - \bar{\theta}_2 \tag{2.4.18}$$

are supersymmetric invariants. This means that the two-point functions of SUSY descendants are fixed by those of the quasi-primary fields.

The superspace dependence of three-point functions of general N=2 quasi-primary operators is not fixed by superconformal invariance: there are too many independent supersymmetric invariants generalizing those that show up in the two-point function. However, the dependence is determined for three-point functions with one general quasi-primary superfield and the remaining two fields being either both chiral, both anti-chiral, or one of each chirality. In addition, three-point functions of supercurrent multiplets are also fixed.

Exercise 2.12 Determine the superspace dependence of a three-point function of quasi-primary fields

$$\langle\Phi_1(z_1)\overline{\Phi}_2(z_2)A(z_3)\rangle\,,$$

where Φ and $\overline{\Phi}$ are, respectively, chiral and anti-chiral. An effective way to do this is to first find a basis for the supersymmetric invariants that can be constructed from z_1, z_2, z_3 subject to the chirality conditions. With these in hand, make an Ansatz for the correlator and then use the remaining constraints from (2.4.16). You should find that the correlation function vanishes unless $q_{\text{tot}} = q_1 + q_2 + q_A \in \{0, 1\}$. When $q_{\text{tot}} = 0$, then A must be an anti-chiral primary, in which case

$$\langle\Phi_1(z_1)\overline{\Phi}_2(z_2)A(z_3)\rangle = \frac{F(\bar{z})}{\xi_{13}^{2h_1}\,\xi_{23}^{2h_2}}\,, \qquad \xi_{ik} = z_{ik} + \theta_i\bar{\theta}_{ik} + \theta_{ki}\bar{\theta}_k\,.$$

Check this result and then derive the expression for the case $q_{\text{tot}} = 1$.   □

2.4.4 The Super-Virasoro Algebra

The global superconformal algebra is a sub-algebra of the infinite-dimensional super-Virasoro algebra encoded in the conserved holomorphic currents $J(z)$, $G^{\pm}(z)$, and $T(z)$, respectively of spin 1, 3/2, and 2, with the OPEs

$$T(z)J(w) \sim \frac{J(w)}{(z-w)^2} + \frac{\partial J(w)}{z-w} \,,$$

$$T(z)G^{\pm}(w) \sim \frac{3/2 G^{\pm}(w)}{(z-w)^2} + \frac{\partial G^{\pm}(w)}{(z-w)} \,,$$

$$J(z)G^{\pm}(w) \sim \frac{\pm G^{\pm}(w)}{z-w} \,. \tag{2.4.19}$$

The remaining OPEs involve central terms that are fixed by closure of the algebra in terms of the central charge c. The non-trivial OPEs are

$$J(z)J(w) \sim \frac{c/3}{(z-w)^2} \,,$$

$$G^+(z)G^-(w) \sim \frac{2c/3}{(z-w)^3} + \frac{2J(w)}{(z-w)^2} + \frac{2T(w) + \partial J(w)}{z-w} \,,$$

$$T(z)T(w) \sim \frac{c/2}{(z-w)^4} + \frac{2T(w)}{(z-w)^2} + \frac{\partial T(w)}{z-w} \,. \tag{2.4.20}$$

Exercise 2.13 Show that these fields assemble into a supercurrent multiplet as in (1.7.5). With our superspace conventions it takes the form

$$\mathcal{S} = J + \theta G^- - \bar{\theta} G^+ + 2\theta\bar{\theta} T \,. \tag{2.4.21}$$

Check that the two-point function is given by

$$\langle \mathcal{S}(z_1)\mathcal{S}(z_2) \rangle = \frac{c/3}{\zeta_{12}^2} \,. \tag{2.4.22}$$

$\square$

Using the mode expansions

$$T(z) = \sum_{n\in\mathbb{Z}} L_n z^{-n-2} \,, \quad J(z) = \sum_{n\in\mathbb{Z}} J_n z^{-n-1} \,, \quad G^{\pm}(z) = \sum_{r\in\mathbb{Z}\pm 1/2\pm\eta} G_r^{\pm} z^{-r-3/2}$$

$$\tag{2.4.23}$$

then leads to the mode algebra that includes the Virasoro and U(1) KM algebras, together with additional (anti)commutators

$$[L_n, G_r^{\pm}] = (\tfrac{m}{2} - r)G_{r+n}^{\pm} \,, \qquad [J_n, G_r^{\pm}] = \pm G_{r+n}^{\pm} \,, \qquad \{G_r^{\pm}, G_s^{\pm}\} = 0 \,,$$

$$\{G_r^+, G_s^-\} = 2L_{r+s} + (r-s)J_{r+s} + \tfrac{c}{12}(4r^2 - 1)\delta_{r,-s} \,. \tag{2.4.24}$$

We introduced the parameter η in the moding of $G^\pm$, anticipating a twist by the KM algebra. The untwisted sector, where $G^\pm(z)|0\rangle$ is single-valued, has $\eta = 0$. As in the discussion of the twisted fermion above, this is the NS sector. The Hermitian adjoints of the operators take the form shown in (2.4.5), with n, r now running over the full range of values.

Sticking for now to the NS sector, we can organize the states of the CFT into multiplets of this algebra. N=2 primary states are defined to be annihilated by all the lowering modes $L_{n>0}$, $J_{n>0}$, and $G^\pm_{r>0}$, and all other states are obtained by acting on the primary states with raising modes. The chiral/anti-chiral quasi-primary states identified above remain distinguished in the full algebra.

Exercise 2.14 Use the N=2 super-Virasoro algebra to show that in a unitary theory a chiral (anti-chiral) state with $q = 2h$ $(-q = 2h)$ is N=2 primary.   □

The N=2 super-Virasoro algebra has some further implications for unitarity. For instance, the norm of $G^\mp_{-r}$ on an N=2 primary state of weight h and charge q follows from the algebra:

$$\| G^\mp_{-r} |\Phi\rangle \|^2 = \left(2h \pm 2rq + \tfrac{c}{12}(4r^2 - 1) \right) \| |\Phi\rangle \|^2 . \tag{2.4.25}$$

Setting $r = 3/2$, we find $2h \pm 3q + 2c/3 \geq 0$, so that for chiral or anti-chiral primary states we find $h \leq c/6$.

Further implications of unitarity are discussed in [95, 159]. As in the case of the Virasoro algebra, when the central charge is sufficiently small (it turns out the value is $c < 3$) these unitarity bounds restrict the possible representations to the N=2 minimal models that exist for

$$c = \frac{3m}{m + 2} , \qquad\qquad m = 1, 2, \ldots . \tag{2.4.26}$$

2.4.5 The Chiral Ring

Consider the OPE of two chiral primary operators,

$$\Phi_1(z_1, \bar{z}_1)\Phi_2(z_2, \bar{z}_2) \sim \sum_s C^s_{12} z_{12}^{h_s - h_1 - h_2} \bar{z}_{12}^{\bar{h}_s - \bar{h}_1 - \bar{h}_2} \Psi_s(z_2, \bar{z}_2) . \tag{2.4.27}$$

$U(1)_R$ charge conservation implies that only operators with $q_s = q_1 + q_2$ contribute, and this allows us to rewrite the OPE as

$$\Phi_1(z_1, \bar{z}_1)\Phi_2(z_2, \bar{z}_2) \sim \sum_{s | q_s = q_1 + q_2} C^s_{12} z_{12}^{h_s - q_s/2} \bar{z}_{12}^{\bar{h}_s - \bar{h}_1 - \bar{h}_2} \Psi_s(z_2, \bar{z}_2) . \tag{2.4.28}$$

In a theory with (2,2) superconformal invariance we can make a further restriction to operators $\Phi \in H_{cc}$, where

$$H_{cc} = \{\Phi \in \mathcal{H} \mid h = q/2 \text{ and } \overline{h} = \overline{q}/2\}, \tag{2.4.29}$$

i.e. Φ is left-chiral-primary and right-chiral-primary. Since $h \leq c/6$ for a chiral primary operator, it follows that $\dim H_{cc} < \infty$ in a compact CFT. Furthermore, we can grade this subspace by the charges of left- and right-moving R-symmetries $u(1)_L \oplus u(1)_R$:

$$H_{cc} = \bigoplus_{q,\overline{q}} H_{cc}^{q,\overline{q}}. \tag{2.4.30}$$

The OPE of operators in H_{cc} is non-singular:

$$\Phi_1(z_1, \overline{z}_1)\Phi_2(z_2, \overline{z}_2) \sim \sum_{\substack{q_s = q_1 + q_2 \\ \overline{q}_s = \overline{q}_1 + \overline{q}_2}} C_{12}{}^s z_{12}^{h_s - q_s/2} \overline{z}_{12}^{\overline{h}_s - \overline{q}_s/2} \Psi_s(z_2, \overline{z}_2).$$

$$\tag{2.4.31}$$

In this case the limit $z_1 \to z_2$ yields another operator in H_{cc}:

$$\lim_{z_1 \to z_2} \Phi_i(z_1)\Phi_j(z_2) = C_{ij}{}^k \Phi_k(z_2). \tag{2.4.32}$$

This endows H_{cc} with a ring structure that respects the $u(1)_L \oplus u(1)_R$ grading. In an analogous fashion we can also define H_{ac}, consisting of left-anti-chiral primary and right-chiral primary operators, as well as the complex-conjugate versions H_{aa} and H_{ca}.

2.4.6 Hodge Decomposition

As we will explore in greater detail in Chap. 4, there are many parallels between notions of complex geometry and the N=2 algebra. Perhaps the most basic is the statement that any state $|\phi\rangle$ can be uniquely written as a linear combination

$$|\phi\rangle = |\psi\rangle + G^+_{-1/2}|\chi\rangle + G^-_{+1/2}|\lambda\rangle, \tag{2.4.33}$$

where $|\psi\rangle$ is a chiral primary state. To see this, observe that our assumption of compactness guarantees that the space of chiral primary states with a fixed value of $\overline{h}$ is finite-dimensional. Working in the Hilbert space of states with a fixed $\overline{h}$, this means we have an orthogonal projector $\Pi = \Pi^\dagger$ from the Hilbert space $\mathcal{H}_{\overline{h}}$ to the subspace of chiral primary states. It follows that

$$\mathbb{1} - \Pi = (2L_0 - J_0)\Sigma \,, \tag{2.4.34}$$

where the operator Σ annihilates all chiral primary states and otherwise acts by multiplication: $\Sigma|\phi_{h,q}\rangle = (2h - q)^{-1}|\phi_{h,q}\rangle$, but this means

$$\mathbb{1} = \Pi + G^+_{-1/2}G^-_{1/2}\Sigma + G^-_{1/2}G^+_{-1/2}\Sigma \,. \tag{2.4.35}$$

Hence (2.4.33) holds with

$$|\psi\rangle = \Pi|\phi\rangle \,, \qquad |\chi\rangle = G^-_{1/2}\Sigma|\phi\rangle \,, \qquad |\lambda\rangle = G^+_{-1/2}\Sigma|\phi\rangle \,. \tag{2.4.36}$$

Observe that this decomposition into three mutually orthogonal components respects the grading by q. The reader should compare this to the Hodge decomposition in Sect. B.3.1 of the geometric appendix.

Another parallel with geometry is in the relationship between the chiral primary states and the $G^+_{-1/2}$ cohomology. The latter can be defined as follows. Let

$$H_{c,\bar{h}} = \bigoplus_q H^q_{c,\bar{h}} \,, \qquad H^q_{c,\bar{h}} = \left.\frac{\{\ker G^+_{-1/2} \cap \mathcal{H}_{\bar{h}}\}}{\{\operatorname{im} G^+_{-1/2} \cap \mathcal{H}_{\bar{h}}\}}\right|_{\text{R charge } q} \,. \tag{2.4.37}$$

While $\mathcal{H}_{\bar{h}} \cap \ker G^+_{-1/2}$ is infinite-dimensional, the cohomology groups are finite-dimensional and in fact isomorphic to the space of chiral-primary operators with right-moving weight $\bar{h}$ and R-charge q. The geometric analogue is the cohomology of the Dolbeault operator $\bar{\partial}$ and harmonic forms of the corresponding Laplacian. In either case, the result is a straightforward consequence of the Hodge decomposition, as the reader can verify.

In the special case of (2,2) supersymmetry we can obtain decompositions like this on both the left and right. In that case we have a stronger statement: any state $|\phi\rangle$ can be decomposed as

$$|\phi\rangle = |\psi_{c,c}\rangle + (G^+_{-1/2} + \overline{G}^+_{-1/2})|\chi\rangle + (G^-_{1/2} + \overline{G}^-_{1/2})|\lambda\rangle \,, \tag{2.4.38}$$

where $|\psi_{c,c}\rangle$ belongs to the cc ring. This follows from exactly analogous arguments applied to the diagonal (left-right symmetric) N=2 algebra. The geometric parallel is to the Hodge–De Rham decomposition on a compact Kähler manifold.

2.4.7 Spectral Flow

The U(1) KM symmetry can be "bosonized" as above by writing $J = i\sqrt{\frac{c}{3}}\partial\phi$. This allows us to introduce the twisted ground states

$$|\eta\rangle = \lim_{z\to 0} e^{-i\phi\eta\sqrt{c/3}}|0\rangle, \tag{2.4.39}$$

and since the supercurrents $G^\pm$ are charged with respect to U(1), their moding is shifted as indicated in (2.4.23). For $\eta = 0$, (the NS sector), $G^\pm(z)$ are single-valued. More generally, we have an isomorphism of the different twisted Hilbert spaces $\mathcal{H}^\nu$, and this isomorphism respects the N=2 algebra: a simple generalization of Exercise 2.7 shows that we can represent the action of the N=2 algebra in a twisted sector labeled by η by an isomorphic algebra defined on the untwisted sector with generators

$$L_n^\eta = L_n - \eta J_n + \frac{\eta^2 c}{6}\delta_{n,0}, \quad J_n^\eta = J_n - \frac{\eta c}{3}\delta_{n,0}, \quad G_r^{\eta\pm} = G_{r\mp\eta}^\pm, \quad r \in \mathbb{Z} \pm \eta \pm 1/2. \tag{2.4.40}$$

This equivalence goes by the name of "spectral flow" [283, 340].

As we saw in our general discussion of twisting, quantization of the U(1) charges has important implications. So it is here. Consider, first, the situation where the charges with respect to J of the N=2 algebra are quantized. In this case we have a natural isomorphism from the NS sector to itself obtained by "1 unit of spectral flow," i.e. $\eta = 1$. Under this identification the chiral primary states with charge q and weight $h = q/2$ are mapped to

$$(q, q/2) \xrightarrow{\quad\eta=1\quad} (q - c/3, -(q - c/3)/2). \tag{2.4.41}$$

Thus, we observe that

1. in such theories $c/3 \in \mathbb{Z}$;
2. the space of chiral primary states is isomorphic to that of anti-chiral primary states, with an "order-reversing" isomorphism $q \to q - c/3$;
3. there exists a unique holomorphic anti-chiral primary operator $\overline{\Omega}(z)$ with $q = -c/3$ that is the image of the vacuum. Its conjugate is the chiral primary $\Omega(z)$ with $q = c/3$.

The last point implies that the theory has an extended chiral algebra that is explored further in [160].

The operators are $\mathbb{Z}_2$ graded with respect to the ± 1 eigenvalues of $e^{i\pi J_0}$, a grading reminiscent of that by a fermion number operator $(-1)^F$. This is more than a coincidence: in most theories that we will discuss we will be able to identify $e^{i\pi J_0}$

as a "left-moving" contribution to the fermion number. In these sorts of theories, the flow by 1/2 unit of spectral flow leads to the Ramond sector, which has some important features. For example, the supercurrents have zero-modes $G_0^\pm$, which satisfy

$$\{G_0^+, G_0^-\} = 2\left(L_0 - \frac{c}{24}\right) = 2E_0 \,. \tag{2.4.42}$$

We define the Ramond ground states as the states annihilated by $G_0^\pm$.[15] Obviously these states have a vanishing ground state energy E_0, and we can identify them via (2.4.40) with the corresponding states in the NS sector: since $G_0^{\eta\pm} = G_{\mp 1/2}^\pm$, the Ramond ground states are in 1:1 correspondence with the chiral primary states: a chiral primary state with charge q corresponds to a Ramond ground state with charge $q - c/6$.

Rational Charges

More generally, we might have a situation where the J charges are rational, so that for some choice of level $r \in \mathbb{Z}$, not necessarily with $r = c/3$, we have a $\widehat{u(1)}_r$ KM algebra of the sort defined above. In this case, we will still have the natural identification of the untwisted sector with $\mathcal{H}^1$, as well as the corresponding holomorphic operators with charge $\pm r$.

2.4.8 N = 1, 3, 4

The N=2 algebra, whether superconformal or merely supersymmetric, obviously plays a key role in (0,2) QFT. However, other two-dimensional supersymmetries are also occasionally useful, and in this section we will make a brief tour of these structures.

Every N=2 algebra has an N=1 sub-algebra with a supercurrent

$$G(z) = \frac{1}{\sqrt{2}}G^+(z) + \frac{1}{\sqrt{2}}G^-(z) \,, \tag{2.4.43}$$

---

[15]The off-set of $-c/24$ in the ground state energy E_0 has a nice interpretation: if we make a conformal transformation to go from the plane to the cylinder via $z = e^w$, we obtain a standard Hamiltonian formulation of our theory on $S^1 \times \mathbb{R}$, with $\mathbb{R}$ specifying the Euclidean time. In making this transformation, $T(z)$ picks up an inhomogeneous shift by a Schwartzian derivative which leads to a shift of the modes: $(L_0)_{\text{cyl}} = L_0 - c/24$.

and therefore OPE

$$G(z)G(w) \sim \frac{2c/3}{(z-w)^3} + \frac{2T(w)}{z-w} . \qquad (2.4.44)$$

We can be a bit more general and take $G(z) = \frac{1}{\sqrt{2}} \left[e^{i\alpha} G^+(z) + e^{-i\alpha} G^-(z) \right]$ for any phase α. The N=1 algebra is less constraining than the N=2 algebra but deserves our attention as a fundamental symmetry of the RNS formulation in superstrings and heterotic strings: the most general background of the critical perturbative heterotic string is an N=1 superconformal theory. That being said, it is perhaps not surprising that it is not an easy class of theories to tackle in full generality.

The remaining possibilities are classified once we make some reasonable assumptions:

1. the energy momentum tensor T is the unique holomorphic field with spin 2;
2. the remaining fields have spins $s \in \{1/2, 1, 3/2\}$;
3. the algebra contains the N-extended supersymmetry algebra

$$\{G^A_{-1/2}, G^B_{-1/2}\} = 2\delta^{AB} L_{-1}$$

as a subalgebra;
4. the Virasoro algebra is a sub-algebra as well.

Remarkably, there is just one possibility without "extra" spin 1/2 fields: the (small) N=4 algebra. This contains a $\widehat{\mathfrak{su}(2)}_k$ KM algebra and has central charge $c = 6k$. The 4 supercharges transform as $\mathbf{2} \oplus \mathbf{2}$ of the $\mathfrak{su}(2)$.

In addition, there are two possibilities with spin 1/2 holomorphic fields ψ: the N=3 and the large N=4 algebra. The former has an $\widehat{\mathfrak{su}(2)}_k$ KM and $c = 3k/2$, with supercharges in $\mathbf{3}$ of $\mathfrak{su}(2)$ and one $\mathfrak{su}(2)$-invariant free fermion. The large N=4 algebra has the KM symmetry

$$\widehat{\mathfrak{su}(2)}_{k_1} \oplus \widehat{\mathfrak{su}(2)}_{k_2} \oplus \mathfrak{u}(1) ,$$

and central charge $c = 6k_1 k_2/(k_1 + k_2)$. The supercharges and the fermions ψ transform in $(\mathbf{2}, \mathbf{2})_0$ representation. We will not consider these possibilities in more detail here, but more information can be found in [331, 340, 351]. On the other hand, we will return to the (small) N=4 algebra in due course and will drop the diminutive terminology from here on.

Exercise 2.15 In this exercise we take a look at extended supersymmetry realizations via free fields. Consider a Lagrangian for a free real compact scalar ϕ and a Majorana-Weyl fermion λ

$$\mathcal{L} = \partial\phi\bar{\partial}\phi + \lambda\bar{\partial}\lambda .$$

Verify that this is invariant under the SUSY transformation

$$Q \cdot \phi = i\lambda \qquad\qquad Q \cdot \lambda = -i\partial\phi \,.$$

Now consider n copies of the same theory, with

$$\mathcal{L} = \partial\phi^T \bar{\partial}\phi + \lambda^T \bar{\partial}\lambda \,.$$

Suppose there is an additional invariance

$$Q' \cdot \phi = i\mathcal{J}\lambda \,, \qquad\qquad Q' \cdot \lambda = i\mathcal{K}\partial\phi \,, \qquad\qquad (2.4.45)$$

such that $(Q')^2 = \partial$ and $\{Q', Q\} = 0$. Show that $\mathcal{K} = \mathcal{J}$, and $\mathcal{J}$ is an anti-symmetric matrix satisfying $\mathcal{J}^2 = -\mathbb{1}_n$. Such a $\mathcal{J}$ defines a Hermitian structure on the Euclidean target-space T^n where ϕ takes values; evidently, such a $\mathcal{J}$ exists if and only if $n = 2m$. Thus, we obtain N=2 (more precisely (2,0)) supersymmetry on a complex manifold.

Next, consider a more general situation, where we have a number $N - 1$ of such extra supersymmetries, Q_A, corresponding to $N-1$ complex structures $\mathcal{J}_A$. In order to obtain a standard SUSY algebra

$$\{Q_A, Q_B\} = 2\delta_{AB}\partial \,,$$

we need the complex structures to satisfy $\{\mathcal{J}_A, \mathcal{J}_B\} = -2\delta_{AB}\mathbb{1}_n$. Show that given two complex structures $\mathcal{J}_1$ and $\mathcal{J}_2$, we automatically obtain a third linearly independent one, $\mathcal{J}_3 = \mathcal{J}_1\mathcal{J}_2$. Thus, in this case, $N = 3$ implies $N = 4$, and it corresponds to a hyper-Hermitian structure, characterized by $\mathcal{J}_A$, $A = 1, 2, 3$ that obey

$$\mathcal{J}_A\mathcal{J}_B = -\delta_{AB} + \sum_C \epsilon_{ABC}\mathcal{J}_C \,.$$

Show that such a structure is possible if and only if $n = 4k$.

Finally, show that $N \geq 5$ is too nice to exist: there is no $\mathcal{J}_4$. This can be accomplished by evaluating

$$(\mathcal{J}_1\mathcal{J}_2\mathcal{J}_3\mathcal{J}_4)^2$$

in two different ways, once using the anti-commutation relations of all the $\mathcal{J}$, and once using $\mathcal{J}_3 = \mathcal{J}_1\mathcal{J}_2$.

The example has one more little lesson. While there is at most one N=4 supersymmetry algebra that closes to translations, the action is invariant under many more fermionic transformations simply because it is a sum of free supersymmetric theories. We will briefly touch on this again in Chap. 4.   □

2.4.9 Superconformal Kac-Moody Algebras

Suppose an N=1 superconformal theory with central charge c has a continuous symmetry with a simple Lie algebra $\mathfrak{g}$ of dimension n that commutes with the supercharge. In this case there exists a supersymmetric generalization of the Sugawara decomposition. We summarize the results, most of which can be found in Chapter 18 of [327], and discuss a few details in the appendix.

The theory has a $\widehat{\mathfrak{g}}_k$ KM symmetry with currents $\{J^a\}_{a=1,...,n}$, and each J^a is the upper component of an N=1 superconformal multiplet. The corresponding lowest components are $h = 1/2$ N=1 primary operators Ψ^a that transform in the adjoint representation of $\mathfrak{g}$. The presence of the fermions leads to a larger KM symmetry that includes a $\widehat{\mathfrak{so}(n)}_1$ KM algebra constructed from the fermion bilinears. This larger KM symmetry has a sub-algebra $\widehat{\mathfrak{g}}_{k_b}$ with level $k_b = k - h(\mathfrak{g})$ and currents J^a_{bos} that have non-singular OPE with the fermions Ψ^a. Thus, the energy momentum tensor can be decomposed into three commuting terms: (1) T_ψ with $c_\psi = n/2$ for the fermions; (2) T_{bos} for the $\widehat{\mathfrak{g}}_{k_b}$ with $c_b = k_b n/(k_b + h(\mathfrak{g}))$; and (3) T' for the remaining degrees of freedom with $c' = c - c_\psi - c_b$. There is a super-Sugawara decomposition: the N=1 superconformal algebra decomposes into two commuting N=1 subalgebras: the first has energy-momentum tensor $T_\psi + T_b$ and is the super-KM algebra, and the second has T'.

Unitarity leads to constraints on the levels and central charges: $k \geq h(\mathfrak{g})$ (with equality if and only if T_b is trivial), and $c \geq c_{\text{SKM}}$, where the super-Kac-Moody central charge is

$$c_{\text{SKM}} = c_\psi + c_b = \frac{k_b n}{k_b + h(\mathfrak{g})} + \frac{n}{2} = \left(\frac{k_b}{k} + \frac{1}{2}\right) \dim \mathfrak{g} \ . \tag{2.4.46}$$

The original theory may also have symmetries that do not commute with the supercharge but rather have an OPE

$$J^A(z)G(w) = \frac{\mathcal{X}^A(w)}{z - w} \ , \tag{2.4.47}$$

where the $\mathcal{X}^A$ are Virasoro primary operators with $h = 3/2$. The Jacobi identity can be used to show that these extend the superconformal algebra, and if the irreducibility assumptions stated above hold, the currents J^A must generate the R-symmetry algebra of the extended superconformal algebra.

The structure of KM algebras in N=2 SCFTs is just a little bit more restricted. As discussed in [346, 351], the key requirement is that the compact simply connected Lie group of $\mathfrak{g}$ must admit an integrable complex structure compatible with the metric. We will return to this point again when we study (0,2) non-linear sigma models, but for now we will just state the a geometric fact: any even dimensional compact Lie group admits an integrable complex structure [337, 374]. This means that an N=1 SKM algebra corresponding to an even-dimensional compact Lie group is automatically an N=2 SKM. Finally, just as the KM algebra has a Sugawara construction, so do the N=1 and N=2 SKMs.[133, 149, 346]. That means that any time we have a theory with an $N = 2$ SKM, the superconformal algebra decomposes into two commuting factors:

$$\mathcal{A}_c \doteq \mathcal{A}_{c-c_{\text{SKM}}} \oplus \mathcal{A}_{c_{\text{SKM}}} . \tag{2.4.48}$$

Exercise 2.16 The simplest N=2 SKM algebra has $c = 3$ and consists of a free Weyl fermion ψ, its conjugate $\overline{\psi}$, and two bosonic currents combined into complex combinations K and $K^\dagger$. The non-vanishing OPEs of these fields take the form

$$\psi(z)\psi^\dagger(w) \sim \frac{1}{z-w} , \qquad\qquad K(z)K^\dagger(w) \sim \frac{1}{(z-w)^2} .$$

Show that the following fields satisfy the $c = 3$ $N = 2$ algebra:

$$J_{\text{skm}} =: \psi\psi^\dagger : , \qquad\qquad G_{\text{skm}} = \sqrt{2}\psi K^\dagger, \overline{G}_{\text{skm}} = \sqrt{2}\psi^\dagger K,$$

$$T_{\text{skm}} =: KK^\dagger : -\tfrac{1}{2}(: \psi^\dagger \partial\psi : + : \psi\partial\psi^\dagger :).$$

Next, show that $J' = J - J_{\text{skm}}$ and similarly defined G' $\overline{G}'$ and T' commute with the N=2 SKM fields. $\qquad\qquad\qquad\qquad\qquad\qquad\qquad\qquad\qquad\qquad\qquad\square$

2.5 Background Fields

The properties of a QFT are often elucidated by coupling it to background fields. This is particularly powerful when the coupling is made by gauging a continuous symmetry of the theory. In this section we will explore such couplings for a CFT.

2.5.1 Gauging a KM Symmetry

Suppose we have a CFT with a $\widehat{u(1)}_k \oplus \widehat{u(1)}_{\overline{k}}$ symmetry and corresponding currents $J(z)$ and $\overline{J}(\overline{z})$ satisfying

$$\langle J(z)J(0)\rangle = \frac{k}{z^2} , \qquad\qquad \langle \overline{J}(\overline{z})\overline{J}(0)\rangle = \frac{\overline{k}}{\overline{z}^2} . \tag{2.5.1}$$

We introduce a background gauge field, with components A and $\overline{A}$, and we define the QFT in the background via

$$\langle\cdots\rangle_{A,\overline{A}} = \langle\cdots e^{-S[A,\overline{A}]}\rangle , \tag{2.5.2}$$

where $\cdots$ represents the insertions of any local operators and

$$S[A, \overline{A}] = \frac{1}{2\pi} \int d^2z \left\{ iJ\overline{A} + i\overline{J}A \right\} . \qquad (2.5.3)$$

The correlation functions to any order in the background are then obtained by expanding the exponential. The right-hand-side is not well-defined as written. Since the action involves an integral over the world-sheet there will be singularities when the interaction terms approach the insertions or each other; since the OPE is only well-defined for separated points, there will also be ambiguities and corresponding choices of local counter-terms. These issues are addressed by fixing some regularization scheme and renormalizing the operators and S.

In our example the story is rather simple, since the correlators of the currents are determined by the OPE. For example, denoting $J(z_i)$ by J_i, we have

$$\langle J_1 J_2 \rangle = \frac{k}{z_{12}^2} , \qquad \langle J_1 J_2 J_3 \rangle = 0 ,$$

$$\langle J_1 J_2 J_3 J_4 \rangle = \langle J_1 J_2 \rangle \langle J_3 J_4 \rangle + \langle J_1 J_3 \rangle \langle J_2 J_4 \rangle + \langle J_1 J_4 \rangle \langle J_2 J_3 \rangle . \qquad (2.5.4)$$

So, if we define the partition function

$$Z[A, \overline{A}] = \langle 1 \rangle_{A, \overline{A}} = \langle e^{-S[A, \overline{A}]} \rangle , \qquad (2.5.5)$$

we have $Z = e^{-W}$ with

$$W[A, \overline{A}] = \frac{1}{8\pi^2} \int d^2z_1 d^2z_2 \left\{ k \frac{\overline{A}_1 \overline{A}_2}{z_{12}^2} + \overline{k} \frac{A_1 A_2}{\overline{z}_{12}^2} \right\} . \qquad (2.5.6)$$

The integrand exhibits a mild singularity as $z_1 \to z_2$, which we can dispose of by defining the improper integral as $\lim_{\ell \to 0} \int_{|z_{12}| \geq \ell}$. While simple, W is not gauge-invariant. The original action $S[A, \overline{A}]$ is invariant under

$$\delta_f A = -\partial f , \qquad \delta_f \overline{A} = -\overline{\partial} f \qquad (2.5.7)$$

for any function $f(z, \overline{z})$ because J and $\overline{J}$ are (separately) conserved currents. On the other hand, using (1.8.2) we have

$$\delta W = -\frac{1}{2\pi} \int d^2z \, f \left\{ k\partial \overline{A} + \overline{k}\overline{\partial} A \right\}$$

$$= -\frac{1}{4\pi}(k + \overline{k})\delta_f \int d^2z A\overline{A} - \frac{1}{4\pi}(k - \overline{k}) \int d^2z \, f(\partial \overline{A} - \overline{\partial} A)$$

$$= -\frac{k + \overline{k}}{4\pi}\delta_f \int d^2z A\overline{A} - \frac{i(k - \overline{k})}{4\pi} \int fF , \qquad (2.5.8)$$

where $F = F_{12}dy^1 \wedge dy^2 = -2i(\partial\overline{A} - \overline{\partial}A)dy^1 \wedge dy^2$ is the gauge-invariant field strength.

Exercise 2.17 The quick way to make this computation is simply by ignoring the ℓ regulator and using (1.8.2). Show that the boundary terms in the regulated integral lead to the same form for the gauge variation. This had to be: the gauge variation is guaranteed to be local and finite.   □

While the first contribution is manifestly the gauge variation of a local operator, the second contribution cannot be put in this form. Hence, gauge invariance can only be maintained if $k = \overline{k}$. If that holds, we add the counter-term to obtain the gauge-invariant

$$W[A, \overline{A}] = \frac{k}{8\pi^2} \int d^2z_1 d^2z_2 \left\{ \frac{\overline{A}_1\overline{A}_2}{z_{12}^2} + \frac{A_1 A_2}{\overline{z}_{12}^2} \right\} + \frac{k}{2\pi} \int d^2z A\overline{A} . \qquad (2.5.9)$$

A direct way to check the gauge invariance is to check the conservation of the current. We compute

$$\langle J_1 \rangle_A = -2\pi i \frac{\delta W}{\delta \overline{A}_1} = \frac{ik}{2\pi} \int d^2z_2 \frac{\partial_2 \overline{A}_2}{z_{12}} - ikA_1 ,$$

$$\langle \overline{J}_1 \rangle_A = -2\pi i \frac{\delta W}{\delta A_1} = \frac{ik}{2\pi} \int d^2z_2 \frac{\overline{\partial}_2 A_2}{\overline{z}_{12}} - ik\overline{A}_1 . \qquad (2.5.10)$$

So, the current has a non-zero 1-point function in the background, but it remains conserved:

$$\overline{\partial}_1 \langle J_1 \rangle_A + \partial_1 \langle \overline{J}_1 \rangle_A = 0 . \qquad (2.5.11)$$

Although the partition function does not have the same simple form, the same conclusion holds true for non-abelian KM algebras as well: we can preserve gauge invariance by adding a local counter-term if and only if $k = \overline{k}$.

There is another, equivalent, perspective on the origin of this contact term: the OPEs we used in the computation are only determined for separated points, and there are ambiguities in the contact terms that arise when operators collide. These are constrained by physical requirements, such as gauge invariance. The analysis above amounts to the observation that if $k = \overline{k}$, then we can preserve gauge invariance by introducing a contact term

$$\langle J(z_1)\overline{J}(\overline{z}_2) \rangle = k\delta^2(z_{12}, \overline{z}_{12}) . \qquad (2.5.12)$$

This is consistent with the scale invariance of the CFT; more generally, the absence of a scale in the original theory constrains the possible contact terms. When we regulate a non-chiral theory, for example when we gauge the rotation symmetry of a free complex scalar, such a contact term will be "automatically" provided in the regulated theory.

Exercise 2.18 Check gauge invariance in the non-abelian case by expanding the partition function up to $O(A^4, \overline{A}^4)$ corrections for a general KM algebra with currents satisfying (2.3.13). □

Free Fermions and Fujikawa

In a gauge theory a global symmetry may suffer from an Adler-Bardeen-Jackiw type anomaly, where the current with components J and $\overline{J}$ satisfies $\overline{\partial} J + \partial \overline{J} \propto F_{12}$. In a weakly coupled theory with fermions there is a standard result that we can understand the right-hand-side as arising from a non-invariance of the fermion measure [375]. This will be a useful perspective for us when we study two-dimensional gauge theories, so let us recover that result. Rather than going through the details of determinants, we can use the CFT results we just developed.

We start with a free fermion action for n_γ left-moving fermions γ^a and n_ψ right-moving fermions ψ^i, and we couple these to background gauge fields:

$$S_{\min} = \frac{1}{2\pi} \int d^2z \left[\sum_a \overline{\gamma}^a (\overline{\partial} + i\overline{A}^a)\gamma^a + \sum_i \psi^i (\partial + iA^i)\psi^i \right]. \qquad (2.5.13)$$

We are interested in studying global symmetries of the theory in the presence of a dynamical gauge field $(A_g, \overline{A}_g)$ that corresponds to setting $A^a = 0, \overline{A}^i = 0$ and

$$\overline{A}^a = Q_\gamma^a \overline{A}_g, \qquad\qquad A^i = Q_\psi^i A_g. \qquad (2.5.14)$$

By the previous discussion this symmetry will be non-anomalous provided $k = \overline{k}$ with $\overline{k} = \sum_i (Q_\psi^i)^2$ and $k = \sum_a (Q_\gamma^a)^2$. We will assume that holds, so that by adding a counter-term we can render the effective action gauge-invariant. We take the counter-term to be

$$S_{\text{c.t.}} = \frac{1}{2\pi} \int d^2z \sum_{a,i} A^a C^{ai} \overline{A}^i, \qquad (2.5.15)$$

where the matrix is

$$C^{ai} = \tfrac{1}{k} Q_\gamma^a Q_\psi^i + D^{ai}, \qquad (2.5.16)$$

and $\sum_a Q_\gamma^a D^{ai} = 0$, and $\sum_i D^{ai} Q_\psi^i = 0$. By design it reduces to the usual counter-term when evaluated on (2.5.14).

The free theory has conserved currents $J^a =: \overline{\gamma}^a \gamma^a :$ and $\overline{J}^i =: \overline{\psi}^i \psi^i :$ that correspond to the obvious $U(1)^{n_\gamma} \times U(1)^{n_\psi}$ symmetries $\delta_\epsilon \gamma^a = i\epsilon^a \gamma^a$

and $\delta_\epsilon \psi^i = i\epsilon^i \psi^i$. The general set of background fields gauge fields and the accompanying counter-term allow us to construct improved currents via

$$J^a_{\text{imp}} = -2\pi i \left. \frac{\delta(S_{\min} + S_{\text{ct}})}{\delta \overline{A}^a} \right|_{A=A_g} = J^a - iQ^a_\gamma A_g \,,$$

$$\overline{J}^i_{\text{imp}} = -2\pi i \left. \frac{\delta(S_{\min} + S_{\text{ct}})}{\delta A^i} \right|_{A=A_g} = \overline{J}^i - iQ^i_\psi \overline{A}_g \,. \qquad (2.5.17)$$

By computing the one-point functions of these improved currents it is easy to derive the operator relations

$$\bar{\partial} J^a_{\text{imp}} + Q^a_\gamma \tfrac{1}{2} F_{12} = 0 \,, \qquad \partial \overline{J}^i_{\text{imp}} - Q^i_\psi \tfrac{1}{2} F_{12} = 0 \,. \qquad (2.5.18)$$

Finally, we can give a Fujikawa interpretation of this result. Under a change of variables $\delta_\epsilon \gamma^a = i\epsilon^a \gamma^a$ and $\delta_\epsilon \psi^i = i\epsilon^i \psi^i$ with non-constant parameters, the change in the classical action is

$$\delta_\epsilon S_{\min} = \frac{1}{2\pi} \int d^2z \left\{ \sum_a \epsilon^a (-i\bar{\partial} J^a_\gamma) + \sum_i \epsilon^i (-i\partial \overline{J}^i_\psi) \right\} \qquad (2.5.19)$$

The path-integral interpretation of (2.5.18) is then that the terms proportional to F_{12} arise from the transformation of the measure, which, when interpreted as a shift of the action, takes the form

$$\delta_\epsilon S_{\text{eff}} = \frac{1}{2\pi} \int d^2z \left\{ \sum_a \epsilon^a Q^a_\gamma - \sum_i \epsilon^i Q^i_\psi \right\} \tfrac{i}{2} F_{12} \,. \qquad (2.5.20)$$

Finally, setting ϵ^a and ϵ^i to be constant, we conclude that under the transformation $\delta_\epsilon \gamma^a = i\epsilon^a \gamma^a$ and $\delta_\epsilon \psi^i = i\epsilon^i \psi^i$ the transformation of the measure leads to a shift of the action by

$$\delta_\epsilon S_{\text{eff}} = -i\left(\sum_a \epsilon^a Q^a_\gamma - \sum_i \epsilon^i Q^i_\psi \right) \tfrac{1}{2\pi} \int F \,. \qquad (2.5.21)$$

If our action includes a ϑ-angle term

$$S_\vartheta = i \frac{\vartheta}{2\pi} \int F \,, \qquad (2.5.22)$$

then the effective action is invariant provided that we shift ϑ:

$$\delta_\epsilon \vartheta = +i\left(\sum_a \epsilon^a Q^a_\gamma - \sum_i \epsilon^i Q^i_\psi \right) \,. \qquad (2.5.23)$$

If ϑ is dynamical, then the symmetry is maintained, but if ϑ is a background this leads to selection rules. Although we only performed the computation to leading order in the gauge field, this result is exact. As we will see further in Chap. 5 , or as the reader can find in the references, the transformation of the measure is determined by an index theorem, so that to the extent that the path integral makes sense, the result is exact.

2.5.2 A Background Metric

So far we discussed the theory defined on the Euclidean plane with a flat metric. More generally, we can use the energy-momentum tensor to couple the theory to background metric $g = \delta + h$, where h is a small perturbation, by following the same strategy as we used for the gauge field. For any theory, not necessarily a CFT, we write

$$S[h] = -\frac{1}{8\pi} \int d^2z \, h^{\mu\nu} T_{\mu\nu} \,, \tag{2.5.24}$$

where h is the metric perturbation, and indices are raised/lowered with the background flat metric. Writing this in terms of the Θ, T and $\overline{T}$ components as in (2.1.5), we obtain

$$
\begin{aligned}
S[h] &= -\frac{1}{8\pi} \int d^2z \Big[\tfrac{1}{2}(h_{11}+h_{22})(T_{11}+T_{22}) + \tfrac{1}{2}(h_{11}-h_{22})(T_{11}-T_{22}) + 2h_{12}T_{12} \Big] \\
&= -\frac{1}{8\pi} \int d^2z \Big[\sigma\Theta + \overline{h}\,T + h\overline{T} \Big] \,,
\end{aligned}
\tag{2.5.25}
$$

where, repeating (2.1.5) for convenience

$$\Theta = \tfrac{1}{4}(T_{11}+T_{22}) \,, \quad T = \tfrac{1}{4}(T_{11}-T_{22}) - \tfrac{i}{2}T_{12} \,, \quad \overline{T} = \tfrac{1}{4}(T_{11}-T_{22}) + \tfrac{i}{2}T_{12} \,,$$

and

$$\sigma = 2(h_{11}+h_{22}) \,, \quad \overline{h} = h_{11} - h_{22} + 2ih_{12} \,, \quad h = h_{11} - h_{22} - 2ih_{12} \,. \tag{2.5.26}$$

We now define the perturbed correlators via

$$\langle \cdots \rangle_h = \langle \cdots e^{-S[h]} \rangle \,. \tag{2.5.27}$$

In a CFT, where $\Theta = 0$, we use the T–T OPE to obtain to leading order in the perturbation

$$\langle T_1 \rangle_h = \frac{c}{96\pi} \int d^2z_2 \frac{\partial_2^3 \overline{h}_2}{z_{12}} \,, \qquad \langle \overline{T}_1 \rangle_h = \frac{\overline{c}}{96\pi} \int d^2z_2 \frac{\overline{\partial}_2^3 h_2}{\overline{z}_{12}} \,. \tag{2.5.28}$$

This does not bode well for conservation of the energy-momentum tensor in a general background, but we have yet to consider the possible local counter-terms. These must be Lorentz-invariant and contain two derivatives; we restrict attention to terms quadratic in the perturbation since we work to first order in the 1-point functions. This leads to the following possible terms:

$$S_{c.t.} = -\frac{1}{8\pi} \int d^2z \left[a_1 \bar{h} \partial^2 \sigma + a_2 \bar{h} \partial \bar{\partial} h + a_3 h \bar{\partial}^2 \sigma + a_4 \sigma \partial \bar{\partial} \sigma \right] \tag{2.5.29}$$

for some to-be-determined coefficients a_1, a_2, a_3, a_4.

Exercise 2.19 Show that the inclusion of $S_{c.t.}$ leads to

$$\langle T \rangle_h = \frac{c}{96\pi} \int d^2 z_2 \frac{\partial_2^3 \bar{h}_2}{z - z_2} + a_1 \partial^2 \sigma + a_2 \partial \bar{\partial} h \, ,$$

$$\langle \Theta \rangle_h = a_1 \partial^2 \bar{h} + a_3 \bar{\partial}^2 h + 2 a_4 \partial \bar{\partial} \sigma \, ,$$

$$\langle \bar{T} \rangle_h = \frac{\bar{c}}{96\pi} \int d^2 z_2 \frac{\bar{\partial}_2^3 h_2}{\bar{z} - \bar{z}_2} + a_3 \bar{\partial}^2 \sigma + a_2 \partial \bar{\partial} \bar{h} \, .$$

Show that these satisfy the conservation equations (2.1.6) if and only if $c = \bar{c}$, in which case

$$a_1 = -a_2 = a_3 = -2a_4 = -c/48 \, ,$$

which in turn leads to

$$\langle \Theta \rangle_h = \frac{c}{48} \left[\partial \bar{\partial} \sigma - \partial^2 \bar{h} - \bar{\partial}^2 h \right] \, . \qquad \square$$

From the exercise it follows that $c = \bar{c}$ is necessary in order to preserve diffeomorphism invariance, and, unless $c = \bar{c} = 0$, conformal invariance is then violated. The violation is proportional to the linearized Ricci-scalar, so that the fully covariant term is uniquely determined (the derivative order is fixed by scale invariance) to be[16]

$$\langle \Theta \rangle_g = -\frac{c}{48} R(g) \, . \tag{2.5.30}$$

This violation is the two-dimensional version of the famous conformal anomaly, and it is the price for preserving diffeomorphism invariance. More generally, in any even dimension there are two types of contributions to the conformal anomaly, denoted as A and B. The former is topological in character and arises from a part of the effective

---

[16]We recall that for $g = \delta + h$ the linearized form of the Ricci scalar is $R(g) = -\partial^\mu \partial_\mu h^\alpha_\alpha - \partial^\alpha \partial^\beta h_{\alpha\beta}$.

action W that does not involve any scale μ; the latter is associated to terms in W that involve a renormalization scale explicitly. There are no type B anomalies in two dimensions. A lucid account and classification of A and B anomalies is provided in [130].

Incidentally, there is another kind of central term that occasionally shows up in discussing world-sheet anomalies. It may be that a (1,0) current J has OPE with T

$$T(z)J(w) \sim \frac{a}{(z-w)^3} + \frac{J(w)}{(z-w)^2} + \frac{\partial J(w)}{z-w} , \tag{2.5.31}$$

and if we were to couple T and J to background fields, then the coefficient a would lead to a mixed anomaly. Such an anomaly can always be removed by improving the energy-momentum tensor. Namely, if $J(z)J(w) \sim r(z-w)^{-2}$, then setting $T_{\text{new}} = T + \frac{a}{2r}\partial J$ leads to a standard Virasoro-KM algebra for J and T_{new} with central charge $c_{\text{new}} = c + 3a^2/2r$. We will see a variation on this theme in the last chapter when we discuss topological field theories.

2.6 Conformal Perturbation Theory

We now turn to one of the most important concepts in quantum field theory: the renormalization group flow and QFTs defined by conformal perturbation theory. This is a central feature in our conceptual understanding of QFT, but it is technically challenging to provide a precise and computable definition of this framework. We will give a little bit of a discussion that will guide much of our thinking about the theories that we will encounter in the next chapters. The discussion will focus on two dimensional Lorentz-invariant theories, but the essential points apply in any dimension and to less symmetric situations.

The idea, in a nut-shell, is that given a CFT for which we know the spectrum of local operators $\mathcal{O}_i(z, \bar{z})$ with spins $s_i = h_i - \bar{h}_i$ and scaling dimensions $\Delta_i = h_i + \bar{h}_i$, and their correlation functions, we can attempt to define a perturbed theory by

$$\langle \cdots \rangle_\lambda = \langle \cdots e^{-S[\lambda]} \rangle , \qquad S[\lambda] = \int d^2z \sum_i \lambda^i \mathcal{O}_i(z, \bar{z}) . \tag{2.6.1}$$

To preserve Lorentz invariance we restrict the sum in $S[\lambda]$ to operators with $s_i = 0$. Since the action is dimensionless, the coupling λ^i has mass dimension $\eta_i = 2 - \Delta_i$. If we know all of the correlation functions of the unperturbed theory, then we can obtain all correlation functions in the perturbed theory by expanding the exponential order by order in λ and computing the integrals.

The nut-shell neglects a crucial subtlety: the right-hand side is not defined due to singularities as the integrated operators collide with each other and with the insertions (the operators in $\cdots$). A concrete regulator is provided by cutting out "little disks" of radius a from the domain of integration. In general we must also address potential IR divergences. This can be done in a number of ways, for instance by placing the theory on a compact round sphere S^2 of radius L. Alternatively, we may take the couplings to be background fields $\lambda^i(z, \overline{z})$ with compact support, or we can introduce an explicit position space cut-off L and restrict the integrated operators to $|z| \le L$.

While the regulated correlation functions are well-defined (though ponderous to compute even at relatively low orders in perturbation theory), they also depend on the UV regulator. To obtain a local QFT from this description, we must renormalize the theory and send $a \to 0$; we are also interested in sending $L \to \infty$. In the next section we will take a look at what the result of renormalization should be.

2.6.1 Some Axioms of Renormalized QFT

In this section we will remind the reader what sort of structure we would like for a renormalized conformal perturbation theory. Our discussion closely follows the incredible reference [397]. An alternative (and more powerful) approach makes stronger assumptions based on a local renormalization group formulation [56, 173, 174, 314, 320]. Aspects of the problem can also be attacked in an axiomatic yet useful fashion [230]. While our main interest is in two-dimensional theories, much of what we say here is dimension-independent. The reader will find the discussion in [173, 174, 179, 345] useful and in some ways complementary to the presentation given here.

We assume that a renormalized QFT includes the following ingredients.

1. We consider (the Euclidean continuation of) a unitary, Lorentz-invariant QFT. The Lorentz charges are constructed from a symmetric energy momentum tensor.
2. There is a set of local operators $\widehat{\Phi}_A(w, \overline{w})$ of mass dimension Δ_A. The m-point correlation functions $\langle \widehat{\Phi}_{A_1}(w_1, \overline{w}_1) \cdots \widehat{\Phi}_{A_m}(w_m, \overline{w}_m) \rangle$ are the basic observables. Whenever it is not confusing, we will use the shorthand $\widehat{\Phi}_{A_i} = \widehat{\Phi}_{A_i}(w_i, \overline{w}_i)$. For simplicity we will assume that the $\widehat{\Phi}_A$ transform as primary operators with respect to scale transformations: under $(w, \overline{w}) \to (sw, s\overline{w})$ we have $\widehat{\Phi}_A(w, \overline{w}) \to s^{-\Delta_A} \widehat{\Phi}_A(sw, s\overline{w})$.
3. The correlation functions of the local operators depend on a renormalization length-scale ℓ and a (possibly infinite) set of dimensionless couplings g^i, but the dependence is subject to the Callan-Symanzik equation:

$$\left\{ -\ell \frac{\partial}{\partial \ell} + \sum_i \beta^i(g) \frac{\partial}{\partial g^i} \right\} \langle \widehat{\Phi}_{A_1} \cdots \widehat{\Phi}_{A_m} \rangle = \Gamma \cdot \langle \widehat{\Phi}_{A_1} \cdots \widehat{\Phi}_{A_m} \rangle, \qquad (2.6.2)$$

where the β function is

$$\beta^i(g) = -\ell \frac{dg^i}{d\ell}, \qquad (2.6.3)$$

and

$$\Gamma \cdot \langle \widehat{\Phi}_{A_1} \cdots \widehat{\Phi}_{A_m} \rangle = \Gamma^B_{A_1} \langle \widehat{\Phi}_B(z_1) \widehat{\Phi}_{A_2} \cdots \widehat{\Phi}_{A_m} \rangle$$
$$+ \Gamma^B_{A_2} \langle \widehat{\Phi}_{A_1} \widehat{\Phi}_B(z_2) \widehat{\Phi}_{A_3} \cdots \widehat{\Phi}_{A_m} \rangle$$
$$+ \cdots + \Gamma^B_{A_m} \langle \widehat{\Phi}_{A_1} \widehat{\Phi}_{A_m} \cdots \widehat{\Phi}_B(z_m) \rangle . \qquad (2.6.4)$$

$\Gamma^A_B(g)$ is the matrix of "anomalous dimensions."[17] Thus, a change in ℓ is equivalent to a motion in the coupling space controlled by the beta function (a misnomer—it is really a vector field) $\beta^i(g)$ and a rotation on the basis of local operators according to $\Gamma(g)$; by dimensional analysis $\Gamma^B_A(g) = 0$ whenever $\Delta_A \neq \Delta_B$; by Lorentz-invariance $\Gamma^B_A = 0$ whenever the operators $\widehat{\Phi}_A$ and $\widehat{\Phi}_B$ have different spins.

4. The trace of the energy-momentum tensor, denoted by Θ, generates scale transformations. That is, if we set

$$S = \sum_{s=1}^m \left[w_s \frac{\partial}{\partial w_s} + \overline{w}_s \frac{\partial}{\partial \overline{w}_s} \right] , \qquad \Delta_{\text{tot}} = \sum_{s=1}^m \Delta_{A_s} , \qquad (2.6.5)$$

then

$$\left\{ S + \Delta_{\text{tot}} - \widehat{\Delta} \right\} \langle \widehat{\Phi}_{A_1} \cdots \widehat{\Phi}_{A_m} \rangle = -\frac{1}{\pi} \int_{\text{reg}} d^2 z \langle \Theta(z, \overline{z}) \widehat{\Phi}_{A_1} \cdots \widehat{\Phi}_{A_m} \rangle . \qquad (2.6.6)$$

The $\widehat{\Delta}$ is a linear operator that acts in the same way as Γ. Its motivation comes from the divergences in the integral on the right-hand-side: the integral needs to be regulated, and $\widehat{\Delta}$ provides the counter-terms necessary to remove the cut-off dependence. Even when $\Theta = 0$, so that the renormalized theory is a conformal theory, there can be a non-trivial $\widehat{\Delta}$ contribution.

The correlation functions obey a tautological identity. Since we are free to change our reference lengthscale (i.e. our meter stick), we have

$$\left\{ S + \Delta_{\text{tot}} + \ell \frac{\partial}{\partial \ell} \right\} \langle \widehat{\Phi}_{A_1} \cdots \widehat{\Phi}_{A_m} \rangle = 0 . \qquad (2.6.7)$$

This leads to a useful equation when combined with (2.6.6):

$$-\ell \frac{\partial}{\partial \ell} \langle \widehat{\Phi}_{A_1} \cdots \widehat{\Phi}_{A_m} \rangle = \widehat{\Delta} \cdot \langle \widehat{\Phi}_{A_1} \cdots \widehat{\Phi}_{A_m} \rangle - \frac{1}{\pi} \int d^2 z \langle \Theta(z, \overline{z}) \widehat{\Phi}_{A_1} \cdots \widehat{\Phi}_{A_m} \rangle . \qquad (2.6.8)$$

---

[17] We will frequently use the summation convention: repeated indices are summed.

We can also combine this with the Callan-Symanzik equation to obtain

$$\beta^i \frac{\partial}{\partial g^i} \langle \widehat{\Phi}_{A_1} \cdots \widehat{\Phi}_{A_m} \rangle - \frac{1}{\pi} \int_{\mathrm{reg}} d^2 z \langle \Theta(z, \bar{z}) \widehat{\Phi}_{A_1} \cdots \widehat{\Phi}_{A_m} \rangle$$
$$= (\Gamma - \widehat{\Delta}) \cdot \langle \widehat{\Phi}_{A_1} \cdots \widehat{\Phi}_{A_m} \rangle . \tag{2.6.9}$$

A Family of Conformal Field Theories

To illustrate some of this structure, suppose we deform a CFT_0 by exactly marginal operators to produce a family of theories CFT_g that depend on the dimensionless couplings g^i. The compact boson of Exercise 2.8 is the prototypical example of this scenario—the radius is the exactly marginal coupling—and shows what we are to expect: the marginal operator remains exactly marginal, while the dimensions of other operators depend on the coupling(s). By assumption we then have $\beta = 0$ and $\Theta = 0$. Our tautological statement then requires $\widehat{\Delta} = \Gamma$.

The interpretation of Γ or $\widehat{\Delta}$ is made clear by looking at the two-point functions of operators in the renormalized QFT. By dimensional analysis we have

$$\langle \widehat{\Phi}_A(x) \widehat{\Phi}_B(0) \rangle = x^{-\Delta_A - \Delta_B} G_{AB}(x/\ell; g) , \tag{2.6.10}$$

and the Callan-Symanzik equation

$$-\ell \frac{\partial}{\partial \ell} G_{AB} = \Gamma_A^C G_{CB} + G_{AC} \Gamma_B^C \tag{2.6.11}$$

has the solution, written in matrix form,

$$G(\ell/x; g) = e^{-\Gamma \log \ell/x} \widetilde{G}(g) e^{-\Gamma^T \log \ell/x} , \tag{2.6.12}$$

where the symmetric matrix $\widetilde{G}_{AB}$ is

$$\widetilde{G}_{AB}(g) = G_{AB}(1; g) = \ell^{\Delta_A + \Delta_B} \langle \widehat{\Phi}_A(\ell) \widehat{\Phi}_B(0) \rangle . \tag{2.6.13}$$

Suppose that Γ is diagonalizable:

$$\Gamma = S^{-1} D S , \tag{2.6.14}$$

where for each block of operators with fixed Δ_A—say there are N of them—we have

$$D = \mathrm{diag}(\gamma_1, \gamma_2, \ldots, \gamma_N) . \tag{2.6.15}$$

Let $\mathcal{G} = SGS^T$. From the general solution for G we then find

$$\mathcal{G}_{AB}(\ell/x; g) = \left(\frac{x}{\ell}\right)^{\gamma_A + \gamma_B} \mathcal{G}_{AB}(1; g), \tag{2.6.16}$$

Therefore, the ℓ dependence of these two-point functions can be absorbed into a rescaling of the fields. More precisely, if we set $\Psi_A = \ell^{\gamma_A} S_A^B \widehat{\Phi}_B$, we obtain

$$\langle \Psi_A(x) \Psi_B(0) \rangle = x^{-\Delta_A - \Delta_B + \gamma_A + \gamma_B} \mathcal{G}_{AB}(1; g). \tag{2.6.17}$$

So, these correlation functions take an explicitly scale-invariant form as expected. There is more to conformal invariance than scale invariance, and the reader can discover the remaining consequences by writing Ward identities for T and $\overline{T}$, which are separately conserved because $\Theta = 0$.

If Γ is not diagonalizable, then the correlation function will inevitably have $\log(\ell/x)$ terms. This is a pathology for a unitary CFT. There is a body of literature on "log" CFTs of this sort—see, e.g. [123], but we will not consider them here. Another way of seeing why Γ should be diagonalizable, at least in this CFT context, is that it should be a correction to the dilatation operator, and in a unitary theory that operator should be Hermitian.

Once we write the two point-functions in the Ψ_A basis, the Callan-Symanzik equation becomes trivial: there is no ℓ dependence at all, and $\Gamma = 0$. Equation (2.6.6) then holds with $\Delta_A \to \Delta_A - \gamma_A(g)$. These results should extend to general n-point functions, but the structure is complicated beyond $n = 3$.

The Action Principle and Geometry

In this section we add a few more axioms that we will require of a renormalized QFT. Motivated by sources and fields of path integral formulation of Lagrangian QFTs, we postulate that the coupling dependence of correlation functions is constrained by the "action principle":

$$\frac{\partial}{\partial g^i} \langle \widehat{\Phi}_{A_1} \cdots \widehat{\Phi}_{A_m} \rangle = +B_i(g, \ell/a) \cdot \langle \widehat{\Phi}_{A_1} \cdots \widehat{\Phi}_{A_m} \rangle - \int_{\text{reg}} d^2 z \langle \widehat{\mathcal{O}}_i(z) \widehat{\Phi}_{A_1} \cdots \widehat{\Phi}_{A_m} \rangle.$$
$$\tag{2.6.18}$$

That is, for every coupling there is a local operator $\widehat{\mathcal{O}}_i$, and there is a connection B_i such that (2.6.18) holds for every correlation function.[18] We will call the $\widehat{\mathcal{O}}_i$ the deforming operators and will assume that the set of deforming operators is complete

---

[18] As in the integrated correlation function with Θ, this integrated correlator is senseless without a regularization.

in the sense that $(B_i)^A_j = 0$ and $\Gamma^A_j = 0$ unless $A = k$, i.e. deforming operators only mix among themselves.

We assume that the space parametrized by g^i is, at least locally, a manifold $\mathcal{M}$. In general, $\mathcal{M}$ may be infinite-dimensional, but in favorable cases, where the theory is renormalizable, it is finite dimensional. We assume that the correlation functions transform under diffeomorphisms on $\mathcal{M}$ as sections of various bundles over $\mathcal{M}$. For instance, m-point correlators of the deforming operators have a natural interpretation as sections of $(T^*_{\mathcal{M}})^{\otimes m}$, and if we think of $\langle \widehat{\Phi}_{A_1} \cdots \widehat{\Phi}_{A_m} \rangle$ as a section of a bundle $\mathcal{E} \to \mathcal{M}$, then we have an interpretation for the action principle, which we rewrite as

$$\mathcal{D}_i \langle \widehat{\Phi}_{A_1} \cdots \widehat{\Phi}_{A_m} \rangle = \left(\frac{\partial}{\partial g^i} - B_i \right) \langle \widehat{\Phi}_{A_1} \cdots \widehat{\Phi}_{A_m} \rangle = - \int_{\text{reg}} d^2 z \langle \widehat{\mathcal{O}}_i \widehat{\Phi}_{A_1} \cdots \widehat{\Phi}_{A_m} \rangle .$$

$$(2.6.19)$$

That is, we now have a derivative operator $\mathcal{D}$ that is a map $\mathcal{E} \to \mathcal{E} \otimes T^*_{\mathcal{M}}$.

The other assumption necessary for the geometric interpretation is that $\beta = \beta^i \frac{\partial}{\partial g^i}$ behaves as a vector field under diffeomorphisms, and the derivative operators extend in the usual way that is consistent with Leibniz rule and contractions:

$$\mathcal{D}_i \beta^j = \beta^j_{,i} + B^j_{ik} \beta^k .$$

$$(2.6.20)$$

We see that B has a natural interpretation as a connection, and if one wants to allow for general diffeomorphisms on $\mathcal{M}$, then a B_i is inevitable. As we will see below, the interpretation of B is more complicated because it may also have an explicit ℓ dependence.

The action principle can be combined with the statement that Θ generates scale transformations in (2.6.6) to produce

$$\int_{\text{reg}} d^2 z \, \langle (\beta^i \widehat{\mathcal{O}}_i(z) + \tfrac{1}{\pi} \Theta(z)) \widehat{\Phi}_{A_1} \cdots \widehat{\Phi}_{A_m} \rangle = (\widehat{\Delta} + \beta^i B_i - \Gamma) \cdot \langle \widehat{\Phi}_{A_1} \cdots \widehat{\Phi}_{A_m} \rangle .$$

$$(2.6.21)$$

There is a more general formulation of a renormalized QFT in terms of a local renormalization group—see, for instance, [56, 320, 321], where it is assumed that the couplings g^i can be promoted to background fields. In that formulation it is possible to show a relation

$$\Theta = -\pi \beta^i \widehat{\mathcal{O}}_i + \cdots ,$$

$$(2.6.22)$$

where $\cdots$ involves redundant operators, i.e. operators that vanish in correlation functions unless placed at a coincident point with another operator—a detailed discussion relevant to two dimensions can be found in [63]. The integrated form (2.6.19) we obtain with our weaker axioms is consistent with the stronger (2.6.22).

Properties of the B Connection

We will now discuss some aspects of the connection B on the coupling space. Our first point is cautionary: the divergent integrals that occur in the relevant definitions must be treated carefully. For example, one might naively conclude that the connection is flat, i.e. the covariant derivative $\mathcal{D}_i$ satisfies $[\mathcal{D}_i, \mathcal{D}_j] = 0$. After all, formally

$$
\mathcal{D}_i \mathcal{D}_j \langle \widehat{\Phi}_{A_1} \cdots \widehat{\Phi}_{A_m} \rangle = -\mathcal{D}_i \int d^2 z \langle \widehat{\mathcal{O}}_j(z, \bar{z}) \widehat{\Phi}_{A_1} \cdots \widehat{\Phi}_{A_m} \rangle
$$

$$
= \int d^2 y \, d^2 z \langle \widehat{\mathcal{O}}_i(y, \bar{y}) \widehat{\mathcal{O}}_j(z, \bar{z}) \widehat{\Phi}_{A_1} \cdots \widehat{\Phi}_{A_m} \rangle , \qquad (2.6.23)
$$

and the last line appears to be symmetric in i, j. This is not the case precisely because the integral is formal, and in general the connection has curvature. For instance, the moduli space of a CFT is generally not flat [174, 279].[19]

We will now discuss some properties of the B connection; the reader may wish to review some of the geometric concepts in Appendix B. To simplify the presentation, we will abbreviate the ponderous $\widehat{\Phi}_{A_1} \cdots \widehat{\Phi}_{A_m}$ by X_A; the results extend over multiple indices in an obvious tensorial form. We will also often abbreviate $\frac{\partial}{\partial g^i}$ by ∂_i.

First, we define the torsion tensor

$$
T_{ij}^k = B_{ij}^k - B_{ji}^k . \qquad (2.6.24)
$$

By our assumption of the completeness of the deforming operators, we have no occasion to speak of T_{ij}^A with $A \neq k$ for some deforming operator $\widehat{\mathcal{O}}_k$. The commutator of the derivatives yields the curvature tensor F:

$$
[\mathcal{D}_m, \mathcal{D}_i] \langle X_A \rangle = \left(\partial_i B_{mA}^C - \partial_m B_{iA}^C + B_{mA}^B B_{iB}^C - B_{iA}^B B_{mB}^C \right) \langle X_C \rangle
$$

$$
+ (B_{im}^k - B_{mi}^k) \mathcal{D}_k \langle X_A \rangle
$$

$$
= (F_{im})_A^C \langle X_C \rangle + T_{im}^k \mathcal{D}_k \langle X_A \rangle , \qquad (2.6.25)
$$

where the curvature tensor is

$$
(F_{im})_A^C = \partial_i B_{mA}^C - \partial_m B_{iA}^C + B_{mA}^B B_{iB}^C - B_{iA}^B B_{mB}^C . \qquad (2.6.26)
$$

---

[19]In the context of (2,2) SCFT topological field theory techniques can be used to evaluate the moduli space metric [237], and at least locally the metric is known in many non-trivial examples. A pedagogical general discussion of moduli space geometry from a CFT point of view and applications to moduli spaces of (2,2) SCFTs can be found in [126]; in the same context it has been shown that sphere partition functions can be used to compute the moduli space metric and also derive general constraints on the global moduli space geometry [201].

The curvature satisfies a Bianchi identity:

$$(F_{im})^k_j + (F_{ji})^k_m + (F_{mj})^k_i = \mathcal{D}_i T^k_{mj} + \mathcal{D}_j T^k_{im} + \mathcal{D}_m T^k_{ji} . \tag{2.6.27}$$

Exercise 2.20 Use the covariant derivative to rewrite the Callan-Symanzik equation in a covariant form:

$$-\ell \frac{\partial}{\partial \ell} \langle X_A \rangle = -\beta^m \mathcal{D}_m \langle X_A \rangle + S^B_A \langle X_B \rangle , \tag{2.6.28}$$

where S is given by

$$S^B_A = \Gamma^B_A - \beta^m B^B_{mA} . \tag{2.6.29}$$

Argue that for a sensible geometric interpretation S should transform as a tensor under coupling-space diffeomorphisms (this is not the case for either Γ or B separately.).   □

We require that the partial derivatives with respect to ℓ and g^i commute, i.e.

$$[-\ell \frac{\partial}{\partial \ell}, \frac{\partial}{\partial g^i}]\langle X_A \rangle = 0 . \tag{2.6.30}$$

This leads to a consistency requirement that is explored in the next exercise.

Exercise 2.21 Let

$$\mathcal{X}^i_j = \Gamma^i_j + \partial_j \beta^i . \tag{2.6.31}$$

Show that $\mathcal{X}^i_j$ is a tensor by rewriting it as

$$\mathcal{X}^i_j = S^i_j + \mathcal{D}_j \beta^i + \beta^m T^i_{mj} . \tag{2.6.32}$$

Next, show that setting $[-\ell \frac{\partial}{\partial \ell}, \frac{\partial}{\partial g^i}]\langle X_A \rangle = 0$ leads to

$$\mathcal{X}^j_i \mathcal{D}_j \langle X_A \rangle = \left\{ \beta^m (F_{im})^B_A + \mathcal{D}_i (S^B_A) + \left[\ell \frac{\partial}{\partial \ell} B^B_{iA} \right] \right\} \langle X_B \rangle . \tag{2.6.33}$$

Finally, check that $\mathcal{X}^j_i = 0$ implies that the Callan-Symanzik equation for the deforming operators can be written in terms of a Lie derivative with respect to the β-function:

$$-\ell \frac{\partial}{\partial \ell} \langle X_i \rangle = -\mathcal{L}_\beta \langle X_i \rangle = - \left[\beta^j \partial_j \langle X_i \rangle + \beta^j_{,i} \langle X_j \rangle \right] . \tag{2.6.34}$$

□

If $\mathcal{X}^i_j = 0$ and the torsion tensor $T = 0$ as well, we obtain an equation for the connection[20]:

$$-\ell \frac{\partial}{\partial \ell} B^k_{ij} = \beta^m F^k_{imj} - \mathcal{D}_i \mathcal{D}_j \beta^k \, . \tag{2.6.35}$$

We will see that this expression is satisfied in conformal perturbation theory.

2.6.2 Bare and Renormalized Couplings

Conformal perturbation theory (CPT) is a general method for obtaining renormalized QFTs. Conventional perturbation theory around a free field theory is a limit of the more general situation, and in fact a somewhat pathological limit.

In describing a family of QFTs over some coupling space $\mathcal{M}$, at best we might hope that there is a point $p \in \mathcal{M}$ where a perturbation expansion has a non-zero radius of convergence, with radius given by the distance to the nearest singular point. It is well-known that this hope does not hold when p is a free field theory. On the other hand, this hope is believed to be realized in conformal perturbation theory around a compact unitary CFT. A general argument for this is sketched out in [263], and many exact results suggest this to be the case, but there is no proof of this remarkable conjecture. As the next few paragraphs will indicate, there is probably a good reason for this: our (or at least the author's) understanding of the technical aspects of CPT is still rather limited.

To describe CPT, we must make two aspects of the problem more precise: we must define regulated correlation functions and then give a renormalization prescription. To ease up on the notation, we only discuss correlation functions of the deforming operators in the text. A study of the non-deforming operators and a few technical points are relegated to the Appendix A.3.

A theory CFT_0 has local quasi-primary spin zero operators $\mathcal{O}_I$ with mass dimension $2 \geq \Delta_I \geq 0$, and we suppose we know all of their correlation functions at separated insertion points. We assume there is a unique identity operator, which we will write as $\mathcal{O}_0 = 1$ in some of the manipulations that follow.

We define renormalized correlators for a deformed theory by the following procedure.

1. Define the bare action by

$$S(\lambda_B) = \int d^2z \left[\lambda^i_B \mathcal{O}_i(z, \bar{z}) + \lambda^0_B \right] \, , \tag{2.6.36}$$

---

[20]The reader can verify that with these assumptions the right-hand-side of the equation is symmetric in the i and j indices.

where the λ_B^i are the bare coupling constants with mass dimension $\eta_i = 2 - \Delta_i$, and λ_B^0 is the cosmological constant.

2. Fix n distinct points $w_i \in \mathbb{C}$ and m points $z_\alpha \in \mathbb{C}$ and let $\mathbb{D}_{n;m}^{\mathrm{uv}}(w_1, w_2, \ldots, w_n) \subset \mathbb{C}^m$ be

$$\mathbb{D}_{n;m}^{\mathrm{uv}} = \left\{ (z_1, \ldots, z_m) \in \mathbb{C}^m \mid |z_{\alpha\beta}| \geq a , \ |z_\alpha - w_s| \geq a \right\} . \qquad (2.6.37)$$

This domain will regulate UV divergences due to operator collisions: the smallest distance between any two operators will be the cut-off a. Next, restrict the domain further to

$$\mathbb{D}_{n;m} = \left\{ (z_1, \ldots, z_m) \in \mathbb{D}_{n;m}^{\mathrm{uv}} \mid |z_{\alpha\beta}| \leq L , \ |z_\alpha - w_s| \leq L \right\} , \qquad (2.6.38)$$

where L is the IR cut-off length. We take it to be larger than the maximum separation between the fixed operators in the correlation function.

We denote the regulated correlation functions of the bare operators by

$$\langle \mathcal{O}_{i_1}(w_1) \cdots \mathcal{O}_{i_n}(w_n) e^{-S(\lambda_B)} \rangle_{\mathrm{reg}} . \qquad (2.6.39)$$

These are defined by taking the formal power-series expansion of e^{-S} in the coupling constants, and at each order m replacing the integration domain $\mathbb{C}^m \to \mathbb{D}_{n;m}$.

The regulated correlation functions with at least one insertion of $\mathcal{O}_i$ are finite and Lorentz-invariant. If the inserted operators carry spin, the regulated correlation functions will be Lorentz-covariant. All correlation functions are also manifestly translation-invariant. The partition function $Z = \langle e^{-S(\lambda_B)} \rangle_{\mathrm{reg}}$ is IR-divergent, but the divergence is just the infinite volume $\int d^2 z$ factor: we write the free energy as $W = -\log Z = -\int d^2 z f$, and the regulated free energy density f will be finite.

3. We express the bare coupling constants in terms of dimensionless renormalized couplings g^i via

$$\lambda_B^i = \ell^{-\eta_i} \Lambda_B^i(g, a/\ell) = \ell^{-\eta_i} g^i + O(g^2) , \qquad (2.6.40)$$

Now we define the renormalized operators by

$$\widehat{\mathcal{O}}_i = Z_i^j(g, a/\ell) \ell^{-\eta_j} \mathcal{O}_j + \ell^{-2} Z_i^0(g, a/\ell), \iff \mathcal{O}_j = \ell^{\eta_j} (Z^{-1})_j^k \left[\widehat{\mathcal{O}}_k - \ell^{-2} Z_k^0 \right]. \qquad (2.6.41)$$

where the dimensionless wavefunction renormalization coefficients Z_i^j and Z_i^0 are to be determined by a renormalization prescription: a specification for the values of a finite set of correlation functions at some chosen kinematics (e.g. particular insertion points for the local operators). The correlation functions are also constrained by demanding that CPT obeys the axioms of a renormalized QFT sketched above. The renormalized correlation functions are denoted by the double brackets:

$$\langle\!\langle \widehat{\mathcal{O}}_{k_1} \cdots \widehat{\mathcal{O}}_{k_n} \rangle\!\rangle = \lim_{a \to 0} \langle \widehat{\mathcal{O}}_{k_1} \cdots \widehat{\mathcal{O}}_{k_n} \rangle_{\text{reg}} . \tag{2.6.42}$$

Note that the $\widehat{\mathcal{O}}_i$ have mass dimension 2.

The definition we give deserves a few more comments.

Descendants and Renormalizability

The quasi-primary operators $\mathcal{O}_i$ with $\Delta_i \leq 2$ do not exhaust the set of local spin zero operators. There are irrelevant operators, and there may also be some relevant/marginal descendants given by $\mathcal{O}_i^{p,\overline{p}}(z, \overline{z}) = \partial^p \overline{\partial}^{\overline{p}} \mathcal{O}_i(z, \overline{z})$. Such operators can in principle appear in the renormalized operators since Lorentz invariance allows for

$$\widehat{\mathcal{O}}_i = \cdots + \ell^{\Delta_\Phi + p + \overline{p} - 2} \partial^p \overline{\partial}^{\overline{p}} \Phi + \cdots , \tag{2.6.43}$$

where Φ is a quasi-primary operator with spin $s_\Phi = h_\Phi - \overline{h}_\Phi = p - \overline{p}$.

We are interested in Lorentz-preserving deformations by relevant and marginal operators, and we will make a standard renormalizability assumption: wavefunction renormalization of an operator with dimension Δ_i should only mix operators of dimension $\leq \Delta_i$. Therefore a descendant could only appear in the renormalized operator if

$$2 \geq \Delta_\Phi + p + \overline{p} \implies 1 \geq h_\Phi + p . \tag{2.6.44}$$

The solutions are $\Delta_i = 2$, and, either $h_\Phi = 0$, $\overline{h}_\Phi = 1$ and $p = 1$, or $h_\Phi = 1$, $\overline{h}_\Phi = 0$, and $\overline{p} = 1$. In a unitary compact CFT such operators Φ are conserved anti-holomorphic or holomorphic currents that satisfy $\overline{\partial}\Phi = 0$ or $\partial\Phi = 0$. Therefore CFT_0 has no candidate descendants that could mix with the $\mathcal{O}_i$.

IR Divergences

Already at first order in perturbation theory we encounter the following integral in the computation of the renormalized two-point function:

$$\langle \mathcal{O}_i(x) \mathcal{O}_j(0) \rangle \sim \int d^2 z \frac{C_{ijk}}{|z|^{\Delta_i + \Delta_k - \Delta_j} |z - x|^{\Delta_j + \Delta_k - \Delta_i} |x|^{\Delta_i + \Delta_j - \Delta_k}} . \tag{2.6.45}$$

For large $|z|$ the integrand scales as $|z|^{-2\Delta_k}$, so if $\Delta_k \leq 1$ there will be IR divergences. At least in some situations these IR divergences are harmless once they are summed up to all orders in perturbation theory. For instance, whenever the deformation leads to a gapped theory, we expect sensible IR behavior. A simple example is where CFT_0 is a system of N massless Dirac fermions deformed by a mass term.

To avoid these IR divergences, we will make an "IR safety" assumption: the deforming operators $\mathcal{O}_i$ have $2 \geq \Delta_i > 1$, and their OPE closes on the identity and any operators with $\Delta > 1$. This is a strong assumption: for instance, it is easy to violate for a large radius compact boson. While this guarantees IR safety at the first few orders in perturbation theory, it is not completely clear that it does so to all orders.

With those caveats, we are ready to explore how the general renormalized QFT is exemplified by conformal perturbation theory (CPT). We will see that there are consistency requirements on renormalization coefficients that CPT must satisfy in order to be a renormalized QFT. At low orders in perturbation theory, we will see in explicit computations that these consistency conditions are fulfilled. Showing that they hold to all orders is a key open problem in CPT.

The β Function

We take bare couplings to have the form

$$\lambda_B^i(g) = \ell^{-\eta_i} \left[g^i + \tfrac{1}{2} A_{2jk}^i g^j g^k + \tfrac{1}{3!} A_{3jkl}^i g^j g^k g^l + O(g^4) \right] ,$$

$$\lambda_B^0(g) = \ell^{-2} \left[\tfrac{1}{2} A_{2jk}^0 g^j g^k + \tfrac{1}{3!} A_{3jkl}^0 g^j g^k g^l + O(g^4) \right] , \qquad (2.6.46)$$

where the dimensionless coefficients A_{2jk}^i are functions of the ratio a/ℓ.

The β function in CPT is determined by demanding that the λ_B^i are independent of the renormalization scale[21]:

$$0 = \ell \frac{d\lambda_B^i}{d\ell} = \ell \frac{\partial \lambda_B^i}{\partial \ell} + \frac{\partial \lambda_B^i}{\partial g^k} \ell \frac{dg^k}{d\ell} = \ell \frac{\partial \lambda_B^i}{\partial \ell} - \frac{\partial \lambda_B^i}{\partial g^k} \beta^k . \qquad (2.6.47)$$

We solve this for the β function:

$$\beta^k = \left(\frac{\partial g}{\partial \lambda_B} \right)_i^k \ell \frac{\partial \lambda_B^i}{\partial \ell} . \qquad (2.6.48)$$

---

[21] We define the β function as $\beta = -\ell \frac{dg}{d\ell}$. When β is negative, then the coupling becomes small as $\ell \to 0$, i.e. in the UV.

We explore its perturbative expansion below. We also require that λ_B^0 is ℓ-independent, which means

$$\left(-\ell\frac{\partial}{\partial\ell} + \beta^i\frac{\partial}{\partial g^i}\right)\lambda_B^0 = 0. \tag{2.6.49}$$

The Callan-Symanzik Equation

Since our wavefunction renormalization mixes the operators $\mathcal{O}_i$ with the identity, it is useful to introduce an index I that collectively refers to $\{1, \mathcal{O}_i\}$. With that notation, in order for CPT to have the axiomatic properties we desire, we need

$$\left(-\ell\frac{\partial}{\partial\ell} + \beta^j\frac{\partial}{\partial g^j}\right)\langle\widehat{\mathcal{O}}_{i_1}\cdots\widehat{\mathcal{O}}_{i_n}\rangle = -\ell\frac{d}{d\ell}\left[Z_{i_1}^{J_1}\cdots Z_{i_n}^{J_n}\ell^{-\eta_{J_1}-\cdots-\eta_{J_n}}\right]\langle\mathcal{O}_{J_1}\cdots\mathcal{O}_{J_n}\rangle$$

$$= \Gamma\cdot\langle\widehat{\mathcal{O}}_{i_1}\cdots\widehat{\mathcal{O}}_{i_n}\rangle, \tag{2.6.50}$$

where

$$\Gamma_i^J\widehat{\mathcal{O}}_J = -\ell\frac{d}{d\ell}\left(Z_i^J\ell^{-\eta_J}\right)\mathcal{O}_J. \tag{2.6.51}$$

Using (2.6.41), we expand both sides as

$$\Gamma_i^j\widehat{\mathcal{O}}_j + \Gamma_i^0 = -\ell\frac{d}{d\ell}\left(Z_i^k\ell^{-\eta_k}\right)\ell^{\eta_k}(Z^{-1})_k^j\widehat{\mathcal{O}}_j - \ell\frac{d}{d\ell}\left(Z_i^0\ell^{-2}\right)$$

$$+ \ell\frac{d}{d\ell}\left(Z_i^k\ell^{-\eta_k}\right)\ell^{\eta_k-2}(Z^{-1})_k^j Z_j^0, \tag{2.6.52}$$

so that

$$\Gamma_i^j = -\ell\frac{d}{d\ell}\left(Z_i^k\ell^{-\eta_k}\right)\ell^{\eta_k}(Z^{-1})_k^j,$$

$$\Gamma_i^0 = -\ell\frac{d}{d\ell}\left(Z_i^0\ell^{-2}\right) + \ell\frac{d}{d\ell}\left(Z_i^k\ell^{-\eta_k}\right)\ell^{\eta_k-2}(Z^{-1})_k^j Z_j^0$$

$$= -\ell\frac{d}{d\ell}\left(Z_i^0\ell^{-2}\right) - \Gamma_i^j Z_j^0\ell^{-2}. \tag{2.6.53}$$

The Action Principle

In the same way we compute

$$\frac{\partial}{\partial g^j} \langle \widehat{\mathcal{O}}_{i_1} \cdots \widehat{\mathcal{O}}_{i_n} \rangle = B_j \cdot \langle \widehat{\mathcal{O}}_{i_1} \cdots \widehat{\mathcal{O}}_{i_n} \rangle$$

$$- \int d^2z \, \langle \left(\frac{\partial \lambda_B^k}{\partial g^j} \mathcal{O}_k(z) + \frac{\partial \lambda_B^0}{\partial g^j} \right) \widehat{\mathcal{O}}_{i_1} \cdots \widehat{\mathcal{O}}_{i_n} \rangle \,, \qquad (2.6.54)$$

where the connection B arises through Z relating the bare and renormalized quantities:

$$(B_j)_i^k \widehat{\mathcal{O}}_k + (B_j)_i^0 = \frac{\partial}{\partial g^j} \left[Z_i^k \ell^{-\eta_k} \mathcal{O}_k + Z_i^0 \ell^{-2} \right] . \qquad (2.6.55)$$

Working out the coefficients, we obtain

$$(B_j)_i^k = \frac{\partial Z_i^m}{\partial g^j} (Z^{-1})_m^k \,, \qquad (B_j)_i^0 = \left[\frac{\partial Z_i^0}{\partial g^j} - (B_j)_i^m Z_m^0 \right] \ell^{-2} . \qquad (2.6.56)$$

For the action principle to hold, we need

$$\frac{\partial \lambda_B^k}{\partial g^j} \mathcal{O}_k + \frac{\partial \lambda_B^0}{\partial g^j} = \widehat{\mathcal{O}}_j = Z_j^k \ell^{-\eta_k} \mathcal{O}_k + Z_j^0 \ell^{-2} \implies Z_j^k = \ell^{\eta_k} \frac{\partial \lambda_B^k}{\partial g^j} \,,$$

$$Z_j^0 = \frac{\partial \lambda_B^0}{\partial g^j} \ell^2 . \qquad (2.6.57)$$

This has important consequences:

1. $Z_{i,j}^m = Z_{j,i}^m$ and $Z_{i,j}^0 = Z_{j,i}^0$, and that means B is a torsion-free connection;
2. Γ_i^m is determined by the beta function. To prove this, we use (2.6.57) in (2.6.51):

$$\Gamma_j^i = -\ell \frac{d}{d\ell} \left(\frac{\partial \lambda_B^k}{\partial g^j} \right) \ell^{\eta_k} (Z^{-1})_k^i . \qquad (2.6.58)$$

Now we commute derivatives:

$$-\ell \frac{d}{d\ell} \frac{\partial}{\partial g^j} = \left(-\ell \frac{\partial}{\partial \ell} + \beta^m \frac{\partial}{\partial g^m} \right) \frac{\partial}{\partial g^j} = -\beta_{,j}^m \frac{\partial}{\partial g^m} + \frac{\partial}{\partial g^j} \left(-\ell \frac{d}{d\ell} \right) .$$

$$(2.6.59)$$

Since $d\lambda_B/d\ell = 0$, it now follows

$$\Gamma^i_j = -\beta^m_{\cdot j}\frac{\partial \lambda^k_B}{\partial g^m}\ell^{\eta k}(Z^{-1})^i_k \implies \Gamma^i_j = -\beta^i_{\cdot j}. \qquad (2.6.60)$$

Thus, the matrix of anomalous dimensions for the deforming operators is determined by the β function: the tensor $\mathcal{X}$ of the previous section is automatically zero.

3. $\Gamma^0_i = 0$. This follows from the previous statement, (2.6.57), and the fact that λ^0_B is ℓ-independent.

Finally, we can check that the connection B satisfies (2.6.35).

Exercise 2.22 Use

$$B^k_{ji} = \frac{\partial Z^m_i}{\partial g^j}(Z^{-1})^k_m = \frac{\partial^2 \lambda^m_B}{\partial g^i \partial g^j}\left(\frac{\partial g}{\partial \lambda_B}\right)^k_m \qquad (2.6.61)$$

to show that, as promised,

$$-\ell\frac{\partial}{\partial \ell}B^k_{ij} = \beta^m (F_{im})^k_j - \mathcal{D}_i\mathcal{D}_j\beta^k. \qquad (2.6.62)$$

□

2.6.3 The c-Theorem

Before we delve any further into intricacies of the conformal perturbation theory expansion, we remind the reader of a general result that constrains the RG flow in any renormalizable two-dimensional QFT that satisfies our assumptions. This is Zamolodchikov's c-theorem, and we present the argument following [327]. Rotational invariance implies that the functions

$$F = z^4 \langle T(z,\bar{z})T(0)\rangle,$$

$$G = 4z^3\bar{z}\langle T(z,\bar{z})\Theta(0)\rangle = 4z^3\bar{z}\langle\Theta(z,\bar{z})T(0)\rangle,$$

$$H = 16z^2\bar{z}^2\langle\Theta(z,\bar{z})\Theta(0)\rangle \qquad (2.6.63)$$

only depend on $r^2 = z\bar{z}$.

Exercise 2.23 On rotationally invariant functions $\mathcal{S} = 2z\partial_z = 2\bar{z}\partial_{\bar{z}} = 2\frac{d}{d\log r^2}$. Denote derivatives with respect to $\log r^2$ by a dot. Show that rotational invariance and conservation of the energy-momentum tensor imply the relations

$$4\dot{F} + \dot{G} - 3G = 0, \qquad 4\dot{G} - 4G + \dot{H} - 2H = 0, \qquad (2.6.64)$$

and therefore

$$C = 2F - G - \frac{3}{8}H \tag{2.6.65}$$

satisfies

$$r^2 \frac{dC}{dr^2} = -\frac{3}{4}H . \tag{2.6.66}$$

$\square$

If we set $r = \ell$, our renormalization scale, and use the relation (2.6.22), we therefore obtain

$$\ell \frac{\partial C}{\partial \ell} = -24\pi^2 \sum_{i,j} \beta^i \beta^j \ell^4 \langle \widehat{\mathcal{O}}_i(\ell) \widehat{\mathcal{O}}_j(0) \rangle . \tag{2.6.67}$$

Unitarity implies that the Zamolodchikov metric $\ell^4 \langle \widehat{\mathcal{O}}_i(\ell) \widehat{\mathcal{O}}_j(0) \rangle$ is positive-definite. Hence, as ℓ increases the C function decreases monotonically along the RG flow, and the decrease terminates if and only if $\beta^i = 0$ for all couplings. In a unitary and compact theory the flow must terminate, and the resulting scale-invariant theory is conformal [326], with central charge given by the critical value of C.

2.6.4 Formal g-Expansion of λ^i_B and Beta Function

Consider the expansion of the bare coupling in g^i:

$$\lambda^i_B = \ell^{-\eta_i} \left[g^i + \frac{1}{2} A^i_{2jk} g^j g^k + \frac{1}{3!} A^i_{3jkl} g^j g^k g^l + O(g^4) \right] , \tag{2.6.68}$$

where the coefficients A are functions of the dimensionless ratio ℓ/a. We will find it useful to work with $t = \log \ell/a$ and denote df/dt derivatives by a $\dot{f}$. The A coefficients are completely symmetric in the lower indices.

Taking a derivative, we obtain

$$\frac{\partial \lambda^i_B}{\partial g^j} = \ell^{-\eta_i} \left[\delta^i_j + A^i_{2jk} g^k + \frac{1}{2} A^i_{3jkl} g^k g^l + O(g^3) \right] , \tag{2.6.69}$$

and, with a tiny bit of strain, we invert the matrix

$$\frac{\partial g^j}{\partial \lambda^k_B} = \ell^{\eta_k} \left[\delta^j_k - A^j_{2kn} g^n + (A^j_{2mn} A^m_{2kp} - \frac{1}{2} A^j_{3knp}) g^n g^p + O(g^3) \right] \tag{2.6.70}$$

and compute the derivative

$$
\dot{\lambda}_B^j = \ell^{-\eta_j} \left[-\eta_j g^j + \frac{1}{2} \left(\dot{A}_{2qr}^j - \eta_j A_{2qr}^j \right) g^q g^r \right.
$$
$$
\left. + \frac{1}{3!} \left(\dot{A}_{3qrs}^j - \eta_j A_{3qrs}^j \right) g^q g^r g^s + O(g^4) \right]. \quad (2.6.71)
$$

With this, we can now expand the β-function:

$$
\beta^i = \frac{\partial g^i}{\partial \lambda_B^j} \dot{\lambda}_B^j = -\eta_i g^i + \tfrac{1}{2}\beta_{2qr}^i g^q g^r + \tfrac{1}{3!}\beta_{3qrs}^i g^q g^r g^s + O(g^4) , \quad (2.6.72)
$$

where the coefficients are

$$
\beta_{2qr}^i = \dot{A}_{2qr}^i - (\eta_i - \eta_q - \eta_r) A_{2qr}^i ,
$$
$$
\beta_{3qrs}^i = \dot{A}_{3qrs}^i - (\eta_i - \eta_q - \eta_r - \eta_s) A_{3qrs}^i - A_{2qm}^i \beta_{2rs}^m - A_{2sm}^i \beta_{2qr}^m - A_{2rm}^i \beta_{2sq}^m .
$$
$$
(2.6.73)
$$

The β function cannot have explicit ℓ dependence, which leads to integrability conditions. Let $\sigma_{qr}^i = \eta_i - \eta_q - \eta_r$. If $\sigma_{qr}^i \neq 0$, then

$$
A_{2qr}^i = N_{2qr}^i e^{\sigma_{qr}^i t} - \frac{\beta_{2qr}^i}{\sigma_{qr}^i} = N_{2qr}^i \left(\frac{\ell}{a} \right)^{\sigma_{qr}^i} - \frac{\beta_{2qr}^i}{\sigma_{qr}^i} , \quad (2.6.74)
$$

where N_{2qr}^i is some so-far-undetermined constant. On the other hand, if $\sigma_{qr}^i = 0$ (this is known as a "resonance condition" in the literature), then

$$
A_{2qr}^i = N_{2qr}^i + \beta_{2qr}^i \log \ell/a . \quad (2.6.75)
$$

We can also solve for A_{3qrs}^i in a similar fashion, and for simplicity we will just consider the simple case of just marginal operators, i.e. $\eta_i = 0$ for all i. In this case

$$
A_{3qrs}^i = \tfrac{1}{2} \left(\beta_{2qm}^i \beta_{2rs}^m + \beta_{2rm}^i \beta_{2qs}^m + \beta_{2sm}^i \beta_{2qr}^m \right) \log^2(\ell/a)
$$
$$
+ \widehat{\beta}_{3qrs}^i \log(\ell/a) + N_{3qrs}^i , \quad (2.6.76)
$$

where

$$
\widehat{\beta}_{3qrs}^i = \beta_{3qrs}^i + (N_{2qm}^i \beta_{2rs}^m + N_{2rm}^i \beta_{2qs}^m + N_{2sm}^i \beta_{2qr}^m) . \quad (2.6.77)
$$

The "non-local" $\log^2 \ell/a$ term in A^i_{3qrs} is determined by terms of lower order. This is reminiscent of the cancellation of overlapping divergences in perturbative QFT computations.

There are similar constraints on the cosmological constant, but this time the equations satisfied by the A^0 coefficients are homogeneous, since the β function coefficients are already fixed by the previous computation. At leading order we obtain

$$\dot{A}^0_{2jk} = (2 - \eta_j - \eta_k)A^0_{jk} \, , \qquad (2.6.78)$$

which has the solution

$$A^0_{2jk} = N^0_{jk} \left(\frac{\ell}{a}\right)^{2-\eta_j-\eta_k} \, . \qquad (2.6.79)$$

At the next order we obtain the ODE

$$\dot{A}^0_{3ijk} - (2 - \eta_i - \eta_j - \eta_k)A^0_{3ijk} = 3\beta^m_{2(ij}A^0_{2k)m} \, . \qquad (2.6.80)$$

We obtain a family of solutions

$$A^0_{3ijk} = (\text{particular solution}) + N^0_{3ijk} \left(\frac{\ell}{a}\right)^{2-\eta_i-\eta_j-\eta_k} \, . \qquad (2.6.81)$$

Finally, we can also compute the B connection explicitly:

$$B^k_{ij} = \frac{\partial^2 \lambda^m_B}{\partial g^i \partial g^j} \left(\frac{\partial g}{\partial \lambda_B}\right)^k_m = -A^k_{2ij} + \left(A^k_{3ijn} + A^m_{2ij}A^k_{2mn}\right) g^n + O(g^2) \, .$$
$$(2.6.82)$$

This confirms that B will in general depend on ℓ since there is such a dependence for the A coefficients.

2.6.5 Basic Renormalization in CPT

We will now study renormalization of the zero-, one-, and two-point functions of the deforming operators up to $O(g^2)$ corrections; we will also work out the explicit $O(g^2)$ terms in a simple example. The combinatorics and divergence structures are much simpler for the connected correlation functions, and we will work with those.

A Review of Connected Correlation Functions

Consider a Lagrangian field theory with an operator ϕ. The partition function with a source J is defined as

$$Z[J] = \int D\phi \exp[-S + J \cdot \phi], \qquad (2.6.83)$$

where $J \cdot \phi = \int d^2 z J \phi$. By construction $Z[J]$ is the generating function for the correlation functions of ϕ in the theory. Similarly, $-W[J] = \log Z[J]$ is the generator of connected correlation functions; the first few are then given by

$$\langle \phi \rangle^c = N^{-1} \langle \phi \rangle,$$

$$\langle \phi_1 \phi_2 \rangle^c = N^{-1} \langle \phi_1 \phi_2 \rangle - N^{-2} \langle \phi_1 \rangle \langle \phi_2 \rangle,$$

$$\langle \phi_1 \phi_2 \phi_3 \rangle^c = N^{-1} \langle \phi_1 \phi_2 \phi_3 \rangle - N^{-2} \langle \phi_1 \rangle \langle \phi_2 \phi_3 \rangle - N^{-2} \langle \phi_2 \rangle \langle \phi_3 \phi_1 \rangle$$

$$- N^{-2} \langle \phi_3 \rangle \langle \phi_1 \phi_2 \rangle + 2 N^{-3} \langle \phi_1 \rangle \langle \phi_2 \rangle \langle \phi_3 \rangle, \qquad (2.6.84)$$

and $N = Z[0] = e^{-\int d^2 z f}$, where f is the free energy density.

Exactly the same construction can be given in conformal perturbation theory, with the normalization $N = \langle e^{-S_B} \rangle_{\text{reg}}$. In general RG flows $Z[0]$ is a scheme-dependent unphysical quantity.[22] For our purposes, there are two main properties of the connected correlators: (i) they are independent of the over-all normalization; (ii) with the exception of one-point functions, they are independent of shifts of the fields by c-numbers.

Exercise 2.24 While the first point is obvious, maybe the second is not immediately clear. To show it, let $\widetilde{Z}[J]$ be the generating function for correlators of $\psi = \phi + x$, where x is a c-number, and prove $\widetilde{W}[J] = J \cdot x + W[J]$. $\qquad\qquad$ □

Normalization and One-Point Functions

We now proceed to explicit computations of the connected correlation functions in CPT. We begin with

$$- \log N = \langle S_B \rangle_{\text{reg}} + \frac{1}{2} \left(\langle S_B \rangle^2_{\text{reg}} - \langle S_B^2 \rangle_{\text{reg}} \right) + O(g^3). \qquad (2.6.85)$$

---

[22]One can try to define a partition function for a conformal field theory by placing the theory on a sphere; in a general CFT the resulting object is still scheme-dependent [195].

Inserting the explicit expressions and using $\langle \mathcal{O}_i \rangle = 0$ in CFT$_0$, we obtain

$$-\log N = \int d^2z \left[\lambda_B^0 + \tfrac{1}{2} \int_{\text{reg}} d^2w \, \lambda_B^i \lambda_B^j \langle \mathcal{O}_i(0) \mathcal{O}_j(w) \rangle \right] + O(g^3)$$

$$= \tfrac{1}{2} g^i g^j \ell^{-2} \int d^2z \left[A_{ij}^0 + \frac{2\pi \delta_{ij}}{\Delta_i - 1} \left(\frac{\ell}{a} \right)^{2\Delta_i - 2} \right] + O(g^3) . \qquad (2.6.86)$$

Note the denominator factors of $\Delta_i - 1$; these are rendered harmless our IR safety assumption. We choose A_{ij}^0 to cancel the divergence, so that $N = 1 + O(g^3)$. This is consistent with (2.6.79), and it will lead to simplifications in the computations that follow.

The one-point functions $\langle\!\langle \widehat{\mathcal{O}}_i \rangle\!\rangle^c$ are rather special because they are determined by the action principle: write $N = \exp[-\int d^2z f]$; the action principle implies

$$\int d^2z \, \langle\!\langle \widehat{\mathcal{O}}_i \rangle\!\rangle = -\frac{\partial}{\partial g^i} N \implies \langle\!\langle \widehat{\mathcal{O}}_i \rangle\!\rangle^c = \frac{\partial f}{\partial g^i} . \qquad (2.6.87)$$

If we can choose the cosmological constant so that $N = 1$, we will automatically have $\langle\!\langle \widehat{\mathcal{O}}_i \rangle\!\rangle^c = 0$. We have shown that $N = 1 + O(g^3)$, so that $\langle\!\langle \widehat{\mathcal{O}}_i \rangle\!\rangle = O(g^2)$, and to this order we satisfy the consistency requirement (2.6.79). To push the result to the next order, we would have to prove that we can choose A_{3ijk}^0 so that $N = 1 + O(g^4)$ and (2.6.81) is obeyed.

The Two-Point Functions

We focus on the connected two-point functions; their computation is simplified by noting that they are independent of λ_B^0, and they are also independent of the Z_i^0 factors in the renormalized $\widehat{\mathcal{O}}_i$. Thus (the right-hand-side is to be understood in the $a \to 0$ limit),

$$\langle\!\langle \widehat{\mathcal{O}}_i(x) \widehat{\mathcal{O}}_j(0) \rangle\!\rangle^c = \frac{\langle \widehat{\mathcal{O}}_i'(x) \widehat{\mathcal{O}}'(0) e^{-S_B'} \rangle_{\text{reg}}}{\langle e^{-S_B'} \rangle_{\text{reg}}} - \frac{\langle \widehat{\mathcal{O}}_i'(x) e^{-S_B'} \rangle_{\text{reg}} \langle \widehat{\mathcal{O}}_j'(0) e^{-S_B'} \rangle_{\text{reg}}}{\langle e^{-S_B'} \rangle_{\text{reg}}^2} ,$$

$$(2.6.88)$$

where

$$\widehat{\mathcal{O}}_i'(x) = \ell^{-\eta_i} \mathcal{O}_i(x) + \delta Z_i^k \ell^{-\eta_k} \mathcal{O}_k , \qquad S_B' = \int d^2z \lambda_B^i \mathcal{O}_i . \qquad (2.6.89)$$

Not a whiff of λ_B^0 remains in the computation, and the renormalization coefficients are

$$\delta Z_i^k = \ell^{\eta_k} \frac{\partial \lambda_B^k}{\partial g^i} - \delta_i^k = A_{2im}^k g^m + \tfrac{1}{2} A_{3imn}^k g^m g^n + O(g^3) . \qquad (2.6.90)$$

Let $G_{ij}(x) = \langle \mathcal{O}_i(x)\mathcal{O}_j(0)\rangle$. In the unperturbed CFT we choose a basis of operators so that

$$G_{ij}(x) = \frac{\delta_{ij}}{|x|^{2\Delta_i}}. \tag{2.6.91}$$

Since $\langle \mathcal{O}_i\rangle = 0$, which also implies $\langle S'_B\rangle = 0$, we obtain

$$
\begin{aligned}
\langle\!\langle \widehat{\mathcal{O}}_i(x)\widehat{\mathcal{O}}_j(0)\rangle\!\rangle^c ={}& \ell^{-\eta_i-\eta_j}G_{ij}(x) \\
& - \ell^{-\eta_i-\eta_j}\langle \mathcal{O}_i(x)\mathcal{O}_j(0)S'_B\rangle_{\mathrm{reg}} + \delta Z_i^k \ell^{-\eta_j-\eta_k}G_{kj}(x) \\
& + \delta Z_j^k \ell^{-\eta_i-\eta_k}G_{ik}(x) \\
& + \tfrac{1}{2}\ell^{-\eta_i-\eta_j}\Big[\langle \mathcal{O}_i(x)\mathcal{O}_j(0)(S'_B)^2\rangle_{\mathrm{reg}} - \langle \mathcal{O}_i(x)\mathcal{O}_j(0)\rangle\langle (S'_B)^2\rangle_{\mathrm{reg}} \\
& \qquad\qquad - 2\langle \mathcal{O}_i(x)S'_B\rangle_{\mathrm{reg}}\langle \mathcal{O}_j(0)S'B\rangle_{\mathrm{reg}}\Big] \\
& - \delta Z_i^k \ell^{-\eta_k-\eta_j}\langle \mathcal{O}_k(x)\mathcal{O}_j(0)S'B\rangle \\
& - \delta Z_j^k \ell^{-\eta_k-\eta_i}\langle \mathcal{O}_i(x)\mathcal{O}_k(0)S'B\rangle \\
& + \delta Z_i^k \delta Z_j^m \ell^{-\eta_k-\eta_m}G_{km}(x) \\
& + O(g^3). \tag{2.6.92}
\end{aligned}
$$

We write this as

Exercise 2.25 Use $\langle \mathcal{O}_i\rangle = 0$ and $\langle 1\rangle = 1$ in CFT_0 to show that

$$\langle\!\langle \widehat{\mathcal{O}}_i(x)\widehat{\mathcal{O}}_j(0)\rangle\!\rangle^c = \ell^{-\eta_i-\eta_j}\left[G_{ij} + g^k\delta_k^1 G_{ij} + \tfrac{1}{2}g^k g^m \delta_{km}^2 G_{ij} + O(g^3)\right]. \tag{2.6.93}$$

with

$$
\begin{aligned}
\delta_k^1 G_{ij}(x) ={}& -\ell^{-\eta_k}\int_{\mathrm{reg}} d^2z\langle \mathcal{O}_i(x)\mathcal{O}_j(0)\mathcal{O}_k(z)\rangle + \ell^{\eta_i-\eta_m}A_{2ik}^m\langle \mathcal{O}_m(x)\mathcal{O}_j(0)\rangle \\
& + \ell^{\eta_j-\eta_m}A_{2jk}^m\langle \mathcal{O}_i(x)\mathcal{O}_m(0)\rangle \\
={}& -\ell^{-\eta_k}\int_{\mathrm{reg}} d^2z\langle \mathcal{O}_i(x)\mathcal{O}_j(0)\mathcal{O}_k(z)\rangle \\
& + \ell^{\eta_i-\eta_m}A_{2ik}^m G_{mj} + \ell^{\eta_j-\eta_m}A_{2jk}^m G_{im}. \tag{2.6.94}
\end{aligned}
$$

and (with a bit more algebra)

$$\delta^2_{km} G_{ij}(x) = \ell^{-\eta_k - \eta_m} \int_{\text{reg}} d^2 z_1 d^2 z_2 \, \langle \mathcal{O}_i(x) \mathcal{O}_j(0) \mathcal{O}_k(z_1) \mathcal{O}_m(z_2) \rangle^c$$

$$+ A^n_{2km} \delta^1_n G_{ij} + 2\ell^{\eta_j - \eta_n} A^n_{2j(k} \delta^1_{m)} G_{in} + 2\ell^{\eta_i - \eta_n} A^n_{2i(k} \delta^1_{m)} G_{nj}$$

$$+ \ell^{\eta_i - \eta_p} (A^p_{3ikm} - A^n_{2km} A^p_{2in} - A^n_{2ik} A^p_{2nm} - A^n_{2im} A^p_{2nk}) G_{pj}$$

$$+ \ell^{\eta_j - \eta_p} (A^p_{3jkm} - A^n_{2km} A^p_{2jn} - A^n_{2jk} A^p_{2nm} - A^n_{2jm} A^p_{2nk}) G_{ip}$$

$$- \ell^{\eta_i + \eta_j - \eta_n - \eta_p} (A^n_{2jk} A^p_{2im} + A^n_{2jm} A^p_{2ik}) G_{np} \, . \tag{2.6.95}$$

The right-hand-side is manifestly symmetric in $i \leftrightarrow j$ and in $k \leftrightarrow m$. $\qquad\qquad\square$

Renormalization at First Order

We now examine the first order terms in detail. Our first task is to investigate the integral

$$J_{1ijk} = \int_{\text{reg}} d^2 z \langle \mathcal{O}_i(x) \mathcal{O}_j(0) \mathcal{O}_k(z) \rangle = C_{ijk} |x|^{\eta_k - \Delta_i - \Delta_j} \mathcal{I}_{ijk} \, , \tag{2.6.96}$$

where

$$\mathcal{I}_{ijk} = \int_{\text{reg}} \frac{|x|^{\Delta_k - \eta_k} d^2 z}{|x - z|^{2\alpha + 2} |z|^{2\beta + 2}} \, , \qquad 2\alpha = \Delta_i + \Delta_k - \Delta_j - 2 \, ,$$

$$2\beta = \Delta_j + \Delta_k - \Delta_i - 2 \, . \tag{2.6.97}$$

We can also write

$$2\alpha = \sigma^j_{ik} = \eta_j - \eta_i - \eta_k \, , \qquad 2\beta = \sigma^i_{jk} = \eta_i - \eta_j - \eta_k \, . \tag{2.6.98}$$

The IR safety assumption means the integral converges in the IR, so we will send the IR cut-off to infinity. Making the substitution $z = xu$, we then find that the dimensionless integral $\mathcal{I}_{ijk}$ only depends on the dimensionless parameter $\epsilon = a/x$:

$$\mathcal{I}_{ijk}(\epsilon) = \int_{\text{reg}} \frac{d^2 u}{|1 - u|^{2\alpha + 2} |u|^{2\beta + 2}} \, . \tag{2.6.99}$$

The integration domain is $\mathbb{D}^{\text{uv}}_{2,1} = \{ u \in \mathbb{C} \mid |u| \geq \epsilon, \quad |u - 1| \geq \epsilon \}$. While $\mathcal{I}_{ijk} = \mathcal{I}_{jik}$, there is no symmetry that exchanges j and k. This is in marked contrast to the totally symmetric C_{ijk}. Using

$$\alpha + \beta = -\eta_k \qquad \Longrightarrow \quad -1 < \alpha + \beta \leq 0 ,$$

$$\beta - \alpha = \eta_i - \eta_j \qquad \Longrightarrow \quad -1 < \beta - \alpha < 1 , \qquad (2.6.100)$$

we characterize the possible divergences as follows.

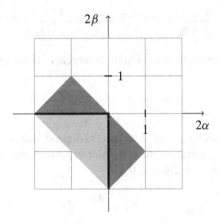

The parameters α, β lie in the interior of the shaded region; the integral will be finite as $\epsilon \to 0$ when the parameters are in the light-shaded region; the integral will diverge as $\epsilon \to 0$ when the parameters are in the dark-shaded region. The thick lines along the axes are where the integral will diverge logarithmically, and the origin $\alpha = \beta = 0$ is the only value of the parameters where there is a divergence both as $u \to 0$ and $u \to 1$.

In the convergent region the integral is given by a classic formula[23]:

$$\mathcal{I}_{ijk}(\epsilon, \alpha, \beta) = 2\pi \frac{\Gamma(-\alpha)\Gamma(-\beta)\Gamma(\alpha + \beta + 1)}{\Gamma(1 + \alpha)\Gamma(1 + \beta)\Gamma(-\alpha - \beta)} + O(\epsilon^2) . \qquad (2.6.101)$$

The $O(\epsilon^2)$ correction just follows by removing the small discs of area proportional to ϵ^2. Some authors use this form to define the regulated correlation functions at all α, β by analytic continuation. We do not know how to define conformal perturbation theory in this fashion, so we do not follow that route. Instead, we compute[24]

$$-\epsilon \frac{d}{d\epsilon} \mathcal{I}_{ijk} = 2\epsilon^{-2\alpha} \int_0^{2\pi} d\theta |1 - \epsilon e^{i\theta}|^{-2\beta - 2} + 2\epsilon^{-2\beta} \int_0^{2\pi} d\theta |1 - \epsilon e^{i\theta}|^{-2\alpha - 2} .$$

$$(2.6.102)$$

---

[23] For completeness we present the derivation in the appendix.

[24] Note the minus sign: increasing ϵ decreases the integral because the integrand is positive, and with a larger ϵ the integration domain is smaller. More details on this computation are in the appendix.

The angular integrals are easy to expand in a series in ϵ, but, because $\alpha < 1$, we only need the leading term to describe the divergent terms in $\mathcal{I}_{ijk}$ (we study the integral more generally in the appendix). In other words, when α and β are non-zero

$$
\begin{aligned}
\mathcal{I}_{ijk} &= \mathcal{I}_{ijk}^{\text{fin}} + \frac{2\pi}{\alpha}\epsilon^{-2\alpha} + \frac{2\pi}{\beta}\epsilon^{-2\beta} \\
&= \mathcal{I}_{ijk}^{\text{fin}} + \frac{4\pi}{\eta_j - \eta_i - \eta_k}\epsilon^{-\eta_j+\eta_i+\eta_k} + \frac{4\pi}{\eta_i - \eta_j - \eta_k}\epsilon^{-\eta_i+\eta_j+\eta_k}
\end{aligned} \tag{2.6.103}
$$

This expression also covers the logarithmic divergences if we simply take the limit

$$
\lim_{\alpha \to 0}\frac{1}{\alpha}\epsilon^{-2\alpha} = -2\log\epsilon . \tag{2.6.104}
$$

By dimensional analysis $\lim_{a \to 0}\mathcal{I}_{ijk}^{\text{fin}}$ has no $|x|$ dependence .

We now study renormalization to first order in perturbation theory. We will line up notation with that of Sect. 2.6.4 so that

$$
\mathcal{I}_{ijk} = \mathcal{I}_{ijk}^{\text{fin}} + \frac{4\pi}{\sigma_{ik}^j}\epsilon^{-\sigma_{ik}^j} + \frac{4\pi}{\sigma_{jk}^i}\epsilon^{-\sigma_{jk}^i} . \tag{2.6.105}
$$

With that, we have

$$
\begin{aligned}
|x|^{\Delta_i+\Delta_j}\delta_k^1 G_{ij} &= -C_{ijk}\left(\frac{|x|}{\ell}\right)^{\eta_k}\mathcal{I}_{ijk} + \left(\frac{|x|}{\ell}\right)^{\eta_m-\eta_i}A_{2ik}^m\delta_{mj} \\
&\quad + \left(\frac{|x|}{\ell}\right)^{\eta_m-\eta_j}A_{2jk}^m\delta_{mi} \\
&= -C_{ijk}\left(\frac{|x|}{\ell}\right)^{\eta_k}\left(\mathcal{I}_{ijk}^{\text{fin}} + \frac{4\pi}{\sigma_{ik}^j}\epsilon^{-\sigma_{ik}^j} + \frac{4\pi}{\sigma_{jk}^i}\epsilon^{-\sigma_{jk}^i}\right) \\
&\quad + \left(\frac{|x|}{\ell}\right)^{\eta_m-\eta_i}A_{2ik}^m\delta_{mj} + \left(\frac{|x|}{\ell}\right)^{\eta_m-\eta_j}A_{2jk}^m\delta_{mi} .
\end{aligned} \tag{2.6.106}
$$

As a final step, we write

$$
\left(\frac{\ell}{|x|}\right)^{\eta_k}|x|^{\Delta_i+\Delta_j}\delta_k^1 G_{ij} = -C_{ijk}\mathcal{I}_{ijk} + \left(\frac{|x|}{\ell}\right)^{\sigma_{ik}^m}A_{2ik}^m\delta_{mj} + \left(\frac{|x|}{\ell}\right)^{\sigma_{jk}^m}A_{2jk}^m\delta_{mi} . \tag{2.6.107}
$$

Following [398], we impose renormalization conditions on

$$
h_{ij}(g) = \ell^4\langle\!\langle\widehat{\mathcal{O}}_i(\ell)\widehat{\mathcal{O}}_j(0)\rangle\!\rangle^c = \delta_{ij} + g^k h_{1ijk} + O(g^2) \tag{2.6.108}
$$

The motivation for this form is a geometric interpretation of the two-point function as a metric. In a family of conformal theories $h_{ij}(g)$ will be ℓ-independent; more generally, the right-hand-side can have an ℓ-dependence consistent with the Callan-Symanzik equation and the other axioms of renormalized QFT. Families of conformal field theories exist, and the moduli space is locally a Riemannian manifold. This means we have no hope to trivialize the metric as a function of g, but we can ask that the perturbative expansion has a canonical form. If the geometric interpretation is to be sensible, then we should be able to understand any ambiguity in the renormalization prescription is an ambiguity in the choice of coordinates, i.e. a local diffeomorphism. We will see this is the case at low orders in g.

Our renormalization tale will be incomplete for at least two reasons. Not only will we stick to low orders in g, but we also will not explicitly fix the values of the couplings $g^i(\ell)$. In principle this can be fixed by computing the three-point functions at some fiducial kinematics at a scale ℓ. We will instead focus on the ambiguities in the wavefunction renormalization coefficients.

Since

$$\ell^4 \langle\!\langle \widehat{\mathcal{O}}_i(\ell)\widehat{\mathcal{O}}_j(0) \rangle\!\rangle^c = \ell^{4-\eta_i-\eta_j}\left[\langle \mathcal{O}_i(\ell)\mathcal{O}_j(0) \rangle + \delta G_{ij} \right]$$
$$= \ell^{\Delta_i+\Delta_j}\left[\langle \mathcal{O}_i(\ell)\mathcal{O}_j(0) \rangle + \delta G_{ij} \right] , \qquad (2.6.109)$$

we have an explicit expression for h_{1ijk}:

$$h_{1ijk} = -C_{ijk}\left(\mathcal{I}^{\mathrm{fin}}_{ijk} + \frac{4\pi}{\sigma^j_{ik}}\left(\frac{\ell}{a}\right)^{\sigma^j_{ik}} + \frac{4\pi}{\sigma^i_{jk}}\left(\frac{\ell}{a}\right)^{\sigma^i_{jk}} \right) + A^j_{2ik} + A^i_{2jk} .$$

$$(2.6.110)$$

We will now show that the wavefunction renormalization coefficients A^j_{2ik} are uniquely determined by the prescription

$$h_{1ijk} = 0 \qquad (2.6.111)$$

and are consistent with constraints from Sect. 2.6.4. We have to treat several different cases. Because of the $i \leftrightarrow j$ symmetry, we can just consider cases with $\sigma^i_{jk} \geq \sigma^j_{ik}$.

1. $\sigma^i_{jk} < 0$. Here the integral is finite, and we have

$$h_{1ijk} = -C_{ijk}\mathcal{I}_{ijk} + A^j_{2ik} + A^i_{2jk} . \qquad (2.6.112)$$

Because $C_{ijk}\mathcal{I}_{ijk}$ is symmetric under $i \leftrightarrow j$, it is always possible to choose the renormalization coefficients A so that $h_{1ijk} = 0$. This is a familiar geometric fact: we can always choose coordinates near any point p in a manifold such that the first derivatives of the metric vanish at p. Since h_{1ijk} has the same symmetries as the Christoffel connection with all lowered indices, $h_{1ijk} = 0$ uniquely determines

$$A^j_{2ik} = \tfrac{1}{2} C_{ijk} \left(\mathcal{I}_{ijk} + \mathcal{I}_{kji} - \mathcal{I}_{ikj} \right) . \tag{2.6.113}$$

This case was studied by Zamolodchikov, who used it to construct perturbative RG flows between large level minimal models, which have primary operators with small positive η. From Sect. 2.6.4, the β-function is of the form

$$\beta^i = -\eta_i g^i - \tfrac{1}{2} \sigma^i_{qr} A^i_{2qr} g^q g^r + O(g^3) , \tag{2.6.114}$$

and when η_i are taken parametrically small there is a fixed point g^i_* with parametrically small couplings. It is a nice exercise to verify that in the small η limit $\beta^j_{2ik} = 2\pi C_{ijk}$.

The $O(g)$ correction to the renormalized two-point function is now given by

$$\left(\frac{\ell}{|x|} \right)^{\eta_k} |x|^{\Delta_i + \Delta_j} \delta^1_k G_{ij} = \frac{C_{ijk}}{2} \left[-2\mathcal{I}^{\text{fin}}_{ijk} + \left(\frac{|x|}{\ell} \right)^{\sigma^j_{ik}} \left(\mathcal{I}_{ijk} + \mathcal{I}_{kji} - \mathcal{I}_{ikj} \right) \right.$$

$$\left. + \left(\frac{|x|}{\ell} \right)^{\sigma^i_{jk}} \left(\mathcal{I}_{jik} + \mathcal{I}_{kij} - \mathcal{I}_{jki} \right) \right] . \tag{2.6.115}$$

2. $\sigma^i_{jk} > 0$. In this case $\sigma^j_{ik} < 0$, so there is one power-law divergence that must be subtracted off. The form of the divergence is consistent with Sect. 2.6.4 and (2.6.74). The correction to the renormalized two-point function is then identical in form to the one obtained above, but with $\mathcal{I}_{ijk} \to \mathcal{I}^{\text{fin}}_{ijk}$.

3. $\sigma^i_{jk} = 0$ and $\sigma^j_{ik} < 0$. In this case we have to subtract off a logarithmic divergence; again, the divergent terms are consistent with Sect. 2.6.4. However, the finite renormalized correlation function is now qualitatively different from the two previous cases because it involves $\log(|x|/\ell)$ terms. We leave the details of this case as an exercise since the main features will also emerge in the last case.

4. $\sigma^i_{jk} = 0$ and $\sigma^j_{ik} = 0$. Now the renormalization condition reads

$$h_{1ijk} = -C_{ijk} \left[\mathcal{I}^{\text{fin}}_{ijk} - 8\pi \log(a/\ell) \right] + A^j_{2ik} + A^i_{2jk} . \tag{2.6.116}$$

We show in the appendix that $\mathcal{I}^{\text{fin}}_{ijk} = 0$, and $h_{1ijk} = 0$ determines

$$A^j_{2ik} = -4\pi C_{ijk} \log a/\ell . \tag{2.6.117}$$

In this case $\eta_k = 0$ and $\Delta_i = \Delta_j$, so the correction to the correlation function is

$$|x|^{\Delta_i+\Delta_j} \delta_k^1 G_{ij} = -8\pi C_{ijk} \log(|x|/\ell) . \qquad (2.6.118)$$

There is also a non-zero β-function: from (2.6.75) with $\sigma_{ik}^j = 0$ we find

$$\beta_{2ik}^j = 4\pi C_{ijk} . \qquad (2.6.119)$$

This is a famous result: non-zero structure constants of marginal operators' 3-point functions lead to the breaking of conformal invariance. By diagonalizing the matrix $\sum_k g^k C_{ijk}$ we can interpret these corrections as shifts in the anomalous dimensions of the deforming operators. In general these shifts have no definite sign, so that the marginal operators may become either marginally irrelevant or marginally relevant. A classic illustration is provided by the compact boson at self-dual radius.

Summary

We have shown that, with some technical assumptions to eliminate IR divergences, at leading order CPT makes sense and leads to results consistent with axioms of a renormalizable QFT. A worth emphasizing is that by imposing those consistency requirements on CPT, a first order computation determines the β-function to second order in the couplings. We also saw examples of some standard results familiar from perturbative QFT computations: power-law divergences are relatively boring and just get subtracted off, while logarithmic divergences lead to universal terms.

The first order computation is a misleading guide to the complexity of CPT. It relies on the simple presentation of the three-point functions, and, more importantly, does not involve the real issues of overlapping divergences and possible IR/UV divergences at higher order in the expansion. We will now examine those issues at second order.

Renormalization at Second Order: An Example

We consider the free CFT of n compact bosons. As is well known [328], this theory has a moduli space of marginal deformations described by the coset

$$\mathcal{M} = \frac{SO(n, n)}{SO(n) \times SO(n)} \qquad (2.6.120)$$

modulo the action of T-duality. We will recover some features of the local geometry of $\mathcal{M}$ by a second order computation in CPT.

The marginal operators are of the form

$$\mathcal{O}_{i\bar{i}} = J_i \bar{J}_{\bar{i}} \,, \tag{2.6.121}$$

where the holomorphic currents J_i, $i = 1, \ldots, n$, and their conjugates $\bar{J}_{\bar{i}}$ satisfy

$$\langle J_{i1} J_{j2} \rangle = \frac{\delta_{ij}}{z_{12}^2} \,, \qquad\qquad \langle \bar{J}_{\bar{i}1} \bar{J}_{\bar{j}2} \rangle = \frac{\delta_{\bar{i}\bar{j}}}{\bar{z}_{12}^2} \,. \tag{2.6.122}$$

This implies

$$\langle \mathcal{O}_{i\bar{i}}(x) \mathcal{O}_{j\bar{j}}(0) \rangle = \frac{\delta_{ij} \delta_{\bar{i}\bar{j}}}{|x|^4} \,. \tag{2.6.123}$$

The three-point functions of the $\mathcal{O}_{i\bar{i}}$ vanish, so that it is consistent to set $A_{2ij}^k = 0$ in our general expressions. The four-point function is non-zero, but it factorizes into a product of the current four-point functions given by

$$\langle J_{i1} J_{j2} J_{k3} J_{m4} \rangle = G_{ijkm} \,, \qquad G_{ijkm} = \frac{\delta_{ij} \delta_{km}}{z_{12}^2 z_{34}^2} + \frac{\delta_{ik} \delta_{jm}}{z_{13}^2 z_{24}^2} + \frac{\delta_{im} \delta_{jk}}{z_{14}^2 z_{23}^2} \,, \tag{2.6.124}$$

and

$$\langle \mathcal{O}_{i\bar{i}1} \mathcal{O}_{j\bar{j}2} \mathcal{O}_{k\bar{k}3} \mathcal{O}_{m\bar{m}4} \rangle = G_{ijkm} \overline{G_{\bar{i}\bar{j}\bar{k}\bar{m}}} \,. \tag{2.6.125}$$

The connected four-point function takes a simple form because the three-point functions are zero. To describe it, we introduce a compact notation

$$(ij) = \delta_{ij} z_{12}^{-2} \,, \quad (ik) = \delta_{ik} z_{13}^{-2} \,, \quad (\bar{i}\bar{j}) = \delta_{\bar{i}\bar{j}} \bar{z}_{12}^{-2} \,, \quad (\bar{i}\bar{k}) = \delta_{\bar{i}\bar{k}} \bar{z}_{13}^{-2} \,, \quad \text{etc.}\,, \tag{2.6.126}$$

so that

$$\langle \mathcal{O}_{i\bar{i}1} \mathcal{O}_{j\bar{j}2} \mathcal{O}_{k\bar{k}3} \mathcal{O}_{m\bar{m}4} \rangle^c = (ij)(km)(\bar{i}\bar{k})(\bar{j}\bar{m}) + (ij)(km)(\bar{i}\bar{m})(\bar{j}\bar{k})$$

$$+ (ik)(jm)(\bar{i}\bar{m})(\bar{j}\bar{k}) + \text{c.c.} \,, \tag{2.6.127}$$

where the "c.c." means to take the complex conjugate of the coordinates and bar/unbar the indices in the obvious fashion.

The integrated four-point function reduces to two integrals that we need to evaluate for $z_{12} \neq 0$.

$$K_1 = \int_{\mathbb{D}_{2,2}} \frac{d^2 z_3 d^2 z_4}{z_{34}^2 \bar{z}_{13}^2 \bar{z}_{24}^2} \,, \qquad K_2 = \int_{\mathbb{D}_{2,2}} \frac{d^2 z_3 d^2 z_4}{z_{13}^2 z_{24}^2 \bar{z}_{14}^2 \bar{z}_{23}^2} \,. \tag{2.6.128}$$

It is easy to see that they are IR finite; we claim that they are also UV finite and given by

$$\mathcal{K}_1 = \frac{(2\pi)^2}{\bar{z}_{12}^2}, \qquad\qquad \mathcal{K}_2 = 0. \qquad (2.6.129)$$

The claim is derived in Appendix (A.3.4).

With that, we have a complete answer for the integrated four-point function:

$$\int d^2z_3 d^2z_4 \langle \mathcal{O}_{i\bar{\imath}1} \mathcal{O}_{j\bar{\jmath}2} \mathcal{O}_{k\bar{k}3} \mathcal{O}_{m\bar{m}4} \rangle^c$$

$$= \frac{(2\pi)^2}{|z_{12}|^4} \left[\delta_{ij}\delta_{km}(\delta_{\bar{\imath}\bar{k}}\delta_{\bar{\jmath}\bar{m}} + \delta_{\bar{\imath}\bar{m}}\delta_{\bar{\jmath}\bar{k}}) + \delta_{\bar{\imath}\bar{\jmath}}\delta_{\bar{k}\bar{m}}(\delta_{ik}\delta_{jm} + \delta_{im}\delta_{jk}) \right]. \qquad (2.6.130)$$

Zamolodchikov Metric Curvature and Renormalization

The torus that we just examined is an example of a more general structure exhibited by a CFT with a moduli space, i.e. a family of CFTs described by a space of exactly marginal couplings.[25] When conformal invariance is preserved, we will obtain the expansion for the two-point function of marginal operators as above:

$$|x|^4 \langle\!\langle \widehat{\mathcal{O}}_i(x) \widehat{\mathcal{O}}_j(0) \rangle\!\rangle = \delta_{ij} + \tfrac{1}{2} g^k g^m \delta_{km}^2 G_{ij} + O(g^3), \qquad (2.6.131)$$

where

$$\delta_{km}^2 G_{ij} = T_{ij,km} + A_{3ikm}^p \delta_{jp} + A_{3jkm}^p \delta_{ip}, \qquad (2.6.132)$$

and

$$T_{ij,km} = \int d^2z_3 d^2z_4 \langle \mathcal{O}_{i1} \mathcal{O}_{j2} \mathcal{O}_{k3} \mathcal{O}_{m4} \rangle^c. \qquad (2.6.133)$$

The question "Can we choose A_{3ikm}^p so that $\delta_{km}^2 G_{ij} = 0$?" was answered by Riemann: the obstruction to doing so is the Riemann tensor of the Zamolodchikov metric evaluated at $g = 0$.

---

[25]It is fashionable to call this space "the conformal manifold." Since the moduli space is typically singular—we will see vivid examples of this below—we will not use this terminology lest we be tempted to apply results from smooth global geometry to the CFT moduli space. We expect that for a well-behaved point there is an open neighborhood that can be treated as a manifold.

Riemann Normal Coordinates

Let us recall some basic facts about normal coordinates. Suppose $\mathbb{R}^n$ parametrizes an open neighborhood of a Riemannian manifold M and a point p corresponds to the origin of $\mathbb{R}^n$ in coordinates x^i. Suppose further that the metric G near p takes the form

$$G = (\delta_{ij} + \tfrac{1}{2}T_{ij,km}x^k x^m + O(x^3))dx^i dx^j ,\qquad(2.6.134)$$

and we would like to improve this by a coordinate transformation

$$x^i = y^i + \tfrac{1}{2}A^i_{2jk}y^j y^k + \tfrac{1}{3!}A^i_{jkm}y^j y^k y^m + O(y^4) .\qquad(2.6.135)$$

The metric in the y coordinates is written as $g = \overline{G}_{ij}dy^i dy^j$, and if we set $A_2 = 0$, we obtain

$$\overline{G}_{ij} - \delta_{ij} = \tfrac{1}{2}y^k y^m X_{ijkm} + O(y^3) ,\qquad(2.6.136)$$

where

$$X_{ijkm} = T_{ij,km} + A^i_{3jkm} + A^j_{3ikm} .\qquad(2.6.137)$$

If $A_2 \neq 0$, then the transformation will inevitably produce y-linear terms in $\overline{G}$; conversely, if the metric in the x coordinates had linear terms to start, then we can always choose A_2 uniquely to eliminate them.[26]

Exercise 2.26 Follow in the steps of Riemann and prove that A_3 is uniquely determined by requiring

$$X_{ijkm} + X_{imjk} + X_{ikmj} = 0 .\qquad(2.6.138)$$

You should find that the unique solution is

$$A^i_{3jkm} = -\tfrac{1}{3}\left(T_{ij,km} + T_{im,jk} + T_{ik,mj}\right) + \tfrac{1}{6}\left(T_{mj,ik} + T_{jk,im} + T_{km,ij}\right) .$$
$$(2.6.139)$$

$\square$

Using the result of the exercise and fixing the condition (2.6.138) leads to

$$X_{ijkm} = \tfrac{1}{3}\left(R_{imkj} + R_{ikmj}\right) ,\qquad(2.6.140)$$

---

[26]It is a good exercise to show this; the diligent reader already met this issue in (2.6.113).

where

$$R_{ikmj} = \tfrac{1}{2}(T_{ij,km} - T_{im,jk} - T_{jk,mi} + T_{km,ji}) \tag{2.6.141}$$

is the Riemann tensor for G evaluated at $x = 0$; the components of R are in one-to-one correspondence with those of X, so that $X = 0$ if and only if $R = 0$. At any rate, now the coordinate transformation (2.6.135) is determined up to $O(y^4)$ corrections and linear orthogonal rotations on the x^i, and the metric takes the form

$$G = \left[\delta_{ij} + \tfrac{1}{3}R_{ikmj}y^k y^m + O(y^3)\right]dy^i dy^j . \tag{2.6.142}$$

We could keep the expansion going to any desired order with the same result: the freedom in the diffeomorphisms at $O(y^{n+1})$ can be used to bring the $O(y^n)$ term in G into a canonical form. In Riemann normal coordinates (also known as geodesic coordinates), the coefficient is fixed, by a generalization of (2.6.138), to be a specific combination of $n - 2$ derivatives of the Riemann tensor evaluated at $x = 0$.

Equipped with that reminder, we return to the Zamolodchikov metric. We fix A_3 uniquely by the renormalization condition

$$\delta^2_{km}G_{ij} + \delta^2_{jk}G_{im} + \delta^2_{mj}G_{ik} = 0 , \tag{2.6.143}$$

and higher order in perturbation theory we could fix the A_{n+1} renormalization coefficients by requiring that $\delta^n G$ has the symmetries of the normal coordinate expansion. There is nothing particularly sacred about this choice, but it does give a clear statement for the geometric quantities encoded in the Zamolodchikov metric.

Sticking to $n = 2$, we therefore find that there is a particular combination of the integrated connected four-point functions that yields the Riemann tensor of the Zamolodchikov metric evaluated at $g = 0$. In the torus example we find

$$2R_{i\bar{i}k\bar{k}m\bar{m}j\bar{j}} = T_{i\bar{i}j\bar{j},k\bar{k}m\bar{m}} + T_{k\bar{k}m\bar{m},i\bar{i}j\bar{j}} - T_{i\bar{i}m\bar{m},j\bar{j}k\bar{k}} - T_{j\bar{j}k\bar{k},m\bar{m}i\bar{i}} , \tag{2.6.144}$$

with

$$T_{i\bar{i}j\bar{j},k\bar{k}m\bar{m}} = (2\pi)^2 \left[\delta_{ij}\delta_{km}(\delta_{\bar{i}\bar{k}}\delta_{\bar{j}\bar{m}} + \delta_{\bar{i}\bar{m}}\delta_{\bar{j}\bar{k}}) + \delta_{\bar{i}\bar{j}}\delta_{\bar{k}\bar{m}}(\delta_{ik}\delta_{jm} + \delta_{im}\delta_{jk})\right] . \tag{2.6.145}$$

The result is not completely transparent, but there is a simplified case we can understand easily [279]. Suppose we only consider deformations with $\bar{J}_{\bar{1}}$, i.e. all barred indices just take the value 1. In this case, we should recover $\mathrm{SO}(n, 1)/\mathrm{SO}(n)$, which is the familiar symmetric hyperbolic space $\mathbb{H}^n$. This is the case:

$$T_{ij,km} = (2\pi)^2 \left[2\delta_{ij}\delta_{km} + \delta_{ik}\delta_{jm} + \delta_{im}\delta_{jk}\right] , \tag{2.6.146}$$

so that

$$R_{ikmj} = -(2\pi)^2 \left[\delta_{im}\delta_{jk} - \delta_{ij}\delta_{km} \right] . \tag{2.6.147}$$

This is consistent with a negatively curved symmetric space, since the Ricci tensor is

$$\mathrm{Ric}_{im} = R_{ijm}{}^{j} = -(2\pi)^2 (n-1)\delta_{im} . \tag{2.6.148}$$

2.6.6 A Few More Remarks on Renormalization

We verified that conformal perturbation theory is renormalizable at first order in the couplings, at least as far as the two-point function goes. This first order result should extend to higher point functions simply because the divergence structure is the same—an integrated operator meets the inserted operators one at a time. From our discussion of the formal properties of the expansion it is clear that if the theory is renormalizable to some order $O(g^p)$, then the resulting renormalized theory will satisfy the formal properties of a renormalized quantum field theory, and various familiar relations like $\Gamma^i_j = -\beta^i_{,j}$ will hold.

To the best of the author's knowledge there is still no framework to show that conformal perturbation theory is renormalizable, i.e. that all divergences can be absorbed into counter-terms for the bare couplings and wavefunction renormalization of the operators. It would be very interesting to construct such a framework, but to the author it seems that some more powerful tool is needed. Perhaps such a tool could come from the geometry of Riemann surfaces and stable maps. The moduli space of a punctured two-sphere seems a natural geometric setting for defining correlation functions in CFT, and it should be possible to describe the divergent structures in terms of the OPE and some compactification of the punctured moduli space.

Tensorial Obstructions to Renormalizability

There is one aspect of that ambitious program that we can already glimpse from the preceding discussion. We just recalled the fundamental fact of Riemannian geometry that the Riemann tensor is an obstruction to "improving" the metric to take its trivial form at second order in the deformation from a given point. There are similar tensorial obstructions to renormalizability as well, so a proof thereof would involve showing that these are zero. In general the obstructions are as complicated as those encountered in constructing Riemann normal coordinates, which involve the Riemann tensor and its covariant derivatives.

Consider a perturbation by marginal operators $\mathcal{O}_i$ with $C_{ijk} = 0$. This means $\beta^i = O(g^3)$, and the first non-trivial divergence in the deformed two-point function is encountered at second order:

$$\delta^2_{km} G_{ij}(x) = \mathcal{J}_{2ij,km} + A^j_{3ikm} + A^i_{3jkm} \,, \tag{2.6.149}$$

where

$$\mathcal{J}_{2ij,km} = \int_{\mathbb{D}_{2,2}} d^2z_3 d^2z_4 \langle \mathcal{O}_i(x)\mathcal{O}_j(0)\mathcal{O}_k(z_3)\mathcal{O}_m(z_4)\rangle^c \,, \tag{2.6.150}$$

and, the only divergent term in A_3 is

$$A^{i,\text{div}}_{3jkm} = \beta^i_{3jkm} \log \ell/a \,. \tag{2.6.151}$$

So, in order for the divergences to be renormalizable we need to be able to solve

$$\mathcal{J}^{\text{div}}_{2ij,km} + (\beta^i_{3jkm} + \beta^j_{3ikm}) \log \ell/a = 0 \,. \tag{2.6.152}$$

There are two potential issues:

1. $\mathcal{J}^{\text{div}}_{2ij,km}$ must be proportional to $\log(|x|/a)$, i.e.

$$\mathcal{J}^{\text{div}}_{2ij,km} = \mathcal{J}^{\text{div},1}_{2ij,km} \log(|x|/a) \,; \tag{2.6.153}$$

2. because β^i_{3jkm} is symmetric in the lower indices, the tensorial structure must be compatible .

Even when the first requirement is met, the second point leads to an obstruction, just as in Riemann's case: we can cancel the divergences by choosing β^i_{3jkm} if and only if

$$R^{\text{div}}_{ikmj} = \frac{1}{2} \left[\mathcal{J}^{\text{div},1}_{2ij,km} - \mathcal{J}^{\text{div},1}_{2im,jk} - \mathcal{J}^{\text{div},1}_{2jk,mi} + \mathcal{J}^{\text{div},1}_{2km,ji} \right] = 0 \,. \tag{2.6.154}$$

When this is non-zero, CPT will not be renormalizable. This is a delicate requirement, since, as we have seen, the finite counter-part of this term may well be non-zero.

Radius of Convergence of Conformal Perturbation Theory

One of the main reasons to establish necessary and sufficient conditions for renormalizability of conformal perturbation theory involves another one of its conjectured properties. It is believed that with a suitable number of conditions on the initial theory, conformal perturbation theory will have a non-zero radius of convergence!

This means that, for example, the β function can be obtained as an analytic function of g around $g = 0$. This is incredibly powerful, since it means that any features we encounter before we run into a pole are completely captured by summing up a perturbation series. A toy example is a beta function of the form

$$\beta(g) = \frac{g(g - g_*)}{1 - g/g_p} \, . \tag{2.6.155}$$

If $g_* < g_p$, this could describe a flow by a relevant operator from the fixed point with $g = 0$ to that at $g = g_*$ (concrete examples include integrable flows based on deformations of minimal models); if $g_* > g_p$, then we cannot describe the low energy behavior in terms of the original degrees of freedom.

It is also interesting to consider the conjecture when applied to an irrelevant perturbation. Once we turn on an irrelevant operator, we expect that we will need to specify an infinite number of couplings to renormalize the theory. It still may be that the perturbative expansion would converge with some UV fixed points within the radius of convergence. Each of would then offer a UV completion to the RG flow.

There is considerable evidence that the conjecture is true in examples, but we do not know the necessary and sufficient conditions. As suggested in [263], the necessary conditions should involve bounds on the growth of OPE coefficients, but a precise statement is not known. The complexity of the problem is already illustrated by a non-example: a free scalar field ϕ in $d > 2$ dimensions is a very nice example of a conformal field theory; from a Lagrangian point of view it is obvious that a perturbation by $g\phi^p$ with $p > 2$ has zero radius of convergence, but to see that from a perturbative expansion of the sort just discussed is not immediate.

Perhaps the most important point to keep in mind is that while such a formal point of view can clarify the structure of difficult problems in QFT, it does not solve them. For instance, there may be strong coupling singularities in the coupling space $\mathcal{M}$ that cannot be described in terms of a perturbation theory around a smooth point $p \in \mathcal{M}$. In some cases the singularity may have an interpretation of a coordinate singularity, but in others it may be genuine, indicating that the theory exhibits a singular behavior: renormalized correlation functions diverge, and the divergence cannot be eliminated by a change of coordinates.

2.7   CPT for (0,2) CFTs

Having plumbed the depths of the author's ignorance of conformal perturbation theory in general, we can now turn to a more specific application to supersymmetric perturbations of superconformal theories.[27]

---

[27]This section follows the discussion given in [83], which is in turn based on ideas to be found in [208], as well as the much earlier [137].

2.7.1 Supersymmetric Deformations at First Order

In making a CPT expansion in a non-supersymmetric theory we considered perturbations defined by

$$S[\lambda] = \int d^2z \sum_i \lambda^i \mathcal{O}_i(z, \bar{z}) , \tag{2.7.1}$$

where $\mathcal{O}_i$ are spin 0 operators; the latter restriction ensures that the first order deformation preserves Lorentz invariance; we saw above that we can easily find Lorentz-invariant regularization schemes, so that CPT is assured to remain Lorentz-invariant at higher order as well.

In a supersymmetric theory it is natural to consider deformations that also preserve some amount of supersymmetry. We will now show that if we wish to preserve the full (0,2) supersymmetry and reality of the action, then the deformations take the form[28]

$$\mathcal{O}(z, \bar{z}) = \left[\{G^-_{-1/2}, \mathcal{U}(z, \bar{z})\} + \text{h.c.} \right] + \{G^+_{-1/2}, [G^-_{-1/2}, \mathcal{K}(z, \bar{z})]\} , \tag{2.7.2}$$

where $\mathcal{U}$ is a fermionic chiral primary operator with

$$\bar{h}_\mathcal{U} = 1/2 + h_\mathcal{U} , \qquad\qquad h_\mathcal{U} = q_\mathcal{U}/2 , \tag{2.7.3}$$

and $\mathcal{K}$ is a real operator with $\bar{h}_\mathcal{K} - h_\mathcal{K} = 1$. Our proof applies to any compact and unitary CFT with N=2 supersymmetry.

To prove the claim it will be convenient to work with the state $|\mathcal{O}\rangle$. $|\mathcal{O}\rangle$ can be taken to be Virasoro quasi-primary; any other term will be a total derivative and will not contribute to the integrated action; $|\mathcal{O}\rangle$ will have 0 spin to preserve Lorentz invariance and will be bosonic. For the deformation to be supersymmetric we need

$$G^\mp_{-1/2}|\mathcal{O}\rangle = L_{-1}|\mathcal{M}^\mp\rangle \tag{2.7.4}$$

for some states $|\mathcal{M}^\mp\rangle$. Applying $G^\pm_{-1/2}$ to both sides of the equation and using the N=2 algebra leads to

$$L_{-1}\left[|\mathcal{O}\rangle - G^+_{-1/2}|\mathcal{M}^-\rangle - G^-_{-1/2}|\mathcal{M}^+\rangle \right] = 0 . \tag{2.7.5}$$

---

[28] In this section we still keep the N=2 algebra on the holomorphic side of the worldsheet, i.e. we are really speaking of a (2,0) theory; we will switch it to the anti-holomorphic side when it will be more convenient to do so.

But that in turn means that up to a constant multiple of the identity operator, which would lead to a trivial deformation of the theory, we can write

$$|\mathcal{O}\rangle = G^-_{-1/2}|\mathcal{M}^+\rangle + G^+_{-1/2}|\mathcal{M}^-\rangle$$

$$= G^-_{-1/2}|\mathcal{U}\rangle + G^+_{-1/2}|\mathcal{V}\rangle + \left[G^+_{-1/2}G^-_{-1/2} - \left(1 + \frac{q\kappa}{2h_\kappa}\right)L_{-1}\right]|\mathcal{K}\rangle ,$$

$$(2.7.6)$$

where $|\mathcal{U}\rangle$, $|\mathcal{V}\rangle$, and $|\mathcal{K}\rangle$ are quasi-primary with respect to the N=2 superconformal algebra i.e. annihilated by the lowering modes L_1, and $G^\pm_{1/2}$. The linear combination of operators in the last term is fixed by the requirement $L_1|\mathcal{O}\rangle = 0$. Lorentz invariance requires $\overline{h}_\kappa = h_\kappa + 1$, $\overline{h}_\mathcal{U} = h_\mathcal{U} + 1/2$, $\overline{h}_\mathcal{V} = h_\mathcal{V} + 1/2$. Supersymmetry does not constrain the operator $\mathcal{K}$.

Because it is the integral of $\mathcal{O}$ that appears in the deformation, the remaining constraints from supersymmetry are

$$G^+_{-1/2}G^-_{-1/2}|\mathcal{U}\rangle = L_{-1}|X\rangle , \qquad G^-_{-1/2}G^+_{-1/2}|\mathcal{V}\rangle = L_{-1}|Y\rangle \qquad (2.7.7)$$

for some states $|X\rangle$ and $|Y\rangle$.

We will now show that the only solution to these conditions is to take $|\mathcal{U}\rangle$ to be chiral primary and $|\mathcal{V}\rangle$ to be anti-chiral primary. It suffices to show the first statement, as the second follows simply by exchanging $|\mathcal{U}\rangle$ with $|\mathcal{V}\rangle$ and G^+ with G^-. Without loss of generality we decompose

$$|X\rangle = a|\mathcal{U}\rangle + |\chi\rangle , \qquad (2.7.8)$$

where a is a constant and $|\chi\rangle$ is orthogonal to $|\mathcal{U}\rangle$. Using this decomposition we obtain

$$\left[G^+_{-1/2}G^-_{-1/2} - aL_{-1}\right]|\mathcal{U}\rangle = L_{-1}|\chi\rangle . \qquad (2.7.9)$$

$\langle \mathcal{U}|L_1 L_{-1}|\chi\rangle = 0$ because $|\mathcal{U}\rangle$ is quasi-primary and orthogonal to $|\chi\rangle$. Therefore we apply $\langle \mathcal{U}|L_1$ to both sides of (2.7.9) to find

$$\langle \mathcal{U}|L_1 \left[G^+_{-1/2}G^-_{-1/2} - aL_{-1}\right]|\mathcal{U}\rangle = 0 \qquad \Longleftrightarrow \qquad a = 1 + \frac{q_\mathcal{U}}{2h_\mathcal{U}} . \qquad (2.7.10)$$

Exercise 2.27 Prove $L_{-1}|\chi\rangle = 0$ by applying $\langle \chi|L_1$ to both sides of (2.7.9). □

Using the result of the exercise, we now have

$$\left[G^+_{-1/2}G^-_{-1/2} - \left(1 + \frac{q_\mathcal{U}}{2h_\mathcal{U}}\right)L_{-1}\right]|\mathcal{U}\rangle = 0 . \qquad (2.7.11)$$

We now apply $G^-_{1/2}$, and the N=2 algebra yields

$$\left[2L_0 - J_0 + \left(1 + \frac{q_\mathcal{U}}{2h_\mathcal{U}}\right)\right] G^-_{-1/2}|\mathcal{U}\rangle = 0. \qquad (2.7.12)$$

So, either $G^-_{-1/2}|\mathcal{U}\rangle = 0$, or

$$2\left(h_\mathcal{U} + \frac{1}{2}\right) - (q - 1) - \left(1 + \frac{q_\mathcal{U}}{2h_\mathcal{U}}\right) = 0. \qquad (2.7.13)$$

The former possibility leads to a trivial deformation; the latter leads is equivalent to

$$(1 + 2h_\mathcal{U})\left(1 - \frac{q_\mathcal{U}}{2h_\mathcal{U}}\right) = 0, \qquad (2.7.14)$$

and the only solution consistent with unitarity is $h_\mathcal{U} = q_\mathcal{U}/2$. $|\mathcal{U}\rangle$ is a chiral primary operator, as was to be shown.

A parallel treatment shows that $|\mathcal{V}\rangle$ must be an anti-chiral primary state. To keep the deformation real requires the $\mathcal{U}$ and $\mathcal{V}$ terms to be conjugate. Finally, returning to the corresponding operators, it follows that we have now established the claim (2.7.2).

The deformations have weights and R-charges

operator	h	$\bar{h}$	q	Δ
$\{G^-_{-1/2}, \mathcal{U}\}$	$h_\mathcal{U} + \frac{1}{2}$	$h_\mathcal{U} + \frac{1}{2}$	$q_\mathcal{U} - 1$	$1 + q_\mathcal{U}$
$\{G^+_{-1/2}, [G^-_{-1/2}, \mathcal{K}(z,\bar{z})]\}$	$h_\mathcal{K} + 1$	$h_\mathcal{K} + 1$	$q_\mathcal{K}$	$2 + 2h_\mathcal{K}$

$$(2.7.15)$$

There are no marginal or relevant deformations of the $\mathcal{K}$ type: the borderline case is $h_\mathcal{K} = 0$, i.e. $\mathcal{K}$ is an anti-holomorphic current, but that is a trivial deformation since the corresponding operator would then be annihilated by the supercharges. On the other hand, we can have relevant and marginal $\mathcal{U}$ operators; the former requires $q_\mathcal{U} < 1$, while the latter requires $q_\mathcal{U} = 1$. As expected, marginal deformations will preserve the R-symmetry at first order in the deformation.

Exercise 2.28 Show that a (0,2) marginal deformation of a (0,4) CFT is necessarily (0,4) marginal. Hint: use the SU(2) Kac-Moody algebra. □

2.7.2 Basic Constraints on RG Flow

In this section we consider flows from a relevant supersymmetric deformation of a known UV fixed point to an IR fixed point. In the next chapters we will explore a zoo of such theories where the UV fixed point is a Lagrangian theory. The constraints we will now discuss will be very useful in that context, but in this section we will discuss them from a more general point of view.

Before we go further, we make a notational switch: *we will revert to the convention that supersymmetry is on the right, i.e. anti-holomorphic side of the worldsheet.*

Unitarity Bound and Marginal Operators

Having identified the marginal supersymmetric operators of an SCFT, we can discuss their fate under a finite deformation. As we saw above, in a general CFT this is a difficult question, but in a supersymmetric theory we have additional tools at our disposal. For example, if we have a marginal coupling g such that the $U(1)_R$ charges are g-independent, a marginal supersymmetric operator $\mathcal{O}$ at $g = 0$ can cease to be marginal for $g \neq 0$ in just one of two ways:

1. "F-term" : in this case $\mathcal{O}$ is lifted together with an anti-chiral primary $\overline{\mathcal{F}}$ with $h = 1, \overline{h} = -\overline{q}/2 = 1$, for $\lambda \neq 0$ the two chiral-primary operators combine into a long multiplet with lowest component having $\overline{q} = 1$ and weight $\overline{h} = 1/2 + \epsilon(g)$.
2. "D-term" : in this case $\mathcal{O}$ is lifted together with a left KM current J: for $\lambda \neq 0$ J and $\mathcal{O}$ now combine into a long multiplet with lowest component having $\overline{q} = 0$ and weight $\overline{h} = \epsilon(g)$.

This was discussed many years ago in [137] in the context of heterotic string compactification to four-dimensional N=1 supersymmetric vacua. The "F"-term and "D"-term terminology is due to the spacetime interpretation of these effects: in the first case the four-dimensional massless modes are lifted by superpotential couplings, and in the second case they are lifted by the supersymmetric Higgs mechanism. We will explore this point of view further in the last two chapters.

To apply the argument to a more general (0,2) SCFT, we need to establish the g-independence of the $U(1)_R$ charges. Since the deformation preserves (0,2) superconformal invariance, the deformed theory still has $\partial \overline{J} = 0$, and the level of the R-current, $\overline{c}/3$, is not modified. The charge of an operator $\mathcal{U}$ can then be extracted from a three-point function determined by the KM Ward identity:

$$\langle\!\langle \overline{J}(z)\mathcal{U}(w_1)\overline{\mathcal{U}}(w_2) \rangle\!\rangle_g = \overline{q}_{\mathcal{U}}(g) \frac{\overline{w}_{12}}{(\overline{z} - \overline{w}_1)(\overline{z} - \overline{w}_2)} \langle\!\langle \mathcal{U}(w_1)\overline{\mathcal{U}}(w_2) \rangle\!\rangle_g . \qquad (2.7.16)$$

The compact boson Exercise 2.8 shows that in a general CFT $\overline{q}_{\mathcal{U}}(g)$ need not be constant. The (0,2) SCFT has a markedly different structure: the marginal operator for the compact boson is not KM primary, since it is essentially of the form $J\overline{J}$; by contrast the marginal operator in the deformed (0,2) theory, $\{G^-_{-1/2}, \mathcal{O}\}$, is primary with respect to the R-current's KM algebra. It would be useful to identify sufficient (and perhaps also necessary) conditions on a compact unitary (0,2) SCFT that ensure g-independence of $\overline{q}_{\mathcal{U}}(g)$.

It is tempting to conjecture other general statements about (0,2) superconformal perturbation theory. For instance, we might expect that the unitarity bound $\bar{h} \geq \bar{q}/2$ will continue to hold under a marginal deformation. Together with a statement of $U(1)_R$-charge invariance this would imply that a marginal deformation can at worst become marginally irrelevant. This seems sensible, and there are theories where the $U(1)_R$ charges are g-independent, but there is no proof that this is always the case. It would be useful to have a more precise statement of the unitarity bound in the language of supersymmetric conformal perturbation theory. The basic technical hurdle to overcome is that, unlike the case of chiral operators in $N = 1\, d = 4$ superconformal theories, in (0,2) theories chiral primary operators can and do have singularities in the OPE. The singularities will be holomorphic, but that is not enough to formulate a formal non-renormalization argument of the sort used in [208]. A first step would be to choose a regularization prescription that makes supersymmetry as manifest as possible.

Consequences of Chirality

Consider a relevant deformation of (0,2) UV fixed point with central charges c_{uv}, $\bar{c}_{uv}$ that flows to an IR fixed point with c_{ir}, $\bar{c}_{ir}$. As we discussed above, the local diffeomorphism anomaly in the UV theory is proportional to the difference $c_{uv} - \bar{c}_{uv}$. This anomaly must also be present in the IR theory, whence we learn that

$$c_{uv} - \bar{c}_{uv} = c_{ir} - \bar{c}_{ir} .$$ (2.7.17)

Any time $c_{uv} \neq \bar{c}_{uv}$ we are guaranteed to have a non-trivial IR fixed point of the flow!

The argument for the c-theorem reviewed above uses diffeomorphism invariance, so it is not immediately obvious how to generalize it to the situation when $c_{uv} \neq \bar{c}_{uv}$. In practice, we focus on theories where the difference $c_{uv} - \bar{c}_{uv}$ can be compensated by introducing decoupled spectator degrees of freedom, say a compact unitary CFT consisting of a number of free fermions. In that case we can sensibly define separate C and $\bar{C}$ functions that interpolate between the UV and IR values of the left and right central charges of the original theory and satisfy

$$\ell \frac{\partial}{\partial \ell}(C - \bar{C}) = 0 .$$ (2.7.18)

A similar, in fact simpler, constraint also exists for any conserved current. Let $J, \bar{J}$ be the components of a conserved current, i.e. we have an operator equation in the renormalized theory

$$\bar{\partial} J + \partial \bar{J} = 0 .$$ (2.7.19)

Consider then the following rotationally invariant correlation functions:

$$K = z^2 \langle J(z, \bar{z}) J(0) \rangle , \qquad M = z\bar{z} \langle J(z, \bar{z}) \bar{J}(0) \rangle = z\bar{z} \langle \bar{J}(z, \bar{z}) J(0) \rangle ,$$

$$\overline{K} = \bar{z}^2 \langle \bar{J}(z, \bar{z}) \bar{J}(0) \rangle .$$

(2.7.20)

Conservation and rotational invariance imply

$$\dot{K} + \dot{M} = M , \qquad\qquad \dot{\overline{K}} + \dot{M} = M ,$$

(2.7.21)

where $\dot{K} = r\frac{\partial}{\partial r} K$ and $r^2 = z\bar{z}$, and as a result we have another RG-invariant quantity:

$$\ell \frac{\partial}{\partial \ell} \left[K - \overline{K} \right] = 0 .$$

(2.7.22)

At the conformal points, where (in unitary and compact theories) K and $\overline{K}$ reduce to the levels of, respectively, J and $\bar{J}$, while $M = 0$, so that RG-invariance implies

$$k_{\mathrm{uv}} - \overline{k}_{\mathrm{uv}} = k_{\mathrm{ir}} - \overline{k}_{\mathrm{ir}} .$$

(2.7.23)

The statement also follows from 't Hooft anomaly matching applied to the two-dimensional theory.

It follows that in any RG flow to a compact unitary CFT, if $k_{\mathrm{uv}} - \overline{k}_{\mathrm{uv}} > 0$, then the IR fixed point must have a left-moving chiral symmetry; similarly, if $k_{\mathrm{uv}} - \overline{k}_{\mathrm{ir}} < 0$, then the IR fixed point must have a right-moving chiral symmetry.

Finding the R-symmetry in the IR

Suppose there is a (0,2) supersymmetric flow with an indecomposable supercurrent multiplet from a UV fixed point to some compact unitary (0,2) SCFT. In such a situation we may hope to identify at least some of the SCFT spectrum with operators constructed from the UV fields. For example, we might expect to reliably study the algebra of chiral operators. The starting point in any such analysis is the proper identification of the IR R-symmetry. There are two possibilities regarding this identification, and each realized in (0,2) dynamics. First, it may be that the IR R-symmetry current is obtained as a linear combination of the UV R-symmetry current and UV global U(1) symmetry currents. In that case the technique of *c-extremization* [66] fixes the linear combination of the UV currents. The alternative is that the IR R-symmetry is a linear combination of UV symmetries and an accidental IR symmetry. Not much can be said in general about this situation, but we will present some well-motivated conjectures in some examples below.

We now examine the first possibility in more detail and review the elegant method of c-extremization. Assume that there is a UV theory with an indecomposable supercurrent multiplet $\overline{\mathcal{S}}$ and conserved abelian currents[29]

$$\mathcal{J}^0_{uv} = J^0_{uv}\frac{\partial}{\partial z} + \overline{J}^0_{uv}\frac{\partial}{\partial \overline{z}}, \qquad \mathcal{J}^\alpha_{uv} = J^\alpha_{uv}\frac{\partial}{\partial z} + \overline{J}^\alpha_{uv}\frac{\partial}{\partial \overline{z}}, \qquad (2.7.24)$$

where $\mathcal{J}_0$ is the UV R-symmetry and $\alpha = 1,\dots,k$ label the remaining global symmetries. We assume that these symmetries are preserved by the RG flow, and the IR limit is a unitary compact (0,2) SCFT with an indecomposable supercurrent multiplet whose lowest component, the IR R-current $\mathcal{J}_{ir} = \overline{J}_{ir}\frac{\partial}{\partial \overline{z}}$, is a linear combination of the abelian UV symmetries. More precisely, there is some choice of k real parameters t_α such that the current

$$\mathcal{J}^t = \mathcal{J}^0_{ir} + \sum_{\alpha=1}^{k} t_\alpha \mathcal{J}^\alpha_{uv} \qquad (2.7.25)$$

flows to $\mathcal{J}_{ir}$. We will now show that 't Hooft anomaly matching uniquely determines the parameters t^α and therefore fixes the presentation of the IR R-symmetry in terms of the UV degrees of freedom.

To apply 't Hooft anomaly matching we couple the currents $\mathcal{J}^t_{uv}$ and $\mathcal{J}^\alpha_{uv}$ to background gauge fields A_t and A_α.[30] As we saw in Sect. 2.5, the effective action will in general transform under gauge transformations $A_t \to A_t - df_t$ and $A_\alpha \to A_\alpha - df_\alpha$. We will choose a regularization scheme where the anomaly takes its symmetric form, so that

$$\delta_f W[A] = -\frac{i}{4\pi}\int \left\{ -\frac{\overline{C}(t)}{3}f_t F_t + \sum_\alpha L^\alpha(t)(f_t F_\alpha + f_\alpha F_t) + \sum_{\alpha,\beta} M^{\alpha\beta} f_\alpha F_\beta \right\}. \qquad (2.7.26)$$

Here the F are the background gauge field strengths, while the coefficients $\overline{C}(t)$, $L(t)$ and $M^{\beta\alpha} = M^{\alpha\beta}$ are polynomial in t and determined by the UV current two-point functions:

$$a^{00} = \lim_{z\to 0}\left[z^2\langle J^0_{uv}(z,\overline{z})J^0_{uv}(0)\rangle - \overline{z}^2\langle \overline{J}^0_{uv}(z,\overline{z})\overline{J}^0_{uv}(0)\rangle \right],$$

$$a^{0\beta} = \lim_{z\to 0}\left[z^2\langle J^0_{uv}(z,\overline{z})J^\beta_{uv}(0)\rangle - \overline{z}^2\langle \overline{J}^0_{uv}(z,\overline{z})\overline{J}^\beta_{uv}(0)\rangle \right],$$

$$M^{\alpha\beta} = \lim_{z\to 0}\left[z^2\langle J^\alpha_{uv}(z,\overline{z})J^\beta_{uv}(0)\rangle - \overline{z}^2\langle \overline{J}^\alpha_{uv}(z,\overline{z})\overline{J}^\beta_{uv}(0)\rangle \right], \qquad (2.7.27)$$

---

[29] We restrict attention to abelian currents, since non-abelian currents cannot mix with the R-symmetry.

[30] Note that A_t has no t dependence; it is simply the background gauge field that couples to $\mathcal{J}^t$.

and

$$-\frac{1}{3}\overline{C}(t) = a^{00} + 2\sum_{\beta} t_{\beta} a^{0\beta} + \sum_{\alpha,\beta} t_{\alpha} t_{\beta} M^{\alpha\beta} \,,$$

$$L^{\beta}(t) = a^{0\beta} + \sum_{\gamma} t_{\gamma} M^{\gamma\beta} = -\frac{1}{6}\frac{d\overline{C}}{dt^{\beta}}(t) \,. \tag{2.7.28}$$

The essence of 't Hooft anomaly matching is that this gauge variation must be reproduced by IR degrees of freedom. Let us now discuss how this might happen. As a first step, we derive some constraints on the symmetric matrix M.

The $k \times k$ symmetric matrix M can be diagonalized and decomposed according to the sign of the eigenvalues: $M = M^- \oplus M^0 \oplus M^+$. With our assumptions M^- must be absent: if it were not, then the IR theory necessarily has an additional anti-holomorphic current that will reproduce the corresponding term in the anomaly. However, an additional anti-holomorphic current means that the IR supercurrent multiplet is reducible because the additional current must either be the bottom component of a separate supercurrent multiplet, or it resides in an N=2 SKM multiplet, in which case the IR supercurrent multiplet is decomposable by the N=2 super-Sugawara construction. So, any theory with a non-trivial M^- must have an accident, and we will not consider it.

A non-trivial M^0 means that there is some UV current with zero anomaly. Such a current must choose between two fates in the IR: it must either decouple or flow to two conserved IR currents with equal level: one holomorphic and one anti-holomorphic. The second fate means again that we have an accidental symmetry; the first fate means that we can ignore every current that corresponds to an element of $\ker M$.

These observations imply that we may as well restrict to currents $\mathcal{J}_{uv}^{\alpha}$ that correspond to the M^+ block. To keep notation simple the index α will now run over just such currents. Every $\mathcal{J}_{\alpha}^{uv}$ must flow to a holomorphic current $\mathcal{J}_{\alpha}^{ir} = J_{\alpha}^{ir}\frac{\partial}{\partial z}$ in the IR. Therefore, the gauge variation of the effective action $W[A]$ in terms of the IR degrees of freedom is

$$\delta_f W[A] = -\frac{i}{4\pi} \int \left\{ -\frac{\overline{c}^{ir}}{3} f_t F_t + \sum_{\alpha,\beta} f_{\alpha} \mathcal{F}_{\beta} M^{\alpha\beta} \right\} \,. \tag{2.7.29}$$

Comparing (2.7.29) and (2.7.26), we see that we must choose the values $t_{\alpha} = t_{\alpha}^{ir}$ so that $L(t^{ir}) = 0$, in which case $\overline{C}(t^{ir}) = \overline{c}_{ir}$. Since M is by assumption invertible, $t_{\alpha} = t_{\alpha}^{ir}$ is the unique solution to $L(t) = 0$. This is nicely summarized by the statement that "the IR R-symmetry maximizes the trial central charge $\overline{C}(t)$."[31]

---

[31] The reader might be surprised that in reviewing c-extremization we arrived at c-maximization. The reason is that our argument is a little bit more refined than the one originally given in [66] and

Thus, with the assumptions we made, all we need to obtain the IR R-symmetry is the set of current two-point function coefficients. These coefficients are familiar when the UV theory is a weakly coupled Lagrangian field theory: the contributions to the anomaly coefficients simply come from the fermion degrees of freedom. Suppose the charge assignments for the Fermi and chiral multiplets are as follows.

symmetry	Φ^a	Γ^A
$U(1)_R$	r_a	R_A
$U(1)^\alpha$	q_a^α	Q_A^α

We then have[32]

$$a^{00} = \sum_A R_A^2 - \sum_a (r_a - 1)^2 , \qquad a^{0\beta} = \sum_A R_A Q_A^\beta - \sum_a (r_a - 1) q_a^\beta ,$$

$$M^{\alpha\beta} = \sum_A Q_A^\alpha Q_A^\beta - \sum_a q_a^\alpha q_a^\beta . \tag{2.7.30}$$

Another simple case where these coefficients are easily determined is when the UV theory is a relevant deformation of a compact and unitary (0,2) SCFT. In that case we can express the anomaly coefficients in terms of the normalizations of the UV currents. The R-current $\bar{J}$ is right-moving, and satisfies

$$-a^{00} = \frac{1}{3} \bar{c}^{uv} = \bar{z}^2 \langle \bar{J}^{uv}(\bar{z}) \bar{J}^{uv}(0) \rangle . \tag{2.7.31}$$

The UV CFT may also have some left-moving currents U(1) currents J_α, $\alpha = 1, \ldots, r$, that satisfy for $z \neq 0$

$$M^{\alpha\beta} = z^2 \langle J_\alpha^{uv}(z) J_\beta^{uv}(0) \rangle , \qquad a^{0\beta} = \langle J_\alpha^{uv}(z) \bar{J}^{uv}(0) \rangle = 0 . \tag{2.7.32}$$

It is possible for the UV theory to have additional right-moving currents. These arise whenever the supercurrent multiplet $\bar{S}$ defined in exercise (2.13) is decomposable:

$$\bar{S} = \sum_{\dot{\alpha}=1}^{\bar{r}} \bar{S}_{\dot{\alpha}} ,$$

---

allows us to dismiss the possibility of extra right-moving currents in the IR. The argument in [66] has a loophole as far as right-moving currents are concerned: it is argued that the R-symmetry current must commute with all the currents in an N=2 SKM multiplet. While that is true for the top components of the SKM multiplet, it is not the case for the R-current of the N=2 SKM multiplet!

[32] The superspace coordinate θ has R-charge $+1$; this accounts for the $(r_a - 1)$ factors in the coefficients.

where each $\overline{S}_{\dot\alpha}$ generates a superconformal algebra with $\overline{c} = \overline{c}_{\dot\alpha}$, and the $\overline{S}_{\dot\alpha}\overline{S}_{\dot\beta}$ OPE is trivial for $\alpha \neq \beta$. This includes the case that the UV CFT has an N=2 SKM algebra—an extension of the N=1 SKM algebra reviewed above. Suppose we now turn on a relevant deformation of this UV theory by a set of chiral primary operators $\mathcal{U}$ with $h_{\mathcal{U}} = 1/2 + \overline{h}_{\mathcal{U}}$ and $\overline{h}_{\mathcal{U}} = \overline{q}_{\mathcal{U}}/2 < 1/2$.

A relevant deformation necessarily preserves every N=2 SKM algebra because there are no N=2 SKM representations with $0 < \overline{q}_{\mathcal{U}} < 1$. To understand that point, note that an abelian N=2 SKM necessarily contains an anti-holomorphic chiral primary field ψ of charge $\overline{q} = 1$, and any $\mathcal{U}$ must have an integer $U(1)_R$ charge in order to have a local OPE with ψ. For a non-abelian N=2 SKM the relevant deformation must involve a primary operator of the current algebra generated by the free fermions, but the dimensions of these satisfy

$$\overline{h} \geq \frac{k}{2} \frac{\ell(r)}{\dim r} \geq 1/2 . \tag{2.7.33}$$

While all the N=2 SKM components of the superconformal algebra must be preserved, it may be that the deformation also preserves a number of other $\overline{S}_{\dot\alpha}$ components. Each such component will lead to a factor in the IR superconformal algebra, and we do not need to work to find the R-symmetry charges for these factors: they are RG-invariant. So, without loss of generality we will now restrict the $\dot\alpha$ to run over the factors that do involve the deformation. More precisely, we can assume that for every $\dot\alpha$ there is some $\mathcal{U}$ charged with respect to $\overline{J}_{\dot\alpha}$. Once this is the case, the most general form for an R-current conserved along the RG flow is of the form

$$\mathcal{J}^{\mathrm{uv}} = \sum_{\alpha=1}^{r} s^{\alpha} J_{\alpha}^{\mathrm{uv}} \frac{\partial}{\partial z} + \overline{J}^{\mathrm{uv}} \frac{\partial}{\partial \overline{z}} . \tag{2.7.34}$$

Here the s^{α} are to-be-determined coefficients, and $\overline{J}^{\mathrm{uv}} = \sum_{\dot\alpha} \overline{J}_{\dot\alpha}$ is the diagonal R-current. This current will be conserved in the presence of the deformation if for every $\mathcal{U}$ we can find a solution to

$$\sum_{\alpha=1}^{r} q_{\mathcal{U}}^{\alpha} s_{\alpha} = 1 - \overline{q}_{\mathcal{U}} . \tag{2.7.35}$$

This is just the condition for the term $\Delta S = \int d^2 z d\theta \, \mathcal{U} + \text{h.c.}$ to be invariant under the $\mathcal{J}^{\mathrm{uv}}$ symmetry.

Suppose that a solution exists, i.e. we can set $s_{\alpha} = \sigma_{\alpha}$ and preserve the current $\mathcal{J}^{\mathrm{uv}}$ along the flow. The solution may be ambiguous: we can shift $\sigma \to \sigma + \omega$, where ω is in the kernel of the matrix of charges $q_{\mathcal{U}}^{\alpha}$, i.e. $\sum_{\alpha} q_{\mathcal{U}}^{\alpha} \omega_{\alpha} = 0$ for all $\mathcal{U}$ involved in the deformation. We normalize the left-moving current two-point functions so that $M^{\alpha\beta} = \delta^{\alpha\beta}$, and fix and orthonormal basis $\{\omega^1, \omega^2, \ldots, \omega^n\}$ for the kernel of $q_{\mathcal{U}}^{\alpha}$. Now the most general form of the coefficients s^{α} can be written as

$$s_{\alpha} = \sigma_{\alpha} + \sum_{i=1}^{n} t_i \omega_{\alpha}^i \qquad (2.7.36)$$

for n real parameters t_i.

If the limit of the RG flow is a compact (0,2) CFT, then from the arguments given above, each ω_i will lead to a left-moving KM current. In addition, if there are no accidents along the flow, then $\mathcal{J}^{\mathrm{uv}}$ must flow to the IR R-symmetry. The latter must have a zero two-point function with the former, and anomaly matching then uniquely determines the coefficients t_i:

$$t_i = -\sum_{\alpha=1}^{r} \omega_{\alpha}^i \sigma_{\alpha} , \qquad (2.7.37)$$

so that the solution for the s_{α} is just given by a projection from σ to the component orthogonal to the ω_i: $s = \sigma_{\perp} = \sigma - \sum_i (\omega^i \cdot \sigma) \omega^i$. This fixes the IR R-charges and determines the central charge from the two-point function of the R-current:

$$\bar{c}^{\mathrm{ir}} = \bar{c}^{\mathrm{uv}} - 3\|\sigma_{\perp}\|^2 . \qquad (2.7.38)$$

A Farewell to Generalities

Having developed the notion of (super)conformal invariance and presented a few useful general tools, we will now turn to specific examples of (0,2) quantum field theories. We leave behind a number of fundamental questions, some technical and some conceptual, on the nature of conformal and superconformal perturbation theory. The examples we tackle in the next chapters will give many glimpses of the beauty and complexity of these RG flows. Perhaps the reader will be inspired by those results to resolve the fundamental questions raised in this chapter.

Chapter 3
Landau-Ginzburg Theories

Abstract In this chapter we study the simplest large class of (0,2) QFTs: the (0,2) Landau-Ginzburg theories. While they are interesting in their own right, the main goal is to introduce useful notions relevant to general (0,2) theories in the context of these simple examples.

3.1 A Class of Lagrangian Theories

Recall that in Chap. 1 we defined the (0,2) Yukawa theories. These are (0,2) Lagrangian theories where the interactions are encoded in terms of the holomorphic potentials E and J. For convenience, we review the construction of the Euclidean Yukawa theory. There are n chiral bosonic superfields Φ^a, $a = 1, \ldots, n$, as well as N fermi E-chiral multiplets Γ^A, $A = 1, \ldots, N$, and the action is

$$S = \frac{1}{2\pi} \int d^2z \left\{ \mathcal{D}\overline{\mathcal{D}}K_z + \frac{1}{\sqrt{2}} \mathcal{D}\,W + \frac{1}{\sqrt{2}} \overline{\mathcal{D}}\,\overline{W} \right\} . \tag{3.1.1}$$

$$K_z = \sum_a \tfrac{1}{4} \left[\overline{\Phi}^a \partial \Phi^a - \Phi^a \partial \overline{\Phi}^a \right] - \tfrac{1}{2} \sum_A \overline{\Gamma}^A \Gamma^A \tag{3.1.2}$$

is the free kinetic term,

$$W = \sum_A \Gamma^A J_A(\Phi) \tag{3.1.3}$$

is the holomorphic superpotential, and the fermi multiplets obey the chirality condition

$$\overline{\mathcal{D}}\Gamma^A = E^A(\Phi) . \tag{3.1.4}$$

The Lagrangian is supersymmetric if and only if $\sum_A E^A J_A = 0$.

© Springer Nature Switzerland AG 2019
I. V. Melnikov, *An Introduction to Two-Dimensional Quantum Field Theory with (0,2) Supersymmetry*, Lecture Notes in Physics 951,
https://doi.org/10.1007/978-3-030-05085-6_3

A Yukawa theory is a Landau-Ginzburg theory (LG) theory if and only if it admits an R-symmetry $U(1)_B$ with charges

	θ	Φ^a	Γ^A	
$U(1)_B$	$+1$	0	$+1$	(3.1.5)

In such a theory $E^A = 0$, and the action is supersymmetric for any J; moreover, the theory automatically admits a supercurrent $\mathcal{R}$-multiplet.

Beyond this, it appears that there is no completely agreed-upon terminology to distinguish different classes of models. Some authors would call any Yukawa theory a LG theory, and many would call any non-linear sigma model with superpotential couplings a Landau-Ginzburg theory. This seems a bit too broad, and there are a number of additional criteria that usefully restrict the landscape of these Landau-Ginzburg theories to a large but perhaps more manageable set. We will now list these criteria along with some comments.

1. We can restrict attention to J_A that are polynomial in the fields. While non-polynomial J_A do play a role, especially in various effective descriptions derived from some more fundamental UV theory, a restriction to polynomials leads to important simplifications because the large field behavior of the potential is simple and universal.

2. We will restrict attention to theories that preserve classical (0,2) SUSY, i.e. there is a point $\phi_* \in \mathbb{C}^n$ such that

$$J_A(\phi_*) = 0 \qquad \text{for all } A.$$

3. We will restrict attention to theories with a compact space of vacua. When this is not the case we must worry about the behavior of the path integral at large field values, and we might suspect that the generic situation will be a lack of normalizable ground state for the theory.

   When the J_A are polynomials the only possible compact holomorphic subset of $\mathbb{C}^n$ defined by the simultaneous vanishing of the J_A is a collections of points.

   We will say a LG theory is "compact" when the space of classical vacua is compact.

   Clearly we must then insist on $N \geq n$, since otherwise there are too few equations to fix the n expectation values of Φ^a.

4. Perhaps the most interesting restriction is to theories that admit another $U(1)$ symmetry, denoted $U(1)_L$, with charges

	θ	Φ^a	Γ^A	
$U(1)_L$	0	q_a	Q_A	(3.1.6)

This the J_A to be quasi-homogeneous:

$$J_A(t^q \Phi) = t^{-Q_A} J_A(\Phi) .$$ (3.1.7)

So far the $U(1)_L$ charges are defined up to an over-all scale. The normalization will be determined by c-extremization. The $U(1)_L$ symmetry will allow us to construct a UV candidate for the R-symmetry of the superconformal IR fixed point. Thus quasi-homogeneous and compact (0,2) LG theories offer a large landscape of theories where we may hope to use the simple UV description to capture properties of non-trivial SCFTs.

An important special case of this construction is provided by (2,2) LG theories. In the (0,2) language these are theories with $N = n$ and

$$J_a = \frac{\partial W}{\partial \Phi_a} ,$$ (3.1.8)

where $W(\Phi)$ is the (2,2) superpotential. The (0,2) multiplets (Φ_a, Γ^a) now combine into (2,2) chiral multiplets X^a, and, introducing left-moving superspace derivatives $\mathcal{D}'$ and $\overline{\mathcal{D}}'$, the action takes the schematic form

$$S_{(2,2)} = \int d^2z \left\{ \mathcal{D}'\overline{\mathcal{D}}'\mathcal{D}\overline{\mathcal{D}} [\textstyle\sum_a \|X_a\|^2] + \mathcal{D}'\mathcal{D}W(X) + \overline{\mathcal{D}\mathcal{D}'}\overline{W}(\overline{X}) \right\} .$$ (3.1.9)

For a quasi-homogeneous W the chiral superfields X_a have equal $U(1)_L$ and $U(1)_R$ charges q_a. By contrast with the (0,2) LG theories, these (2,2) models are rather well understood [274, 283, 289, 372]. A summary of the basic results is as follows.

1. $Q_a = q_a - 1$;
2. the IR central charge is given by

$$c = \bar{c} = 3 \sum_a (1 - 2q_a) ;$$ (3.1.10)

3. the (a,c) ring is trivial;
4. the (c,c) ring is isomorphic to the Jacobian ideal $\mathbb{C}[X_1, \ldots, X_n]/\langle \frac{\partial W}{\partial X_1}, \ldots, \frac{\partial W}{\partial X_n} \rangle$;
5. the $c < 3$ theories are in 1:1 correspondence with the $N = 2$ minimal models;
6. any rational $\bar{c}$ is realized by a finite number of families of quasi-homogeneous W; in other words, for some fixed choice of $\bar{c}$ there is a finite number (possibly zero) of possibilities for the quasi-homogeneity charges $(q_1, \ldots, q_n)$ that realize that central charge.

We will now study the (0,2) generalization of these (2,2) results.

3.2 Currents and Supercurrents

Every LG theory has an $\mathcal{R}$ supercurrent multiplet associated to the $U(1)_B$ R-symmetry with components

$$\mathcal{R}_B = \sum_{A=1}^{N} \Gamma^A \overline{\Gamma}^A \,, \qquad\quad \overline{\mathcal{R}}_B = -\frac{1}{2} \sum_{a=1}^{n} \overline{\mathcal{D}\Phi}^a \mathcal{D}\Phi^a \,,$$

$$\mathcal{T}_B = -\sum_{a=1}^{n} \partial\Phi^a \partial\overline{\Phi}^a - \frac{1}{2}\sum_{A=1}^{N} \left(\overline{\Gamma}^A \partial\Gamma^A + \Gamma^A \partial\overline{\Gamma}^A \right) \,. \tag{3.2.1}$$

The equations of motion (see Exercise 1.12) that follow from the Lagrangian with $E^A = 0$ are

$$\mathcal{D}\Gamma^A = \sqrt{2}\overline{J}_A \,, \qquad\quad \partial\mathcal{D}\Phi^a = \sqrt{2}\sum_{A=1}^{N} \overline{J}_{A,a}\overline{\Gamma}^A \,,$$

$$\overline{\mathcal{D}\Gamma}^A = \sqrt{2}J_A \,, \qquad\quad \partial\overline{\mathcal{D}\Phi}^a = \sqrt{2}\sum_{A=1}^{N} J_{A,a}\Gamma^A \,, \tag{3.2.2}$$

and they imply

$$\bar{\partial}\mathcal{R}_B + \partial\overline{\mathcal{R}}_B = 0 \,, \qquad\quad \overline{\mathcal{D}}\left(\mathcal{T}_B - \tfrac{1}{2}\partial\mathcal{R}_B\right) = 0 \,. \tag{3.2.3}$$

When the superpotential is quasi-homogeneous, so that there is an $U(1)_L$ global symmetry, there is also a current multiplet with components $\mathcal{K}$—a pure imaginary spin 1 operator, and Ψ—a chiral spin $-1/2$ field that satisfy (up to equations of motion)

$$\overline{\mathcal{D}}\mathcal{K} = -2\partial\Psi \,, \quad \mathcal{D}\mathcal{K} = 2\partial\overline{\Psi} \,, \quad\Longrightarrow\quad \bar{\partial}\mathcal{K} + \partial(\mathcal{D}\Psi - \overline{\mathcal{D}\Psi}) = 0 \,. \tag{3.2.4}$$

A possible improvement term for the current multiplet is specified in terms of a pure imaginary superfield U[1]:

$$\mathcal{K} \to \mathcal{K} - 2\partial U \,, \qquad \Psi \to \Psi + \overline{\mathcal{D}}U \,, \qquad \overline{\Psi} \to \overline{\Psi} - \mathcal{D}U \,. \tag{3.2.5}$$

---

[1]Recall that in our Euclidean conventions reality is defined with respect to the action of the charge conjugation operator $\mathcal{C}$ defined in Sect. 1.8.

For the LG theory these take the form

$$\mathcal{K} = \sum_{A=1}^{N} Q_A \Gamma^A \overline{\Gamma}^A - \frac{1}{2} \sum_{a=1}^{n} q_a (\Phi^a \partial \overline{\Phi}^a - \overline{\Phi}^a \partial \Phi^a) \,,$$

$$\Psi = -\frac{1}{4} \sum_{a=1}^{n} q_a \Phi^a \overline{\mathcal{D}\Phi}^a = \overline{\mathcal{D}} \left[-\frac{1}{4} \sum_{a=1}^{n} q_a \Phi^a \overline{\Phi}^a \right] , \qquad (3.2.6)$$

$\Psi = \overline{\mathcal{D}}\mathcal{A}$, with $\mathcal{A}$ a real superfield

$$\mathcal{A} = -\frac{1}{4} \sum_{a=1}^{n} q_a \Phi^a \overline{\Phi}^a \,. \qquad (3.2.7)$$

The existence of $(\mathcal{K}, \mathcal{A})$ means that the $\mathcal{R}$-multiplet is not unique: for any parameter t

$$\mathcal{R}' = \mathcal{R}_B + t\mathcal{K} \,, \quad \overline{\mathcal{R}}' = \overline{\mathcal{R}}_B + t[\mathcal{D}, \overline{\mathcal{D}}]\mathcal{A} \,, \quad \mathcal{T}' = \mathcal{T}_B - t\partial^2 \mathcal{A} \qquad (3.2.8)$$

is another $\mathcal{R}$-multiplet.

When there are several linearly independent quasi-homogeneous conditions labeled by $\alpha = 1, \ldots, r$, there is a U(1)^r symmetry with current multiplets $(\mathcal{K}^\alpha, \mathcal{A}^\alpha)$ and corresponding charges q_a^α and Q_A^α for the Φ^a and Γ^A respectively. Now for any parameters t_α

$$\mathcal{R}' = \mathcal{R}_B + t_\alpha \mathcal{K}^\alpha \,, \quad \overline{\mathcal{R}}' = \overline{\mathcal{R}}_B + t_\alpha [\mathcal{D}, \overline{\mathcal{D}}]\mathcal{A}^\alpha \,, \quad \mathcal{T}' = \mathcal{T}_B - t_\alpha \partial^2 \mathcal{A}^\alpha \qquad (3.2.9)$$

satisfy the properties of a $\mathcal{R}$-multiplet.

All of this is offers a concrete example of our discussion of extremization in the previous chapter, and the resolution of the ambiguity is then much the same. If we assume that there are no accidents along the flow from the UV LG theory to a compact unitary SCFT, then c-extremization determines the coefficients t_α such that the $\mathcal{R}$ multiplet

$$\mathcal{R}^{\text{uv}} = \mathcal{R}_B + \sum_{\alpha=1}^{r} t_\alpha \mathcal{K}^\alpha \,, \qquad \overline{\mathcal{R}}^{\text{uv}} = \overline{\mathcal{R}}_B + \sum_{\alpha=1}^{r} t_\alpha [\mathcal{D}, \overline{\mathcal{D}}]\mathcal{A}^\alpha \,,$$

$$\mathcal{T}^{\text{uv}} = \mathcal{T}_B - \sum_{\alpha=1}^{r} t_\alpha \partial^2 \mathcal{A}^\alpha \,, \qquad (3.2.10)$$

flows in the IR to the generators of a left-moving Virasoro algebra and a right-moving N = 2 super-Virasoro algebra. Moreover, as we argued quite generally above, in the accident-free scenario the remaining current multiplets must either decouple or flow to left-moving KM currents. Since the lowest components of the (super)current multiplets involved are the $\mathfrak{u}(1)^{\oplus r} \oplus \mathfrak{u}(1)_R$ conserved currents, we can use the c-extremization technique to fix the coefficients t_α: we just need to restrict the coefficients in (2.7.30) to $R_A = 1$ and $r_a = 0$. The results are

$$a^{00} = N - n, \quad a^{0\beta} = \sum_{A=1}^{N} Q_A^\beta + \sum_{a=1}^{n} q_a^\beta, \quad M^{\alpha\beta} = \sum_{A=1}^{N} Q_A^\alpha Q_A^\beta - \sum_{a=1}^{n} q_a^\alpha q_a^\beta.$$
(3.2.11)

Recall that in order for the accident-free scenario to hold M must have non-negative eigenvalues; moreover, all currents that lie in the kernel of M must decouple from the IR, so that we can restrict attention to currents in the orthogonal complement of ker M. In order to keep the notation simple we will just assume that α runs over currents in this orthogonal complement, so that M is in fact a positive matrix. The parameters t_β are then fixed

$$t_\beta = -\sum_{\alpha=1}^{r} a^{0\alpha} (M^{-1})_{\alpha\beta},$$
(3.2.12)

which also determines the IR central charge to be

$$\frac{\bar{c}^{ir}}{3} = n - N + \sum_{\alpha,\beta=1}^{r} a^{0\alpha} (M^{-1})_{\alpha\beta} a^{0\beta}.$$
(3.2.13)

Some simple conclusions follow from this extremization exercise. First, to have a putative accident-free IR SCFT r must be positive: otherwise the central charge is non-sensical. If $r = 1$, then $U(1)_L$ is unique up to a rescaling, which is equivalent to the choice of parameter t. We can fix this to $t = 1$ by normalizing the charges so that

$$-\sum_{A=1}^{N} Q_A - \sum_{a=1}^{n} q_a = \sum_{A=1}^{N} Q_A^2 - \sum_{a=1}^{n} q_a^2.$$
(3.2.14)

This condition was first described in (0,2) LG theories obtained from (0,2) gauged linear sigma models in [145].

For what follows it is helpful to think of extremization as determining a distinguished symmetry sub-algebra $\mathfrak{u}(1)_L \oplus \mathfrak{u}(1)_R \subset \mathfrak{u}(1)_B \oplus \mathfrak{u}^{\oplus r}$; $\mathfrak{u}(1)_L$ has the current multiplet

$$\mathcal{K} = \sum_\alpha t_\alpha \mathcal{K}^\alpha, \qquad \mathcal{A} = \sum_\alpha t_\alpha \mathcal{A}^\alpha,$$
(3.2.15)

and $\mathfrak{u}(1)_R$ corresponds to the R-current in the $\mathcal{R}^{uv}$ multiplet.

3.3 The Chiral Algebra of a LG Theory

A remarkable feature of LG theories and (0,2) Lagrangian theories in general is the existence of the chiral algebra of holomorphic operators, essentially a holomorphic CFT. This structure was elucidated in the context of (2,2) LG theories in [389]; earlier explorations were made in [167]. It was applied to (0,2) LG and GLSM models in [145, 259, 266, 349, 350]. The chiral algebra has since been explored at great length in the context of (0,2) non-linear sigma models: see, e.g. [3, 93, 358, 393]. More recently, the structure has been revisited in the context of (0,2) LG theories in [129, 294].

3.3.1 The General Idea

A theory with a (0,2) supersymmetry algebra, such as any theory with an $\mathcal{R}$-multiplet, has the supercharges Q and $\overline{Q}$ that obey $Q^2 = \overline{Q}^2 = 0$ and $\{Q, \overline{Q}\} = -2\bar{\partial}$. The chiral algebra for the theory is obtained by restricting the operators to $\overline{Q}$ cohomology:

$$H_{\overline{Q}} = \frac{\ker \overline{Q}}{\operatorname{im} \overline{Q}} = \frac{\{\mathcal{O} \mid \overline{Q} \cdot \mathcal{O} = 0\}}{\{\mathcal{O} = \overline{Q} \cdot \mathcal{X}\}} . \tag{3.3.1}$$

The OPE defines an algebra on $H_{\overline{Q}}$. Let $\mathcal{O}(z, \bar{z})$ be a $\overline{Q}$-closed local operator and $[\mathcal{O}(z, \bar{z})]$ be its cohomology class. The supersymmetry algebra implies[2]

$$-2\bar{\partial}\mathcal{O}(z, \bar{z}) = \overline{Q} \cdot (Q \cdot \mathcal{O}) + Q \cdot (\overline{Q} \cdot \mathcal{O}) = \overline{Q} \cdot (Q \cdot \mathcal{O}) . \tag{3.3.2}$$

Thus, the cohomology class $[\mathcal{O}(z, \bar{z})]$ is $\bar{z}$-independent, i.e. holomorphic! It follows that the OPE of $\overline{Q}$-closed operators $\mathcal{O}_i$ takes the form

$$\mathcal{O}_1(z, \bar{z})\mathcal{O}_2(0) \sim \sum_i F_{12}{}^i(z)\mathcal{O}_i(0) + \overline{Q}(\cdots) \tag{3.3.3}$$

and thus descends to an OPE in cohomology:

$$[\mathcal{O}_1](z)[\mathcal{O}_2](0) \sim \sum_i F_{12}{}^i(z)[\mathcal{O}_i](0) . \tag{3.3.4}$$

---

[2]We use the short-hand introduced in the first chapter: $\overline{Q} \cdot \mathcal{O}$ stands for $[\overline{Q}, \mathcal{O}]$ or $\{\overline{Q}, \mathcal{O}\}$, depending on whether $\mathcal{O}$ is bosonic or fermionic.

Lorentz invariance implies that the position dependence of the OPE is entirely determined by the spins of the operators:

$$[\mathcal{O}_1](z)[\mathcal{O}_2](0) \sim \sum_i C_{12}{}_i^i z^{s_i - s_1 - s_2} [\mathcal{O}_i](0) . \tag{3.3.5}$$

The (infinite-dimensional) subspace of operators $H_{\overline{Q}}$, together with the holomorphic OPE, define the chiral algebra of our theory. In general the OPE coefficients $C_{12}{}_i^i$ depend on parameters of the theory, leading to a family of chiral algebras.

The chiral algebra is scale-invariant. Although the original QFT need not be scale-invariant, the chiral algebra naturally is scale-invariant, simply because Lorentz invariance and holomorphy fix the OPE position dependence to be power-law. Thus, the chiral algebra is an RG-invariant of the QFT. In the IR it encodes non-trivial properties of an SCFT, but on the other hand, it can be accessed via computations in the UV, which is particularly useful when the UV theory is weakly coupled.

The scale invariance does not imply that the chiral algebra is determined by *classical* computations in the UV. It is possible to have a Lagrangian theory for which the classical chiral algebra determined by the Lagrangian differs from the quantum chiral algebra. An extreme example is offered by the $(0,2)$ $\mathbb{CP}^1$ sigma model, where the classical chiral algebra is infinite dimensional, while the quantum chiral algebra is trivial [358, 393, 394]. Moreover, accidental symmetries can also invalidate the relation between the UV and IR chiral algebras. However, we also have plenty of examples where classical or semi-classical UV computations do determine the chiral algebra of the theory, or at least some finite-dimensional subalgebra thereof.

It is often useful to present the chiral algebra in superspace. For that purpose it is convenient to note that $\overline{Q}$ and $\overline{D}$ are conjugate operators. That is,

$$\overline{Q} = \exp[-2\theta\overline{\theta}\partial]\overline{D}\exp[2\theta\overline{\theta}\partial] . \tag{3.3.6}$$

This means that $H_{\overline{Q}}$ is isomorphic to $\overline{D}$ cohomology: a field $\mathcal{O}$ is in $H_{\overline{Q}}$ if and only if it is the lowest component of a chiral superfield; $\mathcal{O}$ is trivial in $H_{\overline{Q}}$ if and only if it is the lowest component of a $\overline{D}$-exact superfield.

Exercise 3.1 Show that the preceding statement is correct. First argue that every $\mathcal{O}$ is the lowest component of a unique superfield $\mathcal{S}_{\mathcal{O}}$; next, show that $\overline{Q} \cdot \mathcal{O} = 0$ if and only if $\overline{D}\mathcal{S}_{\mathcal{O}} = 0$, and $\mathcal{O} = \overline{Q} \cdot X$ if and only if X is the lowest component of $\mathcal{S}_X$ with $\mathcal{S}_{\mathcal{O}} = \overline{D}\mathcal{S}_X$.[3]   □

---

[3]Further discussion can be found in [129].

3.3.2 KMV Algebra in Cohomology

$H_{\overline{Q}}$ may contain the stress tensor T. More precisely, suppose that the stress tensor components of the QFT satisfy

$$\overline{Q} \cdot T = 0, \qquad \overline{T} = \overline{Q} \cdot \mathcal{X}_1, \qquad \Theta = \overline{Q} \cdot \mathcal{X}_2. \qquad (3.3.7)$$

In that case the cohomology class $[T]$ is holomorphic, and we obtain a holomorphic CFT structure on the chiral algebra; this may be extended by any additional KM symmetries or supersymmetries that belong to the cohomology. While these properties hold for some theories, such as the classical non-linear sigma model, they do not directly apply to the LG theory, essentially because the superpotential couplings have dimension of mass and explicitly break scale invariance. Remarkably, however, in a LG theory there are operators in $H_{\overline{Q}}$ that do generate a KMV algebra. Consider the superfields

$$\mathcal{T}_\chi = \mathcal{T}^{\mathrm{uv}} - \tfrac{1}{2}\partial\mathcal{R}^{\mathrm{uv}}, \qquad \mathcal{K}_\chi^\alpha = \mathcal{K}^\alpha + 2\partial\mathcal{A}^\alpha. \qquad (3.3.8)$$

By construction these satisfy

$$\overline{\mathcal{D}}\mathcal{T}_\chi = 0, \qquad \overline{\mathcal{D}}\mathcal{K}_\chi^\alpha = 0. \qquad (3.3.9)$$

The lowest components are

$$T_\chi = T_0 - \tfrac{1}{2}\sum_{\alpha=1}^{r} t_\alpha \partial K_\chi^\alpha,$$

$$T_0 = -\sum_{a=1}^{n} \partial\phi^a \partial\overline{\phi}^a - \sum_{A=1}^{N} \gamma^A \partial\overline{\gamma}^A,$$

$$K_\chi^\alpha = \sum_{A=1}^{N} Q_A^\alpha \gamma^A \overline{\gamma}^A - \sum_{a=1}^{n} q_a^\alpha \phi^a \partial\overline{\phi}^a, \qquad (3.3.10)$$

where the t_α are fixed by c-extremization as in (3.2.12). These operators have the correct spins and dimensions to be, respectively, the left-moving energy-momentum tensor and left-moving spin 1 currents. However, they do not have the correct reality properties; for instance, K_χ^α is not pure imaginary! Real or not, the OPE of these chiral operators can be computed in the free UV theory. This is simply because the superpotential W has an over-all dimensionful coupling μ, and, as we argued above, the OPE in $H_{\overline{Q}}$ will be μ-independent.

Exercise 3.2 Use the free field OPEs

$$\gamma^A(z_1)\overline{\gamma}^B(z_2) \sim \frac{\delta^{AB}}{z_{12}}, \qquad \phi^a(z_1)\partial\overline{\phi}^b(z_2) \sim \frac{\delta^{ab}}{z_{12}} \qquad (3.3.11)$$

and (3.2.12) to evaluate the OPEs of T_χ and K_χ^α. The result is

$$K_\chi^\alpha(z_1)K_\chi^\beta(z_2) \sim \frac{M^{\alpha\beta}}{z_{12}^2},$$

$$T_\chi(z_1)K_\chi^\alpha(z_2) \sim \frac{K_\chi^\alpha(z_2)}{z_{12}} + \frac{\partial K_\chi^\alpha(z_2)}{z_{12}^2},$$

$$T_\chi(z_1)T_\chi(z_2) \sim \frac{c/2}{z_{12}^4} + \frac{2T_\chi(z_2)}{z_{12}^2} + \frac{\partial T_\chi(z_2)}{z_{12}}, \qquad (3.3.12)$$

where

$$c = N - n + \overline{c}, \qquad \overline{c} = 3(n - N) + 3(a^0)^T M^{-1} a^0, \qquad (3.3.13)$$

and $M^{\alpha\beta}$ and $a^{0\beta}$ are the anomaly coefficients in (3.2.11). This is precisely what we obtained above using c-extremization.   □

It is therefore tempting to conjecture that these OPEs compute the left-moving KMV sub-algebra contained in $H_{\overline{Q}}$ of the IR SCFT. Barring accidents, this is a sound assumption, and it allows us to determine the charges and scaling dimensions for all operators in the IR $H_{\overline{Q}}$ through computations in the weakly coupled (indeed, free) UV theory. There are some basic unitarity constraints that must be satisfied for this picture to be sensible. For example, when $r = 1$ and there is a unique U(1)$_L$ symmetry, there are the necessary conditions

$$0 < q_a \leq \overline{c}/3, \qquad 0 < 1 + Q_A \leq \overline{c}/3, \qquad (3.3.14)$$

but we will see below that these are not sufficient.

3.3.3 Chiral CFT Beyond LG

The existence of the candidate stress tensor in the chiral algebra of a LG theory depends weakly on the details of the Lagrangian. In any UV theory with an $\mathcal{R}$-supercurrent multiplet there is a candidate operator for the energy momentum tensor in $H_{\overline{Q}}$:

$$\mathcal{T}_\chi = \mathcal{T}^{uv} - \tfrac{1}{2}\partial\mathcal{R}^{uv}. \qquad (3.3.15)$$

Similarly, whenever the UV theory has current multiplets $(\mathcal{K}, \mathcal{A})$ that have the structure as above, we expect to find candidate operators for KM currents in $H_{\overline{Q}}$. Barring accidents, these should describe the holomorphic sector of the IR SCFT. The (0,2) $\mathbb{CP}^1$ model mentioned above is not a counter-example: that theory has no $\mathcal{R}$-multiplet due to an anomaly in the R-current. The anomaly is evaluated as a quantum correction in the UV theory.

It may also be possible to find a UV representation for a more elaborate holomorphic structure. For instance, in the case of (2,2) LG theories there is an N = 2 superconformal algebra in $\overline{Q}$ cohomology [389], and this can also be generalized to (2,2) gauge theories.

3.4 Topological Heterotic Rings: General Structure

In this section we identify a class of operators in (0,2) theories introduced in [2, 3] that closely mimic the cc and ac rings of (2,2) theories. The notions coincide in (2,2) theories, and when a (2,2) theory is smoothly deformed to a more general (0,2) theory then under some mild assumptions the cc and ac rings are deformed to topological heterotic rings. The study of these rings will be a central theme for us.

Consider a (2,2) SCFT. The CFT Hilbert space contains subspaces H_{cc} and H_{ac} (as well as their complex conjugates) that are finite dimensional rings graded by $\mathfrak{u}(1)_L \oplus \mathfrak{u}(1)_R$. Using the supercharge $\overline{Q} = \overline{G}^+_{-1/2}$ it is easy to see that these rings are isomorphic to finite-dimensional sub-spaces of $H_{\overline{Q}}$ defined as follows:

$$H_{cc} \simeq H_{B/2} = \{\mathcal{O} \in H_{\overline{Q}} \mid h_{\mathcal{O}} - q_{\mathcal{O}}/2 = 0\} \,,$$

$$H_{ac} \simeq H_{A/2} = \{\mathcal{O} \in H_{\overline{Q}} \mid h_{\mathcal{O}} + q_{\mathcal{O}}/2 = 0\} \,. \tag{3.4.1}$$

We obtain the ring structure as in (2.4.32).[4] For instance, if $\mathcal{O}_i \in H_{B/2}$, then

$$\mathcal{O}_i(z_1)\mathcal{O}_j(0) \sim \sum_s C_{ij}{}^s z^{h_s - q_s/2} \mathcal{O}_s(0) \,, \tag{3.4.2}$$

where $\mathcal{O}_s \in H_{\overline{Q}}$. Since the (2,2) theory has a unitarity bound $h \geq q/2$ for all operators, it follows that the OPE is non-singular, and in the $z \to 0$ limit the only operators that contribute on the right-hand-side are $\mathcal{O}_k \in H_{B/2}$. The result is the ring structure $\mathcal{O}_i \mathcal{O}_j = \sum_k C_{ij}{}^k \mathcal{O}_k$.

So far, we just reworded the familiar cc ring of a (2,2) theory. We will now show that in a large class of theories the ring structure on $H_{B/2}$ will continue to exist for general marginal deformations. That will include any (0,2) marginal deformations

---

[4]In what follows we will omit the distinction between $\overline{Q}$-closed operator $\mathcal{O}$ and its cohomology class $[\mathcal{O}] \in H_{\overline{Q}}$, and we will drop $\overline{Q}$-exact terms in the OPEs.

of (2,2) theories but will also apply to any (0,2) theory with the following properties in the NS sector:

1. the spin is integer or half-integer: $h - \overline{h} \in \frac{1}{2}\mathbb{Z}$;
2. $q - \overline{q} \in \mathbb{Z}$, and there is a fermion number $(-1)^F = (-1)^{q-\overline{q}}$;
3. spin-statistics holds: the spin is half-integer for fermions and integer for bosons;
4. there is a point in the moduli space where the spectrum of local operators satisfies the bound $h \geq \frac{1}{2}q$.

From these assumptions it follows that for any operator we have

$$h - \overline{h} - \frac{1}{2}(q - \overline{q}) = m \in \mathbb{Z}, \tag{3.4.3}$$

and for any operator in $H_{\overline{Q}}$ the quantity $h - \frac{1}{2}q$ is a non-negative integer.

Consider a marginal deformation that preserves (0,2) superconformal invariance and the U(1)$_L$ symmetry. While the dimensions and charges may shift, the integer m will remain unchanged for all operators. Thus, the bound $h \geq \frac{1}{2}q$ of the undeformed theory will continue to hold in $\overline{Q}$-cohomology under any (0,2) marginal deformation that preserves U(1)$_L$. It follows that the ring structure on $H_{B/2}$ will persist for any marginal U(1)$_L$-preserving deformation off the (2,2) locus.

The SCFT may have more structure, where not only is $q - \overline{q} \in \mathbb{Z}$, but also $q \in \mathbb{Z}$. In that case

$$h - \overline{h} + \frac{1}{2}(q + \overline{q}) = m' \in \mathbb{Z}, \tag{3.4.4}$$

and an exactly analogous argument will show that $H_{A/2}$ will persist for any marginal U(1)$_L$-preserving deformation.

More generally, given a family of (0,2) SCFTs with U(1)$_L$ as above and a point p in the moduli space where $m \geq 0$ for all $\mathcal{O} \in H_{\overline{Q}}$, the family will possess the $H_{B/2}$ ring. If, in addition $q \in \mathbb{Z}$, we also have the $H_{A/2}$ ring. The existence of p sounds like a strong assumption, but there is a class of (0,2) SCFTs where unitarity implies the desired bound. Suppose the $\mathfrak{u}(1)_L$ current has level r. The Sugawara unitarity bound

$$h \geq \frac{q^2}{2r} \tag{3.4.5}$$

implies

$$h - \frac{q}{2} \geq \frac{q(q-r)}{2r} \geq -\frac{r}{8}. \tag{3.4.6}$$

Hence $r < 8$ leads the desired bound: $m \geq 0$ for all $\mathcal{O} \in H_{\overline{Q}}$. The same bound holds for m' whenever $q \in \mathbb{Z}$.

The dimensions of $H_{A/2}$ and $H_{B/2}$ may vary over the SCFT moduli space when states leave or descend to $H_{\overline{Q}}$ cohomology. Consider a non-singular point p in the moduli space. It may be that a pair of states in $H_{\overline{Q}}$ pair up and leave the cohomology for any distance $\epsilon > 0$ away from p. However, a finite deformation is required for a non-chiral state at p to decompose into two chiral states in cohomology. Therefore, any p has an open neighborhood U_p such that for all points $s \in U_p$ dim $H_{B/2}(s) \leq$ dim $H_{B/2}(p)$ (and similarly for the A/2 ring). For a generic p the statement will hold with equality.

The statements just made apply to any massive (0,2) theory with an R-multiplet. Any such theory will have a chiral algebra on $H_{\overline{Q}}$ with stress tensor as in (3.3.15). The topological heterotic ring(s) provide an easily accessible class of RG-invariants of such (0,2) theories. The simplest example is afforded by the LG theories, and we will now examine it in detail. Along the way, we will also start to develop some algebraic and geometric notions that will prove useful in the sequel.

Exercise 3.3   Use the left-moving spectral flow to prove the isomorphism

$$H_{B/2}^{q,\overline{q}} \simeq H_{A/2}^{q-r,\overline{q}} \ . \tag{3.4.7}$$

This relationship will be useful when we relate the topological heterotic rings to gauge-charged matter representations in heterotic compactifications.   □

3.5   Topological Heterotic Ring of (0,2) LG Theory

As we saw in Chap. 1, after integrating out the auxiliary fields, we obtain the following Euclidean action for (0,2) LG theory:

$$S = \frac{1}{2\pi} \int d^2z \left\{ \partial \overline{\phi}^a \partial \phi^a + \overline{\psi}^a \partial \psi^a + \overline{\gamma}_A \overline{\partial} \gamma^A - \gamma^A J_{A,a} \psi^a - \overline{\psi}^a J_{,a}^A \overline{\gamma}_A + J_A \overline{J}^A \right\} \ . \tag{3.5.1}$$

We lowered the index $\overline{\gamma}^A \to \overline{\gamma}_A$ for future convenience. Note that we are using the summation convention to simplify the notation. This action is supersymmetric and in particular invariant under the action of $\overline{Q}$; $\overline{Q}$ leaves ϕ^a, $\overline{\psi}^a$ and γ^A invariant, while acting on the remaining fields—see (1.8.18)—by

$$\overline{Q} \cdot \overline{\phi}^a = \overline{\psi}^a \ , \qquad \overline{Q} \cdot \psi^a = -\overline{\partial} \phi^a \ , \qquad \overline{Q} \cdot \overline{\gamma}_A = -J_A \ . \tag{3.5.2}$$

For any J_A the theory has an R_B-multiplet with a conserved current corresponding to the symmetry with charges

	ϕ^a	ψ^a	γ^A
U(1)$_B$	0	−1	+1

$$\tag{3.5.3}$$

We are interested in the $\overline{Q}$ cohomology $H_{\overline{Q}}$ and more specifically the topological heterotic sub-ring $H_{B/2} \subset H_{\overline{Q}}$. Every cohomology class will have a representative without any $\bar{\partial}$ derivatives; a look at the action of $\overline{Q}$ then also shows that every cohomology class will have a representative without any ψ^a insertions. As we will see shortly, to describe $H_{B/2}$ we can also restrict to operators without any holomorphic derivatives, i.e. every class in $H_{B/2}$ has a representative in the space of operators $\mathcal{H}$ given by linear combinations of

$$\mathcal{O}[\omega^{k,l,s}] = \omega^{A_1\cdots A_k}_{B_1\cdots B_l;a_1\cdots a_s}(\phi,\overline{\phi})\overline{\gamma}_{A_1}\cdots\overline{\gamma}_{A_k}\overline{\psi}^{a_1}\cdots\overline{\psi}^{a_s}\gamma^{B_1}\cdots\gamma^{B_l}. \qquad (3.5.4)$$

The action of $\overline{Q}$ on $\mathcal{H}$ is given by the differential operator

$$\overline{Q} = \left[\sum_{a=1}^{n}\overline{\psi}^a\frac{\partial}{\partial\overline{\phi}^a} - \sum_{A=1}^{N}J_A\frac{\partial}{\partial\overline{\gamma}_A}\right]. \qquad (3.5.5)$$

Since $\overline{Q}$ preserves the number of γ^A insertions, the cohomology can be evaluated at fixed l. By construction $\overline{Q}$ has $\mathfrak{u}(1)_B$ charge $+1$, and we can grade the $\mathcal{H}$ by l and the $\mathfrak{u}(1)_B$ charge $\overline{q}_0$:

$$\mathcal{H} = \bigoplus_{l,\overline{q}_0}\mathcal{H}^{l,\overline{q}_0}. \qquad (3.5.6)$$

So, at every l we are then interested in the cohomology of the complex

$$\cdots \xrightarrow{\overline{Q}} \mathcal{H}^{l,\overline{q}_0-1} \xrightarrow{\overline{Q}} \mathcal{H}^{l,\overline{q}_0} \xrightarrow{\overline{Q}} \mathcal{H}^{l,\overline{q}_0+1} \xrightarrow{\overline{Q}} \cdots \qquad (3.5.7)$$

We will now show that every cohomology class in $H_{\overline{Q}} \cap \mathcal{H}$ has a representative without any $\overline{\psi}$ insertions. It is not too hard to argue for it directly, but we will do it by using a more general set-up of the problem that employs bundle-valued differential forms.[5]

3.5.1 A Geometric Set-Up for the Cohomology

Let $E \to M$ be a holomorphic bundle of rank N over a complex manifold M of dimension $\dim_{\mathbb{C}} M = n$, and suppose E has a global holomorphic section J.[6] Let

$$\Omega^{k,s} = \mathscr{A}^{0,s}(M, \wedge^k E^*), \qquad (3.5.8)$$

---

[5] We will explore these structures in greater detail in the next chapter; here we will just need some of the most basic algebraic concepts and definitions. The exercises provide some practice with these tools.

[6] The specter of geometry appears before us. It may be useful at this point to take a look through the geometry appendix especially Sect. B.3.2. At this point we need little beyond the tensor structure, so the reader should feel free to simply skim the geometry review to get comfortable with some of the conventions.

That is, $\Omega^{k,s}$ is the space of smooth $(0, s)$ differential forms on M valued in the k-th wedge power of the dual bundle E^*.

If we fix a local holomorphic frame $\{e_1, e_2, \ldots, e_N\}$ for E^* and coordinates $(z^i, \bar{z}^i)$ for M, then $\omega \in \Omega^{k,s}$ takes the form

$$\omega = \frac{1}{k!s!} e_{A_1} \wedge \cdots \wedge e_{A_k} \omega^{A_1 \cdots A_k}_{\bar{i}_1 \cdots \bar{i}_s} d\bar{z}^{i_1} \wedge \cdots \wedge d\bar{z}^{i_s} . \tag{3.5.9}$$

We then have the standard definition for the Dolbeault operator: $\bar{\partial} : \Omega^{k,s} \to \Omega^{k,s+1}$:

$$\bar{\partial}\omega = \frac{1}{k!s!} e_{A_1} \wedge \cdots \wedge e_{A_k} \frac{\partial \omega^{A_1 \cdots A_k}_{\bar{i}_1 \cdots \bar{i}_s}}{\partial \bar{z}^{i_0}} d\bar{z}^{i_0} \wedge d\bar{z}^{i_1} \wedge \cdots \wedge d\bar{z}^{i_s} . \tag{3.5.10}$$

We also define the contraction $J \llcorner \omega \in \Omega^{k-1,s}$ via

$$J \llcorner \omega = \frac{1}{(k-1)!s!} e_{A_2} \wedge \cdots \wedge e_{A_k} J_{A_1} \omega^{A_1 A_2 \cdots A_k}_{\bar{i}_1 \cdots \bar{i}_s} d\bar{z}^{i_1} \wedge \cdots \wedge d\bar{z}^{i_s} . \tag{3.5.11}$$

By definition the contraction $J \llcorner$ gives zero on $\Omega^{0,s}$. With these definitions we construct the operator

$$d_J : \Omega^{k,s} \to \Omega^{k,s+1} \oplus \Omega^{k-1,s}$$

$$d_J : \omega \mapsto (-1)^k \bar{\partial}\omega - J \llcorner \omega , \tag{3.5.12}$$

and we observe that $(d_J)^2 = 0$.

Exercise 3.4 Verify that with these definitions $(d_J)^2 = 0$. Also show that if we define for any integer r

$$\Omega^{(r)} = \bigoplus_{\substack{k,s \\ k-s=r}} \Omega^{k,s} \tag{3.5.13}$$

then d_J is a linear operator $d_J : \Omega^{(r)} \to \Omega^{(r-1)}$. Finally, check that if we identify the $\omega \in \Omega^{k,s}$ with operators of the form given in (3.5.4), then the action of d_J is the same as that of $\overline{Q}$ in (3.5.5), and the grading by $\bar{q}_0$ is equivalent to the grading by $-r$. □

The exercise shows that the cohomology of the complex in (3.5.7) is equivalent to the cohomology of the operator d_J on the chain complex built from the $\Omega^{(r)}$[7]

$$H_{\overline{Q}} \cap \mathcal{H}^{\bar{q}_0} = H^{(-\bar{q}_0)}_{d_J} = \frac{\ker\{d_J : \Omega^{(-\bar{q}_0)} \to \Omega^{(-\bar{q}_0-1)}\}}{\operatorname{im}\{d_J : \Omega^{(1-\bar{q}_0)} \to \Omega^{(-\bar{q}_0)}\}} . \tag{3.5.14}$$

---

[7]We suppress the l label, since it does not play any role in the cohomology computation.

The result applies to a wide class of (0,2) theories with a superpotential defined on a non-trivial target space, such as those discussed in [217]. For now, we will just consider the vasty simpler case of LG theories, where $M = \mathbb{C}^n$ and $E = \mathbb{C}^N \times \mathbb{C}^n$. Now the d_J cohomology reduces to an algebraic problem because Dolbeault cohomology of $\mathbb{C}^n$ is very simple: any $\bar{\partial}$-closed form of degree $s > 0$ is exact; a degree $s = 0$ form, i.e. a function, is $\bar{\partial}$-closed if and only if it is holomorphic. Consider now some d_J-closed polyform $\omega^{(r)} \in \Omega^{(r)}$

$$\omega^{(r)} = \sum_{\substack{k,s \\ k-s=r}} \omega^{k,s} = \cdots + \omega_{\text{top}} , \qquad (3.5.15)$$

where $\omega_{\text{top}} \neq 0$ is the component of the polyform $\omega^{(r)}$ with largest value of s, s_{top}. A look at Fig. 3.1 shows that $d_J \omega^{(r)} = 0$ requires $\bar{\partial} \omega_{\text{top}} = 0$, and if $s_{\text{top}} > 0$, then this in turn means that there exists an η such that $\omega_{\text{top}} = \bar{\partial} \eta$. Now let

$$\omega'^{(r)} = \omega^{(r)} - d_J (-1)^{k_{\text{top}}} \eta . \qquad (3.5.16)$$

Since they differ by an exact term, $\omega'^{(r)}$ and $\omega^{(r)}$ belong to the same d_J-cohomology class, but by construction ω' has its largest value of s at most $s_{\text{top}} - 1$. Repeating this construction until the top form has $s = 0$, we therefore show that

$$H_{d_J}^{(r)} = \frac{\ker\{ J_{\llcorner} : \widetilde{\Omega}^r \to \widetilde{\Omega}^{r-1} \}}{\text{im}\{ J_{\llcorner} : \widetilde{\Omega}^{r+1} \to \widetilde{\Omega}^r \}} , \qquad (3.5.17)$$

Fig. 3.1 The spectral sequence for topological heterotic ring of (0,2) LG theory. The solid points label the $\Omega^{k,s}$, and the dashed lines the $\Omega^{(r)}$

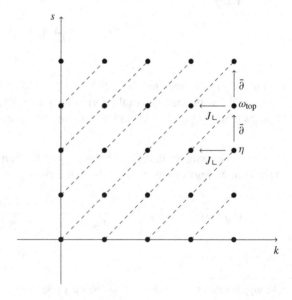

where

$$\widetilde{\Omega}^r = \text{span}\{\omega^{A_1 \cdots A_r}(\phi) e_{A_1} \wedge e_{A_2} \wedge \cdots e_{A_r}\}. \tag{3.5.18}$$

The cohomology $H_{d_J}^{(r)}$ is a standard construction in algebraic geometry [101, 162, 210]. To describe it, we introduce a few more objects. Let $R = \mathbb{C}[\phi_1, \phi_2, \ldots, \phi_n]$ be the ring of polynomials in n variables $\phi_1, \ldots, \phi_n$ with complex coefficients, and take $\mathcal{E} = R^{\oplus N}$ to be the rank N free module over R. It follows that $\widetilde{\Omega}^r = \wedge^r \mathcal{E}$. Finally, let $J = \langle J_1, J_2, \ldots, J_N \rangle \subset R$ be an ideal in R with generators $J_1, J_2, \ldots, J_N$. We now define the Koszul complex for the ideal $J \subset R$ as

$$K_\bullet = \cdots \xrightarrow{\; J_\llcorner \;} \widetilde{\Omega}^{r+1} \xrightarrow{\; J_\llcorner \;} \widetilde{\Omega}^r \xrightarrow{\; J_\llcorner \;} \widetilde{\Omega}^{r-1} \xrightarrow{\; J_\llcorner \;} \cdots \; . \tag{3.5.19}$$

In this terminology the cohomology groups we are after are identified with the homology groups of this complex—$H_k(K_\bullet, J_\llcorner)$. This is the Koszul homology corresponding to ideal $J \subset R$.

Since $\wedge^0 \mathcal{E} = R$, it follows that the complex terminates on the right with

$$\cdots \xrightarrow{\; J_\llcorner \;} \wedge^2 \mathcal{E} \xrightarrow{\; J_\llcorner \;} \mathcal{E} \xrightarrow{\; J_\llcorner \;} R \xrightarrow{\; J_\llcorner \;} 0 \; , \tag{3.5.20}$$

and therefore

$$H_0(K_\bullet, J_\llcorner) = R/J \; . \tag{3.5.21}$$

In other words, $H_0(K_\bullet, J_\llcorner)$ is the quotient ring associated to the ideal.

The Koszul homology groups measure the dimension of the ideal J. It is easy to get an intuitive notion of the dimension of an ideal: the vanishing locus J_A in $\mathbb{C}^n$ may be a smooth sub-manifold, and in that case the dimension of the ideal is simply the dimension of the sub-manifold. More generally the vanishing locus is an algebraic variety and typically is singular as a manifold; however, there are algebraic notions of dimension that generalize to that more general setting [101, 162]. One of these is provided by the Koszul complex:

$$H_k(K_\bullet, J_\llcorner) = \begin{cases} 0 & \text{for } k > N - n + \dim(J) \\ \text{non-zero} & \text{for } k \le N - n + \dim(J) \; . \end{cases} \tag{3.5.22}$$

Additional features are illustrated in the following two exercises.

Exercise 3.5 Show that the $\wedge$ operation gives the space $\bigoplus_k H_k(K_\bullet, J_\llcorner)$ a ring structure. That is, suppose we have classes $[\omega_a] \in H_{k_a}(K_\bullet, J_\llcorner)$ for $a = 1, 2$, with representatives $\omega_a \in \widetilde{\Omega}^{k_a}$. Show that $\omega_1 \wedge \omega_2$ is annihilated by $J_\llcorner$, and that $[\omega_1 \wedge \omega_2] \in H_{k_1+k_2}(K_\bullet, J_\llcorner)$ is independent of representatives, i.e. shifting $\omega_a \to \omega_a + J_\llcorner \eta_a$ for any $\eta_a \in \widetilde{\Omega}^{k_a-1}$ does not change $[\omega_1 \wedge \omega_2]$. □

Exercise 3.6 Let $\epsilon_{A_1 \ldots A_N}$ denote the fully anti-symmetric rank N tensor with $\epsilon_{123 \cdots N} = 1$. Show that this gives an isomorphism $\wedge^k \mathcal{E} = \wedge^{N-k} \mathcal{E}^*$, where $\mathcal{E}^*$ is the dual module to $\mathcal{E}$, where the explicit map is the "Hodge dual"

$$* : \wedge^k \mathcal{E} \quad \to \wedge^{N-k} \mathcal{E}^* ,$$

$$* : \omega^{B_1 \cdots B_k} \mapsto \frac{1}{k!} \omega^{B_1 \cdots B_k} \epsilon_{B_1 \cdots B_k A_1 \cdots A_{N-k}} .$$

Note that $\wedge^k \mathcal{E}$ and $\wedge^k \mathcal{E}^\vee$ are dual in the sense that each element in $\widetilde{\omega} \in \wedge^k \mathcal{E}^*$ yields an R-linear map from $\wedge^k \mathcal{E}$ to R, where $\omega \in \wedge^k \mathcal{E}$ is sent to $\widetilde{\omega} \llcorner \omega \in R$. A more mathematical notation for this statement is $\text{Hom}(\wedge^k \mathcal{E}, R) = \wedge^k \mathcal{E}^*$, and our isomorphism allows us to identify this group with $\wedge^{N-k} \mathcal{E}$.

Now consider the dual Koszul complex

$$K^\bullet = \cdots \xrightarrow{J \wedge} \wedge^{k-1} \mathcal{E}^* \xrightarrow{J \wedge} \wedge^k \mathcal{E}^* \xrightarrow{J \wedge} \wedge^{k+1} \mathcal{E}^* \xrightarrow{J \wedge} \cdots , \qquad (3.5.23)$$

where the map $J \wedge$ is

$$J \wedge : \wedge^k \mathcal{E}^* \to \wedge^{k+1} \mathcal{E}^* ,$$

$$J \wedge : \widetilde{\omega}_{B_1 \cdots B_k} \mapsto (k+1) J_{[B_1} \widetilde{\omega}_{B_2 \cdots B_{k+1}]} ,$$

and the $[\cdots]$ brackets denote total anti-symmetrization, i.e. $T_{[AB]} = \frac{1}{2}(T_{AB} - T_{BA})$ and so forth. Show that $*$ induces an isomorphism of the cohomology groups of (3.5.23), $H^k(K^\bullet, J \wedge)$ to the previously defined $H_k(K_\bullet, J \llcorner)$, with isomorphism given by $*$. To do so, prove the useful identities that $*^2 \wedge^k \mathcal{E} = (-1)^{k(N-k)} \wedge^k \mathcal{E}$ and that for any $\omega \in \wedge^k \mathcal{E}$

$$*(J \llcorner \omega) = (-1)^{k-1} (J \wedge *\omega) . \qquad (3.5.24)$$

Finally, show that the induced map on the cohomology classes is independent of representatives.   □

3.5.2 The LG Topological Heterotic Ring and Koszul Cohomology

Having built up some algebraic machinery, we now return to the LG theory. As shown in the preceding section each of the $\overline{Q}$ cohomology groups $\mathcal{H}^{l, \overline{q}_0}$ is isomorphic to the Koszul homology group $H_{-\overline{q}_0}(K_\bullet, J \llcorner)$. We will now argue that $\mathcal{H}^{0, \overline{q}_0}$ is the topological heterotic ring of LG theory.

Our discussion so far of the $\overline{Q}$ cohomology has been carried out entirely in the classical theory. To work in the quantum theory we need to make sense of the composite operators that appear in (3.5.4). Fortunately, for the purpose of computing the $\overline{Q}$ cohomology, we can use the free UV theory with chiral OPEs of (3.3.11) to define local operators that represent each cohomology class:

$$\mathcal{O}[\omega^{k,l}](z) = \; : \omega^{A_1\cdots A_k}_{B_1\cdots B_l}(\phi)\overline{\gamma}_{A_1}\cdots\overline{\gamma}_{A_k}\gamma^{B_1}\cdots\gamma^{B_l} : (z) \,. \qquad (3.5.25)$$

Since $\phi(z_1)\phi(z_2)$ is non-singular as $z_1 \to z_2$, there are no ordering issues in constructing $\omega^{A_1\cdots A_k}_{B_1\cdots B_l}(\phi)$. Similarly, because the $\overline{\gamma}_A(z_1)\overline{\gamma}_B(z_2)$ is non-singular as $z_1 \to z_2$, normal ordering is only necessary when $l > 0$. The action of $\overline{Q}$ on these operators is represented by

$$\overline{Q} = \oint \frac{dz}{2\pi i} J_A(\phi)\gamma^A \,. \qquad (3.5.26)$$

The symmetry group of our LG theory contains $\mathfrak{u}(1)_B \oplus \mathfrak{u}(1)^{\oplus r}$ as in Sect. 3.2, with charge assignments as follows.

	θ	Φ^a	Γ^A
$\mathfrak{u}(1)_B$	1	0	1
$\mathfrak{u}(1)^\alpha$	0	q^α_a	Q^α_A

$$(3.5.27)$$

We grade the space of operators by these charges and denote the $\mathfrak{u}(1)_B$ and $\mathfrak{u}(1)^\alpha$ charges of an operator $\mathcal{O}$ by, respectively, $\overline{q}_B[\mathcal{O}]$ and $q^\alpha[\mathcal{O}]$. Once we assume the RG flow is accident free, we also find a distinguished sub-algebra $\mathfrak{u}(1)_L \oplus \mathfrak{u}(1)_R$; we denote the $\mathfrak{u}(1)_L \oplus \mathfrak{u}(1)_R$ charges of $\mathcal{O}$ by $q[\mathcal{O}]$ and $\overline{q}[\mathcal{O}]$. These charges are related to $\overline{q}_B$ and q^α by

$$q = t_\alpha q^\alpha \,, \qquad\qquad \overline{q} = \overline{q}_B + q \,. \qquad (3.5.28)$$

To simplify notation for what follows we set

$$q_a = q[\Phi_a] = t_\alpha q^\alpha_a \,, \qquad Q_A = q[\Gamma^A] = t_\alpha Q^\alpha_A \,, \qquad (3.5.29)$$

for the $\mathfrak{u}(1)_L$ charges of the fundamental fields, so that the $\mathfrak{u}(1)_L \oplus \mathfrak{u}(1)_R$ assignments for the superfields of the LG theory are

	θ	Φ^a	Γ^A
$\mathfrak{u}(1)_L$	0	q_a	Q_A
$\mathfrak{u}(1)_R$	1	q_a	$1 + Q_A$

$$(3.5.30)$$

We saw in Sect. 3.3 that the charges q^α are realized by the KMV currents $\mathcal{K}^\alpha_\chi$ in the chiral algebra of the LG theory, with $\mathcal{T}_\chi$ serving as the Virasoro generator. By computing the OPEs of those currents with the fields, we find that ϕ_a, $\overline{\gamma}_A$ and γ^A are all KMV primary with charges and weights that follow from these symmetry assignments.

Exercise 3.7 Compute the OPEs of ϕ_a, $\overline{\gamma}_A$ and γ^A with $\mathcal{T}_\chi$ and $\mathcal{K}_\chi^\alpha$ to verify that they are KMV primary, and determine their charges and weights to verify (3.5.30).

□

Once we know this data, we can also find the right-moving weights and $\mathfrak{u}(1)_R$ charges $\overline{q}$. The latter follows easily since $\overline{q} = \overline{q}_B + q$, and the former is a consequence of the fact that $h - \overline{h}$, the spin of an operator is an RG invariant. Hence, we obtain the table

	ϕ_a	γ^A	$\overline{\gamma}_A$
q^α	q_a^α	$\mathcal{Q}_A^\alpha$	$-\mathcal{Q}_A^\alpha$
q	q_a	$\mathcal{Q}_A$	$-\mathcal{Q}_A$
h	$\frac{1}{2}q_a$	$1+\frac{1}{2}\mathcal{Q}_A$	$-\frac{1}{2}\mathcal{Q}_A$
$\overline{q}$	q_a	$1+\mathcal{Q}_A$	$-1-\mathcal{Q}_A$
$\overline{h}$	$\frac{q_a}{2}$	$\frac{1+\mathcal{Q}_A}{2}$	$\frac{-1-\mathcal{Q}_A}{2}$

(3.5.31)

Note that $\overline{\gamma}^A$ by itself does not belong to the cohomology, but we can use these assignments when $\overline{\gamma}^A$ appears in some composite $\overline{\mathbf{Q}}$-closed field.

We now see that the LG theory satisfies all of the assumptions of Sect. 3.4 necessary for the existence of a B/2 topological heterotic ring. To describe that finite-dimensional space of operators, we consider the $\mathcal{O}[\omega^{k,l}]$ of (3.5.25). We can grade these by $\mathfrak{u}(1)_L \oplus \mathfrak{u}(1)_R$ charges simply by restricting the coefficients $\omega^{A_1\cdots A_k}_{B_1\cdots B_k}(\phi)$ to be quasi-homogeneous polynomials of definite degree

$$q\left[\omega^{A_1\cdots A_k}_{B_1\cdots B_k}(\phi)\right] = d_\omega + \sum_{s=1}^{k}\mathcal{Q}_{A_s} - \sum_{t=1}^{l}\mathcal{Q}_{B_t}.$$

(3.5.32)

With this choice we have

$$q\left[\mathcal{O}[\omega^{k,l}]\right] = d_\omega, \qquad\qquad \overline{q}\left[\mathcal{O}[\omega^{k,l}]\right] = d_\omega + l - k$$

$$h\left[\mathcal{O}[\omega^{k,l}]\right] = \tfrac{1}{2}q\left[\mathcal{O}[\omega^{k,l}]\right] + l, \qquad \overline{h}\left[\mathcal{O}[\omega^{k,l}]\right] = \tfrac{1}{2}\overline{q}\left[\mathcal{O}[\omega^{k,l}]\right].$$

(3.5.33)

As expected, the operators are chiral primary, but also the operators with $l = 0$ satisfy

$$h\left[\mathcal{O}[\omega^{k,0}]\right] = \tfrac{1}{2}q\left[\mathcal{O}[\omega^{k,0}]\right],$$

(3.5.34)

which is the defining relation for operators in the B/2 topological heterotic ring. In describing the operators we did not include any holomorphic derivatives of the fields, such as $\partial\phi_a$ and $\partial\gamma^B$; we now see that is sufficient to completely describe the B/2 ring because an operator with a holomorphic derivative will have $h > q/2$.

These observations imply that our plunge into Koszul cohomology in the previous section has not been in vain. The B/2 ring of the LG theory consists of operators $\mathcal{O}[\omega^k]$, where

$$\mathcal{O}[\omega^k] = \omega^{A_1\cdots A_k}(\phi)\overline{\gamma}_{A_1}\cdots\overline{\gamma}_{A_k}, \tag{3.5.35}$$

and $\omega \in H_k(K_\bullet, J_\llcorner)$. We also see that there are no operator ordering ambiguities for these observables or their OPEs; moreover, the OPE product in the B/2 ring is simply the wedge product on the forms.

Recall that we also wish to restrict attention to compact LG theories, i.e. those where the locus of simultaneous vanishing of the J_A is a collection of points. In that case the dimension of the ideal $J = \langle J_1, \ldots, J_N \rangle$ is zero, and therefore (3.5.22) reduces to

$$H_k(K_\bullet, J_\llcorner) = \begin{cases} 0 & \text{for } k > N - n \\ \text{non-zero} & \text{for } k \leq N - n . \end{cases} \tag{3.5.36}$$

This gives a bound on the largest k (or, equivalently, smallest $\overline{q}_0$ charge) at which the B/2 ring is non-trivial.

3.5.3 Parameter Dependence and the LG Singular Locus

The ideal J will typically depend on complex parameters in the J_A, and some of these may correspond to marginal deformations of the IR SCFT. As we saw in the previous chapter, it is not so easy to determine the dependence of SCFT correlation functions on the marginal couplings. The most remarkable and compelling aspect of the Lagrangians we study here and in subsequent chapters is the technology to find this dependence for a subset of the correlation functions and/or OPE coefficients.

We will soon see that the OPE coefficients of operators in the $H_{B/2}$ topological heterotic ring have a meromorphic dependence on the parameters. More generally, we expect that for small variations around a generic point in the parameter space physical quantities will vary smoothly, but for a sufficiently large deformation the theory may become singular. The subvariety in parameter space where this takes place is known as the singular locus of the theory. In an SCFT constructed as the IR limit of a LG theory a singularity will arise if the ideal J fails to be zero dimensional. That means the superpotential has a classical flat direction, and we expect that the IR SCFT will also exhibit a pathology. We will see below that the half-twisted correlation functions are well-behaved if and only if the ideal J is zero dimensional. The locus in parameter space where this is not the case is known as the B/2 singular locus.

It seems sensible that the only singular points in the LG parameter space (i.e. points where the IR SCFT is pathological) are associated to parameter values where dim $J > 0$, but there is no general proof; the explicit computation of the half-twisted correlation functions is probably the strongest piece of evidence for this conjecture.

The B/2 singular locus has an elegant algebraic characterization. When $N = n$ it is characterized by

$$\det_{A,a} J_{A,a} = 0 \in R/J \,, \tag{3.5.37}$$

i.e. the determinant of the matrix constructed from first derivatives of the ideal generators is zero in the ring. This result is familiar to physicists from (2,2) LG theories [283], but it applies more generally with $N = n$; see, e.g. [368].[8] When $N > n$, the algebraic criterion is provided by (3.5.36): for generic parameters the highest non-vanishing Koszul homology group is $H_{N-n}(K_\bullet, J_L)$, while on the singular locus $H_{N-n+1}(K_\bullet, J_L) \neq 0$. In either situation, we can expect the singular locus to be a complex subvariety of complex co-dimension at least 1.

Exercise 3.8 Consider the case $N = n = 3$ and the ideal generators

$$J_1 = \phi_1^2 - \alpha_1 \phi_2 \phi_3 \,, \quad J_2 = \phi_2^2 - \alpha_2 \phi_3 \phi_1 \,, \quad J_3 = \phi_3^2 - \alpha_3 \phi_1 \phi_2 \,, \tag{3.5.38}$$

which depends on three parameters α_1, α_2, and α_3. Compute the chiral ring $R/J = \mathbb{C}[\phi_1, \phi_2, \phi_3]/J$, and show that

$$\det_{a,b} J_{a,b} = 8(1 - \alpha_1 \alpha_2 \alpha_3) \phi_1 \phi_2 \phi_3 \in R/J \,. \tag{3.5.39}$$

Verify directly that J is zero dimensional if and only if $\alpha_1 \alpha_2 \alpha_3 \neq 1$. Thus, $\alpha_1 \alpha_2 \alpha_3 - 1 \subset \mathbb{C}^3$ is the singular locus in this parameter space, and it is itself a non-singular hypersurface. □

Exercise 3.9 Compute the B/2 ring for the simplest ideal of all: $J_A = \sum_a M_{Aa} \phi^a$, where M is a constant full-rank matrix. [Hint: use an orthogonal transformation to choose a nice basis of fields so that $J_a = m_a \phi^a$ for $A = a$ and $J_A = 0$ for $A > n$.] You should find $H_k \simeq \wedge^k \mathbb{C}^{N-n}$. □

3.6 B/2 Correlation Functions via Localization

The chiral algebra leads to a scale-invariant OPE, and we might expect an interesting structure to be contained in the correlation functions

$$\langle \mathcal{O}_1(z_1) \cdots \mathcal{O}_k(z_k) \rangle \,. \tag{3.6.1}$$

---

[8]The latter reference, devoted to residue currents, contains many algebraic notions that should be useful to a (0,2) LG explorer.

A moment's thought shows that these correlation functions are, after all, not so interesting: the $\mathcal{O}_i$ must have $\overline{q} \geq 0$, so as long as at least one operator has non-zero $\mathfrak{u}(1)_R$ charge, this correlation function will vanish.

In order to obtain a non-zero answer for correlation functions we can perform a twist of the theory. This notion, introduced in [385] in four dimensions, and in [386] in two dimensions, led to a revolution in quantum field theory and especially in its connections to mathematics. There are many reviews of the subject, for example [237, 387]. The ideas relevant for our discussion are presented clearly and concisely in [390]. We will now illustrate them in the LG setting; later, we will revisit this construction from a more general point of view.

The starting point goes back to the basic notions from Chap. 1: upon continuation to Euclidean space the Lorentz symmetry becomes the rotation symmetry, and its action on any field is determined by the field's spin. Any LG theory also has the $U(1)_B$ symmetry, and we therefore have the following assignments for the basic fields:

	ϕ_a	$\overline{\phi}_a$	ψ_a	$\overline{\psi}_a$	γ^A	$\overline{\gamma}_A$
spin	0	0	$-1/2$	$-1/2$	$1/2$	$1/2$
$\mathfrak{u}(1)_B$	0	0	-1	1	1	-1
spin^B	0	0	-1	0	1	0

$$(3.6.2)$$

The last row in the table is just a linear combination of the first two, i.e. it corresponds to the twisted Lorentz current $\mathcal{J}^{\mu}_{\Lambda'}$ defined by (here μ is a worldsheet vector index)

$$\mathcal{J}^{\mu}_{\Lambda'} = \mathcal{J}^{\mu}_{\Lambda} + \tfrac{1}{2}\mathcal{J}^{\mu}_{B} .$$

$$(3.6.3)$$

Up to an overall sign this is the unique combination of Lorentz and $\mathfrak{u}(1)_B$ symmetries that assigns charge 0 to the supercharge $\overline{Q}$ and integral/half-integral charges to the remaining operators.

The half-twisted theory is obtained by declaring $\mathcal{J}^{\mu}_{\Lambda'}$ to be the Lorentz current. Equivalently, we define the theory on a curved worldsheet by turning on a $\mathfrak{u}(1)_B$ background gauge field proportional to the spin connection. From either perspective, we now have a theory equipped with a nilpotent scalar (with respect to spin^B) operator $\overline{Q}$, and we project onto its cohomology to obtain the (infinite-dimensional) spectrum of the half-twisted theory. The preceding discussion shows that we can further truncate the cohomology to the finite-dimensional B/2 ring. We call the half-twisted theory with this B/2 projection the B/2 twisted theory.[9]

Anticipating the BRST projection onto $\overline{Q}$ cohomology, we rewrite the LG action as

---

[9]Since the twist and BRST operator are identical to that of the full half-twisted theory, a better but longer term might be the "B/2 heterotic ring projection of the half-twisted model."

$$S_t = \frac{1}{2\pi} \int d^2z \left\{ t \partial \overline{\phi}^a \partial \phi^a + t \overline{\psi}^a \partial \psi^a + \overline{\gamma}_A \overline{\partial} \gamma^A - \gamma^A J_{A,a} \psi^a \right.$$

$$\left. - \overline{\psi}^a \overline{J}^A_{,a} \overline{\gamma}_A + J_A \overline{J}^A \right\}$$

$$= \frac{1}{2\pi} \int d^2z \left\{ \gamma^A_z \overline{\partial} \overline{\gamma}_A - \gamma^A_z J_{A,a} \psi^a_{\overline{z}} \right\} + \overline{Q} \cdot \frac{1}{2\pi} \int d^2z \left\{ -t \psi^a_{\overline{z}} \partial \overline{\phi}^a - \overline{\gamma}_A \overline{J}^A \right\}.$$

$$\text{(3.6.4)}$$

This defines a family of actions depending on a parameter t; the original action is obtained at $t = 1$. The subscripts z and $\overline{z}$ on γ^A_z and $\psi^a_{\overline{z}}$ emphasize that on a curved worldsheet we should think of these as components of fermionic holomorphic (γ^A_z) and anti-holomorphic ($\psi^a_{\overline{z}}$) 1-forms. The action S_t, although originally defined on a flat worldsheet, makes sense and remains $\overline{Q}$-closed on a curved worldsheet.

The last sentence deserves a contemplative pause. We propose here that we can define the theory on a curved worldsheet. That in itself is perhaps not so remarkable, at least as far as the classical theory is concerned: there is a conserved energy-momentum tensor, and there should be no trouble in minimally coupling this to worldsheet gravity. The remarkable feature of the half-twisted theory is that the coupling preserves some global supersymmetry! That is a new and perhaps unexpected feature: the action will be supersymmetric for any worldsheet metric g—the supercharge simply leaves the metric invariant. By twisting the theory we got rid of all of the spinors, and, more importantly, the $\overline{Q}$ supercharge transforms as a scalar with respect to the twisted Lorentz symmetry.[10]

To make this explicit, we use coordinates $z, \overline{z}$ on the worldsheet and recall our definition $d^2z = i dz \wedge d\overline{z}$, and we write $\psi^a = \psi^a_{\overline{z}} d\overline{z}$ and $\gamma^A = \gamma^A_z dz$ as our fermionic 1-forms. We can then write the following action on the curved worldsheet equipped with a Kähler metric g:

$$S_t = \frac{i}{2\pi} \int_\Sigma \left\{ \gamma^A \wedge \overline{\partial} \overline{\gamma}_A - \gamma^A \wedge J_{A,a} \psi^a \right\}$$

$$+ \overline{Q} \cdot \int_\Sigma \left[-\frac{t}{2\pi} \psi^a \wedge *_g \partial \overline{\phi}^a - \frac{2}{\pi} *_g \overline{\gamma}_A \overline{J}^A \right], \qquad \text{(3.6.5)}$$

where $*_g$ is the Hodge dual taken with respect to metric g on the worldsheet.[11]

The action is therefore explicitly $\overline{Q}$-closed for all t and all g; moreover, it has a $\overline{Q}$-exact dependence on t, $\overline{J}$, and the Weyl mode of the worldsheet metric.

---

[10]The supercharge Q transforms as a 1-form under the twisted Lorentz symmetry, and the corresponding invariance, if it is to be maintained, must be promoted to a local one on a curved worldsheet.

[11]If this is unclear, the reader may want to review the appendix on geometry and especially Exercise B.23.

We therefore expect that correlation functions of $\overline{Q}$-closed operators will be independent of $t, \overline{J}$ and any Weyl rescaling of the worldsheet metric.

The first term in the action has a holomorphic dependence on the coefficients of monomials in the J_A. In addition, since it also explicitly involves the Dolbeault operator $\overline{\partial}$, it also depends holomorphically on the worldsheet complex structure. This is in line with the structure of the chiral algebra of the theory, which defines a holomorphic CFT. The half-twisted theory allows us to define this holomorphic CFT on any Riemann surface, and we expect its correlation functions of local operators to have a holomorphic dependence on the Riemann surface complex structure, as well as on the insertions of the operators.

Do we expect these properties to hold in the quantum theory? In order to define the path integral some regularization must be introduced, and this will inevitably violate the chiral structure of the classical half-twisted theory.[12] We also know that in general our theory will suffer from global and, if $c \neq \overline{c}$, local gravitational anomalies, and these will have to be reflected in the half-twisted theory. As emphasized in [315, 393], these issues are subtle in half-twisted non-linear sigma models, where global properties of the target space lead to obstructions to quantum conformal invariance. We suspect that these issues do not arise in the regularization of the super-renormalizable LG Lagrangian, but the author is not aware of a complete treatment of these subtleties. It would be interesting to study these questions, even in specific examples, for Riemann surfaces of genus 1 or higher, where the worldsheet metric has a non-trivial complex structure moduli space.

We will restrict attention to computations of correlation functions in the B/2 theory where the worldsheet is taken to be a sphere $\mathbb{P}^1$, and we will assume that we can regulate the theory while preserving $\overline{Q}$-invariance. This means that the B/2 theory correlation functions will be constants: on one hand, they have a meromorphic dependence on the operator insertions, but on the other hand, since they also have a non-singular OPE, each such correlation function must be a holomorphic function on the sphere, in other words, a constant. This is the origin of the term "topological heterotic ring" [3]: the half-twisted theory, when restricted to B/2 observables on the sphere yields correlation functions that are constants.

We can compute these B/2 correlation functions on $\mathbb{P}^1$ by supersymmetric localization. Localization computations have now been carried out in diverse dimensions with various amounts of supersymmetry, and there is an extensive literature on the subject [324]. We will content ourselves with a more heuristic treatment here and in subsequent chapters.

At its heart, supersymmetric localization relies on a beautifully simple idea [390]. Consider an integral over some domain $\mathcal{X}$ of the form

$$Z = \int_{\mathcal{X}} d\phi \, e^{-S[\phi]}, \tag{3.6.6}$$

---

[12]The inevitability is easy to understand—a regularization introduces a lengthscale, and that choice is not compatible with the classical scale invariance.

and suppose $S[\phi]$ and the measure are invariant under the action of a bosonic symmetry group G. If G acts without fixed points on $\mathcal{X}$, then we can reduce the integration to the (perhaps) simpler integral over a smaller domain

$$Z = \mathrm{Vol}(G) \int_{\mathcal{X}/G} d\phi\, e^{-S[\phi]} . \tag{3.6.7}$$

If the action of G on $\mathcal{X}$ has fixed points, we need to treat their neighborhoods more carefully. Now suppose, instead, that our symmetry is fermionic. In that case, $\mathrm{Vol}(G) = 0$, and field configurations that are not fixed by the symmetry do not contribute to the integral. This means that the integral will localize onto the fixed points of the G action on $\mathcal{X}$. If we can arrange things so that the fixed point set is simple, for instance a set of isolated points, then we can often perform the integral by a saddle-point method. In supersymmetric field theory localization is particularly powerful when it can reduce a path integral computation to a finite-dimensional integral; in that happy situation we can use a path integral to obtain exact results.

Let us now see how these general ideas are manifested in the example of B/2 half-twisted LG theory on the worldsheet $\Sigma = \mathbb{P}^1$. The path integral we wish to compute is

$$\langle \mathcal{O}[\omega^{k_1}](z_1) \cdots \mathcal{O}[\omega^{k_n}](z_n) \rangle = \mathcal{N} \int D[\text{fields}] e^{-S_t} \mathcal{O}[\omega^{k_1}](z_1) \cdots \mathcal{O}[\omega^{k_n}](z_n) .$$

$$\tag{3.6.8}$$

Here $\mathcal{N}$ is a normalization constant, S_t is our t-deformed action, and $d[\text{fields}]$ is the measure. The latter is the source of all the subtleties in the computation. In theories with a complicated field space, like non-linear sigma models, the path integral is defined patch by patch in the target space. A global definition of the measure is required, and there can be global obstructions to the existence of a measure with particular properties (say target space diffeomorphism covariance); similar but simpler issues are faced in gauge theories, where they provide the path integral interpretation of anomalous global symmetries. These issues do not arise in the half-twisted LG theory—the target space has a trivial topology, and the UV definition of the theory allows us to work with free fields.

3.6.1 The Two-Dimensional Laplacian and a Measure

As we will now show, we can be quite concrete about the regularization of the measure for the LG theory. Let g be a Kähler metric on the worldsheet $\Sigma = \mathbb{P}^1$ with the usual Laplace operators $\Delta_{\bar{\partial}} = \Delta_{\partial} = \frac{1}{2}\Delta_d$.[13] The non-zero-mode normalized eigenfunctions f_α can be taken to be real and are defined by

$$\Delta_{\bar{\partial}} f_\alpha = \lambda_\alpha^2 f_\alpha , \qquad \int_\Sigma 1 \wedge *_g f_\beta = 0 , \qquad \int_\Sigma f_\alpha \wedge *_g f_\beta = \delta_{\alpha\beta} , \qquad (3.6.9)$$

with $\lambda_\alpha > 0$. Together with the zero modes they yield a complete set of eigenfunctions that can be used to expand all of the fields. Since $H^1(\Sigma, \mathbb{C}) = 0$, there are no harmonic 1-forms, and that means we can use the f_α to expand any 1-form as well: $\lambda_\alpha^{-1} \partial f_\alpha$ and $\lambda_\alpha^{-1} \bar{\partial} f_\alpha$ provide orthonormal bases for the $(1, 0)$ and $(0, 1)$ forms respectively, since

$$\int_\Sigma \lambda_\alpha^{-1} \partial f_\alpha \wedge *_g \overline{\lambda_\beta^{-1} \partial f_\beta} = \lambda_\alpha^{-1} \lambda_\beta^{-1} \int_\Sigma f_\alpha \wedge *_g \overline{\Delta_{\bar{\partial}} f_\beta} = \delta_{\alpha\beta} . \qquad (3.6.10)$$

Thus, we expand the fields as

$$\phi = \phi_0 + \sum_\alpha \phi_\alpha f_\alpha ,$$

$$\overline{\psi} = \overline{\psi}_0 + \sum_\alpha \overline{\psi}_\alpha f_\alpha ,$$

$$\overline{\gamma} = \overline{\gamma}_0 + \sum_\alpha \overline{\gamma}_\alpha f_\alpha ,$$

$$\gamma = \sum_\alpha \gamma_\alpha \lambda_\alpha^{-1} \partial f_\alpha ,$$

$$\psi = \sum_\alpha \psi_\alpha \lambda_\alpha^{-1} \bar{\partial} f_\alpha , \qquad (3.6.11)$$

and we insert this mode expansion into the action. First, we rewrite the action in a slightly different form from above by using properties of the Hodge star operator in two dimensions. Since for $\alpha \in \mathscr{A}^{0,1}(\Sigma)$ $*\alpha = i\alpha$, we can write the action as

$$S_t = S_t^{\text{kin}} + S^J , \qquad (3.6.12)$$

where the free kinetic term is

$$S_t^{\text{kin}} = \int_\Sigma \left\{ t\bar{\partial}\phi^a \wedge *_g \partial\overline{\phi}^a + t\psi^a \wedge *_g \partial\overline{\psi}^a + \gamma^A \wedge *_g \bar{\partial}\overline{\gamma}_A \right\}$$

$$= \sum_\alpha \left\{ t(\lambda_\alpha^2 \phi_\alpha^a \overline{\phi}_\alpha^a + \lambda_\alpha \psi_\alpha^a \overline{\psi}_\alpha^a) + \lambda_\alpha \gamma_\alpha^A \overline{\gamma}_{A\alpha} \right\} . \qquad (3.6.13)$$

---

[13]The reader may want to consult the appendix on geometry to recall some notions and review our conventions for Dolbeault and De Rham cohomologies.

The interaction term will be more complicated since

$$S^J = \int_\Sigma \left\{ J_{A,a} \psi^a \wedge *_g \gamma^A + 2 *_g \left[J_A \overline{J}^A - \overline{\psi}^a \overline{J}^A_{,a} \overline{\gamma}_A \right] \right\} \tag{3.6.14}$$

will mix the modes. Let $V = \int_\Sigma *_g 1$ be the volume of Σ in the metric g. We can then write

$$S^J = \frac{V}{2\pi} \left[J_A(\phi_0) \overline{J}^A(\overline{\phi}_0) - \overline{\psi}_0^a \overline{J}^A_{,a} \overline{\gamma}_{A0} \right]$$
$$+ \{\text{terms with at least two non-zero modes}\} \ . \tag{3.6.15}$$

As a warm-up, we study the exactly solvable theory with

$$J_A = M_{Aa} \phi^a \ , \tag{3.6.16}$$

where M_{Aa} is a full-rank (i.e. rank n) matrix. Now the interaction term is also diagonal in the modes:

$$S^J = \frac{V}{\pi} \left[M_{Aa} \overline{M}_{Ab} \phi_0^a \overline{\phi}_0^b - \overline{\psi}_0^a \overline{M}_{Aa} \overline{\gamma}_{A0} \right]$$
$$+ \sum_\alpha \left[2 M_{Aa} \overline{M}_{Ab} \phi_\alpha^a \overline{\phi}_\alpha^b - 2 \overline{\psi}_\alpha^a \overline{M}_{Aa} \overline{\gamma}_{A\alpha} + M_{Aa} \psi_\alpha^a \gamma^A{}_\alpha \right] \ . \tag{3.6.17}$$

A regulated path integral measure is easy to define. We have

$$D[\text{fields}] = D[\text{fields}]_0 \prod_{\alpha, \lambda_\alpha < \Lambda} \prod_a \frac{i}{2\pi} d\phi_\alpha^a \wedge d\overline{\phi}_\alpha^a d\psi_\alpha^a d\overline{\psi}_\alpha^a \prod_A d\gamma_\alpha^A d\overline{\gamma}_{A\alpha} \ ,$$
$$\tag{3.6.18}$$

and

$$D[\text{fields}]_0 = \prod_a \frac{i}{2\pi} d\phi_0^a \wedge d\overline{\phi}_0^a d\psi_0^a \prod_A d\overline{\gamma}_{A0} \ . \tag{3.6.19}$$

The measure has $U(1)_B$ charge $n - N$ and is $\overline{Q}$-invariant.

It is not hard to evaluate the non-vanishing correlation functions directly, but we will follow a route that will generalize beyond this simple example by using $\overline{Q}$-invariance and therefore t-independence. We observe that the measure is invariant under the redefinition

$$\phi_\alpha^a \to t^{-1/2} \phi_\alpha^a \ , \qquad \psi_\alpha^a \to t^{-1/2} \psi_\alpha^a \ , \qquad \overline{\psi}_\alpha^a \to t^{-1/2} \overline{\psi}_\alpha^a \ . \tag{3.6.20}$$

Using this in the action and taking the $t \to \infty$ limit, we find

$$S_t^{\text{kin}} = \frac{1}{2\pi} \sum_\alpha \left\{ t(\lambda_\alpha^2 \phi_\alpha^a \overline{\phi}_\alpha^a + \lambda_\alpha \psi_\alpha^a \overline{\psi}_\alpha^a) + \lambda_\alpha \gamma_\alpha^A \overline{\gamma}_{A\alpha} \right\} ,$$

$$S^J = \frac{V}{\pi} \left[J_A(\phi_0) \overline{J}^A(\overline{\phi}_0) - \overline{\psi}_0^a \overline{J}_{,a}^A \overline{\gamma}_{A0} \right] + O(t^{-1}). \tag{3.6.21}$$

Now consider also the insertions of the $\overline{Q}$-closed operators that saturate the $U(1)_B$ charge of the measure: $\omega^{A_1 \cdots A_r} \overline{\gamma}_{A_1} \cdots \overline{\gamma}_{A_r}$, where $r = N - n$, and ω is a constant. The non-zero modes make a universal, parameter-independent contribution, and the parameter dependence arises through the integral over the zero modes. That is, up to constants, we have

$$\langle \mathcal{O}[\omega] \rangle = \mathcal{N} \prod_{\alpha, \lambda_\alpha < \Lambda} \lambda^{N-n} \int D[\text{fields}]_0 \, \omega^{A_1 \cdots A_r} \overline{\gamma}_{A_1} \cdots \overline{\gamma}_{A_r} e^{-S_0}$$

$$= \mathcal{N} \prod_{\alpha, \lambda_\alpha < \Lambda} \lambda^{N-n}$$

$$\times V^n \int_{\mathbb{C}^n} d\mu \, \omega^{A_1 \cdots A_r} \epsilon_{A_1 \cdots A_r B_1 \cdots B_n} \overline{J}_{,1}^{B_1} \overline{J}_{,2}^{B_2} \cdots \overline{J}_{,n}^{B_n} e^{-V J_C \overline{J}^C} , \tag{3.6.22}$$

where the bosonic measure on $\mathbb{C}^n$ is (we drop the zero mode subscript)

$$d\mu = \frac{i}{2\pi} d\phi^1 \wedge d\overline{\phi}^1 \wedge \frac{i}{2\pi} d\phi^2 \wedge d\overline{\phi}^2 \wedge \cdots \wedge \frac{i}{2\pi} d\phi^n \wedge d\overline{\phi}^n . \tag{3.6.23}$$

The free theory has the special property that $J_A = M_{Aa} \phi^a$, but we wrote the terms in a more general fashion because we will soon generalize the result.

Exercise 3.10 Evaluate the Gaussian integrals and verify (3.6.22). □

We find that unless $N = n$, the non-zero modes lead to a parameter-independent factor that diverges in the $\Lambda \to \infty$ limit. We can absorb this factor into the normalization $\mathcal{N}$. The volume dependence of the correlation function can be absorbed into a rescaling $\overline{J} \to \frac{\pi}{V} \overline{J}$, and we will soon argue that this is a sensible thing to do.

This free theory is a little boring: we can always choose a basis for our fields so that $M_{Aa} = m_a \delta_{A,a}$ for $A \le n$ and zero otherwise, in which case the non-zero correlation functions will be

$$\langle \mathcal{O}[\omega] \rangle = \int_{\mathbb{C}^n} d\mu \, \widetilde{\omega}_{12 \cdots n} \overline{m}_1 \overline{m}_2 \cdots \overline{m}_n e^{-\sum_a |m_a|^2 \phi^a \overline{\phi}^a} , \tag{3.6.24}$$

where the dual to ω is as defined above:

$$\widetilde{\omega}_{B_1 \cdots B_n} = \frac{1}{r!} \omega^{A_1 \cdots A_r} \epsilon_{A_1 \cdots A_r B_1 \cdots B_n} . \tag{3.6.25}$$

After we perform the Gaussian integral on the bosonic zero modes, we obtain

$$\langle \mathcal{O}[\omega] \rangle = \frac{\widetilde{\omega}_{12 \cdots n}}{m_1 m_2 \cdots m_n} . \tag{3.6.26}$$

The answer has a meromorphic dependence on the parameters in J—in this case the m_a, and the correlation function diverges whenever one of the m_a vanishes. The divergence has a straightforward interpretation—the J potential now has a flat direction, and the divergence is due to the integral over the bosonic zero mode of the corresponding field.

Having gone through the simple exercise, we now see that the large t limit will yield similar simplifications in case of more general J_A; at least formally the contributions from the non-zero modes will once again just yield a parameter-independent divergence which we can absorb into the normalization $\mathcal{N}$. This is the case as long as we are only interested in operators that belong to the $B/2$ ring. As we already saw, these have a non-singular OPE, and it is sensible that the path integral computation of the B/2 correlation functions just reduces to the zero modes. The non-zero half-twisted correlators are those that saturate $U(1)_B$ charge of the measure, i.e. $\langle \mathcal{O}[\omega_1] \cdots \mathcal{O}[\omega_k] \rangle$ with $\omega \in H_\bullet(K_\bullet, J_\llcorner)$ such that $\omega_1 \wedge \cdots \wedge \omega_k \in H_{N-n}(K, J_\llcorner)$, and the explicit form is

$$\langle \mathcal{O}[\omega_1] \cdots \mathcal{O}[\omega_k] \rangle = \int d\mu \, \widetilde{\omega}_{A_1 \cdots A_n}(\phi) \overline{J}_{,1}^{A_1} \cdots \overline{J}_{,n}^{A_n} e^{-\|J\|^2} . \tag{3.6.27}$$

Here $\widetilde{\omega} = *(\omega_1 \wedge \omega_2 \wedge \cdots \wedge \omega_k) \in H^n(K^\bullet, J_\wedge)$, and $\|J\|^2 = \sum_A |J_A|^2$.

This formula gives a complete result for B/2 half-twisted correlation functions in LG theories in terms of a single finite-dimensional integral. This is essentially the only class of theories for which such a general result is known. There are also expressions for the B/2 correlation functions in so-called hybrid theories, but, as of this writing, it is not understood when those expressions yield exact results [82]. We will briefly return to this when we touch on hybrid models. There are also some partial results for A/2 half-twisted gauged linear sigma models, but it is not known how to extend them to arbitrary (0,2) gauge theories. Perhaps the enterprising reader can take these next steps!

3.6.2 Properties of the Half-Twisted Correlation Functions

The B/2 correlation functions should have a meromorphic dependence on the parameters encoded in the potential J_A, and they should not depend on particular representatives of the cohomology classes. These important properties hold for (3.6.27) as long as the integral converges.

Exercise 3.11 Verify these claims. First, show that the correlation function is invariant under a small change $\delta \overline{J}$; you can use $J \wedge \widetilde{\omega} = 0$ to show that the variation yields a total derivative that vanishes whenever the integral converges. Second, show that the correlator is zero whenever $\widetilde{\omega} = J \wedge \widetilde{\eta}$, i.e. whenever the insertion is $\overline{Q}$-exact. As a special case, check that $J_A = 0$ is an operator equation, i.e.

$$\langle J_A \mathcal{O}[\omega_1] \cdots \mathcal{O}[\omega_k] \rangle = 0 . \tag{3.6.28}$$

□

The first statement justifies absorbing the Σ volume factor into $\overline{J}$, as was done above.

There are important special cases of this formula. Consider the case that $N = n$. Now the Koszul cohomology collapses to just the bosonic chiral ring, i.e. R/J, and the general formula simplifies to

$$\langle \mathcal{O}[\omega_1] \cdots \mathcal{O}[\omega_k] \rangle = \int d\mu \; \omega_1(\phi) \cdots \omega_k(\phi) \det_{A,a} \overline{J}^A_{,a} e^{-\|J\|^2} . \tag{3.6.29}$$

Now suppose the zero locus of the ideal J consists of a collection of isolated non-degenerate points in $\mathbb{C}^n$ with coordinates ϕ^a_*. For every zero-dimensional ideal the zero locus will consist of isolated points, but we also assume that the points are non-degenerate, i.e. the $n \times n$ matrix $J^A_a(\phi_*)$ is full rank for every zero. In this case we compute the integral by stationary phase in the limit $\overline{J} \to \infty$ and find an elegant result:

$$\langle \mathcal{O}[\omega_1] \cdots \mathcal{O}[\omega_k] \rangle = \sum_{\phi_*} \frac{\omega_1(\phi_*) \cdots \omega_k(\phi_*)}{\det_{A,a} J^A_a(\phi_*)} . \tag{3.6.30}$$

In simple cases, such as $n \leq 2$, this gives a tractable computational method for evaluating the correlation functions.[14] The summand is a rational function of the roots ϕ_*, and since it is being summed over all of the roots, we will obtain a result that will be a meromorphic function of the parameters that specify the ideal J.

It may be the case that while J defines a zero-dimensional ideal (i.e. the ϕ_* are isolated points), some of the roots are degenerate. This is necessarily the case

---

[14]The author has often used this formula together with Maple's "RootOf" command to painlessly obtain explicit expressions for correlation functions.

whenever J is quasi-homogeneous, since in that case a non-zero root is necessarily part of a 1-dimensional family of zeroes. However, we can always perturb the ideal to split the roots, for instance by adding a term $\Delta J_A = \sum_A t_{Aa}\phi^a$. For suitably generic t_{Aa} the roots will now be isolated, and we can apply the formula (3.6.30). Once the sum is computed, it is safe to take the $t_{Aa} \to 0$ limit. When the undeformed ideal has $\phi_* = 0$ as the unique isolated root, then the number of isolated vacua for generic t_{Aa} has an intrinsic definition—it is given by $\dim_{\mathbb{C}} R/J$ [121].

In the special case that $J_A = \frac{\partial W}{\partial \phi^A}$ for some holomorphic function W—i.e. our LG theory actually has $(2,2)$ supersymmetry and a superpotential $W(\phi)$—the ideal J is known as the Jacobian ideal associated to W. We can now think of $W = 0$ as defining a singular hypersurface in $\mathbb{C}^n$, a subject of much interest in singularity theory. We say that W has an isolated singularity at the origin if $\frac{\partial W}{\partial \phi^a}$ vanish simultaneously only at the origin. In this case $\dim_{\mathbb{C}} R/J$ is known as the Milnor number of the singularity, and R/J is isomorphic to the chiral (c,c) ring of the $(2,2)$ LG theory. The correlation function we determined are then the topological B-model correlation functions of the LG theory.

Whether the J_A can be derived from a superpotential W or not, there is yet another way we can recast the correlation functions when $N = n$, which is known as the Grothendieck residue. We leave this to the references [210, 371], and simply quote the result here:

$$\langle \mathcal{O}[\omega_1] \cdots \mathcal{O}[\omega_k] \rangle = \int_{\Gamma} \frac{\omega_1(\phi) \cdots \omega_k(\phi) d\mu}{J_1(\phi) J_2(\phi) \cdots J_n(\phi)} , \tag{3.6.31}$$

where Γ is a cycle

$$\Gamma = \{\phi \mid |J_a|^2 = \epsilon_a > 0\} \tag{3.6.32}$$

oriented by $d \arg J_1 \wedge \cdots \wedge d \arg J_n > 0$. This residue integral is in general formidable to compute, but it has the advantage that the properties of the correlation functions, such as those explored in Exercise 3.11, are now manifest.

The author is not aware of a manifestly holomorphic form of the correlation functions in the general $N > n$ case. However, there is an important a case where we can compute the integral in (3.6.27) by stationary phase. Consider the case of a quasi-homogeneous ideal J. Suppose that there exists a set

$$\sigma = \{B_1, B_2, \ldots, B_n\} \subseteq \{1, \ldots, N\} , \tag{3.6.33}$$

such that the ideal $K = \langle K_1, \ldots, K_n \rangle$, with $K_i = J_{B_i}$ has an isolated zero at the origin. Since the correlation functions are independent of $\overline{J}$, we can introduce a parameter λ by replacing $\overline{J}_A \to \lambda \overline{J}_A$ for all $A \notin \sigma$, so that (3.6.27) takes the form

$$\langle \mathcal{O}[\omega_1] \cdots \mathcal{O}[\omega_k] \rangle = \int d\mu \left[\widetilde{\omega}_{B_1 \cdots B_n}(\phi) \det_{a,b}(\overline{K}_b^a) + O(\lambda) \right] e^{-\|K\|^2 + O(\lambda)} . \tag{3.6.34}$$

We can now safely take $\lambda \to 0$ because by assumption the integral will converge in that limit, and we recover a formula identical to (3.6.29). We can then obtain (3.6.30) and (3.6.31) by exactly the same logic as in the $N = n$ case, but with the replacement $J \to K$. This simplification is important for "practical" B/2 computations, since in many theories the required subset σ will exist. The λ-independence of the result guarantees that the final answer is independent of a choice of σ, but it may also be that a σ does not exist. For instance, consider the $n = 2$ and $N = 3$ example

$$J_1 = \phi_1 \phi_2^3 , \qquad J_2 = \phi_2(\phi_1^2 - \phi_2) , \qquad J_3 = \phi_1(\phi_1^2 - \phi_2) . \qquad (3.6.35)$$

While the zero locus of $\langle J_1, J_2, J_3 \rangle$ is just the origin, the zero locus of any pair is one-dimensional.

Exercise 3.12 Argue that the B/2 singular locus for a quasi-homogeneous ideal J is the set of parameters where the ideal fails to be zero-dimensional: this is the only way that the integral in (3.6.27) can fail to converge.

More generally, if the ideal is not quasi-homogeneous, argue that the B/2 singular locus is the set of parameters where one or more of the isolated roots wander off to infinity.   □

3.7 Parameters, Redefinitions, and Accidents

Whenever a (0,2) LG theory flows to a (0,2) SCFT the chiral algebra structure and the topological heterotic ring, which can be studied in the UV theory, provide insights into the IR dynamics as long as we can identify the R-symmetry of the SCFT with a UV symmetry. In this section we will show that the qualification is not an idle concern: in many examples a naive identification of the R-symmetry is incorrect.

A (0,2) LG theory with $N > n$ must have some non-trivial IR dynamics simply because $c - \bar{c} = N - n$ is an RG invariant that encodes the local gravitational anomaly. Given a LG theory with a potential with isolated vacua and a manifestly unitary UV theory, we expect the endpoint of the RG flow to be a compact unitary CFT with $c - \bar{c} = N - n$. Does it carry a representation of the chiral algebra and heterotic ring of the UV theory? We will now explore this question and show that the answer can be negative. In a more constructive direction, we will sketch out some conditions under which we expect to have some control on the IR.

3.7.1 Newton Polytopes

Before we dive into the physics, we will make a small interlude to define a useful concept in polynomial algebra, that of a Newton polytope [121, 354]. Consider two polynomials in two variables, say

$$f(x, y) = x^2 + y^2 + xy^2 + xy^3 , \quad g(x, y) = x^3 + y^3 + xy^2 + x^2y . \qquad (3.7.1)$$

Each monomial $x^{m_1} y^{m_2}$ corresponds to a point in a two-dimensional lattice $\mathbb{Z}^2 \subset \mathbb{R}^2$, and we can therefore encode the monomials in f and g by marking all of the lattice points with a non-zero coefficient:

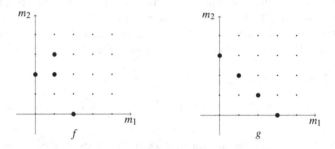

The Newton polytope of the polynomial is obtained by taking the convex hull of all of the lattice points corresponding to non-zero monomials:

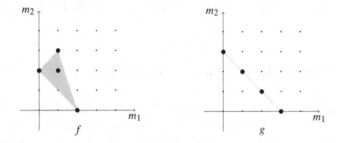

That is, if $\boldsymbol{m}_i$ denote the lattice points with non-zero monomials in a polynomial f, then the Newton polytope of f is the set of all points

$$\Delta(f) = \{y = \sum_i t_i \boldsymbol{m}_i \mid t_i \geq 0, \sum_i t_i = 1\}. \tag{3.7.2}$$

Each polytope is convex by construction, and is in fact the convex hull of its vertices, i.e. the sum on i in the definition can be restricted to the vertices. The polytope may be full-dimensional like $\Delta(f)$, or it may be degenerate, such as $\Delta(g)$. Our examples illustrate a quasi-homogeneous polynomial—g, and a non-quasi-homogeneous polynomial f. A moment's reflection shows that any quasi-homogeneous polynomial is necessarily degenerate, since all of the monomials must lie in a hyperplane defined by the charges. For instance, if $J_A(\phi)$ is a quasi-homogeneous polynomial as in our LG theories, then the lattice points corresponding to the monomials in J_A satisfy

$$\sum_a q_a \boldsymbol{m}_i^a = -Q_A. \tag{3.7.3}$$

It is often useful to think of the charge q_a as a vector $\boldsymbol{q} \in (\mathbb{R}^n)^*$, where $(\mathbb{R}^n)^*$ is the dual space to $\mathbb{R}^n$. Denoting the natural dual pairing by $\langle \cdot, \cdot \rangle$, we express the quasi-homogeneity of J_A by the statement that $\langle \boldsymbol{q}, \boldsymbol{m} \rangle = -Q_A$ for every $\boldsymbol{m} \in \Delta_A \cap \mathbb{Z}^n$.

3.7.2 Naive Parameter Counting

Consider a quasi-homogenous ideal $\boldsymbol{J}$ with an isolated zero at the origin as above. Once the charges q_a and Q_A are fixed, then for each A we can find all of the integer lattice points $\boldsymbol{m} \in \mathbb{R}^n$ that lie in the positive orthant and satisfy $\langle \boldsymbol{q}, \boldsymbol{m} \rangle = -Q_A$. For each such $\boldsymbol{m}$ we define the monomial $\mathsf{M}_{\boldsymbol{m}} = \prod_a (\phi^a)^{m^a}$. The convex hull of these points defines the Newton polytope Δ_A corresponding to the generic J_A, and the full superpotential is

$$\mathcal{W} = \sum_A \Gamma^A \sum_{\boldsymbol{m} \in \Delta_A \cap \mathbb{Z}^n} \alpha_{Am} \mathsf{M}_{\boldsymbol{m}}, \qquad (3.7.4)$$

where the α_{Am} are complex coefficients.

Let $|\Delta_A|$ denote the total number of points in $\Delta_A \cap \mathbb{Z}^n$. The superpotential then depends on $\sum_A |\Delta_A|$ complex parameters. Not all of these will affect the IR dynamics of the theory. To see that, recall the equations of motion for our theory with some fixed choice of J_A (3.2.2), which include

$$\overline{\mathcal{D}} \Gamma^A = \sqrt{2} J_A(\Phi), \qquad \overline{\mathcal{D}} \partial \overline{\Phi}^a = \sqrt{2} \sum_A \Gamma^A J_{A,a}. \qquad (3.7.5)$$

Now consider a small deformation of the superpotential of the form

$$\delta J_A = \sqrt{2} \sum_A \delta M_A^B(\Phi) J_B(\Phi) + \sqrt{2} \sum_a J_{A,a} \delta F^a(\Phi). \qquad (3.7.6)$$

Up to equations of motion the corresponding deformation of the superpotential is therefore a D-term, since

$$\delta \mathcal{W} = \overline{\mathcal{D}} \left[\sum_{A,B} \overline{\Gamma}^A \delta M_A^B J_B + \sum_{A,a} \partial \overline{\Phi}^a \delta F^a(\Phi) \right]. \qquad (3.7.7)$$

Given our discussion of marginal and relevant deformations of (0,2) SCFTs in the previous chapter, it is reasonable to conjecture that the D-term couplings in the LG Lagrangian are irrelevant. Similarly, an F-term that is equivalent up to equations of motion to a D-term, where the D-term coupling is of the form $\mathcal{D} \overline{\mathcal{D}} X$ for some well-defined operator X should also be an irrelevant deformation.[15]

---

[15]We will see applications of similar ideas to (0,2) non-linear sigma models and gauged linear sigma models in the next chapters. Closely related ideas also arise in four-dimensional gauge theories [58].

Thus we expect the δJ_A of the form (3.7.6) will not affect the IR dynamics. This implies that two theories with UV superpotentials $\mathcal{W}(\Gamma, \Phi)$ and $\widetilde{\mathcal{W}}(\widetilde{\Gamma}, \widetilde{\Phi})$ that are related by a holomorphic field redefinition of the form

$$\Phi^a = F^a(\widetilde{\Phi}), \qquad\qquad \Gamma^A = \sum_B M_B^A(\widetilde{\Phi})\widetilde{\Gamma}^B \qquad\qquad (3.7.8)$$

should lead to the same IR dynamics. If we perform the redefinition in the D-terms and the F-terms, then we have done nothing. However, because we assume the D-term couplings are irrelevant, it follows that $\mathcal{W}$ and $\widetilde{\mathcal{W}}$ should flow to the same fixed point.

We can now formulate a first guess at a UV description of the IR SCFT's parameter space. We start with

$$\mathcal{M}_0 = \mathbb{C}^{\sum_A |\Delta_A|} \setminus \{\text{singular locus}\} . \qquad\qquad (3.7.9)$$

The singular locus will in general be a complicated variety, so the algebraic geometry of $\mathcal{M}_0$ will also be formidable. Given the discussion above, we should then also take a quotient by the field redefinitions to obtain a UV description of (a subset of) marginal deformations of the IR SCFT:

$$\mathcal{M}_{\text{uv}} = \frac{\mathcal{M}_0}{\text{group of field redefinitions}} . \qquad\qquad (3.7.10)$$

The geometric meaning of the quotient is far from obvious because the quotient topology will fail to separate points (the quotient will typically not be Hausdorff). When the quotient is by a reductive group, the mathematical technology suited to handle such problems goes by the name of Geometric Invariant Theory (GIT) [313]; for more general groups it may not be possible to construct a moduli space from the quotient, and we may have to work with a moduli stack [363]. We leave these details to the references and the reader's fortitude. It is fair to say that the author's understanding of the physics of these theories and in particular global aspects of the parameter spaces have not yet reached a level where these details begin to play a significant role. Perhaps the reader will do better!

We can also consider additional deformations of the superpotential that break the $U(1)_R$ symmetry. If we established that a superpotential J_A leads to some IR SCFT, then we expect that a deformation

$$\Delta \mathcal{W} = \epsilon \Gamma^A F(\phi) \qquad\qquad (3.7.11)$$

will be relevant if $Q_A + \overline{q}[F] < 0$ and irrelevant if $Q_A + \overline{q}[F] > 0$.

3.7.3 Accidents

One reason that it may be premature to get lost in details of the quotient in $\mathcal{M}_{uv}$ is that the IR SCFT can be very different from what one might naively expect based on c-extremization based on the charges preserved by a generic J_A. One source of this disparity between the naive UV description and the IR is an accidental symmetry [83].

Consider an $N = 3, n = 2$ example from [83]:

$$\mathcal{W}_0 = \begin{pmatrix} \Gamma^1 & \Gamma^2 & \Gamma^3 \end{pmatrix} \begin{pmatrix} \alpha_{11} & \alpha_{12} & \alpha_{13} \\ \alpha_{21} & \alpha_{22} & \alpha_{23} \\ \alpha_{31} & \alpha_{32} & \alpha_{33} \end{pmatrix} \begin{pmatrix} \Phi_1^6 \\ \Phi_2^2 \\ \Phi_1^3 \Phi_2 \end{pmatrix} . \tag{3.7.12}$$

For generic values of the 9 parameters α the potential preserves a unique $U(1)_L$ symmetry, and normalizing the charges as in (3.2.14) leads to $\bar{c} = 3$ and charge assignments

$$\begin{array}{cccc} & \Phi_1 & \Phi_2 & \Gamma^{1,2,3} \\ q & \frac{1}{7} & \frac{3}{7} & -\frac{6}{7} \\ \bar{q} & \frac{1}{7} & \frac{3}{7} & \frac{1}{7} \end{array} \tag{3.7.13}$$

To obtain a description of the parameter space of the IR theory we consider the α modulo field redefinitions consistent with (0,2) SUSY and the $U(1)_L$ symmetry:

$$\Gamma^A \to \sum_B \Gamma^B M_B^A , \qquad \Phi_1 \to x\Phi_1 , \qquad \Phi_2 \to y\Phi_2 + z\Phi_1^3 . \tag{3.7.14}$$

These transformations are invertible if and only if $M \in GL(3, \mathbb{C})$ and $x, y \in \mathbb{C}^*$. The induced action on the Φ monomials is then

$$\begin{pmatrix} \Phi_1^6 \\ \Phi_2^2 \\ \Phi_1^3 \Phi_2 \end{pmatrix} \to S \begin{pmatrix} \Phi_1^6 \\ \Phi_2^2 \\ \Phi_1^3 \Phi_2 \end{pmatrix} , \qquad S = \begin{pmatrix} x^6 & 0 & 0 \\ x^3 z & x^3 y & 0 \\ z^2 & 2yz & y^2 \end{pmatrix} , \tag{3.7.15}$$

and hence the action on the parameters α is $\alpha \mapsto M\alpha S$.

A bit of algebra shows that every non-singular ideal J described by α is equivalent by a field redefinition to one of three superpotentials:

$$\mathcal{W}_1 = \Gamma^1 \Phi_1^6 + \Gamma^2 \Phi_2^2 + \Gamma^3 \Phi_1^3 \Phi_2 ,$$
$$\mathcal{W}_2 = \Gamma^1 (\Phi_1^6 + \Phi_2^2) + \Gamma^2 \Phi_1^3 \Phi_2 ,$$
$$\mathcal{W}_3 = \Gamma^1 \Phi_1^6 + \Gamma^2 \Phi_2^2 . \tag{3.7.16}$$

The UV parameter space is stratified to three points, and we consider each in turn.

1. $\mathcal{W}_1$ has a $U(1)^2$ global symmetry that acts independently on Φ_1 and Φ_2; extremization picks out the following charges.

	θ	Φ_1	Φ_2	Γ^1	Γ^2	Γ^3	
q	0	$\frac{26}{167}$	$\frac{64}{167}$	$-\frac{156}{167}$	$-\frac{128}{167}$	$-\frac{142}{167}$	$\bar{c} = 3\left(1+\frac{2}{167}\right)$
$\bar{q}$	1	$\frac{26}{167}$	$\frac{64}{167}$	$\frac{11}{167}$	$\frac{39}{167}$	$\frac{25}{167}$	

2. $\mathcal{W}_2$ has a free Γ^3 multiplet. The interacting part of the theory has no extra global symmetries and $U(1)_L^{ir} \times U(1)_R^{ir}$ charges

	θ	Φ_1	Φ_2	Γ^1	Γ^2	
q	0	$\frac{4}{31}$	$\frac{12}{31}$	$-\frac{24}{31}$	$-\frac{24}{31}$	$\bar{c} = 3\left(1+\frac{1}{31}\right)$.
$\bar{q}$	1	$\frac{4}{31}$	$\frac{12}{31}$	$\frac{7}{31}$	$\frac{7}{31}$	

3. $\mathcal{W}_3$ has a free Γ^3 multiplet, and the interacting part of the theory is a product of (2,2) minimal models with (2,2) superpotential $W = X_1^7 + X_2^3$ and charges

	θ	Φ_1	Φ_2	Γ^1	Γ^2	
q	0	$\frac{1}{7}$	$\frac{1}{3}$	$-\frac{6}{7}$	$-\frac{2}{3}$	$\bar{c} = 3\left(1+\frac{1}{21}\right)$.
$\bar{q}$	1	$\frac{1}{7}$	$\frac{1}{3}$	$\frac{1}{7}$	$\frac{1}{3}$	

If we assume that there are no accidental symmetries for the $\mathcal{W}_1$, $\mathcal{W}_2$ and $\mathcal{W}_3$ theories, we obtain a consistent picture of the RG flows starting with the UV theory in (3.7.12). There are three basins of attraction; each has a central charge $\bar{c} > 3$, a set of charges consistent with unitarity bounds and no marginal deformations. Moreover, we can construct interpolating RG flows $\mathcal{W}_3 \to \mathcal{W}_2 \to \mathcal{W}_1$ by adding relevant deformations to the superpotentials. However, $\mathcal{W}_1$ has no $U(1)_L$-invariant relevant deformations that make it flow to a putative $\bar{c} = 3$ theory described by $\mathcal{W}_0$.

We conclude that (0,2) LG RG flows have accidental symmetries, and identifying these is key in order to correctly pinpoint even basic properties of the IR theory. In the example at hand no point in the UV parameter space leads to an IR theory with $\bar{c} = 3$. Whenever a theory exhibits this sort of accident, where no point in $\mathcal{M}$ flows to an IR SCFT with the naive chiral algebra, the computations based thereon have no bearing on the IR physics.

A family of LG theories (over)parameterized by $\mathcal{M}$, where a generic point preserves some particular $U(1)_L \times U(1)_R$ symmetry and has a putative fixed value for the central charge $\bar{c}$, will in general admit loci of enhanced symmetry. Suppose we fix to a locus $\mathcal{S} \subset \mathcal{M}$ of enhanced symmetry and use c-extremization to find a candidate central charge $\bar{c}'$. If $\bar{c}' < \bar{c}$, then we know that at least small deformations away from $\mathcal{S}$ are irrelevant. As far as the IR dynamics is concerned, we can then restrict attention to $\mathcal{S}$. So, let us instead assume that $\bar{c}' > \bar{c}$. In this case, the deformation away from $\mathcal{S}$ is relevant, and the endpoint of the flow may be the IR SCFT with central charge $\bar{c}$ and chiral ring determined by the original theory. The precise conditions for when this is the case are not known. A conjecture that, if proven, would partially address this question is put forward in [83]. In [119] an intriguing suggestion is made that the notion of K-stability may be an appropriate condition.

The accidents just described are associated to loci of enhanced symmetry and can be probed by c-extremization. This does not exhaust all possible accidents in IR SCFTs, and we will see examples of that in further chapters; however, it is rather remarkable that we are able to have even this test. After all, IR accidents are, by definition, hard to identify from a UV perspective. Unitarity bounds provide a familiar check for accidents, and their violation has been used to great effect to unravel non-trivial IR dynamics, e.g. [341]. Enhanced symmetry loci and c-extremization give us another tool to explore this complicated phenomenon in two-dimensional QFT, and it would be useful to develop these ideas further.

It is not unreasonable to have the expectation that "in a sufficiently complicated" (0,2) LG theory, where, among other things, the naive parameter dimension is much larger than the dimension of the group of field redefinitions, there should be some locus in the parameter space that does flow to an IR SCFT with the naive $\bar{c}$ and chiral algebra structure. It would be useful to quantify "sufficiently complicated."

We leave this discussion with a final comment about the simpler world of (2,2) LG theories. Those too have loci of enhanced symmetry, but for compact theories they turn out to be rather benign: a non-singular (2,2) superpotential $W(X)$ has an enhanced symmetry if and only if we can divide the fields into two groups $\{X\} \to \{Y\} \cup \{Z\}$ such that $W(X) = W(Y, Z) = W_1(Y) + W_2(Z)$, and in this case the central charge works just as expected: $\bar{c} = \bar{c}_1 + \bar{c}_2$. Thus, the accidents we discuss are a common feature of (0,2) theories but do not afflict the tamer (2,2) world [83].

Chapter 4
Heterotic Non-linear Sigma Models

Abstract In this chapter we discuss the basic features of non-linear sigma models with (0,2) supersymmetry. This is a large universe, and to circumscribe our explorations we mainly stick to the theories relevant to compactifications of the heterotic string. To elucidate the geometric structures it turns out easiest to start with (0,1) supersymmetry. The reader may find it useful to skim through the geometry appendix before diving into the details of this chapter.

4.1 Quick Review of the Bosonic Case

The physics of the non-linear sigma model is intimately connected with the geometry of the target space manifold, and the most important features are already apparent in the bosonic theory. Consider such a theory with n bosonic fields ϕ^μ and action

$$S = \frac{1}{4\pi} \int d^2z \left[g_{\mu\nu}(\phi) + B_{\mu\nu}(\phi) \right] \partial \phi^\mu \bar\partial \phi^\nu . \qquad (4.1.1)$$

This is the form of the action for a flat Euclidean world-sheet $\Sigma = \mathbb{R}^2$; the ϕ^μ are coordinates on the target space $M = \mathbb{R}^n$, or, more precisely, the collection of the ϕ fields determines a map

$$\pi : \Sigma \to M$$

$$\pi : (z, \bar z) \mapsto (\phi^1(z, \bar z), \phi^2(z, \bar z), \ldots, \phi^n(z, \bar z)) . \qquad (4.1.2)$$

The $g_{\mu\nu}$ and $B_{\mu\nu}$ are, respectively, symmetric and anti-symmetric in their indices, and their combination is just the most general rank 2 matrix that can couple to a standard two-derivative term. To have a well-defined kinetic term $g + B$ must be positive for every ϕ, and this will be the case if and only if g is positive. We can

© Springer Nature Switzerland AG 2019

I. V. Melnikov, *An Introduction to Two-Dimensional Quantum Field Theory with (0,2) Supersymmetry*, Lecture Notes in Physics 951,
https://doi.org/10.1007/978-3-030-05085-6_4

attempt to define a quantum non-linear sigma model through the path integral

$$Z = \int [D\Phi] e^{-S} \qquad (4.1.3)$$

over "all maps" from $\Sigma \to M$. In order to be well-defined without extra data, this formulation requires that the measure and the action are invariant under changes of coordinates on the target space. The invariance of the action will follow if g and B transform as rank 2 tensors on M; more precisely, the g and B that appear in the action are given by pulling back said tensors from M by the map π; we will treat the pull-back as understood unless it is likely to cause confusion.

Having made this construction for $M = \mathbb{R}^n$, we easily extend it to a more general target space, which we take to be a smooth compact manifold M.[1] We should then define the Lagrangian for each open patch $U \subset M$, $U \simeq \mathbb{R}^n$. The action (and measure) will then be well-defined on an overlap $U \cap V$ between any two such patches because it is diffeomorphism invariant on each patch.

We can also replace the flat worldsheet Σ with a genus g Riemann surface.[2] This requires a choice of worldsheet metric. The classical non-linear sigma model is conformally invariant on a flat worldsheet and is independent of the Weyl mode of the worldsheet metric on a curved Σ. Quantum corrections can spoil this independence and lead to β-functions for the metric, B-field, and the dilaton [172, 267, 328].[3] The leading order terms in the beta function(al)s are [107, 328, 367]

$$\beta_{\mu\nu}^g = \alpha' \text{Ric}_{\mu\nu} + 2\alpha' \nabla_\mu \nabla_\nu \text{Dil} - \frac{\alpha'}{4} H_{\mu\lambda\rho} H_\nu^{\ \lambda\rho} + O(\alpha'^2) \, ,$$

$$\beta_{\mu\nu}^B = -\frac{\alpha'}{2} \nabla^\lambda H_{\lambda\mu\nu} + \alpha' (\nabla^\lambda \text{Dil}) H_{\lambda\mu\nu} + O(\alpha'^2) \, ,$$

$$\beta^{\text{Dil}} = \frac{n-26}{6} - \frac{\alpha'}{2} \nabla^2 \text{Dil} + \alpha' (\nabla \text{Dil})^2 - \frac{\alpha'}{24} H^2 + O(\alpha'^2) \, . \qquad (4.1.4)$$

The string tension α' is introduced by multiplying the action in (4.1.1) by $\frac{2}{\alpha'}$ and emphasizes that the beta functions will be small provided that curvature terms are small in units of α'. Indices are raised and lowered with the metric g, and ∇ is the Levi-Civita connection for g. Ric is the Ricci tensor for g, and $H = dB$ is the 3-form field strength associated to the B-field. The dilaton beta function includes the constant term from the Weyl anomaly. As discussed in [107], if β^g and β^B

---

[1] We will follow a standard abuse of terminology: for us, unless otherwise noted, *a compact manifold* will mean *a compact manifold without boundary*.

[2] We will always stick to closed Riemann surfaces, but those interested in D-branes and open strings will want to add boundaries and boundary conditions for the fields.

[3] Our predilection for a flat worldsheet has caused us to leave out the dilaton coupling that can be added to the action on a curved worldsheet; qualitatively it is of the form $\Delta S = \int_\Sigma \text{Dil}(\phi) \boldsymbol{R}_\Sigma$, where $\text{Dil}(\phi)$ is the dilaton field and $\boldsymbol{R}_\Sigma$ is the Ricci form on Σ.

vanish to leading order in α', then $\beta^{\text{Dil}} = \text{const} + O(\alpha'^2)$: non-zero β^g and β^B act as obstructions to conformal invariance on a flat worldsheet, and once they vanish $\beta^{\text{Dil}} = \frac{c_{\text{eff}} - 26}{6}$, where c_{eff} is the central charge of the resulting conformal field theory (and which may not equal n, the target space dimension).[4] This result is expected to hold to all orders in the α' expansion.

A Ricci-flat background with $H = 0$ and $\text{Dil} = \text{const}$ is a good starting point for describing a conformal field theory as a weakly coupled non-linear sigma model, but without further assumptions it is not clear when such a non-linear sigma model flows to a nearby conformal field theory. Indeed, since the α' expansion, as any expansion around a free field theory, only defines an asymptotic series, it is not clear in what sense the non-linear sigma model is defined at all. It is sometimes said that we can let the RG flow of the NLSM define the IR fixed point, whether it is strongly-coupled or not. To do so, we define the NLSM with an explicit distance cut-off a and study the theory at scales $\ell \gg a$. This requires us to specify more than just the classical continuum Lagrangian just discussed. The situation is somewhat better for an asymptotically free NLSM, such as the famous $O(N)$ model, where the UV theory can be described by a classical continuum theory. That perspective will be developed further in the next chapter, where we will generate geometries that arise along the RG flow from an asymptotically free gauge theory to the IR.

In the rest of this chapter we will take the pragmatic attitude that the Lagrangian description of a putative fixed point will be useful whenever there is a parameter (it may be continuous, such as the volume of the torus, or discrete, such as the level of a WZW model), which when taken large leads to a weakly coupled theory. When this is not the case, the non-linear sigma model is really a formal construction, and any conclusions about a putative CFT "defined" by such a Lagrangian should be made with care.

In what follows we will discuss circumstances when a classical geometry can be corrected so that the resulting theory (M,g,B,Dil) has zero beta functions. Finding the general set of such geometries is a fundamental and formidable problem in string theory. As a sense of the difficulty, we point out that it is very likely that there are geometries that preserve conformal invariance to all orders in the α' expansion, yet fail to define SCFTs due to non-perturbative corrections. Nevertheless, there is strong evidence that many geometries do describe SCFTs. We will describe many examples in this chapter and next in the context of heterotic string theory, but it is fair to say that while we have many qualitative insights and quantitative computations in special examples, the general structure is still out of reach. Perhaps the enthusiastic reader will unravel it.

---

[4]More precisely, the β function is an obstruction to scale invariance; it is possible to find non-linear sigma models that are scale invariant but not conformally invariant[244], but the examples given are all non-compact. This is consistent with the theorem that a two-dimensional unitary and compact scale-invariant theory is automatically conformal [326]. We will not discuss non-compact non-linear sigma models and so will conflate conformal invariance with scale invariance.

Clearly we glossed over most technical points and details, but one basic one deserves explicit mention. The computation of the beta functions and construction of other physical quantities in the theory relies on being able to regulate the theory while preserving target space diffeomorphism invariance. With this invariance it is sensible and possible to describe the beta functions for the background fields: even though there is formally an infinite number of couplings, they are organized into well-defined tensors by the diffeomorphism invariance. In what follows, we will therefore require that target space diffeomorphism invariance is preserved at the quantum level. We will see that in heterotic sigma models this leads to non-trivial constraints.

4.1.1 Gerbes

We observe that under $B_{\mu\nu} \to B_{\mu\nu} - (\partial_\mu A_\nu - \partial_\nu A_\mu)$ the classical Lagrangian transforms by a total derivative. This means that when we construct the non-linear sigma model on M, the B-field need not be a tensor, and on an overlap it can patch as a connection on an abelian gerbe: on $U \cap V$

$$B^U = B^V - dA^{UV} , \qquad (4.1.5)$$

where A^{UV} is locally a 1-form. The curvature $H = dB$ satisfies

$$H^U = H^V . \qquad (4.1.6)$$

In contrast to B, which we can call a "local" 2-form, H is an honest (or "global") 3-form, and $dH = 0$.

This leads to two developments. First, under a shift of the B-field by a closed 2-form ω on M the action will shift by an additive constant. This is because we may write the B-field action as (verifying this is a good exercise)

$$S_B = \frac{i}{4\pi} \int_\Sigma \pi^*(B) , \qquad (4.1.7)$$

and under a shift $B \to B + \omega$, where $d\omega = 0$, we find

$$S_B \to S_B + \frac{i}{4\pi} \int_\Sigma \pi^*([\omega]) , \qquad (4.1.8)$$

where $[\omega] \in H^2(M, \mathbb{R})$ is the cohomology class of the closed form ω. Thus, we find that the shift only depends on the cohomology class of ω; the exponentiated action e^{-S_B} therefore transforms by a phase, and the phase is trivial if $\omega \in H^2(M, 8\pi^2\mathbb{Z})$. The latter notation is a little bit imprecise, but its meaning is hopefully clear: the cohomology class of $\omega/8\pi^2$ is integral, i.e. $\int_Y \frac{\omega}{8\pi^2} \in \mathbb{Z}$ for any compact (recall that for us that also means boundary-less) submanifold $Y \subset M$.

The second point is that globally B should be thought of as a connection on a $U(1)$ gerbe. The topological data of this gerbe includes the cohomology class of the 3-form H. In any open patch we have $H|_U = dB|_U$, but H need not be globally exact; instead it satisfies the quantization condition

$$[H] \in H^3(M, 8\pi^2\mathbb{Z}) . \tag{4.1.9}$$

This quantization can be understood intrinsically as a Dirac quantization condition for strings in a background B or from the topology of the gerbe [228, 327]. We will content ourselves with a quick argument from [334, 381] that is perhaps familiar to those who pondered the definition of Wilson lines in gauge theory. The idea is to use the global form H to give an alternative definition of S_B. Let T be a 3-manifold with boundary $\partial T = \Sigma$. By extending the map $\pi : \Sigma \to M$ to $\widehat{\pi} : T \to M$, we construct an alternative form of the B-field action that is explicitly invariant under the gerbe shifts:

$$S_{B,T,\widehat{\pi}} = \frac{i}{4\pi} \int_{\widehat{\pi}(T)} H . \tag{4.1.10}$$

While this is gauge-invariant, the action may carry a dependence on the choice of $\widehat{\pi}(T) \subset M$. Since $dH = 0$, small deformations of the map $\widehat{\pi}$ leave the action invariant. However, there may be topologically distinct manifolds $\widehat{\pi}(T)$ and $\widehat{\pi}'(T')$ which share the boundary $\pi(\Sigma)$. The difference in the action is then

$$S_{B,T',\widehat{\pi}'} - S_{B,T,\widehat{\pi}} = \frac{i}{4\pi} \int_X H , \tag{4.1.11}$$

where $X \in M$ is a submanifold in M constructed by gluing $\widehat{\pi}(T)$ and $\widehat{\pi}'(T)$ along the common boundary.[5] This will in general be non-zero, but, if (4.1.9) holds, then the difference is $2\pi i \times$ integer, and therefore the definition is independent of the choice of $\widehat{\pi}$.

Exercise 4.1 Study the classical equations of motion and the B-field. Show that the equations of motion derived from the bosonic NLSM action are

$$\nabla_z \bar{\partial}\phi^\mu - \tfrac{1}{2} H^\mu_{\nu\rho} \partial\phi^\nu \bar{\partial}\phi^\rho = 0 , \tag{4.1.12}$$

---

[5]More precisely, X is a submanifold that represents the homology cycle $H_3(M, \mathbb{Z})$ obtained by gluing $\widehat{\pi}(T)$ and $\widehat{\pi}'(T)$ along the common boundary. It is always possible to represent such a cycle by a smooth submanifold [92].

where indices are raised/lowered with $g^{\mu\nu}$ and $g_{\mu\nu}$, and

$$\nabla_z \bar{\partial}\phi^\mu = \partial\bar{\partial}\phi^\mu + \Gamma^\mu_{\nu\rho}\partial X^\nu \bar{\partial}X^\rho \,,$$

$$H_{\mu\nu\rho} = B_{\mu\nu,\rho} + B_{\rho\mu,\nu} + B_{\nu\rho,\mu} \,. \tag{4.1.13}$$

Γ is the Christoffel connection for the metric g.

Show that it is *not* possible to think of the H coupling as twisting the Christoffel connection for the simple reason that $\nabla_z \bar{\partial}\phi^\mu$ and $\frac{1}{2}H^\mu_{\nu\rho}\partial\phi^\nu \bar{\partial}\phi^\rho$ have opposite reality properties: the first term is manifestly real, while the second is imaginary! This will have implications for topologically non-trivial solutions to the equations of motion, also known as worldsheet instantons.   □

Exercise 4.2 Consider the bosonic sigma model with $H = 0$ on a compact Riemannian manifold M with metric g. Suppose that g has an isometry generated by a Killing vector V, i.e. the Lie derivative of the metric with respect to V is zero: $\mathcal{L}_V g = 0$. Recall that in general the Lie derivative of the metric is defined by[6]

$$(\mathcal{L}_V g)_{\mu\nu} = V^\alpha g_{\mu\nu,\alpha} + V^\alpha_{,\mu}g_{\alpha\nu} + V^\alpha_{,\nu}g_{\mu\alpha} \,. \tag{4.1.14}$$

If this is unfamiliar, then show that $\mathcal{L}_V g = 0$ is equivalent to the Killing equation for V: $\nabla_\mu V_\nu + \nabla_\nu V_\mu = 0$. Next, show that the classical action has a conserved current with components $J_V = V^\mu g_{\mu\nu}\partial\phi^\mu$ and $\bar{J}_V = V^\mu g_{\mu\nu}\bar{\partial}\phi^\mu$ that obey

$$\bar{\partial}J_V + \partial\bar{J}_V = 0 + \text{equations of motion} \,. \tag{4.1.15}$$

Now suppose this non-linear sigma model actually defines a compact and unitary conformal field theory. We then know that the left-moving and right-moving components of the current must be separately conserved. Show that this is possible for J_V if and only if V is a parallel vector field, i.e. $\nabla_\mu V_\nu = 0$. This is a very strong statement, since it indicates that M has a finite cover by $N \times S^1$, where the circle direction is generated by the vector field. The statement fits nicely with results on the classical geometry of Ricci-flat manifolds with isometries, e.g. [255].   □

4.2 (0,1) Superspace and Target Space

Having reviewed the basic structure, we now discuss the non-linear sigma models that are relevant to the heterotic string. The starting point for us will be a free theory

---

[6]More details may be found in the geometry appendix.

with bosonic target space $\mathbb{R}^{1,9}$ and action

$$S = \frac{1}{4\pi} \int d^2z \left[\eta_{\mu\nu} \partial \phi^\mu \bar{\partial} \phi^\nu + \eta_{\mu\nu} \psi^\mu \partial \psi^\nu + \lambda^T \bar{\partial} \lambda \right] . \tag{4.2.1}$$

The λ is a vector of 32 holomorphic Majorana–Weyl fermions. This is a $c = 26, \bar{c} = 15$ theory, which, when coupled to (0,1) supergravity leads to the heterotic string [206, 213, 327]. A GSO projection leads to a supersymmetric and tachyon-free string spectrum in $\mathbb{R}^{1,9}$; this is the standard superstring projection onto $e^{i\pi \bar{F}} = 1$ in the right-moving sector, while in the left-moving sector there are two choices, corresponding either to a $\mathbb{Z}_2$ orbifold on the 32 Majorana fermions, or a $\mathbb{Z}_2 \times \mathbb{Z}_2$ orbifold, with each $\mathbb{Z}_2$ factor acting on groups of 16 Majorana–Weyl fermions. In the former case the massless ten-dimensional spectrum includes $\mathfrak{so}(32)$ gauge fields, while in the latter the gauge algebra is $\mathfrak{e}_8 \oplus \mathfrak{e}_8$. Globally, the gauge group is Spin(32)/$\mathbb{Z}_2$ or $E_8 \times E_8$.

The $\mathbb{R}^{1,9}$ solutions are not the only possible vacua of string theory. If we work in the perturbative string, where the string coupling can be taken to be arbitrarily small, then we can consider more general $(0, 1)$ superconformal field theories with $c = 26, \bar{c} = 15$ that we couple to (0,1) supergravity. We will focus on a simple subset of these, where the theory is a product of a free (0,1) theory describing a $\mathbb{R}^{1,D-1}$ spacetime and k free left-moving fermions and an interacting theory that describes the "compact" theory. This internal theory must have the central charges and modular transformation properties that match those of a (0,1) non-linear sigma model on $\mathbb{R}^n$, $n = 10 - D$, with $r = 32 - k$ Majorana–Weyl fermions, so that we can define for the product theory the same GSO projections as for the free ten-dimensional theory. This internal theory will be taken to be unitary and compact, so that the full string theory has a positive norm on its states and a discrete spectrum of strings propagating on the $\mathbb{R}^{1,D-1}$ spacetime.

4.2.1 Superfields

With that motivation from Minkowski vacua of the perturbative heterotic string, we next focus on (0,1) unitary compact SCFTs defined on the Euclidean plane. Classical Lagrangian theories of this sort are then (0,1) supersymmetric non-linear sigma models. To construct these, we fix a (0,1) superspace with coordinates $z = (z, \bar{z}, \vartheta)$ and superspace derivatives and supercharges (apologies for fraktur)

$$\mathfrak{D} = \partial_\vartheta + \vartheta \bar{\partial} , \qquad\qquad \mathfrak{Q} = \partial_\vartheta - \vartheta \bar{\partial} , \tag{4.2.2}$$

which satisfy

$$\mathfrak{D}^2 = \bar{\partial} , \qquad \mathfrak{Q}^2 = -\bar{\partial} , \qquad \{\mathfrak{D}, \mathfrak{Q}\} = 0 . \tag{4.2.3}$$

A supersymmetry transformation with parameter ξ acts as

$$\delta_\xi z = \delta_\xi(z, \overline{z}, \vartheta) = (0, -\xi\vartheta, \xi) \,, \tag{4.2.4}$$

and a $(0,1)$ supercharge $\boldsymbol{Q}_1$ acts on a superfield X by

$$\delta_\xi X = \xi \boldsymbol{Q}_1 \cdot X \equiv -\xi \mathfrak{Q} X \,. \tag{4.2.5}$$

Since we continue to work with theories without gauge fields or any other elementary fields with spin larger than $1/2$, it is sufficient to consider two types of $(0,1)$ multiplets:

$$\Phi^\mu = \phi^\mu + i\vartheta\,\psi^\mu \quad \text{(bosonic)}, \quad \Lambda^A = \lambda^A + \vartheta L^A \quad \text{(fermionic)}. \tag{4.2.6}$$

The λ^A have negative chirality, and the L^A auxiliary fields are Lorentz scalars; it will be convenient to assemble the Λ^A into an r-dimensional vector $\boldsymbol{\Lambda}$. Both of these are real multiplets with respect to the conjugation described in Chap. 1.

These fields describe the field space in an open patch U of the target manifold M, with the ϕ^μ identified with local coordinates on U, or more precisely, describing the map $\pi : \mathbb{R}^2 \to U$ via $(z, \overline{z}) \mapsto (\phi^1(z, \overline{z}), \phi^2(z, \overline{z}), \ldots, \phi^n(z, \overline{z}))$. To build a sensible non-linear sigma model, we must make sure that the action and local operators do not depend on the coordinates we use to describe these maps, and we would like the coordinate changes to be suitably $(0,1)$ supersymmetry covariant. Given a change of coordinates $\phi^\mu = \phi^\mu(\widehat{\phi}^1, \ldots, \widehat{\phi}^n)$, we can accomplish this for the Φ^μ multiplets by setting

$$\Phi^\mu(\widehat{\Phi}) = \Phi^\mu(\widehat{\Phi}^1, \ldots, \widehat{\Phi}^n) \,. \tag{4.2.7}$$

The components of the superfields then satisfy

$$\phi^\mu(\widehat{\phi}) = \phi^\mu(\widehat{\phi}^1, \ldots, \widehat{\phi}^n) \,, \qquad \psi^\mu = \frac{\partial \phi^\mu}{\partial \widehat{\phi}^{\mu'}} \widehat{\psi}^{\mu'} \,. \tag{4.2.8}$$

The transformations are local on the worldsheet, and that of the fermions involves the same matrix as components of a vector field on M. So, while the bosonic fields transform as coordinates, their superpartners transform as sections of the pullback of the tangent bundle on U. This is very nice, since it immediately tells us how to globally think about the fermions ψ. By definition the coordinates on two overlapping patches $U \subset M$ and $V \subset M$ will be related by a diffeomorphism on $U \cap V$, and that means the fermions ψ have a global meaning:

$$\psi \in \mathscr{A}^0(\Sigma, \overline{K}_\Sigma^{1/2} \otimes \pi^*(T_M)) \,. \tag{4.2.9}$$

The notation $\mathscr{A}^k(M, V)$ means smooth sections of $\wedge^k T_M^* \otimes V$, and $\overline{K}$ is the complex conjugate of the canonical bundle on the worldsheet.[7]

The fermi multiplets Λ can also transform non-trivially on the target space. By analogy with the right-moving fermions, we define λ as

$$\lambda \in \mathscr{A}^0(\Sigma, K_\Sigma^{1/2} \otimes \pi^*(E)), \qquad (4.2.10)$$

where E is a rank r real vector bundle $p : E \to M$. This means there is a cover of M with patches U, V, etc. such that whenever $U \cap V \neq 0$, we can relate the λ via transition functions:

$$\lambda^U = G^{UV}(\phi)\lambda^V, \qquad (4.2.11)$$

and the transition functions are smooth on the overlaps, valued in $\mathrm{GL}(r, n)$, and satisfy the usual consistency conditions for a well-defined vector bundle:

$$G^{UV} G^{VU} = \mathbb{1} \quad \text{for } U \cap V \neq \emptyset,$$

$$G^{UV} G^{VW} G^{WU} = \mathbb{1} \quad \text{for } U \cap V \cap W \neq \emptyset. \qquad (4.2.12)$$

We can extend this to the superfields by taking

$$\Lambda^U = G^{UV}(\Phi)\Lambda^V. \qquad (4.2.13)$$

In components the λ transform covariantly, but the auxiliary fields acquire a non-homogeneous shift reminiscent of the non-homogeneous term in the transformation of a connection on the bundle:

$$L^U = G^{UV} \left(L^V + (G^{UV})^{-1} \frac{\partial G^{UV}}{\partial \phi^\rho} i \psi^\rho \lambda^V \right). \qquad (4.2.14)$$

The superspace derivative $\mathfrak{D}\Lambda$ will have a similar non-homogeneous term in its transformation, and its "covariantization" will involve a connection on the bundle.

Exercise 4.3 Consider the transformations of

$$\mathfrak{D}\Lambda + A_\mu(\Phi)\mathfrak{D}\Phi^\mu \Lambda \qquad (4.2.15)$$

---

[7]The canonical bundle is trivial on the plane or cylinder (or torus), but it is generally non-trivial on a Riemann surface Σ; for instance, on $\Sigma = \mathbb{P}^1$, we have $K_\Sigma = \mathcal{O}(-2)$, so that $\overline{K}_\Sigma^{1/2} = \overline{\mathcal{O}(-1)}$. On the plane the notation is simply a reminder that ψ is a right-moving (aka anti-holomorphic) fermion, but we keep the more general notation since we will want to define twisted versions of this non-linear sigma model on a compact Riemann surface.

under (4.2.13). Show that the combination transforms covariantly provided that A_μ transforms as a connection on the bundle. It may be useful to review the basics of vector bundles and connections in the geometric appendix.   □

4.2.2 The Classical Action

Before we write the general supersymmetric form of the action, we anticipate one point: the kinetic terms for the left-moving fermions λ should be invariant under local frame rotations for the vector bundle $p : E \to M$, and they should be positive in each patch on the target space. That is, the action will include a term of the form

$$S \supset \frac{1}{4\pi} \int d^2z \lambda^T \mathcal{G} \left[\bar{\partial}\lambda + A_\mu \bar{\partial}\phi^\mu \lambda \right] , \qquad (4.2.16)$$

where $\mathcal{G}$ and A are, respectively (the pullbacks of), a metric on E, and a connection on E. We can dispense with $\mathcal{G}$ by working in a local frame at the price of shifting the connection, so that we can take $\mathcal{G} = \mathbb{1}$ in the action.[8] This is just a convenient choice of basis for the sections of E, but the fact that we have a metric on E means that the structure group of the bundle is reduced to be a subgroup of $\mathrm{SO}(r)$, as opposed to the more general $\mathrm{GL}(r)$.[9]

An action $S = \int d^2z d\vartheta \mathcal{L}$ will be supersymmetric, i.e. annihilated by Q_1 provided that the Lagrangian density $\mathcal{L}$ is a superfield, i.e. a function of Φ, Λ, and their derivatives with respect to ∂ and $\mathfrak{D}$ (the latter include $\bar{\partial}$). We impose the following requirements on the action:

1. (0,1) supersymmetry,
2. Lorentz invariance,
3. invariance under diffeomorphisms on M,
4. invariance under local frame rotations for the vector bundle $p : E \to M$,
5. no more than two spatial derivatives,
6. worldsheet scale invariance (this eliminates potential terms of the form $\Lambda^A V_A(\Phi)$).

The most general action that satisfies these requirements is [376]

$$S = \frac{1}{4\pi} \int d^2z d\vartheta \left\{ (g_{\mu\nu}(\Phi) + B_{\mu\nu}(\Phi))\partial\Phi^\mu \mathfrak{D}\Phi^\nu \right.$$
$$\left. - \Lambda^T (\mathfrak{D}\Lambda + A_\mu(\Phi)\mathfrak{D}\Phi^\mu \Lambda) \right\} , \qquad (4.2.17)$$

---

[8]If this is confusing, the reader can take a look ahead to where we explicitly make a similar manipulation for the right-moving fermions.

[9]For reasons that will become clear when we discuss sigma model anomalies, E is an oriented vector bundle; without that assumption the general structure group would have been $\mathrm{O}(r)$.

and the equations of motion are

$$\mathfrak{D}\Lambda = -A_\mu \mathfrak{D}\Phi^\mu \Lambda,$$

$$g_{\nu\rho}\partial\mathfrak{D}\Phi^\rho = -(\Gamma_{\nu\lambda\mu} - \tfrac{1}{2}dB_{\nu\lambda\mu})\partial\Phi^\lambda \mathfrak{D}\Phi^\mu + \tfrac{1}{2}\Lambda^T F_{\nu\mu}\Lambda\mathfrak{D}\Phi^\mu. \qquad (4.2.18)$$

Here

$$\Gamma_{\nu\lambda\mu} = \frac{1}{2}\left(g_{\lambda\nu,\mu} + g_{\mu\nu,\lambda} - g_{\lambda\mu,\nu}\right) \qquad (4.2.19)$$

is the Levi-Civita connection for the metric g with index lowered: $\Gamma_{\nu\lambda\mu} = g_{\nu\rho}\Gamma^\rho_{\lambda\mu}$, $dB_{\nu\lambda\mu}$ are the components of the three-form dB, A is a connection on E, and $F = dA + A^2$ is the curvature of the connection; F is valued in the adjoint of $\mathfrak{so}(r)$ and is therefore anti-symmetric in the vector bundle indices.

The component action, with auxiliary fields L eliminated by equations of motion, is

$$S = \frac{1}{4\pi}\int d^2z \Big\{ (g_{\mu\nu} + B_{\mu\nu})\partial\phi^\mu\bar\partial\phi^\nu$$

$$+ g_{\mu\nu}\psi^\mu\partial\psi^\nu + \partial\phi^\lambda\psi^\mu\psi^\nu(\Gamma_{\mu\lambda\nu} - \tfrac{1}{2}dB_{\mu\lambda\nu})$$

$$+ \lambda^T(\bar\partial\lambda + \bar\partial\phi^\mu A_\mu\lambda) - \tfrac{1}{2}\lambda^T F_{\mu\nu}\lambda\psi^\mu\psi^\nu \Big\}. \qquad (4.2.20)$$

While the kinetic terms for the left- and right-moving fermions appear to have a very different form, we can use a vielbein e^a_μ and its inverse $E^{a\mu}$ to express the action in terms of frame bundle fermions $\psi^a \equiv e^a_\mu\psi^\mu$ with the result

$$g_{\mu\nu}\psi^\mu\partial\psi^\nu + \partial\phi^\lambda\psi^\mu\psi^\nu(\Gamma_{\mu\lambda\nu} - \tfrac{1}{2}dB_{\mu\lambda\nu}) = \boldsymbol{\psi}^T(\partial\boldsymbol{\psi} + \partial\phi^\mu S^-_\mu\boldsymbol{\psi}), \qquad (4.2.21)$$

where $S^\pm$ denote the spin connection ω twisted by $H = dB$:

$$S^{\pm ab}_\lambda = \omega^{ab}_\lambda \pm \tfrac{1}{2}E^{a\sigma}E^{b\nu}H_{\sigma\lambda\nu}. \qquad (4.2.22)$$

This will be useful for some further developments, but it is a little bit awkward from the point of view of manifest (0,1) supersymmetry.[10]

Since we motivated (0,1) non-linear sigma models from the heterotic perspective, we should note that the classical action just constructed cannot describe every heterotic compactification, even if the latter has a large radius limit, where the non-linear sigma model description should be reliable, and the underlying geometry

---

[10]The reader should be aware of some notational differences in the literature: for us it is S^- that shows up in the action, and, as we will see shortly S^+ that shows up in the Green–Schwarz anomaly. Sometimes conventions have it just the other way around.

is a smooth compact manifold M. The issue is in the left-moving sector, which we describe in terms of some weakly coupled fermions. Such couplings do not capture every possible Spin(32)/$\mathbb{Z}_2$ or $E_8 \times E_8$ bundle but rather only those that admit connections that have a regular embedding in $\mathfrak{so}(32)$ or $\mathfrak{so}(16) \oplus \mathfrak{so}(16)$, respectively. There are more sophisticated techniques, which involve WZW models fibered over M, that can accommodate more sophisticated structure groups [147]. Fortunately, even the "smaller set", with a weakly-coupled free fermion description, affords plenty of interesting and mysterious two-dimensional phenomena for our exploration. We will stick to it in what follows.

4.3 Sigma Model Anomalies

The classical non-linear sigma model we just introduced has chiral fermions, and therefore may suffer from anomalies. We discussed in Sect. 2.5 the coupling of Kac–Moody currents to a background gauge field A and found that this is in general anomalous: the partition function $Z[A]$ obtained by integrating out the fermions may not be gauge-invariant.

The fermion sector of the non-linear sigma model we just constructed has some qualitatively similar features, since we can think of the fermions as free fields coupled to background gauge fields, and at first blush we need not worry that the gauge fields are themselves complicated composites of the bosonic fields. For instance, we start with the free theory of r free fermions λ^A, which realizes an $\mathfrak{so}(r)$ Kac–Moody symmetry with currents $J^{AB} = i\lambda^A\lambda^B$ and $A > B$. We then couple these currents to the gauge field $A_{\bar{z}}^{AB} = A_\mu^{AB}(\phi)\bar{\partial}\phi^\mu$ and obtain the desired kinetic term of the non-linear sigma model's left-moving fermions. A parallel story may be told for the right-moving fermions.

These fermionic tales are naive since they neglect the non-trivial geometric facts encoded in

$$\lambda \in \mathscr{A}^0(\Sigma, K^{1/2} \otimes \pi^*(E)), \qquad \psi \in \mathscr{A}^0(\Sigma, \overline{K}^{1/2} \otimes \pi^*(T_M)). \qquad (4.3.1)$$

These structures cannot be ignored when $\pi(\Sigma)$ is topologically non-trivial, and we will remark on them further when we discuss worldsheet instantons. Nevertheless, the naive picture is useful in the realm of sigma model perturbation theory around the topologically trivial map where the worldsheet is mapped to a point: $\pi(\Sigma) = x \in M$. We should at least be able to perturbatively define the path integral for this class of maps, and that involves couplings of the sort described above, where to a first approximation we can treat $A_\mu^{AB}(\phi)$ and other connections (and metrics) by expanding ϕ as a constant plus quantum fluctuations. This leads us to the background field method and the identification of sigma model anomalies.

Sigma model anomalies were discussed some time ago in higher dimensional quantum field theories in [308] and have resurfaced in various guises. In recent years, for instance, they have shown up in discussions of quantum field theory considered on curved backgrounds; see, e.g. [342] and references therein. In higher dimensional theories we often find a space of vacua parameterized by the vacuum expectation values of scalar fields, and [308] showed that it is in general not possible to consistently define the fermion partition function as a function over the space of vacua (it is in general a section of a bundle, and that leads to a pathology in the theory's definition).

In two dimensions the situation is more pressing, since there is no such notion as a space of vacua labeled by expectation values of scalar fields. Infrared fluctuations are strong, and we cannot localize the scalar fields in field space. This is a generalization of the famous result that there are no Goldstone bosons in two dimensions [117, 229, 304]. As a result, we must insist that the integral over the fermions produces a function of the background bosonic scalars. This means it must be invariant under gauge transformations that shift the connection A_μ and local Lorentz transformations that shift the spin connection ω.[11]

If we are interested in applications to heterotic string theory, it is essential that whatever choices we make to achieve this, they should preserve $(0,1)$ supersymmetry. Following [239, 245, 246], we will see this is the case when (M, E) satisfy a topological criterion.

4.3.1 The Background Field Expansion

There are three basic steps that lead to the isolation of the sigma model anomaly through the supersymmetric background field expansion. These are nicely presented in [267], and these same tools can be used to compute the perturbative β-functions and various observables in sigma model perturbation theory. The presentation here is taken almost word for word from [301].

First, we split the fields into a background and quantum contributions, using geodesic normal coordinates. We then expand the action about a background that satisfies the classical equations of motion, keeping terms quadratic in the quantum fields. This is sufficient to compute the effective action to quadratic order in A and $\mathcal{S}^+$. The necessary methodology is well described in [267].

Second, we evaluate the quadratic contributions to the effective action. As these are one-loop computations, there is no need for supergraph machinery; instead, we compute directly using superspace OPEs, taking care to regularize divergences and evaluating contributions from certain canonical contact terms. The latter were described in [204].

---

[11] There is a relation between local Lorentz anomalies and diffeomorphism anomalies—there are local counter-terms that can be added to shift the Lorentz anomaly into a diffeomorphism anomaly. The former is simpler to work with, so we will stick to that. A relevant discussion may be found in the timeless [10, 11].

Finally, by using the background equations of motion, we isolate the non-covariant terms. To analyze the anomaly we can drop the explicitly covariant terms, and this is a significant source of simplification. We then check that the gauge variation of these terms can be canceled by adding a local counter-term and redefining the gauge transformation properties of the B-field.

Covariant Background Superfields

Let $\widetilde{\Phi}(s)$ and $\widetilde{\Lambda}(s)$ denote a one-parameter family of $(0,1)$ superfields with derivatives

$$\Sigma_s \equiv \frac{d}{ds}\widetilde{\Phi}(s), \qquad \mathcal{X}_s \equiv \frac{d}{ds}\widetilde{\Lambda} + \Sigma^\mu \mathcal{A}_\mu(\widetilde{\Phi})\widetilde{\Lambda} \tag{4.3.2}$$

that satisfy the parallel transport equations

$$\dot{\Sigma}_s^\lambda + \Gamma_{\mu\nu}^\lambda(\widetilde{\Phi})\Sigma_s^\mu \Sigma_s^\nu = 0, \qquad \nabla_s \mathcal{X}_s = 0, \tag{4.3.3}$$

with ∇_s the covariant derivative constructed with the gauge connection $\mathcal{A}(\widetilde{\Phi})$. The background fields (Φ, Λ) specify the initial values $\widetilde{\Phi}(0) = \Phi$ and $\widetilde{\Lambda}(0) = \Lambda$, and we take the quantum fields to be $\Sigma \equiv \Sigma_{s=0}$ and $\mathcal{X} \equiv \mathcal{X}_{s=0}$. With this in mind, we obtain the action for the fluctuations by solving the geodesic equations in a power-series around $s = 0$ and expanding

$$S(\widetilde{\Phi}, \widetilde{\Lambda}) = \sum_{n=0}^\infty \frac{s^n}{n!} S_n(\Phi, \Lambda; \Sigma, \mathcal{X}). \tag{4.3.4}$$

The nth term is the $O(n)$ term in the expansion of the action in the fluctuating fields. The great virtue of this "geodesic expansion", appreciated early on [12, 172], is that the resulting quantum action is explicitly target space diffeomorphism-invariant. As emphasized in [267], extracting the terms order by order is greatly simplified by using a covariant derivative and not the naive d/ds. If we assume that the background fields satisfy the classical equations of motion (4.2.18), then the $O(s)$ terms vanish, and the leading terms in the expansion of (4.2.17) have the action $S_2 = \frac{1}{4\pi}\int d^2z d\vartheta \mathcal{L}_2$ with

$$\mathcal{L}_2 = g_{\alpha\beta}D_{\bar{z}}^-\Sigma^\alpha D_\vartheta^+\Sigma^\beta$$

$$+\Sigma^\alpha\Sigma^\beta\partial\Phi^\mu\mathfrak{D}\Phi^\nu\left[R_{\mu\alpha\beta\nu} + \frac{1}{2}\nabla_\alpha H_{\beta\mu\nu} + \frac{1}{4}H_{\gamma\mu\alpha}H_{\delta\nu\beta}g^{\gamma\delta}\right]$$

$$-\mathcal{X}^T D_\vartheta \mathcal{X} + 2\mathfrak{D}\Phi^\mu\Sigma^\nu\mathcal{X}^T F_{\nu\mu}\Lambda + \frac{1}{2}\Sigma^\nu D_\vartheta\Sigma^\mu\Lambda^T\mathcal{F}_{\nu\mu}\Lambda$$

$$+\frac{1}{2}\mathfrak{D}\Phi^\mu\Sigma^\nu\Sigma^\lambda\Lambda^T\nabla_\lambda F_{\nu\mu}\Lambda, \tag{4.3.5}$$

where

$$D_\vartheta \mathcal{X} = \mathfrak{D}\mathcal{X} + \mathfrak{D}\Phi^\mu A_\mu \mathcal{X}, \qquad (4.3.6)$$

$H \equiv dB$, and

$$D_{\bar{z}}^- \Sigma^\alpha = \partial \Sigma^\alpha + \partial \Phi^\mu (\Gamma^\alpha_{\mu\gamma} - \tfrac{1}{2} H^\alpha_{\mu\gamma}) \Sigma^\gamma,$$

$$D_\theta^+ \Sigma^\beta = \mathfrak{D}\Sigma^\beta + \mathfrak{D}\Phi^\nu (\Gamma^\beta_{\nu\delta} + \tfrac{1}{2} H^\beta_{\nu\delta}) \Sigma^\delta. \qquad (4.3.7)$$

The final step is to re-express the Σ^μ in terms of the more convenient frame bundle fields Σ^a. We introduce a vielbein e^a_μ and its inverse $E^{a\mu}$ such that $g_{\mu\nu} = e^a_\mu e^a_\nu$ and write the action in terms of $\Sigma^a = e^a_\mu \Sigma^\mu$. The result is

$$\mathcal{L}_2 = (\partial \Sigma^a + \partial \Phi^\lambda S^{-ab}_\lambda \Sigma^b)(\mathfrak{D}\Sigma^a + \mathfrak{D}\Phi^\mu S^{+ac}_\mu \Sigma^c) + \Sigma^a \Sigma^b \partial \Phi^\mu \mathfrak{D}\Phi^\nu R^+_{\mu abv}$$

$$- \mathcal{X}^T (\mathfrak{D}\mathcal{X} + \mathfrak{D}\Phi^\lambda A_\lambda \mathcal{X}) + 2\mathfrak{D}\Phi^\mu \Sigma^a \mathcal{X}^T F_{a\mu} \Lambda$$

$$+ \tfrac{1}{2}\Sigma^a (\mathfrak{D}\Sigma^b + \mathfrak{D}\Phi^\lambda \omega^{bc}_\lambda \Sigma^c) \Lambda^T F_{ab} \Lambda + \tfrac{1}{2}\mathfrak{D}\Phi^\mu \Sigma^a \Sigma^b \Lambda^T \nabla_b F_{a\mu} \Lambda, \qquad (4.3.8)$$

where

$$R^+_{\mu abv} = E^\alpha_a E^\beta_b \left[R_{\mu\alpha\beta\nu} + \frac{1}{2}\nabla_\alpha H_{\beta\mu\nu} + \frac{1}{4} H_{\gamma\mu\alpha} H_{\delta\nu\beta} g^{\gamma\delta} \right], \qquad (4.3.9)$$

ω is the torsion-free, metric compatible spin connection, and, as in (4.2.22),

$$S^{\pm ab}_\lambda = \omega^{ab}_\lambda \pm \tfrac{1}{2} E^{a\sigma} E^{bv} H_{\sigma\lambda\nu}. \qquad (4.3.10)$$

The Quadratic Effective Action

Having written down the quadratic action, we are ready to compute the one-loop corrections to the effective action that are quadratic in the background fields A, $S^\pm$ and ω. This is a very special set of terms because we can compute them just by considering the terms in $\mathcal{L}_2$; we do not need the $O(s^3)$ or higher terms in the quantum action.

We expand around the free theory with action

$$S_{\text{free}} = \frac{1}{4\pi} \int d^2 z d\vartheta \left[\partial \Sigma^a \mathfrak{D}\Sigma^a - \mathcal{X}^T \mathfrak{D}\mathcal{X} \right]. \qquad (4.3.11)$$

The super OPEs

$$\Sigma^a(z_1)\Sigma^b(z_2) \sim -\delta^{ab}\log(z_{12}(\bar{z}_{12}-\vartheta_1\vartheta_2)),\qquad \mathcal{X}^A(z_1)\mathcal{X}^B(z_2) \sim \frac{\delta^{AB}}{z_{12}}$$

$$(4.3.12)$$

determine all correlators by Wick's theorem. It is a familiar fact that sufficiently singular functions of z_{12} are non-holomorphic due to contact terms (e.g. $\bar{\partial}_1 z_{12}^{-1} = 2\pi\delta^2(z_{12},\bar{z}_{12})$); similarly, they also carry a ϑ dependence if we wish them to be supersymmetric [204]. That is, if $\mathfrak{Q}_{1,2}$ denote the supercharge derivatives at $z = z_1$ and $z = z_2$ respectively, then

$$\xi(\mathfrak{Q}_1 + \mathfrak{Q}_2)\frac{1}{z_{12}} = 0 \implies \partial_{\vartheta_1}\frac{1}{z_{12}} = -2\pi\vartheta_2\delta^2(z_{12},\bar{z}_{12}).\qquad(4.3.13)$$

In fact, one can define a ϑ-independent "principal part" of z_{12}^{-1} by

$$\frac{1}{z_{12}} = \mathsf{P}\frac{1}{z_{12}} - 2\pi\vartheta_1\vartheta_2\delta^2(z_{12},\bar{z}_{12}).\qquad(4.3.14)$$

An important consequence for what follows is

$$\mathfrak{D}_1 z_{12}^{-1} = 2\pi(\vartheta_1 - \vartheta_2)\delta^2(z_{12},\bar{z}_{12}).\qquad(4.3.15)$$

Exercise 4.4 Verify the superspace acrobatics in the previous three equations. □

To express the interaction Lagrangian of (4.3.8) succinctly, we introduce a shorthand for various pull-backs from the target space; for example, $\mathcal{S}_\vartheta^{\pm ab} \equiv \mathfrak{D}\Phi^\mu \mathcal{S}_\mu^{\pm ab}$, $\mathcal{S}_z^{\pm ab} \equiv \partial\Phi^\mu \mathcal{S}_\mu^{\pm ab}$, etc. With this notation the interaction terms linear in the background are

$$\mathcal{L}_{\text{int}} = \partial\Sigma^a \mathcal{S}_\vartheta^{+ab}\Sigma^b + \mathfrak{D}\Sigma^a(\mathcal{S}_z^{-ab} - \tfrac{1}{2}\Lambda^T F_{ab}\Lambda)\Sigma^b - \mathcal{X}^T \mathcal{A}_\theta\mathcal{X}$$
$$- 2\Sigma^a\mathcal{X}^T F_{a\theta}\Lambda + \Sigma^a\Sigma^b(R_{z(ab)\vartheta}^+ - \tfrac{1}{2}\Lambda^T\nabla_{(b}F_{a)\vartheta}\Lambda).\qquad(4.3.16)$$

At quadratic order, the terms in the first line have no non-trivial contractions with those in the second line.[12] Since the contractions among terms from the second line yield explicitly covariant terms, we can concentrate on the quadratic terms due to

$$\mathcal{L}_{\text{int}}' = \partial\Sigma^a \mathcal{S}_\vartheta^{+ab}\Sigma^b + \mathfrak{D}\Sigma^a\mathcal{T}^{ab}\Sigma^b - \mathcal{X}^T \mathcal{A}_\theta\mathcal{X},\qquad \mathcal{T}^{ab} \equiv \mathcal{S}_z^{-ab} - \tfrac{1}{2}\Lambda^T F_{ab}\Lambda.$$
$$(4.3.17)$$

---

[12]Either a full contraction is impossible, or it is zero due to symmetry properties under $a \leftrightarrow b$.

At quadratic order in the background, the possible contractions of these interactions yield either $O(A^2)$ or $O(S_+^2)$ terms; we consider these in turn.

The $\mathcal{X}$ Contributions

The $O(A^2)$ correction to the partition function is

$$\Delta Z_{\mathcal{X}} = \frac{1}{2} \int \frac{d^2z_1 d^2z_2 d\vartheta_2 d\vartheta_1}{(4\pi)^2} \langle \mathcal{X}_1^T A_{1\vartheta} \mathcal{X}_1 \times \mathcal{X}_2^T A_{2\vartheta} \mathcal{X}_2 \rangle, \qquad (4.3.18)$$

where the correlator is to be evaluated with free field OPEs.

Exercise 4.5 Evaluate the correlation function and show that this leads to a contribution to the effective action

$$\Delta S_{\mathcal{X}} = -\int \frac{d^2z_1 d^2z_2 d\vartheta_2 d\vartheta_1}{(4\pi)^2} \frac{\mathrm{tr}\{A_{1\vartheta} A_{2\vartheta}\}}{z_{12}^2}. \qquad (4.3.19)$$

Here the notation "tr{...}" refers to taking the trace in the fundamental representation of $\mathfrak{so}(r)$, and the subscripts 1 and 2 refer to the superspace insertion of the field; thus $A_{1\vartheta} \equiv A_\mu(\Phi(z_1))\mathfrak{D}_1\Phi(z_1)$, and $\mathfrak{D}_1 = \partial_{\vartheta^1} + \vartheta^1\partial_1$. □

The reader should compare this to what we obtained in Sect. 2.5; while this result is technically more complicated and written in superspace, it is qualitatively just what we had computed there.

The Σ Contributions

The $O(S_+^2)$ terms are somewhat more involved. The main complication is due to the logarithm in the $\Sigma_1\Sigma_2$ OPE. The resulting logarithms lead to IR divergences in the $z_{1,2}$ integrals. To handle these we regulate the OPE in a supersymmetric manner. Introducing the supersymmetric invariants $\vartheta_{12} \equiv \vartheta_1 - \vartheta_2$ and $\zeta_{12} \equiv \bar{z}_{12} - \vartheta_1\vartheta_2$, we take the regulated two-point function to be

$$\langle \Sigma_1^a \Sigma_2^b \rangle = -\delta^{ab}\Delta_{12}, \qquad \Delta_{12} \equiv \log(z_{12}\bar{\zeta}_{12} + \ell^2), \qquad (4.3.20)$$

where ℓ is a regulating lengthscale. This is still explicitly supersymmetric because

$$R \equiv z_{12}\bar{\zeta}_{12} + \ell^2 \qquad (4.3.21)$$

is annihilated by $(\mathfrak{Q}_1 + \mathfrak{Q}_2)$. With this regulator, we obtain

$$\Delta S_{\Sigma} = \int \frac{d^2z_1 d^2z_2 d\vartheta_2 d\vartheta_1}{(4\pi)^2} \left[\frac{1}{2} \mathrm{tr}\{S_{1\vartheta}^+ S_{2\vartheta}^+\}X + \mathrm{tr}\{S_{1\vartheta}^+ T_2\}Y + \frac{1}{2}\mathrm{tr}\{T_1 T_2\}Z \right],$$

$$(4.3.22)$$

where "tr$\{\cdots\}$" is the trace in the fundamental of $\mathfrak{so}(n)$, the Lie algebra of the structure group of the tangent bundle, and

$$X = \frac{1}{2}\partial_1\partial_2\Delta_{12}^2 - 2\partial_1\Delta_{12}\partial_2\Delta_{12},$$

$$Y = -\frac{1}{2}\partial_1\mathfrak{D}_2\Delta_{12}^2 + 2\Delta_{12}\partial_1\mathfrak{D}_2\Delta_{12},$$

$$Z = \Delta_{12}\mathfrak{D}_1\mathfrak{D}_2\Delta_{12} - \mathfrak{D}_1\Delta_{12}\mathfrak{D}_2\Delta_{12}. \tag{4.3.23}$$

To simplify these terms, we first note that since $\mathfrak{D}_1\Delta_{12} = z_{12}\vartheta_{12}R^{-1}$, the second term in Z vanishes. The second term in X has a simple $\ell \to 0$ limit:

$$-2\partial_1\Delta_{12}\partial_2\Delta_{12} = \frac{2\bar{\zeta}_{12}^2}{(z_{12}\bar{\zeta}_{12} + \ell^2)^2} \xrightarrow{\ell\to 0} \frac{2}{z_{12}^2}; \tag{4.3.24}$$

while the second term in Y is a UV-divergent local term since

$$\partial_1\mathfrak{D}_2\Delta_{12} = \vartheta_{12}\frac{\ell^2}{(z_{12}\bar{z}_{12} + \ell^2)^2} \xrightarrow{\ell\to 0} 2\pi\vartheta_{12}\delta^2(z_{12}). \tag{4.3.25}$$

Thus, up to a local counter-term, we find $\Delta S_\Sigma = \Delta S_1 + \Delta S_2$ with

$$\Delta S_1 = \int \frac{d^2z_1 d^2z_2 d\vartheta_2 d\vartheta_1}{(4\pi z_{12})^2} \, \mathrm{tr}\{\mathcal{S}_{1\vartheta}^+ \mathcal{S}_{2\vartheta}^+\},$$

$$\Delta S_2 = \int \frac{d^2z_1 d^2z_2 d\vartheta_2 d\vartheta_1}{4(4\pi)^2} \, \mathrm{tr}\{(\partial_1\mathcal{S}_{1\vartheta}^+ - \mathfrak{D}_1\mathcal{T}_1)(\partial_2\mathcal{S}_{2\vartheta}^+ - \mathfrak{D}_2\mathcal{T}_2)\}\Delta_{12}^2. \tag{4.3.26}$$

The second contribution looks complicated, but fortunately we need not consider it. Up to terms of higher order in the background and using the classical equations of motion for Φ and Λ, we find

$$\partial S_\theta^{+ab} - \mathfrak{D}\mathcal{T}^{ab} = \mathfrak{D}\Phi^\mu\partial\Phi^\lambda(d\omega_{\lambda\mu}^{ab} + \tfrac{1}{2}H^{a\ b}_{\ \mu\ ,\lambda} + \tfrac{1}{2}H^{a\ b}_{\ \lambda\ ,\mu}). \tag{4.3.27}$$

This is invariant under the linearized Lorentz transformations, and we expect that incorporation of higher order terms in the background will provide a fully covariant form for ΔS_2. So, the non-covariant terms in the $O(\mathcal{S}_+^2)$ contribution to the one-loop effective action have, up to a minus sign, the same form as ΔS_A, and the combined non-covariant terms are

$$\Delta S = \int \frac{d^2z_1 d^2z_2 d\theta_2 d\theta_1}{(4\pi)^2} \frac{\mathrm{tr}\{\mathcal{S}_{1\vartheta}^+ \mathcal{S}_{2\vartheta}^+\} - \mathrm{tr}\{A_{1\vartheta} A_{2\vartheta}\}}{z_{12}^2}. \tag{4.3.28}$$

4.3.2 Interpretation of the Result

We make one more manipulation on the terms by using $z_{12}^{-2} = \partial_2 z_{12}^{-1}$ and rewriting ∂A_ϑ in a more convenient way up to background fields' equations of motion and higher order terms in A:

$$\partial A_\vartheta = \partial \mathfrak{D}\Phi^\lambda A_\lambda = \mathfrak{D}\Phi^\lambda \partial \Phi^\rho A_{\lambda,\rho} = \mathfrak{D}\Phi^\lambda \partial \Phi^\rho dA_{\rho\lambda} + \mathfrak{D}(A_z). \qquad (4.3.29)$$

Using this ΔS, we therefore have the presentation of the non-covariant terms as

$$\Delta S = \Delta S_A + \Delta S_{S^+} - S_{\text{c.t.}}, \qquad (4.3.30)$$

where the "truly non-local" terms are

$$\Delta S_A = -\int \frac{d^2 z_1 d^2 z_2}{(4\pi)^2 z_{12}} d\vartheta_2 d\vartheta_1 \, \mathrm{tr}\{A_{1\mu} dA_{2\lambda\rho}\} \mathfrak{D}_1 \Phi_1^\mu \mathfrak{D}_2 \Phi_2^\lambda \partial_2 \Phi_2^\rho,$$

$$\Delta S_{S^+} = +\int \frac{d^2 z_1 d^2 z_2}{(4\pi)^2 z_{12}} d\vartheta_2 d\vartheta_1 \, \mathrm{tr}\{S_{1\mu}^+ dS_{2\lambda\rho}^+\} \mathfrak{D}_1 \Phi_1^\mu \mathfrak{D}_2 \Phi_2^\lambda \partial_2 \Phi_2^\rho, \qquad (4.3.31)$$

and $S_{\text{c.t.}}$ is a local term

$$S_{\text{c.t.}} = -\frac{1}{8\pi} \int d^2 z d\vartheta \left[\mathrm{tr}\{A_\mu A_\nu\} - \mathrm{tr}\{S_\mu^+ S_\nu^+\} \right] \partial \Phi^\mu \mathfrak{D}\Phi^\nu. \qquad (4.3.32)$$

Note that $\mathrm{tr}\{\cdots\}$ denotes either the fundamental of $\mathfrak{so}(r)$ or $\mathfrak{so}(n)$, depending on whether the argument is a gauge or Lorentz object. As the name suggests, this contribution is canceled by adding $S_{\text{c.t.}}$, a finite local counter-term, to the action.

While the effective action is explicitly $(0,1)$ supersymmetric, it is not gauge-invariant. The classical fermionic action is invariant under infinitesimal gauge transformations

$$\delta_\epsilon \lambda = \epsilon\lambda, \qquad\qquad \delta_\epsilon A = -\nabla\epsilon = -d\epsilon - [A, \epsilon], \qquad (4.3.33)$$

where the gauge parameter ϵ is pulled back from the target space and valued in the adjoint of $\mathfrak{so}(r)$. Similarly, there is a classical invariance under Lorentz transformations

$$\delta_\kappa \psi = \kappa\psi, \qquad\qquad \delta_\kappa \omega = -\nabla\kappa = -d\kappa - [\omega, \kappa]. \qquad (4.3.34)$$

Following the general discussion above, we must require that this invariance holds for the effective action, but the supersymmetry identity

$$\mathfrak{D}_1 z_{12}^{-1} = 2\pi(\vartheta_1 - \vartheta_2)\delta^2(z_{12}, \bar{z}_{12}) \qquad (4.3.35)$$

shows that under linearized transformations $\delta_\epsilon A = -d\epsilon$ and $\delta_\kappa S^+ = -d\kappa$, the effective action transforms by a local term

$$\delta \Delta S = \frac{1}{8\pi} \int d^2z d\theta (\text{tr}\{\epsilon d A_{\mu\nu}\} - \text{tr}\{\kappa d S^+_{\mu\nu}\})\partial \Phi^\mu \mathcal{D}\Phi^\nu. \qquad (4.3.36)$$

This variation can be canceled by postulating a non-trivial gauge transformation for the B-field:

$$\delta B = -\frac{1}{2}\,\text{tr}\{\epsilon d A\} + \frac{1}{2}\,\text{tr}\{\kappa d S^+\}. \qquad (4.3.37)$$

This is the world-sheet realization of the famous Green–Schwarz anomaly cancelation mechanism.

A key consequence is that the gauge-invariant 3-form cannot be dB but is rather

$$\mathcal{H} = dB - \frac{1}{2}\text{CS}_3(A) + \frac{1}{2}\text{CS}_3(S^+), \qquad (4.3.38)$$

where CS_3 denotes the Chern–Simons 3-form

$$\text{CS}_3(A) \equiv \text{tr}\{AdA + \tfrac{2}{3}A^3\}, \qquad (4.3.39)$$

and similarly for $\text{CS}_3(S^+)$. We only obtained the quadratic terms from the explicit computation, but we expect (and will assume) that inclusion of the higher order terms will lead to the non-linear covariant form. Before we discuss its significance, we will make a short foray into characteristic classes.

4.4 Anomalies and the Heterotic B-Field

We will make a few comments on vector bundles and anomalies relevant for heterotic geometry. The reader who wishes to refresh her memory of some vector bundle notions can take a look in the relevant sections of the geometric appendix.

4.4.1 Characteristic Classes and Descent

The characteristic classes that we review are defined axiomatically in algebraic topology [307]. In physics applications to local field theory and especially those in Lagrangian theories, the classes typically appear through their differential form representatives. Often one can compute a representative, and therefore the class, in some approximation, and the topological rigidity of the class then leads to strong physical constraints that hold beyond the approximation. We will see these features in the (0,1) and (0,2) quantum field theories. At any rate, because of our physics motivation, we will approach the classes from the differential form point of view [94, 161].

Given a rank k vector bundle $p : E \to M$ with structure group G equipped with a connection A and field strength $F = dA + A \wedge A$, we construct the gauge-invariant $2m$-forms $P_m = \text{tr}\, F^m$, where F^m denotes taking the wedge product m times, while contracting the vector bundle indices. Since the connection can be taken to be valued in any representation of G, the trace can also be taken in any representation r of G, so that, really, we have $P_{m,r}$. We will use a typical notation that tr denotes the trace in the fundamental representation of G, while Tr will denote the trace in the adjoint representation.

Each $P_{m,r}$ is a closed form and therefore defines a cohomology class $[P_{m,r}] \in H^{2m}(M, \mathbb{R})$. In fact, by normalizing the form appropriately, we obtain integral cohomology classes! This may at first sound strange, since $P_{m,r}$ seems to depend on the choice of connection A, while integrality of a cohomology class would be a topological constraint. Nevertheless this is exactly the magic of Chern–Weil theory, and the invariants constructed from these cohomology classes partly characterize the topology of the bundle.

A Line Bundle

The simplest case is a complex line bundle $\mathcal{L} \to M$, where the fiber is $\mathbb{C}$, and the structure group is U(1). A connection A on $\mathcal{L}$ is locally a 1-form, and the difference of two connections $A - A'$ is a global 1-form. The corresponding field-strengths $F = dA$ and $F' = dA'$ are gauge-invariant, and their $F - F' = d\xi$ for some imaginary 1-form $\xi \in \mathscr{A}^1(M, \mathbb{C})$. Thus, clearly F defines a cohomology class $[F] \in H^2(M, \mathbb{R})$ that is independent of the choice of connection. We then have the standard result, reviewed in the geometry appendix, that for any submanifold $\iota : C \hookrightarrow M$ that represents a cycle $[C] \in H_2(M, \mathbb{Z})$,

$$\int_C \iota^*(c_1(\mathcal{L})) \in \mathbb{Z} . \tag{4.4.1}$$

We will abbreviate this statement by

$$c_1(\mathcal{L}) = \left[\frac{i}{2\pi} F \right] \in H^2(M, \mathbb{Z}) . \tag{4.4.2}$$

This is not a terrible notation, since every class in $H^2(M, \mathbb{Z})$ can be represented by a smooth submanifold.[13]

---

[13]This fundamental result is due to Hopf and should be understood in the larger context of representability of cycles as described by Thom: for $H_k(M, \mathbb{Z})$ with $\dim M = n$, if $k \leq 6$ or

In this case $P_m = F^m$ clearly satisfies all of the quoted properties, and in every patch $P_m = dQ_{2m-1}$, where Q_{2m-1} is the Chern–Simons form

$$Q_{2m-1} = A \wedge F^{m-1} \,, \tag{4.4.3}$$

which has the gauge transformation

$$\delta_\epsilon Q_{2m-1} = -d\epsilon \wedge F^{m-1} = dQ^1_{2m-2} \,. \tag{4.4.4}$$

This chain of reasoning, that relates the forms P_m, Q_{2m-1} and Q^1_{2m-1}, is known as descent, and it generalizes to any other vector bundle, with P_m as defined above, and local forms Q_{2m-1} and Q^1_{2m-2} defined by $P_m = dQ_{2m-1}$, and $\delta_\epsilon Q_{2m-1} = dQ^1_{2m-2}$. A general and explicit solution to the descent equations can be found in [10], but for low m it is easy to carry it out explicitly by hand, as the following exercise shows.

Exercise 4.6 Carry out the descent procedure for $P_2 = \operatorname{tr} F^2$ of a general vector bundle. Show that

$$\operatorname{tr} F^2 = d\mathrm{CS}_3 \,, \qquad \mathrm{CS}_3 = dA + \tfrac{2}{3}A^2 \,, \qquad \delta_\epsilon \mathrm{CS}_3 = d\operatorname{tr}\{\epsilon dA\} \,. \tag{4.4.5}$$

Let $A^g = g(A - g^{-1}dg)g^{-1}$ for some gauge transformation $g \in \mathrm{Aut}(V)$, and show that

$$\mathrm{CS}_3(A^g) - \mathrm{CS}_3(A) = \frac{1}{3}\operatorname{tr}\{\omega_g^3\} - d\operatorname{tr}\{A\omega_g\} \,. \tag{4.4.6}$$

where $\omega_g = g^{-1}dg$. Use $d\omega_g = -\omega_g^2$ to show that this difference of Chern–Simons forms is closed.

Even on a trivial bundle, which admits the connection $A = 0$, the difference is an object that is familiar from standard discussions of instantons in four-dimensional gauge theory—e.g. in [375]: ω_g is the Maurer–Cartan form, and $\frac{1}{3}\operatorname{tr}\{\omega_g^3\}$ encodes the winding number of the gauge transformation g. If the base manifold is $\mathbb{R}^4$, then g restricted to the three-sphere at infinity defines a homotopy class in $\pi_3(G)$, and $\int_{S^3}\frac{1}{3}\operatorname{tr}\{\omega_g^3\}$ is proportional to the instanton number. $\square$

Chern Classes

Suppose $E \to M$ is a rank k complex vector bundle, i.e. the fibers are $\mathbb{C}^k$. The curvature F is then an anti-Hermitian $k \times k$ matrix of 2 forms valued in $E \otimes E^*$,

---

if $n - 2 \le k \le n$, then every cycle can be represented by a submanifold. A readable summary and references may be found in [92].

and we define the total Chern class of E via

$$c(E) = \det\left(\mathbb{1} + \frac{i}{2\pi}F\right) = 1 + \sum_i c_i(E). \tag{4.4.7}$$

where the $c_i \in \mathscr{A}^{2i}(M)$.[14] Using the familiar $\det = \exp\operatorname{tr}\log$ identity we find

$$c(E) = 1 + \frac{i}{2\pi}\operatorname{tr} F + \frac{1}{8\pi^2}\left(\operatorname{tr} F^2 - (\operatorname{tr} F)^2\right)$$
$$- \frac{i}{48\pi^3}\left(2\operatorname{tr} F^3 - 3\operatorname{tr} F\operatorname{tr} F^2 + (\operatorname{tr} F)^3\right) + \dots. \tag{4.4.8}$$

The total Chern class behaves well with respect to vector bundle addition: $c(E_1 \oplus E_2) = c(E_1)c(E_2)$. There is a closely related object, the Chern character $\operatorname{ch}(E)$, which behaves well with respect to $\oplus$ and $\otimes$; this is defined as

$$\operatorname{ch}(E) = \operatorname{tr}\exp(i F/2\pi) \tag{4.4.9}$$

and satisfies $\operatorname{ch}(E_1 \otimes E_2) = \operatorname{ch}(E_1)\operatorname{ch}(E_2)$ and $\operatorname{ch}(E_1 \oplus E_2) = \operatorname{ch}(E_1) + \operatorname{ch}(E_2)$. For a line bundle $\mathcal{L}$ we have $\operatorname{ch}(\mathcal{L}) = \exp c_1(\mathcal{L})$.

Many practical computations are facilitated by the "splitting principle," which states that an identity of the classes that holds for a vector bundle $E = \mathcal{L}_1 \oplus \mathcal{L}_2 \oplus \cdots \oplus \mathcal{L}_r$, where $\mathcal{L}_i$ are line bundles, holds for any rank r complex bundle [94].

Exercise 4.7 Establish the identities

$$\operatorname{ch}_1 = c_1,$$

$$\operatorname{ch}_2 = \frac{1}{2}c_1^2 - c_2,$$

$$\operatorname{ch}_3 = \frac{1}{2}\left[c_3 - c_1 c_2 + \frac{1}{3}c_1^3\right],$$

$$\operatorname{ch}_4 = \frac{1}{24}c_1^4 - \frac{1}{6}c_1^2 c_2 + \frac{1}{6}c_1 c_3 + \frac{1}{12}c_2^2 - \frac{1}{6}c_4. \tag{4.4.10}$$

---

[14]We are again slightly abusing notation here and conflating the topological invariants $c_i \in H^{2i}(M, \mathbb{Z})$, with the specific representatives in terms of the curvature forms. This usually does not cause trouble, but the reader should be aware of this abuse.

Use the splitting principle to prove that for a rank r bundle E

$$\text{ch}_0(\wedge^k E) = \binom{r}{k} \,,$$

$$\text{ch}_1(\wedge^k E) = \binom{r-1}{k-1} \text{ch}_1(E) \,,$$

$$\text{ch}_2(\wedge^k E) = \binom{r-1}{k-1} \text{ch}_2(E) + \binom{r-2}{k-2} \left[\frac{1}{2} \text{ch}_1(E)^2 - \text{ch}_2(E) \right] \,. \qquad (4.4.11)$$

Finally, show that the dual bundle E^* has

$$c_k(E^*) = (-1)^k c_k(E) \,, \qquad (4.4.12)$$

and complex conjugation shows $c_k(\overline{E}) = c_k(E^*)$. □

Pontryagin Classes

For a real rank k vector bundle V with structure group $SO(k)$, the curvature F is anti-symmetric matrix of 2-forms. This means that $\text{tr } F^{2m+1} = 0$, and the expansion

$$p(V) = \det\left(\mathbb{1} - \frac{1}{2\pi} F \right) = 1 + p_1(V) + p_2(V) + \cdots \qquad (4.4.13)$$

only involves $p_i(V) \in \mathscr{A}^{4i}(M)$. This defines the Pontryagin classes of the bundle V. The first non-trivial class is

$$p_1(V) = -\frac{1}{8\pi^2} \text{tr } F^2 \,. \qquad (4.4.14)$$

A complex bundle E of rank k is, in particular, a real bundle V of rank $2k$, and, more precisely, $V \otimes \mathbb{C} = E \oplus \overline{E}$, where $\overline{E}$ is the conjugate complex bundle. As the following exercise shows, this relates the Pontryagin classes of V to the Chern classes of E. M need not be a complex manifold for this discussion.

Exercise 4.8 Suppose we have a complex bundle $E \to M$ with structure group $U(r)$. The connection 1-form on E has locally the components $\mathcal{A}^i_{\ j}$, where $i, j = 1, \ldots, r$, and $\mathbb{C}^r$ is the fiber of the bundle. The structure group $U(r)$ means we have a Hermitian metric $h_{i\bar{j}}$, i.e. a section of $E^* \otimes \overline{E}^*$ such that for any section $s \in E$ we have a non-negative function on M given by $h_{i\bar{j}} s^i \bar{s}^j \geq 0$; moreover, we can choose the connection $\mathcal{A}$ so that h is covariantly constant.

Establish the following relations between the connections and curvatures on E and $\overline{E}$.

1. Define the covariant derivative on E as $(Ds)^i = ds^i + \mathcal{A}^i_j s^j$, and show that $h Ds = D(hs)$ for all s requires the connection on $\overline{E}$ to be

$$\overline{\mathcal{A}}^{\bar{\imath}}_{\bar{\jmath}} = h^{\bar{\imath} i} dh_{i \bar{\jmath}} - h^{\bar{\imath} i} \mathcal{A}^k_i h_{k \bar{\jmath}} \, , \tag{4.4.15}$$

where $h^{\bar{\imath} i}$ is the inverse of the Hermitian metric.
2. Prove that $\mathcal{F} = d\mathcal{A} + \mathcal{A}^2$ and $\overline{\mathcal{F}} = d\overline{\mathcal{A}} + \overline{\mathcal{A}}^2$ are related by

$$\overline{\mathcal{F}} = -h^{-1}\,{}^t\mathcal{F}h \qquad \Longleftrightarrow \qquad \overline{\mathcal{F}}^{\bar{\imath}}_{\bar{\jmath}} = -h^{\bar{\imath} i} \mathcal{F}^j_i h_{j \bar{\jmath}} \, . \tag{4.4.16}$$

Now argue that we can think of the real bundle V with structure group $U(r) \subset SO(2r)$ underlying E as $V \otimes \mathbb{C} = E \oplus \overline{E}$, with curvature

$$F = \begin{pmatrix} \mathcal{F} & 0 \\ 0 & \overline{\mathcal{F}} \end{pmatrix} . \tag{4.4.17}$$

Prove that $\operatorname{tr} F^k = (1 + (-1)^k) \operatorname{tr} \mathcal{F}^k$, and, finally compute $\det(\mathbb{1} + \frac{i}{2\pi} F)$ in two different ways. On one hand, from the definition of the Pontryagin class

$$\det\left(\mathbb{1} + \tfrac{i}{2\pi} F\right) = 1 - p_1(V) + p_2(V) - p_3(V) + \dots, \tag{4.4.18}$$

while from $V \otimes \mathbb{C} = E \oplus \overline{E}$ and properties of the Chern class

$$\det(\mathbb{1} + \tfrac{i}{2\pi} F) = c(E)c(\overline{E}) = c(E)c(E^*) \, . \tag{4.4.19}$$

Now prove that

$$p_1(V) = c_1(E)^2 - 2c_2(E) \, , \quad p_2(V) = -2c_1(E)c_3(E) + c_2^2(E) + 2c_4(E) \, . \tag{4.4.20}$$

$\square$

As another special example of a Pontryagin class, we consider the adjoint vector bundle $\mathfrak{g}_{\mathcal{P}}$ associated to a principal G-bundle $\mathcal{P} \to M$. Its basic topological invariant is the first Pontryagin class [44]

$$p_1(\mathfrak{g}_{\mathcal{P}}) = \frac{1}{8\pi^2 h_{\mathfrak{g}}} \operatorname{Tr} F^2 \in H^4(M, \mathbb{Z}) \, , \tag{4.4.21}$$

where $h_{\mathfrak{g}}$ is the dual Coxeter number of the Lie algebra, and Tr is taken in the adjoint representation normalized so that the highest root has length-squared 2. We will use this form below.

A Comment on Torsion

We leave this brief discussion of Chern–Weil theory and characteristic classes with a topological confession. The Chern and Pontryagin classes that we just introduced are really defined intrinsically by the topology of the bundles and do not rely on the choice of connection. In general these carry more information than the differential forms constructed from the curvature because the differential forms cannot represent the part of the class valued in the torsion subgroup group of the cohomology. To illustrate the issues, let us consider the simplest example: a line bundle $\mathcal{L} \to M$ over a compact manifold and its first Chern class $c_1(\mathcal{L})$. We know from algebraic topology [223] that if we decompose the singular homology groups of M into the free and torsion subgroups as

$$H_1(M, \mathbb{Z}) = \mathbb{Z}^{\oplus b_1} \oplus T_1 , \qquad H_2(M, \mathbb{Z}) = \mathbb{Z}^{\oplus b_2} \oplus T_2 , \qquad (4.4.22)$$

then

$$H^2(M, \mathbb{Z}) = \mathbb{Z}^{\oplus b_2} \oplus T_1 . \qquad (4.4.23)$$

The algebraic topology definition of $c_1(\mathcal{L})$ [210, 224] shows that $c_1(\mathcal{L}) \in H^2(M, \mathbb{Z})$ classifies line bundles in the C^∞ (i.e. smooth) category: there is a smooth isomorphism between two line bundles if and only if they have the same first Chern class. It is perfectly possible to have a line bundle $\mathcal{L}$ that admits a flat connection (i.e. a connection with $F = 0$) but $c_1(\mathcal{L}) \neq 0$. In this case the line bundle is "torsion," and since T_1 is a finite (abelian) group, $\mathcal{L}^{\otimes p}$ is then trivial for a sufficiently large integer p. These topological features will be missed when we represent $c_1(\mathcal{L})$ by the cohomology class $[i F/2\pi]$. The issue does not arise when $T_1 = 0$, for instance if M is simply connected. In that case Chern–Weil theory completely captures the topology of line bundles, and a line bundle is trivial if and only if it admits a flat connection.

The Heterotic B-Field

We are now in a position to discuss the significance of (4.3.38). Taking the exterior derivative of both sides, we obtain

$$0 = d\mathcal{H} = \frac{1}{2} \operatorname{tr} \mathcal{R}_+^2 - \frac{1}{2} \operatorname{tr} F^2 . \qquad (4.4.24)$$

This is the famous Bianchi identity of heterotic compactification. The subscript on the Ricci form reminds us that it is computed with the curvature twisted by $\mathcal{H}$, so that this is a complicated non-linear condition on $\mathcal{H}$. From our discussion in the previous section, however, we recognize that by taking cohomology classes, this leads to

$$\frac{1}{2} p_1(T_M) - \frac{1}{2} p_1(E) = 0 \quad \text{in} \quad H^4(M, \mathbb{R}) . \tag{4.4.25}$$

A stronger condition follows from demanding that there is no global sigma model anomaly [124, 142, 168, 382, 383]. The global anomaly is zero if the Stiefel–Whitney classes w_1 and w_2 vanish for both E and T_M. We will not delve into the Stiefel–Whitney classes but state two basic properties. First, $w_1(E) \in H^1(M, \mathbb{Z}_2)$ is an obstruction to orienting the bundle: E is orientable and has structure group $SO(r)$ if and only if $w_1(E) = 0$. In some cases the associated principal $SO(r)$ bundle can be lifted to a principal $\mathrm{Spin}(r)$ bundle over M: the class $w_2(E) \in H^2(M, \mathbb{Z}_2)$ is the obstruction to the existence of such a lift: a lift exists if and only if $w_2(E) = 0$. Such a lift is called a spin structure on E [281]. These requirements are natural in the context of spacetime–supersymmetric heterotic geometry: they are necessary for the left-moving GSO projection(s) to be consistently defined just as in the free heterotic string in terms of a $\mathbb{Z}_2$ or $\mathbb{Z}_2 \times \mathbb{Z}_2$ action on the left-moving fermions. We will assume that they are satisfied by E and M. It can be shown [307] that when $w_1(E) = 0$

$$p_1(E) = w_2(E)^2 \mod 2 . \tag{4.4.26}$$

Similar considerations apply to the tangent bundle T_M and the right-moving GSO projection, and therefore $p_1(E)$ and $p_1(T_M)$ are both even integral cohomology classes. Thus, the Bianchi identity is a condition in integral cohomology:

$$\frac{1}{2} p_1(T_M) - \frac{1}{2} p_1(E) = 0 \quad \text{in} \quad H^4(M, \mathbb{Z}) . \tag{4.4.27}$$

Suppose we pick the defining topological data for $p : E \to M$. Concretely, this means an open cover $\{U_\alpha\}$ and transition functions on the overlaps for both the frame bundle over M, as well as E. Assuming that (4.4.27) is satisfied, we can ask whether it is possible to find connections and a B-field such that

$$\mathcal{H} = dB + \frac{1}{2} \mathrm{CS}_3(\mathcal{S}^+) - \frac{1}{2} \mathrm{CS}_3(A) \tag{4.4.28}$$

is a global form. This is complicated given that $\mathcal{S}^+$ depends on dB! Before we study this equation in detail, however, we should recall that it was computed in perturbation theory, and while its cohomological content (4.4.27) will not be modified by higher order corrections, the particular differential form representatives may well change. For example, we expect that higher order terms will shift $\mathcal{S}^+$ to be defined with the gauge-invariant $\mathcal{H}$, rather than dB. To think about this more clearly, it then helps to restore the NLSM expansion parameter, the string tension α'

in the equation. With it in place, the condition is

$$\mathcal{H} = dB + \frac{\alpha'}{4}\mathrm{CS}_3(\mathcal{S}^+) - \frac{\alpha'}{4}\mathrm{CS}_3(A) + O(\alpha'^2) . \qquad (4.4.29)$$

We can now ask whether we can find a B-field to leading order in the α' expansion, and now the answer is positive. Since (4.4.27) holds, it follows that there exists some global three form ξ such that

$$\frac{1}{2}d\mathrm{CS}_3(\mathcal{S}^+) - \frac{1}{2}d\mathrm{CS}_3(A) = d\xi . \qquad (4.4.30)$$

So, on each patch $U_\alpha \simeq \mathbb{R}^n$ we can make an Ansatz for the local B-field B_α as a solution to

$$dB_\alpha = \frac{\alpha'}{2}\left[\xi - \frac{1}{2}\mathrm{CS}_3(\mathcal{S}_\alpha^+) + \frac{1}{2}\mathrm{CS}_3(A_\alpha)\right] . \qquad (4.4.31)$$

This has a solution because the right-hand-side is closed by (4.4.27), and the ambiguity in the solution is a gerbe shift of B_α by a local exact form: $B_\alpha \rightarrow B_\alpha + d\Lambda_\alpha$. On an overlap $U_\alpha \cap U_\beta \neq \emptyset$, the solutions will patch by

$$B_\alpha - B_\beta = d\Lambda_{\alpha\beta} + \frac{\alpha'}{2}\Delta_{\alpha\beta}^{\mathrm{gauge}} , \qquad (4.4.32)$$

where $\Delta_{\alpha\beta}^{\mathrm{gauge}}$ is constructed from the differences in Chern–Simons forms defined on the patches, as in Exercise 4.6.

As a consequence, there is in general no invariant meaning to the integral of $\mathcal{H}$ over some three-cycle $Y \subset M$. This should be contrasted with the H-field of bosonic or NSR type II sigma models, which obeys $[H] \in H^3(M, 4\pi^2\alpha'\mathbb{Z})$, as described above.[15] In particular, since $d\mathcal{H} \neq 0$, it is not in cohomology, and therefore the integral $\int_Y i^*(\mathcal{H})$ depends on the details of the embedding $i : Y \subset M$.

Still, $\mathcal{H}$ is not an arbitrary 3-form, and there is a certain amount of rigidity in it [28, 334]. The integral of the Chern–Simons forms over Y is defined modulo integer multiples of $4\pi^2\alpha'$,[16] and the ambiguity has the same quantization as that of the naive gerbe 3-form dB. There is a consistent structure: once we fix the topology for the bundle $E \rightarrow M$ that obeys (4.4.27) and we pick connections $\mathcal{S}^+$ and $\mathcal{A}$, then we can find a solution for a B-field with a gauge-invariant $\mathcal{H}$. The solution does not determine B: we are free to shift B by a gerbe connection B', with $dB' \in H^3(M, 4\pi^2\alpha')$, for instance a connection with $dB' = 0$. Some of these shifts will lead to gauge-equivalent configurations, and others will describe motion in the SCFT moduli space.

---

[15] We restored units of α' in restating the quantization condition.

[16] We work this out explicitly in an abelian example in Sect. B.2.4.

4.5 Heterotic Spacetimes

The $(0,1)$ non-linear sigma model is of fundamental importance in the heterotic string: the covariant definition of the theory, as provided in [175], requires the internal theory in a string compactification to be a $(0,1)$ superconformal theory. The physical constraints on the theory are obtained by coupling the product of the internal and spacetime theories to $(0,1)$ supergravity. The general case is difficult to analyze. When does a particular $(0,1)$ non-linear sigma model that satisfies the topological anomaly cancellation conditions flow to a $(0,1)$ superconformal theory? When is the resulting theory weakly coupled, so that we can use the non-linear sigma model Lagrangian and related geometric concepts to describe the dynamics? These are fundamental questions in string theory, and we are not in a position to answer them in full generality. This is not surprising: such a general understanding would mean that we have a general picture of solutions to the classical heterotic string equations of motion, which would include non-supersymmetric and time-dependent vacua.

We have a better understanding of solutions that preserve some amount of supersymmetry in spacetime. In perturbative heterotic string theory, such solutions in general preserve some extended supersymmetry, typically accompanied by some integrality conditions. The first results of this sort were derived in a number of heterotic classics: [46, 48, 165, 246, 343], and a general analysis can be found in [303]. The general story will not be needed, but it is worthwhile to summarize some specific cases.

1. $\mathcal{N} = (1, 0)$ supersymmetry in $\mathbb{R}^{1,5}$ requires an internal worldsheet theory with $\bar{c} = 6$ and $(0, 4)$ superconformal invariance.
2. $\mathcal{N} = 1$ supersymmetry in $\mathbb{R}^{1,3}$ requires an internal worldsheet theory with $\bar{c} = 9$, $(0, 2)$ superconformal invariance, and integral R-charges in the $\overline{\mathrm{NS}}$ sector.
3. $\mathcal{N} = 2$ supersymmetry in $\mathbb{R}^{1,3}$ requires the $\bar{c} = 9$ N=2 superconformal worldsheet algebra to decompose into a sum of $\bar{c} = 3$ N=2 algebra with integral R-charges and a $\bar{c} = 6$ N=4 algebra.
4. $\mathcal{N} = (2, 0)$ supersymmetry in $\mathbb{R}^{1,1}$ requires an internal worldsheet theory with $\bar{c} = 12$, $(0, 2)$ superconformal invariance, and integral R-charges in the $\overline{\mathrm{NS}}$ sector.
5. $\mathcal{N} = (1, 0)$ supersymmetry in $\mathbb{R}^{1,1}$ requires an internal worldsheet theory with $\bar{c} = 12$, $(0, 1)$ superconformal invariance; the right-moving Virasoro algebra should contain a $\bar{c} = 1/2$ Virasoro algebra (i.e. the antiholomorphic sector of the Ising model) as a sub-algebra.

The last case is included to remind the reader that extended worldsheet supersymmetry may not be necessary for certain amounts of spacetime supersymmetry. On the other hand, as the remaining cases show, there is a nice association between extended worldsheet supersymmetry and spacetime supersymmetry.[17] Worldsheet theories with at least (0,2) superconformal invariance play a starring role, and we now turn to them.

4.5.1 Classical (0,2) Non-linear Sigma Models

Extended chiral supersymmetry in a (0,1) non-linear sigma model requires the target space M to possess additional geometric structure beyond that of a Riemannian manifold. This goes back to [399] and is reviewed in many classic works, e.g. [240, 376]. We will follow here the presentation in [246, 343]. A key ingredient will be a choice of a complex structure on M; the reader may want to review some basic complex geometry described in the geometric appendix and references therein.

Our starting point is the classical (0,1) supersymmetric component action (4.2.20), repeated here for convenience:

$$S = \frac{1}{4\pi} \int d^2z \left\{ (g_{\mu\nu} + B_{\mu\nu}) \partial\phi^\mu \bar{\partial}\phi^\nu \right.$$

$$+ g_{\mu\nu}\psi^\mu \partial\psi^\nu + \partial\phi^\lambda \psi^\mu \psi^\nu (\Gamma_{\mu\lambda\nu} - \tfrac{1}{2}dB_{\mu\lambda\nu})$$

$$\left. + \lambda^T (\bar{\partial}\lambda + \bar{\partial}\phi^\mu A_\mu \lambda) - \tfrac{1}{2}\lambda^T F_{\mu\nu}\lambda \psi^\mu \psi^\nu \right\}. \tag{4.5.1}$$

By construction, the action is invariant under the supersymmetry generated by $\boldsymbol{Q}_1$:

$$\boldsymbol{Q}_1 \cdot \phi^\mu = -i\psi^\mu \,, \quad \boldsymbol{Q}_1 \cdot \psi^\mu = i\bar{\partial}\phi^\mu \,, \quad \boldsymbol{Q}_1 \cdot \lambda = iA_\mu(\phi)\psi^\mu \lambda \,. \tag{4.5.2}$$

The last expression is obtained from the superspace formulas above once we use the $\boldsymbol{L}$ equation of motion: $\boldsymbol{L} = -iA_\mu(\phi)\psi^\mu \lambda$.

Exercise 4.9 Show that supersymmetry closes: up to λ equations of motion $\boldsymbol{Q}_1^2 = \boldsymbol{P}$, where the translation generator $\boldsymbol{P}$ acts as $\bar{\partial}$ on the fields. Observe that the closure only uses the λ equations of motion. Argue that this will be the case for any other chiral supersymmetry satisfying $\boldsymbol{Q}_2^2 = \boldsymbol{P}$; the point is that $\boldsymbol{Q}_2$ has spin $-1/2$, while the equations of motion for ϕ and ψ involve $\partial\bar{\partial}$ and ∂ derivatives respectively. A further useful (though a tad masochistic) exercise is to verify that $\boldsymbol{Q}_1 \cdot S = 0$ in the explicit component formulation.   □

---

[17]There is just one more exception in the case of Minkowski space heterotic string vacua— $\mathbb{R}^{1,2}$ with $\mathcal{N} = 1$ supersymmetry. All other Minkowski compactifications do require some amount of extended supersymmetry for the internal theory. More discussion of the exceptional $\mathbb{R}^{1,2}$ and $\mathbb{R}^{1,1}$ cases can be found in [303, 348]; exotic $\mathbb{R}^{1,1}$ supersymmetries are discussed in [166].

We are interested in non-linear sigma models that have a classical (0,2) super-conformal invariance and a compact target space. This means that we not only seek a second supercharge Q_2, but also a right-moving conserved R-symmetry charge R; this is sufficient for (0,2) superconformal invariance because the classical sigma model Lagrangian is scale-invariant, and the promotion of global symmetries to holomorphic/anti-holomorphic symmetries is automatic. In a weakly-coupled non-linear sigma model realizing a superconformal theory, the bosons are assigned a conformal weight of zero,[18] while the ψ and λ have weight $\Delta = 1/2$ and spins $-1/2$ and $1/2$ respectively. With that assignment the classical action is conformally invariant, and the R-symmetry action on the fields takes the form

$$R \cdot \phi^\mu = 0 , \qquad R \cdot \psi^\mu = -i \mathcal{J}^\mu_\nu(\phi)\psi^\nu , \qquad R \cdot \lambda = 0 , \qquad (4.5.3)$$

where $\mathcal{J}^\mu_\nu(\phi)$ is a tensor pulled back from the target space.

The Ansatz for the R-symmetry action, so far motivated by free fields, deserves one more comment. In principle, we could also consider adding to R the action of a symmetry that corresponds to an isometry of M or of the bundle. The latter is a linear action on the λ: $\delta\lambda = \epsilon M(\phi)\lambda$, and it leads to a purely left-moving conserved current. Such symmetries do play an important role in our story, for instance as a left-moving R-symmetry of a (2,2) theory, but they are not going to be involved in the right-moving R-symmetry. The former require the manifold to have an isometry. As we remarked at the start of this chapter, this is a strong requirement, and it leads to additional left- and right-moving symmetries. However, these symmetries also have the property that they commute with Q_1. Thus, we conclude that (4.5.3) is the most general possibility for the R-action that we need to consider.

A short calculation [246, 343] shows that R-invariance of the (ϕ, ψ) action requires the tensor $\mathcal{J}$ to be compatible with the metric and covariantly constant with respect to the twisted connection ∇^-:

$$0 = \mathcal{J}^\nu_\mu g_{\nu\lambda} + \mathcal{J}^\nu_\lambda g_{\nu\mu} ,$$
$$0 = \nabla^-_\nu \mathcal{J}^\mu_\lambda = \mathcal{J}^\mu_{\lambda,\nu} + (\Gamma^\mu_{\nu\rho} - \tfrac{1}{2}(dB)^\mu{}_{\nu\rho})\mathcal{J}^\rho_\lambda - (\Gamma^\rho_{\nu\lambda} - \tfrac{1}{2}(dB)^\rho{}_{\nu\lambda})\mathcal{J}^\mu_\rho . \qquad (4.5.4)$$

The action for the λ will be invariant under this R-symmetry if

$$\mathcal{J}^\nu_\mu F_{\nu\lambda} + F_{\mu\nu}\mathcal{J}^\nu_\lambda = 0. \qquad (4.5.5)$$

We will explore the geometric significance of these conditions shortly.

---

[18]The presence of these operators of weight zero should not bother the reader, because, just as with a free compact boson, a ϕ by itself is not a well-defined local operator; those are constructed by pulling back tensors valued in the bundle $E \to M$ and contracting the indices with $\partial\phi$, $\bar\partial\phi$, ψ, and λ to construct gauge- and diffeomorphism-invariant operators.

We now want to ensure that the operators P, Q_1, R, and the putative Q_2 satisfy the N=2 algebra

$$[R, Q_A] = i\epsilon_{AB} Q_B \,, \qquad\qquad \{Q_A, Q_B\} = 2\delta_{AB} P \,, \qquad\qquad (4.5.6)$$

with all other (anti-)commutators zero. Since we already know the action of Q_1 and R, we define $Q_2 \equiv i[Q_1, R]$. This leads to the following action on the fields:

$$Q_2 \cdot \phi^\mu = i[Q_1, R] \cdot \phi^\mu = i \mathcal{J}^\mu_\nu \psi^\nu$$
$$Q_2 \cdot \psi^\mu = i[Q_1, R] \cdot \psi^\mu = \mathcal{J}^\mu_\nu \bar\partial \phi^\nu + i \mathcal{J}^\mu_{\nu,\rho} \psi^\nu \psi^\rho \,,$$
$$Q_2 \cdot \lambda = i[Q_1, R] \cdot \lambda = -i \psi^\nu \mathcal{J}^\mu_\nu A_\mu \lambda. \qquad\qquad (4.5.7)$$

Since R and Q_1 are symmetries of the action, this Q_2 is a symmetry as well. The Jacobi identity implies that the operators we just defined will generate the N=2 algebra if and only if $[R, P] = 0$ and $Q_1 = i[R, Q_2]$. Since that action of P on the fields is simply $\bar\partial$, the former condition clearly holds. On the other hand, the latter is non-trivial and leads to

$$Q_1 \cdot \phi^\mu = i[R, Q_2] \cdot \phi^\mu \implies \mathcal{J}^2 = -\mathbb{1},$$
$$Q_1 \cdot \psi^\mu = i[R, Q_2] \cdot \psi^\mu \implies \mathcal{N}^\mu_{\lambda\rho} = 0, \qquad\qquad (4.5.8)$$

where

$$\mathcal{N}^\mu_{\lambda\rho} = \mathcal{J}^\mu_{\nu,[\rho} \mathcal{J}^\nu_{\lambda]} - 2 \mathcal{J}^\nu_{[\lambda} \mathcal{J}^\mu_{\rho],\nu} - \mathcal{J}^\mu_\nu \mathcal{J}^\nu_{[\lambda,\rho]}. \qquad\qquad (4.5.9)$$

The first condition means that $\mathcal{J}$ is an almost complex structure on M; using this in $\mathcal{N}$ allows us to express it in a more familiar form:

$$\mathcal{N}^\mu_{\lambda\rho} = \mathcal{J}^\nu_\lambda (\mathcal{J}^\mu_{\rho,\nu} - \mathcal{J}^\mu_{\nu,\rho}) - \mathcal{J}^\nu_\rho (\mathcal{J}^\mu_{\lambda,\nu} - \mathcal{J}^\mu_{\nu,\lambda}). \qquad\qquad (4.5.10)$$

This is the Nijenhuis tensor for $\mathcal{J}$, and its vanishing implies that $\mathcal{J}$ defines a complex structure on M.

Exercise 4.10 Verify Eqs. (4.5.4), (4.5.5), (4.5.8).   □

Geometric Interpretation

We showed that a classical (0,1) non-linear sigma model with target space M equipped with a metric g and left-moving fermions coupling to bundle $E \to M$ with curvature F has (0,2) invariance if and only if the following three conditions are satisfied:

1. M is a complex manifold with a complex structure $\mathcal{J}$ that is compatible with the metric—that is the first condition in (4.5.4);
2. $\mathcal{J}$ is covariantly constant with respect to the ∇^- connection, which is obtained by twisting the Levi-Civita connection derived from g by adding the torsion term $\Gamma^\mu_{\lambda\rho} \rightarrow \Gamma^\mu_{\lambda\rho} - \frac{1}{2}(dB)^\mu_{\lambda\rho}$;
3. F is a $(1,1)$ form with respect to the $\mathcal{J}$ complex structure—that is the condition in (4.5.5).

The first statement means that $(M, g, \mathcal{J})$ define a Hermitian geometry with Hermitian form $\omega_{\mu\nu} = \mathcal{J}^\lambda_\mu g_{\lambda\nu}$. We can therefore choose complex coordinates ϕ^a, $\overline{\phi}^{\overline{a}}$ on M, and write the Riemannian interval in terms of a Hermitian metric:

$$ds^2 = g_{\mu\nu}d\phi^\mu d\phi^\nu = g_{a\overline{b}}d\phi^a d\overline{\phi}^{\overline{a}} + \text{c.c.} \tag{4.5.11}$$

Rather than expand the alphabet soup of fields and variables, we will overload the notation: thus, g will denote either the Riemannian or Hermitian metric, depending on the context; similar remark holds for ϕ^μ versus ϕ^a.

The advantage of the complex coordinates is that the complex structure takes a simple and constant form: $\mathcal{J}(d\phi^a) = id\phi^a$, and $\mathcal{J}(d\overline{\phi}^{\overline{a}}) = -id\overline{\phi}^{\overline{a}}$, so that the Hermitian form $\omega \in \mathcal{A}^{1,1}(M)$ is

$$\omega = ig_{a\overline{a}}d\phi^a \wedge d\overline{\phi}^{\overline{b}} . \tag{4.5.12}$$

The reader can quickly check that ω is a real form. As shown in Exercises B.21 and B.22, the only non-zero covariant derivatives of $\mathcal{J}$ are

$$\nabla_{\overline{c}}\mathcal{J}^{\overline{a}}_b = g^{\overline{a}a}(\partial\omega)_{ba\overline{c}} , \qquad \nabla_c\mathcal{J}^a_{\overline{b}} = -g^{\overline{a}a}(\overline{\partial}\omega)_{c\overline{a}\overline{b}} . \tag{4.5.13}$$

Since supersymmetry requires $\nabla^-\mathcal{J} = 0$, we therefore relate $\partial\omega$ and $\overline{\partial}\omega$ to dB:

$$dB = i(\overline{\partial} - \partial)\omega . \tag{4.5.14}$$

Thus, dB is determined by the Hermitian metric on M. As we discovered above, dB is not a well-defined object once we consider sigma-model anomalies. There is a natural gauge-invariant modification of this condition:

$$\mathcal{H} = i(\overline{\partial} - \partial)\omega , \tag{4.5.15}$$

and it can be shown that $(0,2)$ supersymmetry variations of the non-local terms in the effective action (4.3.26) induce the necessary terms [245, 301, 343].

Exercise 4.11 Use (4.5.13) to show that $\nabla^-\mathcal{J} = 0$ if and only if (4.5.14) holds.

In addition, show that the non-vanishing components of the ∇^- connection are

$$\Gamma^a_{bc} - \tfrac{1}{2}(dB)^a_{bc} = g^{\bar{a}a} g_{b\bar{a},c} , \qquad\qquad \Gamma^{\bar{a}}_{\bar{b}\bar{c}} - \tfrac{1}{2}(dB)^{\bar{a}}_{\bar{b}\bar{c}} = g^{\bar{a}a} g_{a\bar{b},\bar{c}} ,$$

$$\Gamma^b_{\bar{a}c} - \tfrac{1}{2}(dB)^b_{\bar{a}c} = g^{\bar{b}b}(g_{c\bar{b},\bar{a}} - g_{c\bar{a},\bar{b}}) \qquad \Gamma^{\bar{b}}_{a\bar{c}} - \tfrac{1}{2}(dB)^{\bar{b}}_{a\bar{c}} = g^{\bar{b}b}(g_{b\bar{c},a} - g_{a\bar{c},b}) .$$
$$\tag{4.5.16}$$

Again, it is possible to include corrections from the $(0,2)$ supersymmetry variations of the non-local terms in the effective action to show that dB is replaced by $\mathcal{H}$, as we might expect.   □

Finally, we discuss the significance of the restriction on the curvature, following the standard treatment from, e.g. [207]. Supersymmetry constrains the curvature to take the form $F = F_{a\bar{a}} d\phi^a \wedge d\bar{\phi}^{\bar{a}} \in \mathcal{A}^{1,1}(M, E \otimes E^*)$. Since A is a connection on the real rank r bundle $p : E \to M$, on each patch $U_\alpha \subset M$ we can express it as

$$A_\alpha = \mathcal{A}_\alpha - \mathcal{A}_\alpha^\dagger , \tag{4.5.17}$$

where $\mathcal{A}$ is a (in general complex) $r \times r$ matrix of $(1,0)$ forms on U_α. The $(2,0)$ and $(0,2)$ components of the curvature are

$$F_\alpha^{2,0} = \partial \mathcal{A}_\alpha + \mathcal{A}_\alpha \wedge \mathcal{A}_\alpha = 0 , \quad F_\alpha^{0,2} = -\bar{\partial} \mathcal{A}_\alpha^\dagger + \mathcal{A}_\alpha^\dagger \wedge \mathcal{A}_\alpha^\dagger = 0 . \tag{4.5.18}$$

These equations are readily integrated in terms of a "potential" $\xi_\alpha : U_\alpha \to GL(r, \mathbb{C})$:

$$\mathcal{A}_\alpha = \xi_\alpha^{-1} \partial \xi_\alpha . \tag{4.5.19}$$

This leads to a strong condition on the bundle E: we will show that $E \otimes \mathbb{C}$ is a holomorphic vector bundle, i.e. has holomorphic transition functions.

Suppose that we have a cover $\{U_\alpha\}$ for M, with transition functions $g_{\alpha\beta}$ for $p : E \to M$ on $U_\alpha \cap U_\beta \neq \emptyset$ obeying the usual co-cycle conditions. Clearly the $g_{\alpha\beta}$ also define transition functions for the bundle $E \otimes \mathbb{C}$. Given any $\eta_\alpha : U_\alpha \to GL(r, \mathbb{C})$, we obtain an isomorphic bundle with transition functions $\widetilde{g}_{\alpha\beta} = \eta_\alpha g_{\alpha\beta} \eta_\beta^{-1}$. If A_α defines a connection with transition functions g, then

$$\widetilde{A}_\alpha = \eta_\alpha A_\alpha \eta_\alpha^{-1} - d\eta_\alpha \eta_\alpha^{-1} \tag{4.5.20}$$

is a connection with transition functions $\widetilde{g}$. We apply this to our case by setting $\eta_\alpha = (\xi_\alpha^\dagger)^{-1}$ and obtain a connection on $E \otimes \mathbb{C}$

$$\widetilde{A}_\alpha = (\xi_\alpha \xi_\alpha^\dagger)^{-1} \partial (\xi_\alpha \xi_\alpha^\dagger) . \tag{4.5.21}$$

Since this form is valid for every U_α, the transition functions $\widetilde{g}_{\alpha\beta} : U_\alpha \cap U_\beta \to GL(r, \mathbb{C})$ must be holomorphic.

The explicit form for the connection bears a close resemblance to the Chern connection that exists on any holomorphic bundle $\mathcal{E} \to M$ equipped with a Hermitian metric h, a section of $\mathcal{E}^* \otimes \overline{\mathcal{E}}^*$. With explicit holomorphic fiber indices A, B, we have $h(s, t) = h_{A\overline{B}} s^A \overline{t^B}$ for any two sections s, t of $\mathcal{E}$, and the Chern connection is $A_B^A = h^{\overline{B}A} \partial h_{B\overline{B}}$. Comparing with the expression above, we are tempted to identify $h_\alpha = (\xi_\alpha \xi_\alpha^\dagger)^T$ as a Hermitian metric on $\mathcal{E} = E \otimes \mathbb{C}$. While $\xi_\alpha \xi_\alpha^\dagger$ has the correct positivity and Hermiticity properties, we have not quite shown that it transforms as a section of $\overline{\mathcal{E}}^* \otimes \mathcal{E}^*$. However, the discussion makes it clear that every holomorphic vector bundle does admit a connection with (1,1) curvature form.

The Chern connection naturally emerges when $E \to M$ is isomorphic to a holomorphic bundle $\mathcal{E} \to M$ (the isomorphism is reviewed in Exercise (B.16)). In this case we can, instead of working with r real fermions λ^A work with $r/2$ Weyl fermions and their conjugates. We will develop this further when we work with a (0,2) superfield formulation below. Most work on heterotic geometry has been carried out in this context, but it is just a special case: the more general condition of (0,2) supersymmetry is that $E \otimes \mathbb{C}$ is holomorphic.

4.5.2 Constraints from Spacetime Supersymmetry

The classical conditions for a (0,1) NLSM to exhibit (0,2) supersymmetry are not sufficient to guarantee that the quantum non-linear sigma model is a (0,2) SCFT. At the same time, the geometries that lead to classical (0,2) supersymmetry constitute a large class: any bundle $E \to M$, such that M is complex and $E \otimes \mathbb{C} \to M$ is holomorphic can be used to construct a classical (0,2) non-linear sigma model, provided that we can choose the B-field so that $\mathcal{H} = i(\overline{\partial} - \partial)\omega$. The simplest realization of this is to take M to be a Kähler manifold with Kähler form M and $dB = 0$.

To reduce the geometries to a smaller class, we will impose additional requirements on the theory. We will assume that the weakly coupled geometry is a good guide to the spectrum of an SCFT with spacetime supersymmetry, i.e. a (0,2) SCFT with spectral flow. As such, we need the non-linear sigma model to exhibit a non-anomalous $U(1)_R$ symmetry that assigns integral charges to NS sector operators. For starters, we will focus on the most-studied case of $\overline{c} = 9$ theories.

Since we assume the Lagrangian theory is a good description of the physics, the central charges are easy to describe in general: if $E \to M$ has rank r and M has complex dimension n, then $c = r + 2n$, while $\overline{c} = 3n$. In particular, for $\overline{c} = 9$ we take $n = 3$. M must be compact to ensure a compact SCFT. The $U(1)_R$ symmetry just acts on the right-moving fermions alone and will in general be anomalous since the fermions couple to the connection on the tangent bundle. Just as in Sect. 2.5, we can easily see that the anomaly is proportional to the pull-back of

$$\tfrac{1}{2} \mathcal{J}_\mu^\nu \, \mathrm{tr}\{\mathcal{R}_{+\nu\rho}\} d\phi^\mu \wedge d\phi^\rho \ . \tag{4.5.22}$$

In complex coordinates, this is in turn simply $i \, \mathrm{tr}\{\mathcal{R}_{+a\bar{b}}\} d\phi^a \wedge d\bar{\phi}^{\bar{b}}$, and we recognize this to be a representative of the first Chern class $c_1(T_M)$, computed with the twisted connection. Thus, a necessary condition for the non-anomalous $U(1)_R$, is that $c_1(T_M) = 0$.[19]

Spectral flow implies that our $\bar{c} = 9$ theory has a purely anti-holomorphic anti-chiral primary operator $\mathcal{O}_{\mathrm{bot}}(\bar{z})$ with $\bar{q} = -3$ and $\bar{h} = 3/2$. In the non-linear sigma model the unique candidate is [343]

$$\mathcal{O}_{\mathrm{bot}} = \tfrac{1}{3!} \Omega_{\lambda\mu\nu}(\phi) \psi^\lambda \psi^\mu \psi^\nu \ . \tag{4.5.23}$$

The spectral flow operator itself is constructed as a square root of $\mathcal{O}_{\mathrm{bot}}$, and for this square root to be well-defined on all patches in field space, we require Ω to be a nowhere vanishing form on M. In order for $\mathcal{O}_{\mathrm{bot}}$ to have $U(1)_R$ charge -3 we need Ω to be a section of the canonical bundle $K_M = \wedge^3 T_M^*$. In local complex coordinates

$$\Omega = \frac{1}{3!} \Omega_{abc} d\phi^a \wedge d\phi^b \wedge d\phi^c = \frac{1}{3!} \eta(\phi, \bar{\phi}) \epsilon_{abc} d\phi^a \wedge d\phi^b \wedge d\phi^c \ , \tag{4.5.24}$$

where $\eta(\phi, \bar{\phi})$ is nowhere vanishing. We now observe a little subtlety: by our assumptions the canonical bundle has a nowhere vanishing section, which means the bundle is topologically trivial;[20] that is consistent with $c_1(T_M) = 0$. However, this does not imply that the bundle is holomorphically trivial in the chosen complex structure. There are many complex manifolds that satisfy $c_1(T_M) = 0$ yet have a non-trivial canonical bundle. Examples include $S^3 \times S^3$ with the Calabi–Eckmann complex structure, as well as any even-dimensional semi-simple Lie group.[21]

We need to check that $\mathcal{O}_{\mathrm{bot}}$ is right-moving, i.e. $\partial \mathcal{O}_{\mathrm{bot}} = 0$ up to the equations of motion. Using the ψ equation of motion, we find

$$\partial \mathcal{O}_{\mathrm{bot}} = \frac{1}{3!} \nabla_\rho^- \Omega_{\sigma\mu\nu} \partial \phi^\rho \psi^\sigma \psi^\mu \psi^\nu + \frac{1}{4} \Omega_{\lambda\mu\nu} \lambda^T F^\lambda_\sigma \lambda \psi^\sigma \psi^\mu \psi^\nu \ . \tag{4.5.25}$$

---

[19]Since we are now working with complex M, unless otherwise noted, T_M will denote $T_M^{1,0}$, the holomorphic tangent bundle.

[20]This is discussed in the Sect. B.2.1.

[21]Incidentally, these were some of the first constructions of complex non-Kähler manifolds [336].

The two terms have to vanish separately, which leads to two conditions that we will examine in turn:

$$\nabla^- \Omega = 0, \qquad F_{[\sigma}^{\lambda} \Omega_{\mu\nu]\lambda} = 0. \qquad (4.5.26)$$

The geometric content of these conditions is most transparent in complex coordinates. We begin with the implications of $\nabla^- \Omega = 0$. Given the form of ∇^- determined in (4.5.16), it is not hard to see that the non-trivial conditions arise from $\nabla_d^- \Omega_{abc} = 0$ and $\nabla_{\bar{d}}^- \Omega_{abc} = 0$. This holds if and only if

$$\partial_d \log \eta = g^{\bar{e}e} g_{d\bar{e},e}, \qquad \partial_{\bar{d}} \log \eta = g^{\bar{e}d}(g_{d\bar{e},\bar{d}} - g_{d\bar{d},\bar{e}}). \qquad (4.5.27)$$

While the right-hand-side of the first equation is not a tensor, in the second equation we find a tensor on the right-hand-side. Define the Lee 1-form β by

$$\beta = g^{\bar{e}e}(g_{e\bar{e},d} - g_{d\bar{e},e})d\phi^d + g^{\bar{e}d}(g_{d\bar{e},\bar{d}} - g_{d\bar{d},\bar{e}})d\phi^{\bar{d}}. \qquad (4.5.28)$$

Evidently β is real, and it is not hard to verify[22]

$$d\Omega = \beta \wedge \Omega, \qquad d\overline{\Omega} = \beta \wedge \overline{\Omega}, \qquad d(\omega \wedge \omega) = \beta \wedge \omega \wedge \omega. \qquad (4.5.29)$$

Furthermore, the norm of Ω defines a non-vanishing function on M:

$$\|\Omega\|^2 = \Omega_{abc} \overline{\Omega}_{\bar{a}\bar{b}\bar{c}} g^{\bar{a}a} g^{\bar{b}b} g^{\bar{c}c} = 3! \eta \bar{\eta} (\det g)^{-1}, \qquad (4.5.30)$$

but since $\nabla^- \Omega = \nabla^- \overline{\Omega} = 0$ and $\nabla^- g = 0$, it follows that this function a constant, and we can choose it so that $\eta \bar{\eta} = \det g/3!$. This means that in addition to (4.5.29), the non-degenerate forms Ω, and ω also satisfy the algebraic relations

$$\Omega \wedge \omega = 0, \qquad \overline{\Omega} \wedge \omega = 0, \qquad i\Omega \wedge \overline{\Omega} = \frac{1}{3!} \omega \wedge \omega \wedge \omega. \qquad (4.5.31)$$

These algebraic relations are the defining relations for M to be an SU(3) structure manifold.[23]

---

[22]Basic tensor identities such as $g_{a\bar{a}} g_{b\bar{b}} \epsilon^{abc} = \epsilon_{\bar{a}\bar{b}\bar{c}} g^{\bar{c}c} \det g$ are useful in showing these relations.

[23]G-structures and related notions of pure spinors play an important role in supersymmetric solutions of supergravity, both for type II and heterotic theories. We will not delve too far into these spacetime notions and refer the reader to some basic references. A very readable description of G-structures in the supergravity context is given in [190]; the connection to pure spinors and generalized geometry is discussed in many references; we refer to [203] for an introduction and references.

Exercise 4.12 Consider the second equation in (4.5.26). Show that $F_{[\sigma}{}^\lambda \Omega_{\mu\nu]\lambda} = 0$ if and only if the (1,1) curvature satisfies the zero-slope Hermitian Yang–Mills equation:

$$g^{\bar{a}a} F_{a\bar{a}} = 0 , \qquad \Longleftrightarrow \qquad \omega \wedge \omega \wedge F = 0 . \qquad (4.5.32)$$

In addition, verify that the requirement that F is a $(1, 1)$ form is equivalent to $\Omega \wedge F = 0$. □

The "zero slope" refers to a generalization that allows a right-hand-side linear in the first Chern class of the bundle. Hermitian–Yang–Mills equations with non-zero slope play an important role in type II string theory and gauge fields on D-branes; see, e.g. [30, 40]. The existence of solutions to these equations for holomorphic bundles over compact Kähler manifolds are discussed in the classic work [369]; this was generalized to more general complex manifolds obeying, respectively, $\partial\bar{\partial}\omega = 0$ and $\partial\bar{\partial}\omega^{n-1} = 0$ in [105] and in [285]. A survey of these developments, together with many other references is given in [284]. Briefly put, a holomorphic bundle admits a Hermitian–Yang–Mills connection if and only if it satisfies a stability criterion, which is defined purely in terms of holomorphic data of the bundle. Conceptually this is a huge simplification, since solutions of a non-linear partial differential equation are related to a structure that, in many specific examples, is purely algebraic. At a technical level, the algebraic problem can be quite formidable. A discussion of effective computational approaches for a large class of bundles is given in [13].

A key point for us is that bundle stability is an open condition in the space of Hermitian metrics on M: if $\mathcal{E} \to M$ is a stable holomorphic bundle with respect to a Hermitian form ω_0, then it remains stable under a small deformation $\omega_0 \to \omega_0 + \delta\omega$.

4.5.3 Vanishing of the Leading Order Beta Function

If the non-linear sigma model is to usefully describe a superconformal theory, the β functions for the background fields should vanish order by order in α' perturbation theory. The (0,2) beta functions have been studied in a number of works, going back to [243]; a survey of results may be found in [267], and a modern presentation useful for our purposes is given in [318]. We will not delve into the details but just state the results. For the class of geometries at hand, namely those where the (0,2) non-linear sigma model possesses a non-anomalous $U(1)_R$ and a candidate for the spectral flow operator, the vanishing of the leading order β function amounts to just one additional condition: the Lee form β that appears in $d\Omega$ and $d\omega^2$ is exact:

$$\beta = 2d\varphi , \qquad (4.5.33)$$

where φ is the dilaton profile, i.e. a real function on M.[24]

The exactness of the Lee form leads to strong conditions on the geometry. The first of these is that the Hermitian metric g on M is conformally balanced:

$$d(e^{-2\varphi}\omega \wedge \omega) = 0 .$$ (4.5.34)

More generally, a Hermitian metric on a compact complex n-fold M is conformally balanced if $d(e^{-2\varphi}\omega^{n-1}) = 0$ for some real smooth function φ. There are obstructions to finding a conformally balanced metric on a compact complex manifold [305].[25] A presentation of these and related geometric properties can be found in [21, 177, 188].

Since the Lee form also relates to the closure of Ω,

$$d(e^{-2\varphi}\Omega) = 0 ,$$ (4.5.35)

and this implies that the canonical bundle is holomorphically trivial.

It is perhaps worthwhile to describe this little point in detail to contrast M with other complex 3-folds with vanishing $c_1(T_M)$. Suppose that the complex atlas for M includes two neighborhoods U and V, each isomorphic to $\mathbb{C}^3$, with coordinates z and y respectively, with $U \cap V \neq \emptyset$. It follows that on the overlap $U \cap V$ we have

$$e^{-2\varphi}\Omega = e^{-2\varphi(z,\bar{z})}\eta_U(z,\bar{z})dz^1 \wedge dz^2 \wedge dz^3$$
$$= e^{-2\varphi(y,\bar{y})}\eta_V(y,\bar{y})dy^1 \wedge dy^2 \wedge dy^3 .$$ (4.5.36)

The sections are related by the holomorphic transition function $\mathcal{J}_{VU} = \det_{a,b}\frac{\partial z^a}{\partial y^b}$:

$$\eta_V = \mathcal{J}_{VU}\eta_U .$$ (4.5.37)

This shows that, as promised above, as soon as we know that Ω is non-degenerate, it follows that the canonical bundle is topologically trivial, since every transition function can be written as $\mathcal{J}_{VU} = \eta_V\eta_U^{-1}$. In general η_U is not holomorphic, so it does not yield a holomorphic map from the canonical bundle to the trivial bundle. But, $d(e^{-2\varphi}\Omega) = 0$ if and only if $\bar{\partial}(e^{-2\varphi}\eta_U) = 0$; this is just right—now $e^{-2\varphi}\eta_U$ is holomorphic in every patch, and the bundle is holomorphically trivial.

It is easy to construct complex manifolds where $c_1(T_M) = 0$ but the canonical bundle is holomorphically non-trivial. Perhaps the simplest example is a Hopf surface X, which is diffeomorphic to $S^3 \times S^1$ [271]: since $H^2(X, \mathbb{Z}) = 0$, and

---

[24]It is a familiar string–theoretic fact that the zero mode of the dilaton profile is related to the string coupling via e^φ. We will be able to interpret the geometry as a perturbative string theory vacuum when the dilaton profile remains bounded on M and its zero mode is a modulus.

[25]This can be contrasted with the result of Gauduchon that every compact complex manifold admits a hermitian metric with $\partial\bar{\partial}\omega^{n-1} = 0$.

line bundles are topologically classified by $c_1 \in H^2(X, \mathbb{Z})$, the canonical bundle is topologically trivial, but it is not holomorphically trivial. Put another way, the holomorphic connection is flat but has holonomy. More generally, holomorphic line bundles are classified by the sheaf cohomology group $\text{Pic}(X) = H^1(X, \mathcal{O}^*)$, where $\mathcal{O}^*$ is the sheaf of non-zero holomorphic functions on X [210], and there is a long exact sequence

$$0 \longrightarrow H^1(X, \mathbb{Z}) \longrightarrow H^1(X, \mathcal{O}) \longrightarrow \text{Pic}(X) \xrightarrow{c_1} H^2(X, \mathbb{Z}) \longrightarrow \cdots . \tag{4.5.38}$$

In particular, $\text{Pic}^0(X) = H^1(X, \mathcal{O})/H^1(X, \mathbb{Z}) = \ker c_1$ describes the isomorphism classes of topologically trivial holomorphic line bundles.

It is worthwhile to note that in general the dilaton profile cannot be absorbed into some conformal rescaling of various quantities. For instance, while $\widehat{\omega} = e^{-\varphi}\omega$ and $\widehat{\Omega} = e^{-2\varphi}\Omega$ satisfy $d(\widehat{\omega}^2) = d\widehat{\Omega} = 0$, it is the ω and Ω that satisfy the SU(3) structure relations.

4.5.4 Summary

We pause to summarize the state of affairs so far. In order for a $(0,1)$ heterotic non-linear sigma model to describe an $\mathcal{N} = 1$ supersymmetric $\mathbb{R}^{1,3}$ vacuum of the heterotic string, it should preserve $(0,2)$ supersymmetry and have a candidate for the spectral flow operator, and it should have vanishing leading order beta function. We also need the background to satisfy the consistency conditions that follow by demanding the vanishing of the sigma model anomalies.

All in all, we obtain the following conditions. The non-linear sigma model is specified by a rank r bundle $E \to M$, equipped with a connection A, metric g, B-field B, and the short-shrifted dilaton φ.

1. The bundle has vanishing Stiefel–Whitney classes $w_1(E) = w_2(E) = 0$, as well as $\frac{1}{2}p_1(T_M) - \frac{1}{2}p_1(E) = 0 \in H^4(M, \mathbb{Z})$.
2. The gauge-invariant curvature $\mathcal{H}$ of the B-feld satisfies $d\mathcal{H} = \frac{\alpha'}{4}\left(\text{tr}\,\mathcal{R}_+^2 - \text{tr}\,F^2\right)$.
3. M is complex manifold with Hermitian form ω. The non-closure of ω determines $\mathcal{H} = i(\bar{\partial} - \partial)\omega$.
4. The bundle $E \otimes \mathbb{C} \to M$ is holomorphic, and the connection on E is chosen so that the curvature 2-form F has no $(0,2)$ or $(2,0)$ components.
5. The connection on E satisfies the Hermitian–Yang–Mills equation: $\omega \wedge \omega \wedge F = 0$.
6. M is an SU(3) structure manifold with $c_1(T_M) = 0$ and non-degenerate $(3,0)$ form Ω obeying (4.5.31), (4.5.29).
7. M has a trivial canonical bundle and is conformally balanced, with dilaton chosen to satisfy $d(e^{-2\varphi}\Omega) = 0$ and $d(e^{-2\varphi}\omega \wedge \omega) = 0$.

The first two conditions follow from the absence of local and global sigma model anomalies. The next two are required by (0,2) supersymmetry. The fifth and sixth condition are required for spectral flow, and the last is implied by the vanishing of the leading order beta function.

It is remarkable that most of the conditions can be obtained without a detailed analysis of the (0,2) beta functions. Perhaps even the last condition can be obtained from some further consistency conditions on the putative (0,2) SCFT with spectral flow, but the author is not aware of such an argument.

These conditions satisfy a consistency check: aside from the topological conditions on the bundle, the same conditions arise by demanding that a compactification of heterotic supergravity preserves four supercharges in Minkowski space [352]. A detailed analysis of the ten-dimensional supergravity's supersymmetry transformations shows that the first corrections to these requirements are $O(\alpha'^2)$ [75].[26]

4.6 A (2,2) Calabi–Yau Sketch

The general geometry described in the previous section is intricate, and even the special cases only seem simple through (too) much exposure and by comparison to the general case (and provided one does not ask too many detailed questions). In this section we will give a brief overview of a classic heterotic geometry—the Calabi–Yau compactification described in [109] and masterfully presented in [207]. This will help to set the notation and to prepare for some (0,2) elaborations.

The supersymmetry conditions simplify if $H = 0$ and $E = T_M$. In that case M is a Kähler manifold with trivial canonical bundle. By setting the SU(3) structure connection A on E to be equal to the Levi-Civita connection on the tangent bundle—this is known as "standard embedding" in the literature—the conditions reduce to the requirement that the pair ω, Ω define a metric with holonomy contained in SU(3). Such a metric is Ricci-flat, and by the fundamental result of Yau, such metrics exist on any compact Kähler manifold with trivial canonical bundle [395, 396]. Such a metric is known as a Calabi–Yau metric, and it is unique for a fixed complex structure and Kähler class $[\omega] \in H^{1,1}(M, \mathbb{R})$. For a generic Calabi–Yau metric on M, the solutions come in a moduli space $\mathcal{M}_K \times \mathcal{M}_{\text{c-x}}$, where $\mathcal{M}_K$ is the "Kähler moduli space," and $\mathcal{M}_{\text{c-x}}$ is the "complex structure moduli space [91, 364, 365, 377]. When restricted to some point, i.e. M_0 with a fixed complex structure and Kähler class, there is a canonical identification of the tangent space to the moduli space:

$$T_{\mathcal{M}_K}(M_0) \simeq H^{1,1}(M_0, \mathbb{R}) \, , \quad T_{\mathcal{M}_{\text{c-x}}}(M_0) \simeq H^{0,1}_{\bar{\partial}}(M_0, T_{M_0}) \simeq H^{1,2}(M_0, \mathbb{C}) \, .$$

$$(4.6.1)$$

---

[26] A modern and thorough discussion of these statements is given in [127, 128].

The last isomorphism uses Kodaira–Serre duality and triviality of the canonical bundle. In declaring that the identification is canonical, we are glossing over an important subtlety for manifolds with holonomy contained in $SU(2)$, such as tori, the K3 surface, or their products. For this reason it is often useful to include an additional criterion in the definition of a Calabi–Yau n-fold that excludes those special cases. For instance, it is not uncommon to define a Calabi–Yau n-fold as a compact complex Kähler n-fold with trivial canonical bundle and $H_{\bar\partial}^{0,q}(M) = 0$ for $0 < q < n$. We will follow this definition.

The worldsheet theory obtained from a compact Calabi–Yau target space with standard embedding is a (2,2) supersymmetric non-linear sigma model, and there is by now overwhelming evidence that such a non-linear sigma model defines a (2,2) superconformal field theory with $c = \bar c = 9$. The geometric set-up just described fits beautifully with the structure of the $c = \bar c = 9$ SCFT with integral $U(1)_L \times U(1)_R$ charges. The (2,2) preserving marginal deformations follow the general pattern we described in Chap. 2, except that marginal operators must now satisfy chirality conditions on both the right and the left. The (2,2) chiral ring therefore consists of four sets of operators—(c,c), (a,c), (a,a), and (c,a). Marginal deformations are therefore identified with operators of $U(1)_L \times U(1)_R$ charges $(1, 1)$ in the (c,c) ring and $(-1, 1)$ in the (a,c) ring. These deformations are believed to be unobstructed. An argument that comes close to showing this directly in conformal perturbation theory was given in [148]. There is also a powerful spacetime argument that goes back to [139, 140]: such a superconformal field theory gives rise to a spacetime compactification of the type IIA or IIB string that preserves eight supercharges, and the marginal deformations reside in different massless multiplets of the resulting $\mathcal{N} = 2$ ungauged supergravity theory. For instance, in a IIA string the (a,c) moduli reside in four-dimensional vector multiplets, while the (c,c) moduli reside in the hypermultiplets. Any obstructions to marginality of these deformations must be encoded in spacetime potential terms, but such terms are forbidden by spacetime supersymmetry! Moreover, since the $\mathcal{N} = 2$ moduli space decomposes into the vector- and hyper-moduli spaces, it must be that the Zamolodchikov metric is also a product metric.

The latter point is "almost" clear in the SCFT. After all, $U(1)_L \times U(1)_R$ symmetry and conformal invariance imply that an (a,c) marginal deformation has zero two-point function with either a (c,c) or an (a,a) marginal deformation. That is true, but it does not imply the desired splitting because the operators that show up in the deformation are $U(1)_L \times U(1)_R$-neutral descendants of the chiral ring operators. There are also well-known counter-examples. From Chap. 2 we know that in a (4,4) SCFT every (2,2)-preserving deformation is (4,4) marginal, but the (4,4) moduli space does not split, and is not even complex. It turns out that the culprit is the non-abelian KM symmetry whose currents lead to additional singular terms in the OPE of (c,c) and (a,c) operators [202]. There is a direct geometric parallel for manifolds with holonomy contained in $SU(2)$ and their moduli spaces—this is the subtlety mentioned above. The same features can be seen from the spacetime supergravity theory: when spacetime supersymmetry is enhanced beyond to 16 or more supercharges, the moduli space no longer splits. We will assume, unless otherwise noted, that there is no further enhancement of the superconformal algebra of the SCFT, so that the (2,2) SCFT moduli space has a complex structure and splits as a product $\mathcal{M}_{ac} \times \mathcal{M}_{cc}$.

4.6.1   Closed B-Fields and Topological Sectors

In comparing the classical geometry to the SCFT expectation, there is one curious difference: while the complex structure moduli space can be naturally identified with the (c,c) marginal deformations, the Kähler moduli space is real, and it is not clear how to identify it with $\mathcal{M}_{ac}$. The resolution is simple. Consider the bosonic action of the NLSM for a non-linear sigma model map $\sigma : \Sigma \to M$:

$$S_{bos} = \frac{1}{2\pi} \int d^2z \left[g_{a\bar{a}}(\partial\phi^a \bar{\partial}\overline{\phi}^{\bar{a}} + \bar{\partial}\phi^a \partial\overline{\phi}^{\bar{a}}) \right] + \frac{i}{4\pi} \int_\Sigma \sigma^*(B) . \qquad (4.6.2)$$

Since $\mathcal{H} = 0$ point-wise on M, the B-field is closed. Thus, if $\sigma(\Sigma) = $ point, or, more generally, $\sigma(\Sigma)$ is contractible, then the second term is zero.

More generally, to make sense of the path integral over "all maps" $\sigma : \Sigma \to M$, we break up the integral into a sum over topological sectors. Let us restrict attention to $\Sigma = \mathbb{P}^1 = S^2$, i.e. to the tree-level approximation in string theory, where we only keep the leading term in the string coupling. This is the limit that should be described by the worldsheet superconformal theory. In that case, the maps are distinguished by the homotopy class of the image of the map, $[\sigma] \in \pi_2(M)$. If we suppose further that M is simply connected, i.e. $\pi_1(M) = 1$, then by the Hurewicz theorem [223] $\pi_2(M) \simeq H_2(M, \mathbb{Z})$, so that we can label the topological sectors by the cohomology class $[\sigma] \in H_2(M, \mathbb{Z})$.[27] The dependence of the path integral on the B-field takes the form

$$\sum_{[\sigma]\in H_2(M,\mathbb{Z})} e^{-\frac{i}{4\pi}[\sigma]([B])} \int D\Phi' e^{-S'[\omega]} , \qquad (4.6.3)$$

where $D\Phi'$ is a formal expression for integrating over all maps in the fixed homotopy class, and S' depends on the Kähler form ω. This sum should make sense whenever the non-linear-sigma model is weakly coupled, for which we need M to be smooth and all the cycles to be large, and therefore $[\omega]$ is "deep in the interior

---

[27] As Exercise 4.2 indicates, the restriction to simply connected M is reasonable. When M is a compact Calabi–Yau manifold with exactly SU(3) holonomy, then it cannot have continuous isometries and $\pi_1(M)$ is a finite group. Therefore M has a cover $\tilde{M}$ that is simply connected [255]. Our statements then apply to the finite cover, and in principle computations for M can be obtained by taking an orbifold of $\tilde{M}$. As shown in, for instance, [34, 100], the orbifold can have important ramifications for the resulting SCFT.

of the Kähler cone". In the physics literature this is often called a large radius limit. We will make this phrase more precise when we investigate Kähler moduli spaces in more detail in the next chapter. At any rate, when such an expansion is sensible, then we also observe that the path integral is independent under shifts of $B \to B + B'$, where $[B'] \in H^2(M, 8\pi^2\mathbb{Z})$.

With that basic picture in hand, we then have a clear geometric candidate to qualitatively match $\mathcal{M}_{ac}$: for a Calabi–Yau three-fold with $H^{1,1}(M, \mathbb{R}) = H^2(M, \mathbb{R})$ near a large radius limit, the Kähler class and B-field may be expanded in a basis $\{e_1, \ldots, e_{b_2(M)}\}$ for $H^2(M, \mathbb{Z})$ with some convenient normalizations:

$$[\omega] = 2\pi \sum_\alpha r^\alpha e_\alpha , \qquad [B] = -4\pi \sum_\alpha \vartheta^\alpha e_\alpha , \qquad (4.6.4)$$

and we define the complexified Kähler class by

$$J = \left[\frac{i}{4\pi^2}\omega - \frac{1}{8\pi^2}B\right] = \sum_\alpha \underbrace{\tfrac{1}{2\pi}(r^\alpha + i\vartheta^a)}_{=t^\alpha} e_\alpha . \qquad (4.6.5)$$

With this definition, the bosonic action takes the form

$$e^{-S_{bos}} = e^{2\pi i \int_\Sigma \sigma^*(J)} \times \exp^{-\frac{1}{2\pi}\int d^2z \|\bar{\partial}\phi\|^2} , \qquad (4.6.6)$$

where $\|\bar{\partial}\phi\|^2 = g_{a\bar{a}}\bar{\partial}\phi^a\bar{\partial}\overline{\phi}^{\bar{a}} \geq 0$, with equality if and only if $\bar{\partial}\phi^a = 0$, i.e. the map $\sigma : \mathbb{P}^1 \to M$ is holomorphic. These holomorphic maps are known as worldsheet instantons, and the computation just made shows that they are absolute minima of the action.[28]

The full supersymmetric action is of the form

$$S = -2\pi i \int_\Sigma \sigma^*(J) + \{\overline{Q}, \cdot\} , \qquad (4.6.7)$$

so that the dependence of the chiral sector of the theory on the Kähler form is entirely through the complexified Kähler class $J \in H^2(X, \mathbb{C})$. The story of worldsheet instantons in non-linear sigma models is quite venerable [212, 386] and fairly complicated. We will not delve into these matters now, but we will return to them from a gauged linear sigma model perspective in the next chapter. The impatient reader will be well-rewarded by peeking into classic references including [32, 390]. The bottom line is that near a large radius limit we parameterize a neighborhood of $\mathcal{M}_{ac}$ via $h^{1,1}$ complex coordinates $e^{2\pi i t^a}$. Note that these

---

[28]This is a simple instance of a calibrated cycle: in this case the holomorphic curve $\sigma(\Sigma)$ is said to be calibrated by the Kähler form. A more general discussion of calibrated geometry is given in [256].

are single-valued under $\vartheta^a \rightarrow \vartheta^a + 2\pi$, i.e. shifts of the B-field by integral cohomology classes. More generally, we identify the complexified Kähler moduli space $\mathcal{M}_{cK}(M)$ with $\mathcal{M}_{ac}$.

4.6.2 Singular Loci and Special Geometry

The complex structure moduli space of a Calabi–Yau manifold has a natural metric—the Weil-Petersson metric $\mathcal{G}_{c-x}$. The metric is special Kähler [108, 169, 353], and it coincides with the Zamolodchikov metric on $\mathcal{M}_{cc}(M)$. Given a smooth manifold M_0 which corresponds to a point from p in $\mathcal{M}_{cc}(M)$, there are points in $\mathcal{M}_{cc}(M)$ at finite distance from p where the complex structure on M degenerates, and the curvature of $\mathcal{G}_{c-x}$ blows up. The set of all such points is the discriminant locus—a complex co-dimension one subvariety $\mathcal{Z}_B \subset \mathcal{M}_{cc}$, and they lead to singular SCFTs. Quantum corrections to the non-linear sigma model are controlled by $J \in \mathcal{M}_{cK}(M)$ and therefore do not modify the structure of $\mathcal{M}_{cc}(M)$.

The complexified Kähler moduli space mirrors these features: there is a special Kähler metric $\mathcal{G}_{cK}$ and a singular locus $\mathcal{Z}_A \subset \mathcal{M}_{ac}(M)$. Both structures receive an infinite set of quantum corrections. In addition to a three-loop perturbative shift of the pre-potential proportional to $\zeta(3) \times \chi(M)$,[29] there is an infinite set of non-perturbative corrections that can be organized in a power series expansion in the $e^{2\pi i t^a}$ for the pre-potential for $\mathcal{G}_{cK}$. The sum over the worldsheet instantons fails to converge on the singular locus $\mathcal{Z}_A$, a complex co-dimension 1 subvariety. As a result, there is a notable contrast between the classical geometry of M and the SCFT: the former has singularities in real codimension 1 in the Kähler moduli space and is insensitive to the B-field; the SCFT singularities, on the other hand, really depend on the complexified Kähler class. A succinct and modern discussion of special Kähler geometry in this context is given in [254].

4.6.3 Mirror Symmetry on One Leg

An abstract (2,2) SCFT with integral $U(1)_L \times U(1)_R$ charges and $c = \bar{c} = 9$ has a moduli space $\mathcal{M}_{ac} \times \mathcal{M}_{cc}$; however, there is nothing sacred about the labels (a,c) and (c,c): in fact, the definition is ambiguous, since we can always redefine the $U(1)_L$ charge $q \rightarrow -q$.[30] This observation led to the conjecture that if an SCFT has a large radius limit where $\mathcal{M}_{ac}$ can be interpreted as the complexified Kähler

---

[29]Here $\zeta(3)$ is the Riemann zeta function, and $\chi(M) = 2(h^{1,1} - h^{1,2})$ is the Euler characteristic of M.

[30]There is a similar ambiguity in the right-moving charge; however, if we flip both charges, we merely describe the moduli space in terms of the complex conjugate coordinates.

moduli space of a Calabi–Yau manifold M, while $\mathcal{M}_{cc}$ can be interpreted as the complex structure moduli space of M, then the theory may have another geometric interpretation as a Calabi–Yau manifold M°, whose complexified Kähler moduli space is identified with $\mathcal{M}_{cc}(M)$ [148, 283]. In that case M and M° are a "mirror pair" of Calabi–Yau manifolds with $h^{1,1}(M) = h^{1,2}(M^\circ)$ and $h^{1,2}(M) = h^{1,1}(M^\circ)$, and an isomorphism $\mathcal{M}_{cK}(M) \simeq \mathcal{M}_{c\text{-}x}(M^\circ)$ as special Kähler manifolds.

Shortly after it was made, the mirror conjecture was verified in a class of (2,2) SCFTs built as orbifolds of products of minimal models. These so-called Gepner models were introduced in [192, 193], and strong evidence was given to identify these SCFTs with special points in moduli space of certain Calabi–Yau manifolds.[31] It was shown in [209] that the Gepner construction yielded models in pairs related by a $\mathbb{Z}_2$ isomorphism that exactly implements the flip of the $U(1)_L$ charge. The details of the isomorphism were worked out and applied to make spectacular predictions for enumerative geometry of the Calabi–Yau archetype—the quintic hypersurface in $\mathbb{P}^4$ [110]. The rest, as they say, is history [40, 120, 237, 396].

What then precisely is mirror symmetry? To date there is not a completely straightforward answer to this question. As an illustration of the difficulty, consider the naive proposal—"every Calabi–Yau M has a mirror M°." If this were true, mirror symmetry could be defined on the set of all topological types of Calabi–Yau manifolds. That statement does not hold: there are rigid Calabi–Yau three-folds M with $h^{1,2}(M) = 0$; since $h^{1,1} \geq 1$ for a Calabi-Yau, the mirror of a rigid manifold cannot be given a geometric form.

In the mathematics literature, the question is rephrased in terms of an isomorphism of a pair of open/closed topological string theories and an equivalence of two triangulated categories. In physics literature we would like an isomorphism of a pair of SCFTs. The definitions make sense in either the physics or mathematics parlance, but they are vacuous without a way of producing isomorphic pairs. We will describe a large class of such conjectured pairs in the next chapter.

There is a related fundamental question about (2,2) SCFTs with integral $U(1)_L \times U(1)_R$ charges: which theories have a large radius limit? To the author's knowledge, every construction currently known does have a large radius limit; i.e. by choosing the sign of the $U(1)_L$ charge appropriately, there is always a description of the moduli space with

$$\mathcal{M}_{ac} \simeq \mathcal{M}_{cK}(M) \qquad \text{and} \qquad \mathcal{M}_{cc} \simeq \mathcal{M}_{c\text{-}x}(M) \qquad (4.6.8)$$

for some smooth Calabi–Yau M. It seems likely that there are (2,2) SCFTs without such a limit, but it is not easy to find them. R-charge integrality is a key constraint: without it, it is easy to construct isolated (2,2) SCFTs. It would be most instructive to find examples (or to show that none exist) of isolated theories or, more generally, theories without a large radius limit but with integral charges.

---

[31]The Gepner models were classified in [288].

4.6.4   The Standard Embedding: Heterotic Perspective

A (2,2) SCFT is, in particular, a (0,2) SCFT, and in general it will have deformations
that break the left-moving supersymmetry but preserve the right-moving superconformal algebra, $U(1)_L$, and charge integrality. Such marginal deformations on the
(2,2)-preserving locus take a canonical form:

$$T_{\mathcal{M}_{02}}\big|_{\text{(2,2) locus}} = T_{\mathcal{M}_{ac}} \oplus T_{\mathcal{M}_{cc}} \oplus \mathcal{V}_{02} . \qquad (4.6.9)$$

The first factors are the (2,2)–preserving deformations, and they are, by the
arguments reviewed above, unobstructed. The vector space $\mathcal{V}_{02}$ describes (0,2) and
$U(1)_L$ marginal deformations away from the (2,2) locus. We can understand this
canonical decomposition as follows. By definition $T_{\mathcal{M}_{02}}$ is isomorphic to the set of
right-chiral-primary operators with $\overline{q} = 1$, $q = 0$, and $h = 1$. On the (2,2) locus any
such operator can be uniquely decomposed as

$$\mathcal{O}_{02} = G^-_{-1/2} \cdot \mathcal{O}_{cc} + G^+_{-1/2} \cdot \mathcal{O}_{ac} + \mathcal{O}_{\mathcal{V}} , \qquad (4.6.10)$$

where $\mathcal{O}_{\mathcal{V}}$ is a (2,2) primary operator that is not chiral primary or anti-chiral primary
on the left. Here the $G^{\pm}$ are the left-moving supercurrents of the (2,2) theory.

While the marginal deformations at the (2,2) locus have a canonical decomposition, this structure is not retained away from the locus. We can separate a few
issues:

1. the dimension of $\mathcal{V}_{02}$ jumps at special loci on the (2,2) moduli space;
2. a generic point in $\mathcal{M}_{02}$ need not have a canonical split of the unobstructed
   deformations;
3. in general $\mathcal{M}_{02}$ has distinct branches.

The first of these is very useful, since it enables us to study some features of (0,2)
deformations by moving in the reasonably well-understood (2,2) moduli space [37,
41]. We will meet the next two points in what follows.

The spacetime heterotic theory offers an invaluable perspective and organizing
principle for these features, so we will take a moment to review this familiar story
for the $E_8 \times E_8$ string [327]. As a first step, we work through the details of obtaining
the massless spectrum. This might seem a little bit technically tedious, but it is well-
worth the trouble: while we will just develop the story for (2,2) SCFTs, essentially
identical statements follow for (0,2) SCFTs with integral $U(1)_L \times U(1)_R$ charges.

Massless Spacetime Fermions

The first ingredient in the construction is the "internal theory"—a modular-invariant
(2,2) SCFT with $c = \overline{c} = 9$ and integral $U(1)_L \times U(1)_R$ charges. The theory has
separate left- and right-moving spectral flow, which, starting from the NS-$\overline{\text{NS}}$ sector,

allows us to construct the NS-$\overline{\text{R}}$, R-$\overline{\text{NS}}$, and R-$\overline{\text{R}}$ sectors.[32] We recall that we obtain an isomorphism between the $\overline{\text{NS}}$ chiral primary states with $\overline{\text{R}}$ ground states:

$$|\mathcal{O}^{\overline{q}}_{\text{cc}}; \overline{\text{NS}}\rangle \longleftrightarrow |\mathcal{O}^{\overline{q}}_{\text{cc}}; \overline{\text{R}}\rangle_{\overline{q}-3/2} . \tag{4.6.11}$$

To complete this to a $E_8 \times E_8$ heterotic string theory, we tensor the internal theory with the following ingredients:

1. a left-moving level 1 "hidden" $\mathfrak{e}_8$ current algebra with $c = 8$;
2. 10 left-moving Majorana–Weyl fermions ξ^A with a fermion number operator $(-1)^{F_\xi}$ and $c = 5$;
3. a free $c = 4$, $\overline{c} = 6$ $(0,1)$ SCFT for the uncompactified Minkowski spacetime $\mathbb{R}^{1,3}$ with right-moving fermion number $(-1)^{\overline{F}}$;
4. the superconformal $(0,1)$ supergravity ghosts with $c = -26$ and $\overline{c} = -15$.

A consistent string theory is obtained by constructing the $(\text{NS}, \overline{\text{NS}})$, $(\text{R}, \overline{\text{NS}})$, $(\text{NS}, \overline{\text{R}})$, $(\text{R}, \overline{\text{R}})$ sectors, imposing left and right GSO projections, and restricting to the BRST cohomology. The resulting theory will have $\mathcal{N} = 1$ spacetime supersymmetry on $\mathbb{R}^{1,3}$.[33]

We will be interested in the spacetime massless states, and to describe these it is simplest to work in light-cone gauge, where we drop the ghosts and the longitudinal $\mathbb{R}^{1,1} \subset \mathbb{R}^{1,3}$ worldsheet fields; then the resulting CFT has central charges $c^{\text{tot}} = 24$ and $\overline{c}^{\text{tot}} = 12$. $\mathcal{N} = 1$ supersymmetry means that we can focus on the massless spacetime fermions, which arise in the $\overline{\text{R}}$ sector. Level-matching for the massless states requires

$$\overline{L}_0^{\text{tot}} - \frac{\overline{c}^{\text{tot}}}{24} = 0 \implies \overline{L}_0^{\text{int}} - \frac{\overline{c}^{\text{int}}}{24} = 0 , \quad \text{and} \quad L_0^{\text{tot}} - \frac{c^{\text{tot}}}{24} = 0 . \tag{4.6.12}$$

The first statement means that we restrict to internal $\overline{\text{R}}$ ground states, while the second imposes the restriction to $L_0^{\text{tot}} = 1$.

It remains to discuss the GSO projections. The left-moving GSO is onto states with $e^{i\pi J_0}(-1)^{F_\xi} = 1$, and the right-moving GSO is onto states with $e^{i\pi \overline{J}_0}(-1)^{\overline{F}} = 1$.[34] In the $\overline{\text{R}}$ ground states $(-1)^{\overline{F}} = e^{\pm i\pi/2}$, with $\pm$ sign corresponding to the spacetime chirality and the states $|\pm\rangle$; similarly $(-1)^{F_\xi} = e^{\pm i\pi/2}$, with $+$ sign for

---

[32] The reader may wish to review the spectral flow discussion in Chap. 2.

[33] In pondering modular invariance it is useful to think of the internal $(2,2)$ SCFT as "replacing" an $\mathbb{R}^6$ or T^6 in the usual construction of the ten-dimensional heterotic string. Spectral flow allows us to define NS/R sectors for the internal theory and show that they contribute in the same way to transformations of the worldsheet one-loop partition function as in the free theory. This is explained in detail in [194, 370].

[34] It may be useful to review Exercise 2.9 on twisted fermions to see a convenient definition of fermion numbers and charges in the R sectors of the free theories.

the $|\overline{\mathbf{16}}\rangle$ and $-$ sign for the $|\mathbf{16}\rangle$ representations of $\mathfrak{so}(10)$. Now it is an easy matter to list the states.

First consider the R,$\overline{\text{R}}$ ground states. In the internal theory they are of the form $|\mathcal{O}_{cc}^{q,\bar{q}}; \text{R}, \overline{\text{R}}\rangle_{q-3/2,\bar{q}-3/2}$. The (c,c) ring operators of the internal theory have only a limited range for $q, \bar{q}$:

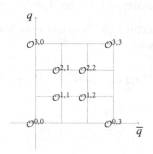

$\mathcal{O}^{0,0}$ is a fancy way of writing the identity operator, while $\mathcal{O}^{3,0}$ and $\mathcal{O}^{0,3}$ are the images of the identity by unit left/right spectral flow, and $\mathcal{O}^{3,3} = \mathcal{O}^{3,0}\mathcal{O}^{0,3}$. We do not include $\mathcal{O}^{s,0}$ or $\mathcal{O}^{0,s}$ operators with $0 < s < 3$. A (2,2) theory with such operators is considerably simpler than the generic one: $\mathcal{O}^{0,1}$ is a free fermion, and $\mathcal{O}^{0,2}$ is a right-moving KM current that will enhance $\text{U}(1)_\text{R}$ to a larger symmetry. The latter is present in (0, 4) theories, and the former is familiar from toroidal compactification. We will assume that such extended symmetry structures are absent, and that implies that their would-be images under unit spectral flow are absent as well. It is now a simple matter to write down the states that survive the GSO projection. We write the states as $|\xi\rangle \otimes |\text{int.}\rangle \otimes |\text{Mink}\rangle$ and obtain[35]

$$|\overline{\mathbf{16}}\rangle \otimes |\mathcal{O}_{cc}^{3,3}; \text{R}, \overline{\text{R}}\rangle_{+3/2,+3/2} \otimes |+\rangle \qquad |\mathbf{16}\rangle \otimes |\mathcal{O}_{cc}^{0,0}; \text{R}, \overline{\text{R}}\rangle_{-3/2,-3/2} \otimes |-\rangle$$

$$|\mathbf{16}\rangle \otimes |\mathcal{O}_{cc}^{0,3}; \text{R}, \overline{\text{R}}\rangle_{-3/2,+3/2} \otimes |+\rangle \qquad |\overline{\mathbf{16}}\rangle \otimes |\mathcal{O}_{cc}^{3,0}; \text{R}, \overline{\text{R}}\rangle_{+3/2,-3/2} \otimes |-\rangle$$

$$|\overline{\mathbf{16}}\rangle \otimes |\mathcal{O}_{cc}^{1,1}; \text{R}, \overline{\text{R}}\rangle_{-1/2,-1/2} \otimes |+\rangle \qquad |\mathbf{16}\rangle \otimes |\mathcal{O}_{cc}^{2,2}; \text{R}, \overline{\text{R}}\rangle_{+1/2,+1/2} \otimes |-\rangle$$

$$|\mathbf{16}\rangle \otimes |\mathcal{O}_{cc}^{2,1}; \text{R}, \overline{\text{R}}\rangle_{+1/2,-1/2} \otimes |+\rangle \qquad |\overline{\mathbf{16}}\rangle \otimes |\mathcal{O}_{cc}^{1,2}; \text{R}, \overline{\text{R}}\rangle_{-1/2,+1/2} \otimes |-\rangle .$$

$$(4.6.13)$$

The first two lines correspond to spacetime fermions in vector multiplets: a simple way to see this is that they arise from half-unit of right-moving spectral flow applied to the R,$\overline{\text{NS}}$ currents. The next two lines describe spacetime fermions in chiral and anti-chiral multiplets; the first column contains the chiral fermions, while the second contains the anti-chiral conjugates. The full set of states is spacetime CPT-invariant: for each state in a representation of $\mathfrak{so}(10) \oplus \mathfrak{u}(1)_L$ there is a state of opposite

---

[35] We will suppress the omni-present momentum-dependent $e^{ik\cdot X}$ factor, and will not write explicit polarization tensors for the gravitino or gauginos.

spacetime chirality and conjugate $\mathfrak{so}(10) \oplus \mathfrak{u}(1)_L$ representation. We see a number of general features:

1. the spectrum is CPT invariant, with CPT implemented by unit spectral flow, and the spacetime spectrum can be determined by describing states with $\bar{q} \leq 0$;
2. $\overline{R}$ states with $\bar{q} = -3/2$ correspond to vector multiplets;
3. $\overline{R}$ states with $\bar{q} = -1/2$ correspond to chiral multiplets.

So, we can organize the spacetime massless spectrum from R,$\overline{R}$ sector according to $\mathfrak{so}(10) \oplus \mathfrak{u}(1)_L$ representations:

$$\text{gauginos:} \quad \mathbf{16}_{-3/2} \oplus \overline{\mathbf{16}}_{3/2}, \qquad \text{chiral:} \quad \mathbf{16}_{1/2}^{\oplus h_{1,1}^{cc}} \oplus \overline{\mathbf{16}}_{-1/2}^{\oplus h_{2,1}^{cc}}, \qquad (4.6.14)$$

where

$$h_{cc}^{q,\bar{q}} = \dim H_{cc}^{q,\bar{q}}. \qquad (4.6.15)$$

Spectral flow implies the isomorphism $H_{cc}^{2,1} \simeq H_{ac}^{1,1}$, so that $h_{cc}^{2,1} = h_{ac}^{1,1}$.

Next, we examine the NS,$\overline{R}$ sector. This contains four types of states.

i. The universal gravitino and dilatino states have $h = 1$ from an oscillator applied to the Minkowski vacuum:

$$|0; \text{NS}\rangle \otimes |\mathcal{O}_{cc}^{0,0}; \text{NS}, \overline{R}\rangle_{0,-3/2} \otimes (\partial X)_{-1}|-\rangle$$

$$|0; \text{NS}\rangle \otimes |\mathcal{O}_{cc}^{0,3}; \text{NS}, \overline{R}\rangle_{0,+3/2} \otimes (\partial X)_{-1}|+\rangle. \qquad (4.6.16)$$

ii. Next, there are the universal $\mathfrak{u}(1)_L$ gaugino states

$$|0; \text{NS}\rangle \otimes J_{-1}|\mathcal{O}_{cc}^{0,0}; \text{NS}, \overline{R}\rangle_{0,-3/2} \otimes |-\rangle$$

$$|0; \text{NS}\rangle \otimes J_{-1}|\mathcal{O}_{cc}^{0,3}; \text{NS}, \overline{R}\rangle_{0,3/2} \otimes |+\rangle, \qquad (4.6.17)$$

and the universal $\mathfrak{so}(10)$ gaugino states

$$\xi\xi|0; \text{NS}\rangle \otimes |\mathcal{O}_{cc}^{0,0}; \text{NS}, \overline{R}\rangle_{0,-3/2} \otimes |-\rangle$$

$$\xi\xi|0; \text{NS}\rangle \otimes |\mathcal{O}_{cc}^{0,3}; \text{NS}, \overline{R}\rangle_{0,3/2} \otimes |+\rangle. \qquad (4.6.18)$$

Here $\xi\xi$ is an informal but useful notation for the oscillators $\xi_{-1/2}^A \xi_{-1/2}^B$.

In this sector we also find the universal "hidden" $\mathfrak{e}_8$ gauginos; they will not play a role in our tale, so we will ignore them.

iii. There are also $\mathfrak{so}(10) \oplus \mathfrak{u}(1)_L$-charged matter states:

$$|0; \text{NS}\rangle \otimes |\mathcal{O}_{cc}^{2,1}; \text{NS}, \overline{R}\rangle_{+2,-1/2} \otimes |+\rangle$$

$$|0; \text{NS}\rangle \otimes |\mathcal{O}_{ac}^{-2,2}; \text{NS}, \overline{R}\rangle_{-2,1/2} \otimes |-\rangle$$

$$\xi|0; \text{NS}\rangle \otimes |\mathcal{O}_{ac}^{-1,1}; \text{NS}, \overline{R}\rangle_{-1,-1/2} \otimes |+\rangle$$

$$\xi |0; NS\rangle \otimes |\mathcal{O}_{cc}^{1,2}; NS, \overline{R}\rangle_{+1,+1/2} \otimes |-\rangle$$

$$|0; NS\rangle \otimes |\mathcal{O}_{ac}^{-2,1}; NS, \overline{R}\rangle_{-2,-1/2} \otimes |+\rangle$$

$$|0; NS\rangle \otimes |\mathcal{O}_{cc}^{2,2}; NS, \overline{R}\rangle_{+2,+1/2} \otimes |-\rangle$$

$$\xi |0; NS\rangle \otimes |\mathcal{O}_{cc}^{1,1}; NS, \overline{R}\rangle_{1,-1/2} \otimes |+\rangle$$

$$\xi |0; NS\rangle \otimes |\mathcal{O}_{ac}^{-1,2}; NS, \overline{R}\rangle_{-1,+1/2} \otimes |-\rangle . \tag{4.6.19}$$

iv. Finally, there are gauge-singlets that have no Minkowski oscillators. These take the form

$$|0, NS\rangle \otimes |\mathcal{V}_c^{0,1}; NS, \overline{R}\rangle_{0,-1/2} \otimes |+\rangle \quad |0, NS\rangle \otimes |\mathcal{V}_c^{0,2}; NS, \overline{R}\rangle_{0,+1/2} \otimes |-\rangle , \tag{4.6.20}$$

where $\mathcal{V}_c^{0,\overline{q}}$ is an NS,$\overline{\text{NS}}$ right-chiral-primary operator with $\overline{q}, h = 1, q = 0$ and trivial OPE with the $\mathfrak{u}(1)_L$ current J. Such operators are in 1:1 correspondence with marginal (0,2) deformations of the internal SCFT. In the special case of (2,2) SCFTs we can canonically decompose these as

$$\mathcal{V}_c^{0,1} = G_{-1/2}^- \cdot \mathcal{O}_{cc}^{1,1} + G_{-1/2}^+ \cdot \mathcal{O}_{ac}^{-1,1} + \mathcal{V}_{02}^{0,1} ,$$

$$\mathcal{V}_c^{0,2} = G_{-1/2}^- \cdot \mathcal{O}_{cc}^{1,2} + G_{-1/2}^+ \cdot \mathcal{O}_{ac}^{-1,2} + \mathcal{V}_{02}^{0,2} . \tag{4.6.21}$$

So, putting all of these features together, we obtain the massless spectrum of the $\mathcal{N} = 1 \, d = 4$ theory that corresponds to the compactification: in addition to the gravity multiplet, the theory has vector multiplets organized according to the canonical decomposition

$$\mathfrak{e}_6 \supset \mathfrak{so}(10) \oplus \mathfrak{u}(1)_L$$

$$\mathbf{78} = \mathbf{45}_0 \oplus \mathbf{16}_{-3/2} \oplus \overline{\mathbf{16}}_{+3/2} \oplus \mathbf{1}_0, \tag{4.6.22}$$

and chiral multiplets. The chiral multiplets include the universal axio-dilaton multiplet and a model-dependent set of gauge-charged and gauge neutral multiplets. The former arrange themselves according to

$$\mathfrak{e}_6 \supset \mathfrak{so}(10) \oplus \mathfrak{u}(1)_L$$

$$\mathbf{27} = \mathbf{16}_{+1/2} \oplus \mathbf{10}_{-1} \oplus \mathbf{1}_{+2}$$

$$\overline{\mathbf{27}} = \overline{\mathbf{16}}_{-1/2} \oplus \mathbf{10}_{+1} \oplus \mathbf{1}_{-2} , \tag{4.6.23}$$

and the charged chiral spectrum is $\mathbf{27}^{\oplus h_{cc}^{1,1}} \oplus \overline{\mathbf{27}}^{\oplus h_{ac}^{-1,1}}$. Finally, there are the gauge singlets

$$\mathbf{1}^{\oplus h_{cc}^{1,1}} \oplus \mathbf{1}^{\oplus h_{ac}^{-1,1}} \oplus \mathbf{1}^{\oplus h\nu} \qquad (4.6.24)$$

that exhibit the "canonical" splitting into the (c,c), (a,c), and "other" moduli.

Exercise 4.13 It is an enlightening exercise to verify the details of the preceding two paragraphs; a good starting point is to derive the massless fermions for the $E_8 \times E_8$ string on $\mathbb{R}^{1,9}$. The computation of the zero-point energies in that basic story then goes over in a straight-forward fashion to the compactification. More details can be found in standard string texts [207, 327].   □

Spacetime and Worldsheet Interplay

Consider now the vacuum structure of the spacetime theory. We focus on perturbative tree-level string theory, so that the gravity multiplet and the axio-dilaton are simply spectators. The supersymmetric deformations of the Minkowski vacuum can then be discussed in perturbative spacetime gauge theory, and they are divided into two categories: there are Higgs branch deformations, where the $\mathbf{27}$ and $\overline{\mathbf{27}}$ scalars obtain vacuum expectation values. The obstructions to these are the $\mathcal{N} = 1$ D-terms and F-terms. In addition, there are also gauge-singlet deformations, and the obstructions are encoded by the F-terms in the $\mathcal{N} = 1$ superpotential W. On the $\mathfrak{e}_6 \oplus \mathfrak{e}_8$-preserving locus we collectively denote the singlets by T_{ac}, Z_{cc}, and U_ν, and the condition for a Minkowski vacuum is

$$W(T, Z, U) = 0\,, \qquad \frac{\partial W}{\partial T} = 0\,, \qquad \frac{\partial W}{\partial Z} = 0\,, \qquad \frac{\partial W}{\partial U} = 0\,. \qquad (4.6.25)$$

Clearly in a "suitably generic" vacuum, i.e. one where W is generic enough, this gives an over-determined system. In other words, the vacua will consist of isolated points. As the preceding discussion has already indicated, finding such generic theories is not at all trivial, and in a (2,2) SCFT, there are always at least $h_{ac}^{-1,1} + h_{cc}^{1,1}$ flat directions, where we set the (0,2) singlets U_ν to 0. This means the singlet superpotential takes the form

$$W = m_{IJ}(T, Z)U^I U^J + g_{IJK}(T, Z)U^I U^J U^K + \ldots . \qquad (4.6.26)$$

As we move around in the (2,2) moduli space, some of the U chiral multiplets can become massive, and in general they can also be obstructed at higher order. Turning the statement around, a $U \neq 0$ vacuum expectation value may only be possible for particular choices of T and Z. As a toy example, we can consider $W = TZU^2$, which has flat directions $T = Z = 0$ and $U = 0$. The two branches have different dimensions, and the moduli space is not a smooth manifold at the origin.[36] The jumping in the U singlets is a characteristic of families of (2,2) SCFTs, and there are a number of techniques for studying it at least at different limiting points in the (2,2) moduli space [41].

The general determination of $W(T, Z, U)$ is a very difficult task, since it is equivalent to finding the full obstruction theory for superconformal perturbation theory. However, we can hope to at least constrain it at points that exhibit some enhanced symmetries, for instance discrete spacetime R-symmetries which can arise at special loci of (2,2) SCFTs. These symmetries become much more powerful when we assume that conformal perturbation theory about a particular vacuum has a finite radius of convergence. In that happy situation, W can be expressed as a power series in the holomorphic superfields with a non-zero radius of convergence [137, 138], so that it is sufficient to study and constrain terms order by order in the expansion to obtain complete results. The very fact that (0,2) obstruction theory is encoded in a holomorphic superpotential is itself a powerful statement, and it would be useful to prove this directly in the SCFT.[37]

The existence worldsheet non-perturbative effects in non-linear sigma models at first sight seems to contradict the previous paragraph. It is crucial to remember that the large radius non-linear sigma model, with its α' expansion, is not a good example of conformal perturbation theory—the free theory (at infinite radius) is at infinite distance from any finite radius theory, and this leads an asymptotic perturbative expansion and the consequent non-perturbative effects. Thus, if one only has a construction of a theory in the large radius limit, then one must worry about non-perturbative destabilization. On the other hand, given a family of theories (that perhaps extends out to the large radius limit) that includes a point at finite distance in the moduli space, all-orders results become exact in an open neighborhood of the point.

Returning to the (2,2) compactification, we now turn to the massless gauge-charged chiral multiplets. In a general $\mathcal{N} = 1$ gauge theory, or, for that matter, in a generic (0,2) compactification, there is no relation between the charged matter sector and the singlets. In this sense the (2,2) compactification is very much non-generic, since there is a correspondence between the (a,c) moduli and matter fields in $\overline{\mathbf{27}}$, as well as between the (c,c) moduli and matter fields in $\mathbf{27}$. This immediately leads to a simple prediction: while gauge invariance allows for a mass term of the form $m_{ij'}(T, Z, U)\mathbf{27}^{j'} \cdot \overline{\mathbf{27}}^{i}$, this mass term is necessarily zero

---

[36]The singularity need not be a singular SCFT: the correlation functions of the SCFT can remain perfectly finite as we approach the origin from either branch, as long as we keep track of the full SCFT spectrum, which will vary smoothly. If we artificially restrict attention to just the marginal operators, there will be a discontinuity at the origin. On the other hand, there are also singularities that cannot be understood in terms of the SCFT degrees of freedom, and in this case, the SCFT will be singular.

[37]We refer to [28] and references therein for results obtained in the large radius limit. D-branes in type II theories also offer an example with some similar features, where the general open-close string deformation theory is encoded in a superpotential, which can be directly studied in the framework of topological field theory [40].

in a (2,2) compactification! In fact, the (2,2) worldsheet structure determines the Zamolodchikov metric on $\mathcal{M}_{ac} \times \mathcal{M}_{cc}$ in terms of holomorphic data encoded in the Yukawa couplings

$$W \supset \mathcal{Y}^{ac}_{ijk}(T)\overline{\mathbf{27}}^i \cdot \overline{\mathbf{27}}^j \cdot \overline{\mathbf{27}}^k + \mathcal{Y}^{cc}_{i'j'k'}(Z)\mathbf{27}^{i'} \cdot \mathbf{27}^{j'} \cdot \mathbf{27}^{k'} , \qquad (4.6.27)$$

This is how special geometry of the (2,2) moduli space arises from the heterotic perspective [150].

4.6.5 Away from Standard Embedding via Higgs

A weakly coupled gauge theory may have Higgs branches, where the gauge group is broken down to a subgroup by non-zero expectation values of the charged fields, and it is interesting to consider the Higgs branches for the theories that arise from (2,2) compactifications.[38] For instance, the embedding we just used to construct the massless spectrum in (4.6.22), (4.6.23) shows that it should be possible to Higgs $\mathfrak{e}_6 \to \mathfrak{so}(10)$. If this can be done while preserving supersymmetry, then the deformation has an interpretation as a (0,2) marginal deformation of the worldsheet SCFT. In the supersymmetric Higgs mechanism a massive vector multiplet has the same degrees of freedom as a massless vector and a chiral multiplet, and the massive gauge bosons have to fit into representations of the unbroken gauge group. Taking a look at the decompositions above and ignoring D- and F-terms for the moment, the charged spectrum should behave as follows:

$$\mathfrak{e}_6 , \quad \mathbf{27}^{\oplus n} \oplus \overline{\mathbf{27}}^{\oplus \bar{n}} \to \mathfrak{so}(10) ,$$

$$\mathbf{16}^{\oplus (n-1)} \oplus \overline{\mathbf{16}}^{\oplus (\bar{n}-1)} \oplus \mathbf{10}^{\oplus (n+\bar{n})} \oplus \mathbf{1}^{\oplus (n+\bar{n}-1)} . \qquad (4.6.28)$$

It is not too hard to argue that a solution to the D-terms exists: if we only turn on $\mathbf{1}_{+2} \subset \mathbf{27}$ and $\mathbf{1}_{-2} \subset \overline{\mathbf{27}}$ expectation values, then the only possibly non-zero D-term is the one for the $\mathfrak{u}(1)_L$ symmetry, and it is always possible to set it to zero by giving expectation values to scalars with both signs of $\mathfrak{u}(1)_L$ charge. The F-terms are more complicated. Already the cubic Yukawa couplings lead to some non-trivial terms, since, qualitatively, we have

$$\mathbf{27} \cdot \mathbf{27} \cdot \mathbf{27} = \mathbf{16}_{+1/2} \cdot \mathbf{10}_{-1} \cdot \mathbf{16}_{+1/2} + \mathbf{10}_{-1} \cdot \mathbf{10}_{-1} \cdot \mathbf{1}_{+2} . \qquad (4.6.29)$$

Thus, expectation values for $\mathbf{1}_{+2}$ will lead to mass terms for the $\mathbf{10}$ representations. If the Yukawa terms and the Higgs branch expectation values are both suitably

---

[38] This was first discussed in the context of large radius compactifications in [384], but the idea applies to general deformations of (0,2) SCFTs.

generic, we might expect that all the **10** chiral multiplets are lifted on the Higgs branch.

More general terms in the superpotential involve powers of the cubic invariants and **27 · $\overline{27}$**. As long as we give expectation values just to the $\mathfrak{so}(10)$ singlets, we see that terms involving at least one cubic invariant cannot lift the Higgs branch. On the other hand, a term of the form $(\mathbf{27} \cdot \overline{\mathbf{27}})^k$ or $U(\mathbf{27} \cdot \overline{\mathbf{27}})^k$, where U is a massless gauge-neutral singlet, could potentially lift the deformation. Remarkably, there are compactifications where discrete R-symmetries can rule out terms of this form [73, 135, 137, 293].

At the infinitesimal level, the construction has a clear interpretation in the worldsheet theory. Organize the 10 left-moving fermions ξ^A into 5 Weyl fermions ξ^α and their conjugates $\overline{\xi}_\alpha$. Now consider the $c = 10$, $\overline{c} = 9$ SCFT that is a product of the $\xi^5, \overline{\xi}_5$ free theory and the internal SCFT. This theory has a $\mathfrak{u}(1)_L^\xi \oplus \mathfrak{u}(1)_L \oplus \mathfrak{u}(1)_R$ KM symmetry, with $U(1)_L^\xi$ corresponding to the current $J_\xi =: \xi^5 \overline{\xi}_5 :$. We can then write, at first order, a deformation of the two-dimensional SCFT with holomorphic parameters g^i and $g^{\prime i'}$ and

$$\Delta S = \int d^2 z \, \overline{G}^-_{-1/2} \cdot \left[g^{\prime i'} \xi \mathcal{O}^{-1,1}_{ac,i'} + g^i \overline{\xi} \mathcal{O}^{1,1}_{cc,i} \right] + \text{h.c.} . \tag{4.6.30}$$

This breaks $\mathfrak{u}(1)_L^\xi \oplus \mathfrak{u}(1)_L \to \mathfrak{u}(1)_L'$ with current $J' = J_\xi + J$. J' defines a level $r = 4$ KM algebra, and as long as the deformation is exactly marginal, which requires at least a D-term constraint on the g' and g couplings, then J' can be used to construct A/2 and B/2 topological heterotic rings, as well as to build the left-moving Ramond sector and GSO projection for the $c = 10$, $\overline{c} = 9$ (0,2) SCFT. The fact that J' does both of these jobs is no accident: as the next exercise shows, there is an intimate relation between the topological heterotic rings and gauge-charged massless spacetime fermions.

Exercise 4.14 The spacetime Higgs $\mathfrak{e}_6 \to \mathfrak{so}(10)$ tale is a special construction of more general $\mathfrak{so}(10)$ compactifications that use a $c = 10$, $\overline{c} = 9$ SCFT with a level $r = 4 \, \mathfrak{u}(1)_L$ symmetry, integral $U(1)_L \times U(1)_R$ charges, and the assumptions on the spectrum spelled out in Sect. 3.4 that also guarantee the existence of A/2 and B/2 topological heterotic rings. Find the massless fermions in such a compactification.[39]

First consider the R,$\overline{\text{R}}$ sector. Using left- and right-moving spectral flow operators, it follows that the massless fermions arise from internal states

$$|\mathcal{O}^{q,\overline{q}}_{B/2}; \text{R}, \overline{\text{R}}\rangle_{q-r/2, \overline{q}-3/2} . \tag{4.6.31}$$

---

[39] A classic reference for this analysis is [144]; although it uses the non-linear sigma model for the internal SCFT, the worldsheet analysis is essentially identical. The reader may find it instructive to take a look and compare the abstract A/2 and B/2 structures to the non-linear sigma model realizations. We will have more to say on this in the next section.

Assume that the SCFT does not have extra holomorphic currents, so that the B/2 cohomology is trivial for $(q, \overline{q}) = (t, 0)$ or $(q, \overline{q}) = (0, s)$ for $0 < t < 4$ and $0 < s < 3$. The right-moving energies and GSO projections work exactly the same way as before, with the result that spacetime anti-chiral gauginos correspond to states with $\overline{q} = -3/2$, while massless fermions in chiral multiplets correspond to states with $\overline{q} = -1/2$. The left-moving GSO projection acts on the 8 fermions ξ^A, $A = 1, \ldots 8$, with $(-1)^{F_\xi}|8^c\rangle = |8^c\rangle$ and $(-1)^{F_\xi}|8^s\rangle = -|8^s\rangle$.[40] Show that the massless fermions in the R, $\overline{\text{R}}$ sector with $\overline{q} < 0$ are therefore

$$\text{gauginos} \begin{cases} |\mathcal{O}_{B/2}^{0,0}; \text{R}, \overline{\text{R}}\rangle_{-2,-3/2} & \to 8_{-2,-3/2}^c \\ |\mathcal{O}_{B/2}^{4,0}; \text{R}, \overline{\text{R}}\rangle_{+2,-3/2} & \to 8_{+2,-3/2}^c \end{cases}$$

$$\text{chiral} \begin{cases} |\mathcal{O}_{B/2}^{1,1}; \text{R}, \overline{\text{R}}\rangle_{-1,-1/2} & \to 8_{-1,-1/2}^s \\ |\mathcal{O}_{B/2}^{2,1}; \text{R}, \overline{\text{R}}\rangle_{0,-1/2} & \to 8_{0,-1/2}^c \\ |\mathcal{O}_{B/2}^{3,1}; \text{R}, \overline{\text{R}}\rangle_{+1,-1/2} & \to 8_{+1,-1/2}^s \end{cases} \qquad (4.6.32)$$

Next, examine the NS, $\overline{\text{R}}$ sector. Use the bounds given in Sect. 3.4 to show that the massless gauge-charged (i.e. charged under the linearly realized $\mathfrak{so}(8) \oplus \mathfrak{u}(1)_L$) states with $\overline{q} < 0$ are classified as follows. In listing these, we omit the explicit tensor product with the Minkowski Ramond ground states (that information is determined by right-moving GSO and the $\overline{q}$ charge), and we also omit the universal gravitino, dilatino, and hidden $\mathfrak{e}_8$ gauginos.

$$\text{gauginos} \begin{cases} \xi\xi|0; \text{NS}\rangle \otimes |\mathcal{O}_{B/2}^{0,0}; \text{NS}, \overline{\text{R}}\rangle_{0,-3/2} & \to 28_{0,-3/2} \\ |0; \text{NS}\rangle \otimes J_{-1}|\mathcal{O}_{B/2}^{0,0}; \text{NS}, \overline{\text{R}}\rangle_{0,-3/2} & \to 1_{0,-3/2} \end{cases}$$

$$\text{chiral} \begin{cases} \xi|0; \text{NS}\rangle \otimes |\mathcal{O}_{B/2}^{1,1}; \text{NS}, \overline{\text{R}}\rangle_{1,-1/2} & \to 8_{1,-1/2}^v \\ \xi|0; \text{NS}\rangle \otimes |\mathcal{O}_{A/2}^{-1,1}; \text{NS}, \overline{\text{R}}\rangle_{-1,-1/2} & \to 8_{-1,-1/2}^v \\ |0; \text{NS}\rangle \otimes |\mathcal{O}_{B/2}^{2,1}; \text{NS}, \overline{\text{R}}\rangle_{2,-1/2} & \to 1_{+2,-1/2} \\ |0; \text{NS}\rangle \otimes |\mathcal{O}_{A/2}^{-2,1}; \text{NS}, \overline{\text{R}}\rangle_{-2,-1/2} & \to 1_{-2,-1/2} \end{cases} \qquad (4.6.33)$$

These fermions fit into representations of $\mathfrak{so}(10)$ according to the decomposition

$$\mathfrak{so}(10) \supset \mathfrak{so}(8) \oplus \mathfrak{u}(1)_L ,$$

$$45 = 28_0 \oplus 8_{-2}^c \oplus 8_{+2}^c \oplus 1_0 ,$$

$$10 = 8_0^c \oplus 1_{-2} \oplus 1_{+2} ,$$

$$16 = 8_{-1}^s \oplus 8_{+1}^v ,$$

$$\overline{16} = 8_{+1}^s \oplus 8_{-1}^v . \qquad (4.6.34)$$

---

[40]This is a convention for the 8^c and 8^s representations.

This decomposition may look a bit unusual, but $\mathfrak{so}(8)$ triality implies that it is isomorphic to the "usual" one where, for instance, $\mathbf{10} = \mathbf{8}^v \oplus \mathbf{1}_{+2} \oplus \mathbf{1}_{-2}$ and $\mathbf{16} = \mathbf{8}^s \oplus \mathbf{8}^c$.

Thus, the gauge-charged massless chiral spectrum organizes into $\mathfrak{so}(10)$ representations

$$\mathfrak{so}(10)\,, \quad \mathbf{16}^{\oplus h_{B/2}^{1,1}} \oplus \mathbf{10}^{\oplus h_{B/2}^{2,1}} \oplus \overline{\mathbf{16}}^{\oplus h_{B/2}^{3,1}}\,. \tag{4.6.35}$$

There may be additional massless gauge-neutral chiral multiplets that arise from $|\mathcal{O}; \text{NS}, \overline{\text{R}}\rangle$ states, where $\mathcal{O}$ is a right-chiral operator with $\bar{q} = 1$, $q = 0$, and $h = 1$. In a general $(0,2)$ theory these have no relation to the topological heterotic rings. As a result, there is also no natural split of the exactly marginal $(0,2)$ deformations into categories like "Kähler," "complex structure," and "bundle." Such terms arise frequently in discussions of heterotic supergravity, but in view of this superconformal perspective, one should apply the terms with care.

On the bright side, as this example demonstrates, the topological heterotic rings that were described in Sect. 3.4 have a natural interpretation in heterotic compactifications, so that the same sort of worldsheet/spacetime interplay that was discussed in the $(2,2)$ setting continues to be useful in the more general $(0,2)$ setting as well. □

Exercise 4.15 It is a nice exercise to repeat the construction for other values for the level of $\mathfrak{u}(1)_L$; $r = 2$ and $r = 5$ are particularly interesting, since they lead to $\mathfrak{e}_7$ and $\mathfrak{su}(5)$ compactifications, respectively. Note that $c = 6 + r$. These have non-trivial gauge and gravitational anomalies (local for $\mathfrak{su}(5)$ and global for $\mathfrak{e}_7$) in four dimensions, and one can check that the matter spectrum obtained is non-anomalous. In addition, in the $r = 2$ case the left-moving KM algebra of the internal theory is automatically enhanced to $\widehat{\mathfrak{su}(2)}_1$, so that $\mathfrak{e}_7$ is realized via the embedding $\mathfrak{e}_7 \supset \mathfrak{su}(2) \oplus \mathfrak{so}(12)$.

The reader may also find it amusing to consider the $r = 1$ case and show that this leads to unbroken $\mathfrak{e}_8$ spacetime gauge group. □

4.7 The α' Expansion for Heterotic Spacetime

$(2,2)$ SCFTs with integral $U(1)_L \times U(1)_R$ charges lead to relatively simple vacua of the $E_8 \times E_8$ heterotic string, and the spacetime perspective illuminates a number of basic features of the SCFTs. Conversely, SCFT results translate rather directly into non-trivial statements in spacetime and lead to a number of conclusions that would not hold in a generic $\mathcal{N} = 1$, $d = 4$ gauge theory. Calabi–Yau geometry and large radius limit non-linear sigma models then provide a way to generate a large class of such SCFTs. Other constructions, such as Gepner models or Landau–Ginzburg theories are also available.

Starting from a (2,2) SCFT, we might try to obtain a nearby (0,2) deformation by perturbing the theory by a (0,2) marginal deformation. The spacetime perspective and flat $\mathfrak{e}_6 \to \mathfrak{so}(10)$ directions in the effective theory offer one such route. Another route would be to turn on a gauge-neutral deformation, i.e. one of the U marginal deformations in the terminology of the previous section.

In this section we will follow a different route: we will retreat to large radius and therefore non-linear sigma models where the target space is a Calabi–Yau manifold equipped with a stable holomorphic bundle. In contrast with the more general heterotic geometries we described above, such backgrounds have a large radius limit, where the volume of M and of any compact cycle in M becomes large. This can act as a starting point for a perturbation theory. One can make a fair bit of progress in analyzing aspects of the resulting theory to all orders in the α' expansion, but non-perturbative corrections can drastically alter qualitative features of the theory. We will also make some comments on non-linear sigma models with $\mathcal{H} \neq 0$.

4.7.1 Gentle (0,2) Sigma Models: Bundles over Calabi–Yau Manifolds

The starting point for the construction is a compact Calabi–Yau 3-fold M equipped with a Ricci-flat Kähler metric and $H^{q,0}_{\bar{\partial}}(M) = 0$ for $q = 1, 2$. To zeroth order in α' we then have $\mathcal{H} = 0$. Let ω_0 and Ω_0 be, respectively, the Kähler and (3,0) form as above. Their conformal closure is then guaranteed and implies that the dilaton $\varphi = \varphi_0$ is constant.

We choose the bundle to be a rank r holomorphic bundle $\mathcal{E} \to M$ and assume it admits a Hermitian–Yang–Mills connection $\mathcal{A}_0$ with respect to ω_0 and Ω_0. Since the bundle is holomorphic, the complex structure on the fibers yields an orientation, and the Stiefel–Whitney classes are $w_1(\mathcal{E}) = 0$ and $w_2(\mathcal{E}) = c_1(\mathcal{E}) \mod 2$. Thus, to ensure absence of global anomalies, we must take $c_1(\mathcal{E})$ to be an even integral class. If $c_1(\mathcal{E}) \neq 0$, then $\mathcal{E}$ has structure group $U(r)$, while with $c_1(\mathcal{E}) = 0$, the structure group is $SU(r)$. The latter case is a bit simpler, so we will start with that. Since $c_1(T_M) = 0$, and we assume $c_1(\mathcal{E}) = 0$, the topological Bianchi identity simplifies to

$$c_2(T_M) - c_2(\mathcal{E}) = 0 \in H^4(M, \mathbb{Z}) . \qquad (4.7.1)$$

Necessity for Non-trivial Bundle

The bundle can be chosen independently from M, and one might wonder if we could simplify the construction by choosing $\mathcal{E}$ to be trivial. The Bianchi identity implies this is not the case, because for M Calabi–Yau with Ricci-flat metric g and Riemann curvature $\mathcal{R}$, we have [396]

$$\int_M c_2(T_M) \wedge \omega_0 = C \int_M \|\mathcal{R}\|^2 d\,\text{Vol}_g\,, \qquad (4.7.2)$$

where $\mathcal{R}$ is the Riemann curvature and C is a positive constant. This is a special case of a more general result given in [369]. In our case it shows that $c_2(T_M)$ is non-trivial in $H^4(M, \mathbb{Z})$ unless M is a flat torus, and the topological Bianchi identity therefore requires a non-trivial bundle $\mathcal{E}$.

4.7.2 Leading Order α' Correction

Supposing that we have a geometry $\mathcal{E} \to M$ that satisfies the topological requirements and the geometric requirements to zeroth order in α', we can ask whether it is possible to systematically correct the connection, metric, and B-field so as to preserve conformal invariance order by order in the α' expansion. In the case of (2,2) non-linear sigma models the answer is yes: although the metric does not remain Ricci-flat [211], it is possible to correct it without changing the Kähler class [316]. Those arguments have been generalized to a large class of (0,2) non-linear sigma models in [252]. There is, in fact, a general spacetime argument [139, 207] for all orders conformal invariance of any (0,2) non-linear sigma model that can be used as a large radius heterotic string compactification. The argument runs as follows: in the large radius limit the overall volume parameter and the component of the B-field proportional to the Kähler class on M reside in a chiral multiplet Φ—compare this to (4.6.5). The vacuum expectation value of the B-field component in Φ decouples to all orders in non-linear sigma model perturbation theory, while any possible obstruction to conformal invariance and therefore existence of an $\mathcal{N} = 1$ supersymmetric vacuum must arise from a superpotential term $W(\Phi)$. Since W is a holomorphic function of Φ, that is independent of the B-field component to all orders in the α' expansion, it must also be independent of the overall volume to all orders in α'.

Because of the spacetime non-renormalization theorem, we will not delve into details of the perturbative expansion, but to get a sense of some of the geometric notions that would be involved, we will take a look at the first non-trivial $O(\alpha')$ correction.[41] The differential and algebraic conditions for conformal invariance take the form

$$\omega \wedge \omega \wedge \omega = 3! i \Omega \wedge \overline{\Omega}\,, \qquad \omega \wedge \Omega = 0\,, \qquad \omega \wedge \overline{\Omega} = 0$$

$$d(e^{-2\varphi}\omega \wedge \omega) = 0\,, \qquad d(e^{-2\varphi}\Omega) = 0\,, \qquad i(\overline{\partial} - \partial)\omega = \mathcal{H}\,,$$

$$\Omega \wedge \mathcal{F} = 0\,, \qquad \overline{\Omega} \wedge \mathcal{F} = 0\,, \qquad \omega \wedge \omega \wedge \mathcal{F} = 0\,, \qquad (4.7.3)$$

---

[41] This discussion closely follows [302].

and

$$d\mathcal{H} = 2i\partial\bar{\partial}\omega = \frac{\alpha'}{4}\left(\operatorname{tr}\mathcal{R}_+^2 - \operatorname{tr}\mathcal{F}^2\right). \qquad (4.7.4)$$

As we mentioned above, these conditions are known up to $O(\alpha'^2)$ corrections, so that, without describing the higher order corrections to the equations, we can sensibly ask whether we can perturb an $O(\alpha'^0)$ solution to an $O(\alpha')$ solution.

To zeroth order in α' the conditions are satisfied by the data ω_0, Ω_0, $\mathcal{A}_0$, φ_0. The differential Bianchi identity indicates that this must be corrected at $O(\alpha')$. We make an Ansatz for the $O(\alpha')$ terms that preserves the complex structure of M:

$$\varphi = \varphi_0 + \alpha'\varphi_1, \qquad \omega = \omega_0 + \alpha'\left[\tfrac{4}{3}\varphi_1\omega_0 + \eta\right], \qquad \Omega = (1 + 2\alpha'\varphi_1)\Omega_0,$$

$$\mathcal{A} = \mathcal{A}_0 + \alpha'\mathcal{A}_1, \qquad (4.7.5)$$

where φ_1 is a correction to the dilaton, η is a $(1,1)$ form, and $\mathcal{A}_1 \in \mathscr{A}^{1,0}(M, \operatorname{End}\mathcal{E})$. The bundle $\operatorname{End}\mathcal{E}$ denotes the traceless endomorphisms of $\mathcal{E}$.

The right-hand-side of the Bianchi identity is

$$\frac{\alpha'}{4}\left[\operatorname{tr}\mathcal{R}_0^2 - \operatorname{tr}\mathcal{F}^2\right] + O(\alpha'^2) = 2i\alpha'\partial\bar{\partial}\sigma + O(\alpha'^2). \qquad (4.7.6)$$

Here $\mathcal{R}_0$ is the curvature of the Calabi–Yau metric. To obtain the right-hand-side, we recall that on a Kähler manifold the curvature $\mathcal{R}$ is a $(1,1)$ form.[42] Therefore, $\left[\operatorname{tr}\mathcal{R}_0^2 - \operatorname{tr}\mathcal{F}^2\right]$ is a d-exact $(2,2)$ form on M, and by the $\partial\bar{\partial}$ lemma, which holds on any Kähler manifold, it is proportional to $i\partial\bar{\partial}\sigma$ for some real $(1,1)$ form σ.

Exercise 4.16 Show that with our Ansatz, the conditions for conformal invariance at $O(\alpha')$ reduce to

$$\omega_0 \wedge \omega_0 \wedge \eta = 0,$$

$$\omega_0 \wedge d\left[\tfrac{1}{3}\varphi_1\omega_0 + \eta\right] = 0,$$

$$i\partial\bar{\partial}\left[\tfrac{4}{3}\varphi_1\omega_0 + \eta - \sigma\right] = 0,$$

$$\omega_0 \wedge \omega_0 \wedge \bar{\partial}\mathcal{A}_1 + 2\omega_0 \wedge \eta \wedge \mathcal{F}_0 = 0. \qquad (4.7.7)$$

$\square$

---

[42]This is not true for a general Hermitian metric, nor for the $\mathcal{R}_+$ curvature of the connection with torsion, and it is another indication that the equations themselves must be corrected in the α' expansion.

We can solve the Bianchi identity constraint (the third equation from the exercise) by writing

$$\eta = \sigma - \frac{4}{3}\varphi_1\omega_0 - \partial\bar{\xi} - \bar{\partial}\xi + i\partial\bar{\partial}f , \qquad (4.7.8)$$

where f is a real function, ξ is a (1,0) form, and $\bar{\xi}$ is its conjugate. Setting aside for the moment the Yang–Mills equation, this leaves us with the first two conditions: the primitivity of η, and the remnant of the conformal balance equation.

Since (M, ω_0, Ω_0) is Kähler, we have a Lefschetz decomposition that involves the operator $L : \mathscr{A}^k(M) \rightarrow \mathscr{A}^{k+2}(M)$, $L(\tau) = \omega_0 \wedge \tau$, and its adjoint $L^\dagger : \mathscr{A}^k(M) \rightarrow \mathscr{A}^{k-2}(M)$.[43] These lead to isomorphisms

$$L^k : \mathscr{A}^{n-k}(M) \simeq \mathscr{A}^{n+k}(M) , \quad \ker L^{k+1} \cap \mathscr{A}^{n-k}(M) \simeq \ker L^\dagger \cap \mathscr{A}^{n-k}(M) . \qquad (4.7.9)$$

Here n is the complex dimension of M (for us $n = 3$), and these isomorphisms can be refined and generalized to $\mathscr{A}^{p,q}(M, \mathcal{E})$ in an obvious fashion. As shown in Exercise B.26, we can always find a real function ψ such that $\sigma = \psi\omega_0 + \kappa$, where the real (1,1) form κ satisfies $\omega_0 \wedge \partial\bar{\partial}\kappa = 0$. We can also take ψ and κ to be orthogonal to all harmonic forms, since any harmonic term will drop out from $\partial\bar{\partial}\sigma$. Now we set $\varphi_1 = 3\psi$. The primitivity condition simplifies to

$$L^\dagger\left[-\frac{1}{3}\psi\omega_0 - \partial\xi - \bar{\partial}\bar{\xi}\right] = -iL^\dagger\partial\bar{\partial}f , \qquad (4.7.10)$$

and the left-hand side is a function without a zero mode, while the right-hand-side is proportional to the Laplacian of f. Therefore, we can solve the equation for f, and f is determined up to a constant. The conformal balance equation is a real equation for the vanishing of a 5-form, and it has (3,2) and (2,3) components. The former is

$$L\partial(\kappa - \bar{\partial}\xi) = 0 . \qquad (4.7.11)$$

Since κ is orthogonal to $\Delta_{\bar{\partial}}$-harmonic forms, we can write

$$\kappa = \bar{\partial}\zeta + \bar{\partial}^\dagger\tau , \qquad (4.7.12)$$

where ζ is a (1,0) form, while τ is a (1,2) form. From Exercise B.24 we have $[L, \partial] = 0$ and $[L, \bar{\partial}^\dagger] = -i\partial$. Since L annihilates (1,2) forms on a complex 3-fold, we have

$$L\partial\bar{\partial}^\dagger\tau = \partial[L, \bar{\partial}^\dagger]\tau = -i\partial\partial\tau = 0 . \qquad (4.7.13)$$

So, setting $\xi = \zeta$, we see that the conformal balance equation is satisfied.

---

[43] The decomposition is discussed in Sect. B.3.4.

Finally, we turn to the correction to the Hermitian Yang–Mills equation. Since we are assuming the connection $\mathcal{A}_0$ has structure group $\mathrm{SU}(r)$, it follows that the bundle of traceless endomorphisms $\mathrm{End}\,\mathcal{E}$ has no non-trivial global section. Therefore $H^{0,0}_{\bar{\partial}}(M, \mathrm{End}\,\mathcal{E}) = 0$, and $H^{3,3}_{\bar{\partial}}(M, \mathrm{End}\,\mathcal{E}) = 0$ as well. Since $\omega_0 \wedge \eta \wedge \mathcal{F}_0$ is a $\bar{\partial}$-closed form in $\mathscr{A}^{3,3}(M, \mathrm{End}\,\mathcal{E})$, it follows that $\omega_0 \wedge \eta \wedge \mathcal{F}_0 = \bar{\partial}\lambda$ for some $\lambda \in \mathscr{A}^{3,2}(M, \mathrm{End}\,\mathcal{E})$. Finally, applying (4.7.9) again, $\lambda = \omega_0^2 \mu$ for some $\mu \in \mathscr{A}^{1,0}(M, \mathrm{End}\,\mathcal{E})$, and therefore $\mathcal{A}_1 = -2\mu$ is a solution to the perturbed Hermitian–Yang–Mills equation.

Exercise 4.17 It is enlightening to study a simpler geometry, where $X = M \times T^2$, and M is a K3 manifold equipped with a hyper-Kähler triplet of 2-forms ω_a, with ω_3 the Kähler form and $\omega_1 + i\omega_2$ the (2,0) form. Taking $\mathcal{E} \to M$ to be a Hermitian–Yang–Mills bundle over the K3, it is not too hard to show that in this case the leading order α' correction is a conformal rescaling of the ω_a (in contrast to the 3-fold case, there is no η shift). $\qquad\qquad\qquad\qquad\qquad\qquad\square$

The argument is not particularly simple, and we might be hard-pressed to make it without the spacetime non-renormalization theorems assuring us that a solution exists. However, there are at least three reasons not to forget it. First, it is important to make consistency checks between the worldsheet and spacetime results. Second, these ideas may be useful in a worldsheet theory where a spacetime interpretation is not available. Finally, if one day a direct and simple worldsheet argument along these lines is available, it will be a measure of significant progress in the physics of non-linear sigma models.

4.7.3 Off-Shell Worldsheet Supersymmetry

We described the conditions for a classical (0,1) non-linear sigma model to possess (0,2) invariance. Another way in which we can proceed to build (0,2) invariant theories with off-shell (0,2) supersymmetry is to work with the familiar chiral multiplets Φ^a, $a = 1, \ldots, n$ and chiral fermi multiplets Λ^A, $A = 1, \ldots, r$.[44] Setting $\alpha' = 2$, the most general two-derivative action without potential terms is then given by [239, 318]—compare to Sect. 1.8

$$S = \frac{1}{2\pi} \int d^2z \, \mathcal{D}\overline{\mathcal{D}} K_z \,,$$

$$K_z = \tfrac{1}{2}\left(\mathcal{K}_a(\Phi, \overline{\Phi})\partial\Phi^a - \mathcal{K}^*_{\bar{a}}(\Phi, \overline{\Phi})\partial\overline{\Phi}^{\bar{a}} \right) - h_{A\overline{B}}(\Phi, \overline{\Phi})\overline{\Lambda}^{\overline{B}}\Lambda^A$$

$$+ \tfrac{1}{2}\tau_{AB}(\Phi, \overline{\Phi})\Lambda^A\Lambda^B - \tfrac{1}{2}\tau^*_{\overline{AB}}(\Phi, \overline{\Phi})\overline{\Lambda}^{\overline{A}}\overline{\Lambda}^{\overline{B}} \,. \qquad (4.7.14)$$

---

[44] We depart from the notation Γ for the fermi multiplets to avoid confusion with the Levi-Civita connection. The chiral fermi multiplets Λ^A will have the components λ^A and L^A that parallel the γ^A and G^A familiar from Sect. 1.8.

This action describes a particular patch in the target space; in order for it to make sense globally, the Λ^A must transform as (a pull-back) of a holomorphic frame for a holomorphic bundle $\mathcal{E} \to M$, and $h_{A\bar{B}}$ must be a Hermitian metric on $\mathcal{E}$. The 1-form "potential" $\mathcal{K}_a$ is a less familiar object. The action is invariant under shifts

$$\mathcal{K} \to \mathcal{K} + \xi + i\partial f, \qquad (4.7.15)$$

where ξ is a holomorphic $(1,0)$ form on M, and f is any real function. This means that on each patch U_α of the target space we specify a 1-form "potential" $\mathcal{K}_\alpha$ and allow these to patch on $U_\alpha \cap U_\beta \neq \emptyset$ as

$$\mathcal{K}_\alpha = \mathcal{K}_\beta + \xi_{\alpha\beta} + i\partial f_{\alpha\beta}. \qquad (4.7.16)$$

This will be consistent on triple overlaps provided

$$\xi_{\alpha\beta} + \xi_{\beta\gamma} + \xi_{\gamma\alpha} = -i\partial f_{\alpha\beta} - i\partial f_{\beta\gamma} - i\partial f_{\gamma\alpha}. \qquad (4.7.17)$$

If M is Kähler with Kähler potential K_α, then $\mathcal{K}_\alpha = \partial K_\alpha$ is an example of such a prepotential, and the ambiguity in shifting $K \to K + \zeta + \bar{\zeta}$ for any holomorphic function ζ translates to $\xi_\alpha = \partial\zeta_\alpha$. More generally, we see from the triple overlap that the real functions $f_{\alpha\beta}$ obey

$$\partial\bar{\partial}(f_{\alpha\beta} + f_{\beta\gamma} + f_{\gamma\alpha}) = 0, \qquad (4.7.18)$$

and since the $\partial\bar{\partial}$-lemma holds for $U_\alpha \cap U_\beta \cap U_\gamma$, this implies that

$$f_{\alpha\beta} + f_{\beta\gamma} + f_{\gamma\alpha} = c_{\alpha\beta\gamma} = \text{constant}. \qquad (4.7.19)$$

The $c_{\alpha\beta\gamma}$ define a class in $\check{H}^2(M, \mathbb{R})$, and by the Čech–de Rham isomorphism a class in $H^2(M, \mathbb{R})$. That is not surprising—this is how the coupling encodes the choice of a closed B-field on M.

The prepotential also determines the Hermitian metric and torsion on M. This torsional "pre-geometry" is of mathematical interest, but it is not immediately clear how to use it beyond the leading order in α'. The Green–Schwarz terms (4.3.37) imply that to preserve manifest $(0,2)$ off-shell supersymmetry, $\mathcal{K}_a$ must be assigned local Lorentz and gauge transformations, and that leads to a target space "metric" that shifts under the transformations. If one wants to work off-shell in a general torsional heterotic geometry, this is a necessity; without including such corrections it is easy to see that $\partial\bar{\partial}\omega = 0$, and that is incompatible with the Bianchi identity except in the case of standard embedding.

It would be useful to explore the precise geometric underpinning of this structure. To the author's knowledge this has not been carried out, although some discussion can be found in [239, 245, 300], and related supergravity questions were considered in [17, 128]. A systematic study may well offer a useful perspective on heterotic geometry. Of course these subtleties are not important when we consider the classical properties of theories with a large-radius limit, in which case $\mathcal{K}_a$ can be derived from a Kähler potential.

The τ coupling also raises some unresolved questions. Locally it is a section of $\wedge^2\mathcal{E}^*$, i.e. on a patch U_α, $\tau_\alpha \in \mathcal{A}^0(U_\alpha, \wedge^2\mathcal{E}^*)$, but the action is unchanged when τ_α is shifted by a holomorphic section $\zeta_\alpha \in \mathcal{O}(U_\alpha, \wedge^2\mathcal{E}^*)$. Thus, to define τ globally we pick $\zeta_{\alpha\beta} \in \mathcal{O}(U_{\alpha\beta}, \wedge^2\mathcal{E}^*)$ satisfying the cocycle condition $\zeta_{\alpha\beta} + \zeta_{\beta\gamma} + \zeta_{\gamma\alpha} = 0$—this is a class in $\check{H}^1(M, \mathcal{O}(\wedge^2\mathcal{E}^*))$, or by the Čech–Dolbeault isomorphism a class in $H^1_{\bar\partial}(M, \wedge^2\mathcal{E}^*)$. It is perhaps worthwhile to explain this in a little more detail. Let ξ be a representative of a class in $H^1_{\bar\partial}(M, \wedge^2\mathcal{E}^*)$. This means that the restriction of ξ to any patch U_α, $\xi_\alpha \in \mathcal{A}^0(U_\alpha, \wedge^2\mathcal{E}^* \otimes \overline{T}^*_M)$, satisfies $\bar\partial\xi_\alpha = 0$. Since $U_\alpha \simeq \mathbb{C}^n$, this means $\xi_\alpha = \bar\partial\tau_\alpha$ for some τ_α, with $\tau_\alpha - \tau_\beta = \zeta_{\alpha\beta}$ on $U_{\alpha\beta}$.

The τ couplings have not been much-studied in the literature. A theory with $\tau = 0$ may have marginal deformations that correspond to turning on $\tau \neq 0$. Although the transition functions used to define the patching of the fermions are specified by the holomorphic bundle $\mathcal{E} \to M$, and the bundle admits a Hermitian metric and therefore a unitary frame with structure group $SU(r)$ (recall we are sticking to $c_1(\mathcal{E}) = 0$ for now), the τ terms lead to a connection with a larger structure group—a subgroup of $SO(2r)$. In a heterotic compactification, where the spacetime gauge group is the commutant of the structure group, it is therefore not surprising that a $\tau \neq 0$ deformation can describe certain spacetime Higgs branches.[45] It would be interesting to study these couplings more generally. For instance one might ask: is every solutions with $\tau \neq 0$ a marginal deformation of a theory with $\tau = 0$?

When $\tau = 0$, the theory has a classical $U(1)_L$ symmetry that assigns charge -1 to Λ^A, $+1$ to $\overline{\Lambda}^{\overline{A}}$, and zero to $\Phi, \overline{\Phi}$. The symmetry is non-anomalous when the structure group of the bundle is $SU(r)$. In such theories it is then natural to restrict the parameters to the $U(1)_L$-preserving locus with $\tau = 0$, and we will stick to that case.

Exercise 4.18 In this exercise we collect some useful results on the $\tau = 0$ (0,2) sigma model with action (4.7.14). First, generalize the computation in Exercise 1.12 to obtain the equations of motion

$$\overline{\mathcal{D}}\Lambda_A = 0, \qquad \overline{\mathcal{D}}\left[2g_{b\bar{a}}\partial\overline{\Phi}^{\bar{a}} + 2h_{A\overline{B},b}\overline{\Lambda}^{\overline{B}}\Lambda^A + (\mathcal{K}_{b,a} - \mathcal{K}_{a,b})\partial\Phi^a\right] = 0.$$

$$(4.7.20)$$

---

[45] Some examples of these were investigated in unpublished work by McOrist and the author. For instance, one concrete non-trivial scenario involves a rank $r = 4$ compactification of the $E_8 \times E_8$ string on a K3 manifold. The structure group is enlarged as $SU(4) \to SO(7)$, and the spacetime Higgs mechanism is $\mathfrak{so}(10) \to \mathfrak{so}(9)$. A discussion of an $r = 3$ case on a Calabi–Yau 3-fold, where the terms give another parametrization of the $\mathfrak{e}_6 \to \mathfrak{so}(10)$ Higgs branch described above, can be found in [1]. Another study of these couplings, in the context of an orbifold compactification, is given in [317].

Next, obtain the component action and integrate out the auxiliary fields in the Λ^A multiplets to obtain

$$
S = \frac{1}{2\pi} \int d^2z \left[g_{a\bar{b}}(\partial\phi^a \bar{\partial}\overline{\phi}^{\bar{b}} + \bar{\partial}\phi^a \partial\overline{\phi}^{\bar{b}}) + B_{a\bar{b}}(\partial\phi^a \bar{\partial}\overline{\phi}^{\bar{b}} - \bar{\partial}\phi^a \partial\overline{\phi}^{\bar{b}}) \right.
$$
$$
+ 2g_{b\bar{b}}\overline{\psi}^{\bar{b}} \left\{ \partial + \partial\phi^c(\Gamma^b_{ca} - \tfrac{1}{2}(dB)^b_{ca}) + \bar{\partial}\overline{\phi}^{\bar{c}}(\Gamma^b_{\bar{c}a} - \tfrac{1}{2}(dB)^b_{\bar{c}a}) \right\} \psi^a
$$
$$
\left. + \overline{\lambda}_A(\bar{\partial} + \bar{\partial}\phi^c \mathcal{A}^A_{Bc})\lambda^B + \overline{\lambda}_A \mathcal{F}^A_{Bb\bar{b}}\lambda^B \psi^b \overline{\psi}^{\bar{b}} \right], \tag{4.7.21}
$$

where

$$
g_{a\bar{b}} = \tfrac{1}{2}(K_{a,\bar{b}} + \overline{K}_{\bar{b},a}), \qquad B_{a\bar{b}} = \tfrac{1}{2}(K_{a,\bar{b}} - \overline{K}_{\bar{b},a}), \qquad \mathcal{A}^A_{Bc} = h^{\overline{A}A}h_{B\overline{A},c},
$$
$$
\mathcal{F}^A_{Bb\bar{b}} = \mathcal{A}^A_{Bb,\bar{b}}, \tag{4.7.22}
$$

and $\overline{\lambda}_A = 2h_{A\overline{B}}\overline{\lambda}^{\overline{B}}$.

Let

$$
\rho_b = 2g_{b\bar{a}}\partial\overline{\phi}^{\bar{a}} + 2h_{A\overline{B},b}\overline{\lambda}^{\overline{B}}\lambda^A + (K_{b,a} - K_{a,b})\partial\phi^a, \tag{4.7.23}
$$

and show that in terms of this field the component action can be rewritten as

$$
S = \frac{1}{2\pi} \int d^2z \left[\rho_b \bar{\partial}\phi^b + \overline{\lambda}_B \bar{\partial}\lambda^B \right.
$$
$$
+ 2g_{b\bar{b}}\overline{\psi}^{\bar{b}} \left\{ \partial + \partial\phi^c(\Gamma^b_{ca} - \tfrac{1}{2}(dB)^b_{ca}) + \bar{\partial}\overline{\phi}^{\bar{c}}(\Gamma^b_{\bar{c}a} - \tfrac{1}{2}(dB)^b_{\bar{c}a}) \right\} \psi^a
$$
$$
\left. + \overline{\lambda}_A \mathcal{F}^A_{Bb\bar{b}}\lambda^B \psi^b \overline{\psi}^{\bar{b}} \right]. \tag{4.7.24}
$$

Recall from Chap. 1, and in particular from (1.8.18), that the supersymmetry transformations act as on the superfields via

$$
\delta_\xi = \frac{1}{\sqrt{2}}\left[\xi Q - \overline{\xi}\overline{Q} \right],
$$

leading to the component transformations

$$
\delta_\xi \phi^a = \xi\psi^a, \quad \delta_\xi \psi^a = -\overline{\xi}\bar{\partial}\phi^a, \quad \delta_\xi \overline{\phi}^{\bar{a}} = \overline{\xi}\overline{\psi}^{\bar{a}}, \quad \delta_\xi \overline{\psi}^{\bar{a}} = -\xi\partial\overline{\phi}^{\bar{a}},
$$
$$
\delta_\xi \lambda^A = \xi L^A, \quad \delta_\xi L^A = -\overline{\xi}\partial\lambda^A, \quad \delta_\xi \overline{\lambda}^{\overline{A}} = -\overline{\xi}\overline{L}^{\overline{A}}, \quad \delta_\xi \overline{L}^{\overline{A}} = \xi\partial\overline{\lambda}^{\overline{A}}. \tag{4.7.25}
$$

Since we define the supercharges Q and $\overline{Q}$ via $\delta_\xi = \overline{\xi}\,\overline{Q} - \xi\,Q$, show that after the L^A and $\overline{L}^A$ are eliminated via their equations of motion, the supercharges act as follows: $\overline{Q}$ annihilates the lowest components of the chiral fields ϕ^a and λ^A, as well as $\overline{\psi}^{\overline{a}}$; similarly Q annihilates $\overline{\phi}^{\overline{a}}$, $\overline{\lambda}^A$ and ψ^a, and the action on the other fields is

$$\overline{Q} \cdot \overline{\phi}^{\overline{a}} = \overline{\psi}^{\overline{a}}\,, \qquad\qquad Q \cdot \phi^a = -\psi^a\,,$$

$$\overline{Q} \cdot \psi^a = -\overline{\partial}\phi^a\,, \qquad\qquad Q \cdot \overline{\psi}^{\overline{a}} = \overline{\partial}\overline{\phi}^{\overline{a}}\,,$$

$$\overline{Q} \cdot \overline{\lambda}^A = \overline{\mathcal{A}}^A_{\overline{B}\overline{b}}\overline{\lambda}^B\overline{\psi}^{\overline{b}}\,, \qquad\qquad Q \cdot \lambda^A = -\mathcal{A}^A_{Bb}\lambda^B\psi^b\,. \qquad (4.7.26)$$

Verify that these obey $Q^2 = \overline{Q}^2 = 0$ without the use of equations of motion, and

$$\{Q, \overline{Q}\} = \overline{\partial} + (\lambda \text{ eom})\,.$$

Finally, prove that $\overline{Q} \cdot \lambda_{\overline{A}} = 0$ and $\overline{Q} \cdot \rho_b = 0$. [Hint: use (4.7.20).]   □

4.7.4 Marginal Deformations

In Chap. 2 we showed that marginal deformations of a (0,2) SCFT are in one-to-one correspondence with right-chiral primary operators $\mathcal{O}$ with $\overline{q} = 1$ and $h = 1$. Such an operator is the lowest component of a chiral superfield, which we denote by the same symbol $\mathcal{O}$, and there is a marginal deformation $\Delta S = \int d^2z\, \mathcal{D}\mathcal{O} + \text{h.c.}$. If we assume that the (0,2) non-linear sigma model describes a (0,2) SCFT, we should therefore be able to classify the deformations by classifying all such operators. In the large radius limit we can construct these out of the classical fields, and it is not hard to carry out this program explicitly [299]. Let us sketch the construction. We seek fermionic operators $\mathcal{O}$ that obey the following properties.

1. $\mathcal{O}$ must be well-defined under diffeomorphisms of the bundle $\mathcal{E} \to M$ and have $(h, \overline{h}) = (1, 1/2)$.
2. It must have $\overline{q} = 1$ with respect to the R-symmetry which assigns charge $+1$ to $\theta, \overline{\mathcal{D}}$ and $\overline{\psi}^{\overline{a}}$.
3. We will also require that $\mathcal{O}$ has vanishing $U(1)_L$ charge; this is done merely to simplify the presentation and can be relaxed along the lines discussed in footnote 45.
4. $\mathcal{O}$ must be chiral up to the equations of motion of the undeformed theory:
   $\overline{\mathcal{D}}\mathcal{O} = 0 + \text{eom}.$

As we will see shortly, the operators obeying these constraints form an infinite-dimensional space. However, these do not always define a genuine "F-term" deformation. In particular, if the difference of two operators is $\mathcal{O} - \mathcal{O}' = \overline{\mathcal{D}}X$

for some well-defined operator X, then these operators correspond to the same deformation of the SCFT. This is simply because $\overline{\mathcal{D}}X$, while chiral, is not a chiral primary operator in theory. Thus, to describe the marginal operators we consider the class of operators satisfying the requirements modulo the equivalence relation[46]

$$\mathcal{O} \sim \mathcal{O} + \overline{\mathcal{D}}X . \qquad (4.7.27)$$

The most general operator that satisfies the first three requirements is

$$\mathcal{O} = \left[\overline{\Lambda}_A \Lambda^B E^A_{B\overline{b}} + Y_{a\overline{b}} \partial \Phi^a + Z^a_{\overline{b}} g_{a\overline{a}} \partial \overline{\Phi}^{\overline{a}} \right] \overline{\mathcal{D} \Phi}^{\overline{b}} . \qquad (4.7.28)$$

The coefficients $E^A_{B\overline{B}}$, $Y_{a\overline{b}}$ and $Z^a_{\overline{b}}$ are all functions of Φ and $\overline{\Phi}$ superfields, and for the operator to be well-defined they must be sections of various bundles:

$$E \in \mathscr{A}^{0,1}(M, \mathrm{End}\,\mathscr{E}) , \qquad Y \in \mathscr{A}^{1,1}(M) , \qquad Z \in \mathscr{A}^{0,1}(M, T_M) . \qquad (4.7.29)$$

Using the equations of motion, $\overline{\mathcal{D}}\mathcal{O} = 0$ if and only if

$$\bar{\partial} Z = 0 , \qquad \bar{\partial} Y = -Z \llcorner \mathcal{H}^{2,1} , \qquad \bar{\partial} E = -Z \llcorner \mathcal{F} . \qquad (4.7.30)$$

Here $\mathcal{H} = dB$ is the classical heterotic torsion, and $\mathcal{H}^{2,1}$ is its projection to $\mathscr{A}^{2,1}(M)$, and the $Z\llcorner$ notation is defined as follows:

$$(Z \llcorner \mathcal{H}^{2,1})_{\overline{a}b\overline{b}} = Z^a_{\overline{a}} \mathcal{H}_{ab\overline{b}} - Z^a_{\overline{b}} \mathcal{H}_{ab\overline{a}} , \qquad (Z \llcorner \mathcal{F})^A_{B\overline{a}b} = Z^a_{\overline{a}} \mathcal{F}^A_{Bab} - Z^a_{\overline{b}} \mathcal{F}^A_{Ba\overline{a}} . \qquad (4.7.31)$$

Next, we consider the possible operators X that determine the equivalence relations on the $\mathcal{O}$. The general form of such an X is

$$X = \overline{\Lambda}_A \Lambda^B \eta^A_B + \mu_a \partial \Phi^a + \zeta^a g_{a\overline{a}} \partial \overline{\Phi}^{\overline{a}} , \qquad (4.7.32)$$

where

$$\eta \in \mathscr{A}^0(M, \mathrm{End}\,\mathscr{E}) , \qquad \mu \in \mathscr{A}^{1,0}(M) , \qquad \zeta \in \mathscr{A}^0(M, T_M) . \qquad (4.7.33)$$

---

[46]There is another equivalence relation possible as well— we can shift $\mathcal{O} \to \mathcal{O} + \partial X'$, where $\partial X'$ is a well-defined chiral operator. The resulting "deformation" is a total derivative. Such operators are intimately tied to the right-moving chiral currents and should be considered in discussing theories with $(0,4)$ or larger superconformal invariance, but they are not present in the SU(3) structure non-linear sigma model of interest here [299].

This leads to equivalence relations

$$Z \sim Z + \bar{\partial}\zeta \,, \quad Y \sim Y + \bar{\partial}\mu - \zeta_\llcorner \mathcal{H}^{2,1} \,, \quad E \sim E + \bar{\partial}\eta - \zeta_\llcorner \mathcal{F} \,. \tag{4.7.34}$$

Equations (4.7.30) and (4.7.33) relate the marginal $U(1)_L$-preserving deformations to cohomology groups on M. Z modulo the ζ equivalence defines a cohomology class in $H_{\bar{\partial}}^{0,1}(M, T_M)$—the group that characterizes the space of infinitesimal deformations of complex structure of M. To describe the remaining conditions, we observe that $\bar{\partial}(Z_\llcorner \mathcal{F}) = 0$, and therefore $Z_\llcorner \mathcal{F}$ defines a class in $\alpha(Z) \in H_{\bar{\partial}}^{0,2}(M, \mathrm{End}\,\mathcal{E})$. Similarly, in the classical (0,2) sigma model $\partial\bar{\partial}\omega = 0$, and therefore $Z_\llcorner \mathcal{H}^{2,1}$ is also $\bar{\partial}$-closed and defines a class $\beta(Z) \in H_{\bar{\partial}}^{1,2}(M)$. In view of the last two equations in (4.7.30), $\alpha(Z)$ and $\beta(Z)$ are obstructions, already at first order, to the complex structure deformation Z. Let $\tilde{Z}$ denote all deformations for which $\alpha(\tilde{Z}) = 0$ and $\beta(\tilde{Z}) = 0$. It follows that for any such deformation

$$\tilde{Z}_\llcorner \mathcal{H}^{2,1} = \bar{\partial}Y_{\tilde{Z}} \,, \qquad\qquad \tilde{Z}_\llcorner \mathcal{F} = \bar{\partial}E_{\tilde{Z}} \tag{4.7.35}$$

for some $Y_{\tilde{Z}} \in \mathscr{A}^{1,1}(M)$ and some $E_{\tilde{Z}} \in \mathscr{A}^0(M)$. The $Y_{\tilde{Z}}$ and $E_{\tilde{Z}}$ shift under $\tilde{Z} \to \tilde{Z} + \bar{\partial}\zeta$:

$$Y_{\tilde{Z}} \to Y_{\tilde{Z}} + \zeta_\llcorner \mathcal{H}^{2,1} \,, \qquad\qquad E_{\tilde{Z}} \to E_{\tilde{Z}} + \zeta_\llcorner \mathcal{F} \,. \tag{4.7.36}$$

Finally, we can untangle (4.7.30) and (4.7.33) by setting $\tilde{Y} = Y + Y_{\tilde{Z}}$ and $\tilde{E} = E + E_{\tilde{Z}}$. We obtain

$$\bar{\partial}\tilde{Z} = 0 \,, \quad \tilde{Z} \sim \tilde{Z} + \bar{\partial}\zeta \,, \quad [\tilde{Z}] \in H_{\bar{\partial}}^{0,1}(M, T_M) \cap \ker\alpha \cap \ker\beta \,,$$

$$\bar{\partial}\tilde{Y} = 0 \,, \quad \tilde{Y} \sim \tilde{Y} + \bar{\partial}\mu \,, \quad [\tilde{Y}] \in H_{\bar{\partial}}^{1,1}(M) \,,$$

$$\bar{\partial}\tilde{E} = 0 \,, \quad \tilde{E} \sim \tilde{E} + \bar{\partial}\lambda \,, \quad [\tilde{E}] \in H_{\bar{\partial}}^{0,1}(M, \mathrm{End}\,\mathcal{E}) \,. \tag{4.7.37}$$

We therefore see that in the classical (0,2) sigma model the infinitesimal deformation space does seem to possess a canonical splitting into three terms, and aside from the restriction on the $\tilde{Z}$ to the kernel of α and β, it has the form of "Kähler+complex structure+bundle" structure familiar from (2,2) compactifications. If we neglect non-perturbative effects, this is entirely reasonable when applied to large radius (0,2) compactifications on Calabi–Yau manifolds with $\mathcal{E} \neq T_M$. In the next chapter we will see that worldsheet instanton corrections modify this structure.

Returning to deformations of (2,2) SCFTs and the (0,2) moduli space $\mathcal{M}_{02}$ discussed above, we can see important similarities with this result. First, since $h^1(M, \mathrm{End}\,\mathcal{E})$ is not a topological invariant, the number of "bundle" deformations will jump as a function of the complex structure moduli space. On the other hand, once the bundle is deformed, some of the complex structure deformations may be lifted, and the moduli space can have distinct branches.

It is less clear how to interpret these results in a torsional geometry with string-scale cycles, even if we formally treat the non-linear sigma model at leading order in α'. Some work on this question has been carried out in [17, 127].[47] The basic issue is that if one simply naively replaces $\mathcal{H}$ with its gauge-invariant form, then $\partial\bar{\partial}\omega \neq 0$, so that $Z_{\mathsf{L}}\mathcal{H}^{2,1}$ is no longer automatically closed, and one cannot perform the simple untangling carried out above. It would be interesting to study this point further directly in the off-shell supersymmetric sigma model; the set-up in [239, 318] should help in taking the first steps.

The restriction to $\ker\alpha$ was pointed out in this context in heterotic supergravity [18]. The geometric underpinning was described in [43]. The total space of a holomorphic vector bundle $\mathcal{E} \to M$ is a complex manifold. For certain deformations of complex structure on M, there is a deformation of the bundle such that the full space remains a complex manifold. The obstruction is encoded by the Atiyah class $\alpha : H_{\bar{\partial}}^{0,1}(M, T_M) \to H_{\bar{\partial}}^{0,2}(M, \mathrm{End}\,\mathcal{E})$, which is exactly the structure we obtained.

Exercise 4.19 Prove that the Atiyah class is zero in the case of $\mathcal{E} = T_M$ and standard embedding. In this case the role of $\mathcal{F}$ is played by the Riemann curvature of the Kähler metric, and its special properties, especially the symmetry $R_{a\bar{a}b\bar{b}} = R_{b\bar{a}a\bar{b}}$, can be used to show $Z_{\mathsf{L}}R = 0$. Details can be found in [299]. □

The Topological Heterotic Ring in the Large Radius Limit

When $c_1(\mathcal{E}) = 0$ and $\tau = 0$, it is a simple matter to deduce the spacetime massless spectrum of a compactification in the large radius limit, and it introduces a useful way to think about the chiral sector of a non-linear sigma model as a "curved bc–$\beta\gamma$ system" [315, 393]. These ideas, which in some sense go back to the application of the Born–Oppenheimer approximation to the seminal relation between supersymmetric non-linear sigma model quantum mechanics and Morse theory [380], were applied to heterotic theories in, for instance, [144, 334] and more recently in the study of $(0,2)$ chiral algebras in [261, 357, 358, 393, 394], but in their modern incarnation they have also been of use in other theories, such as sigma models related to the pure spinor construction for the superstring [315]. In this section we will just touch on this important idea by revisiting the topological heterotic rings in the non-linear sigma model context.

The classical $(0,2)$ action can be recast in a first-order form as in (4.7.24). If we neglect the right-moving fermions, the left-moving sector has a deceptively simple Lagrangian:

$$S \supset \frac{1}{2\pi} \int d^2z \left[\rho_b \bar{\partial}\phi^b + \overline{\gamma}_A \bar{\partial}\gamma^A \right]. \tag{4.7.38}$$

---

[47]Results highlighting the differences in this deformation theory and the more familiar Calabi–Yau case are presented in [251].

The fields shown are annihilated by $\overline{Q}$ and have free OPEs

$$\phi^a(z)\rho_b(w) \sim \frac{\delta^a_b}{z-w} \,, \qquad \gamma^A(z)\overline{\gamma}_B(w) \sim \frac{\delta^A_B}{z-w} \,. \qquad (4.7.39)$$

This is a generalization of the bc–$\beta\gamma$ ghost system, and proceeding very naively, we identify the $U(1)_L$ current and holomorphic energy-momentum tensor

$$J_L =: \overline{\gamma}_A\gamma^A : \,, \qquad T = -: \partial\phi^a\rho_a : -\tfrac{1}{2}\left(: \overline{\gamma}_A\partial\gamma^A : +: \gamma^A\partial\overline{\gamma}_A :\right). \qquad (4.7.40)$$

The reader is invited to compare this to the general chiral algebra discussion in Sect. 3.3 and to check that these are indeed the lowest components for representatives of J_L and T in $\overline{Q}$ cohomology.

The reason this free field structure is deceptively simple is because it only holds patch by patch in the field space of the non-linear sigma model! To study the theory globally, even at just the level of the chiral algebra, requires careful consideration of how the free theories patch together over non-trivial overlaps. This is especially challenging for non-trivial field configurations, where one also has to keep track of the pull-back data. While ϕ^a, γ^A and $\overline{\gamma}_B$ have holomorphic transition functions, and the latter two transform as tensors, this is not the case for ρ_b, which transforms non-homogeneously. Moreover, the transformations for ρ involve both holomorphic and anti-holomorphic fields, so that normal ordering with respect to the free OPE must be taken into account.

The story simplifies substantially when the path integral is expanded in the topologically trivial sector and our interest is in low energy states. For instance, it is a simple matter to describe the topological heterotic rings by using the relation between these operators and the ground states in the R, $\overline{\text{R}}$ sector. Define the right-moving $\overline{\text{R}}$ ground state as the state annihilated by the ψ^a zero modes. In that case the right-moving ground states that are related to the topological heterotic rings are of the form

$$\lambda^{A_1}_{-1/2}\cdots\lambda^{A_k}_{-1/2}(\overline{\lambda}_{B_1})_{-1/2}\cdots(\overline{\lambda}_{B_l})_{-1/2}\Psi^{B_1\cdots B_l}_{A_1\cdots A_k\overline{a}_1\cdots\overline{a}_{\overline{q}}}(\phi,\overline{\phi})\overline{\psi}^{\overline{a}_1}_0\cdots\overline{\psi}^{\overline{a}_{\overline{q}}}_0|\text{R}\rangle_{\overline{q}-3/2} \,. \qquad (4.7.41)$$

These states are well defined if $\Psi \in \mathscr{A}^{0,q}(\wedge^k\mathcal{E} \otimes \wedge^l\mathcal{E}^*)$. Once we restrict to right-moving zero modes, the action of $\overline{Q}$ on the operators is simply via

$$\overline{Q} = \overline{\psi}^{\overline{a}}\frac{\partial}{\partial\overline{\phi}^{\overline{a}}} \,, \qquad (4.7.42)$$

so that $\overline{Q}$ cohomology classes are described by $\bar{\partial}$-closed Ψ: $[\Psi] \in H_{\bar{\partial}}^{0,q}(M, \wedge^k \mathcal{E} \otimes \wedge^l \mathcal{E}^*)$. Now it is easy to see that the topological heterotic rings are described by

$$H_{B/2}^{q,\bar{q}} \simeq H_{\bar{\partial}}^{0,\bar{q}}(M, \wedge^q \mathcal{E}^*) , \qquad H_{A/2}^{-q,\bar{q}} \simeq H_{\bar{\partial}}^{0,\bar{q}}(M, \wedge^q \mathcal{E}) . \qquad (4.7.43)$$

Since $c_1(\mathcal{E}) = 0$, if r is the rank of $\mathcal{E}$, then $\wedge^r \mathcal{E} = \mathcal{O}$, the trivial bundle, and it then also follows $\wedge^q \mathcal{E}^* \simeq \wedge^{q-r} \mathcal{E}$. This is a particular case of the isomorphism

$$H_{B/2}^{q,\bar{q}} \simeq H_{A/2}^{r-q,\bar{q}} . \qquad (4.7.44)$$

Kodaira–Serre duality of the Dolbeault cohomology groups then leads to an isomorphism of these groups to their right-anti-chiral conjugates and leads to CPT invariance of the charged spectrum in the spacetime theory.

It is also possible to describe the marginal deformations from this perspective and recover the results of the previous section. The references [82, 84] discuss this and other aspects of the topological heterotic ring in these models and their generalizations.

4.7.5 Unwieldy Instantons

The large radius discussion given above is incomplete, since it does not include important non-perturbative effects due to non-trivial maps. In the context of Calabi–Yau compactifications this is a tale with a few twists and turns. Briefly, it was realized early on that worldsheet instantons can contribute non-trivially to the spacetime superpotential for the gauge singlets, which include the overall volume modulus [139, 140]. This realization can be a cause for grief and celebration. The grief comes from the potential of these corrections to destabilize the theory: as in the Affleck–Dine–Seiberg potential of super-QCD [7, 24], there may be no spacetime supersymmetric vacuum for any finite radius. A string phenomenologist may take a more optimistic perspective, if he or she seeks a non-perturbative effect to generate just the right top Yukawa and to eliminate unwanted massless scalars. This has obvious difficulties of its own that are familiar from general string phenomenology: if a non-perturbative effect is needed to obtain some desired feature, then one has to worry whether similar effects do not also destabilize the theory entirely; moreover, one must be careful about the regime of validity of the calculations based on a perturbative picture.

On the other hand, it turns out that both responses are a bit premature. It was realized early on that there may be special choices of bundle and manifold for which the destabilizing effects are absent [142]—a sufficient topological condition is that the bundle $\mathcal{E}$, when pulled back to any rational curve in M splits non-trivially. What was at first thought to be a special set of protected theories was vastly expanded in the context of gauged linear sigma models—the topic of the next chapter. Briefly, there is a vast class of (0,2) SCFTs that have large radius limits described as bundle $\mathcal{E} \to M$ over a Calabi–Yau manifold [50, 57, 350]. In these theories there are, in general, many worldsheet instanton contributions, but certain compactness results on gauge instanton moduli spaces guarantee that their contributions to destabilizing superpotential couplings vanish.

Still, one does expect that in a "sufficiently generic" Calabi–Yau compactification the non-renormalization results would fail, but it turns out not so easy to construct explicit examples, because the gauged linear sigma models are essentially the most general constructions that have been developed so far. Nevertheless, examples are known [98, 100, 103, 104].

Finally, it should be stressed that there are many couplings in the spacetime superpotential that receive instanton corrections but do not lead to destabilization. The $\overline{\mathbf{27}}^3$ coupling in a Calabi–Yau compactification is a famous example that can be in principle computed using mirror symmetry. There are other terms as well. For instance, in (2,2) theories the large radius marginal (0,2) deformations can be lifted by worldsheet instantons, though, again, demonstrating this with explicit examples is surprisingly challenging [29, 33, 37, 41]; in particular, the challenge suggests that there may be generalizations of the non-renormalization results quoted above. It is remarkable that this can already be studied in (2,2) theories, and it deserves further study.

4.7.6 (0,4) K3 Sigma Models

There are many heterotic paths to the K3 geometry. The most straightforward goes back to the heroic days and the realization that heterotic compactification on a vector bundle E over a K3 manifold M leads to (1,0) supersymmetric spacetime theories in $\mathbb{R}^{1,5}$ with a non-trivial anomaly cancelation mechanism [205].[48] We leave most details to references [31, 360] and just remind the reader of the massless content of a (1,0) theory. Besides the gravity multiplet, the theory has N_T tensor multiplets, N_H hypermultiplets, and N_V vector multiplets. Anomaly cancelation requires these to satisfy $N_H - N_V = 273 - 29N_T$. Compactification of the perturbative heterotic string yields theories with $N_T = 1$ and $N_H - N_V = 244$; as observed already in [205], the latter constraint is intimately tied to a solution to the Bianchi identity: $\frac{1}{2}p_1(E) - \frac{1}{2}p_1(M) \in H^4(X, \mathbb{Z})$ that we discussed above. There are also additional consistency conditions on a six-dimensional (1,0) theory that go back to [335] having to do with positivity of various gauge kinetic terms; we leave the discussion of these to the references [231, 360].

---

[48] We confront here the confusion between the six-dimensional and two-dimensional supersymmetry labels. We will try to stick to (1,0) as the supersymmetry with eight chiral supercharges in six dimensions.

A K3 non-linear sigma model defines a classical $(0,4)$ superconformal theory with $\bar{c} = 6$, and this is a much more constrained structure than that of $\bar{c} = 9$ SCFTs. First, a six-dimensional version of the spacetime non-renormalization argument guarantees that all marginal deformations that correspond to gauge-neutral massless fields in spacetime are unobstructed simply because there are no neutral scalar potential couplings allowed by $(1,0)$ supersymmetry ; there is also a worldsheet argument given in [47] that applies to the $(4,4)$ case, i.e. a K3 compactification with standard embedding. Spacetime supersymmetry does allow for D-term like potentials that describe various Higgs branches that break the spacetime gauge group. Conversely, at special points in the moduli space the spacetime gauge group may get "un-Higgsed" to a larger group. Second, it is significantly easier to obtain the spacetime massless spectrum, essentially by an index theorem. This is a useful technology, so we will review it briefly before discussing a few additional features of these theories. Further discussion is also provided in, for instance [78, 231].

Massless Spectrum in K3 Compactification: Spacetime Perspective

The essential idea is that since the spectrum is protected by supersymmetry, the supergravity large radius limit computation yields exact results; these may break down at special points in the moduli space where either the SCFT becomes singular (in which case we expect string non-perturbative effects to be important) or where new massless degrees of freedom appear via the Higgs mechanism. In the latter case we can often understand the phenomenon as reduction of the structure group of the vector bundle $\mathcal{E} \to M$.

Let us first discuss the topological conditions for $E \to M$. Since $p_1(M)$ is proportional to the top form on M, we will follow the standard practice and not distinguish between $p_1(M)$ and its integral over M: thus, $p_1(M) = -48$. We suppose that $E \to M$ has structure group G_E, a connected simple group, and associated principal bundle $P \to M$. The topological classification of such bundles on compact connected manifolds is discussed in the appendix of [359]. It is shown that if G_E is simply connected, then the bundle is topologically classified by $p_1(E)$. More generally, one more invariant is needed—a choice of map from the classifying space to $H^2(M, \pi_1(G_E))$. The Stiefel–Whitney class $w_2(E) \in H^2(M, \mathbb{Z}_2)$ is an example of this invariant for $G_E = SO(k)$; global anomaly cancelation requires this class to vanish.

The connection on E must satisfy the Hermitian Yang–Mills equations, just as above, but on K3 these are equivalent to the condition for an anti-self-dual connection $\mathcal{A}$, in other words, an instanton on K3. If E admits an anti-self-dual connection, then the moduli space $\mathcal{M}(\mathcal{A})$ of such connections has quaternionic dimension, determined by an index theorem and some vanishing theorems [44, 155],

$$N_0 \equiv \dim_{\mathbb{H}} \mathcal{M}(\mathcal{A}) = -h_{\mathfrak{g}} \frac{p_1(E)}{2} - \dim \mathfrak{g} \ , \qquad (4.7.45)$$

where $\mathfrak{g}$ is the Lie algebra for G_E, and $h_\mathfrak{g}$ is its dual Coxeter number.

By analogy with the usual Yang–Mills instantons on $\mathbb{R}^4$ we define the instanton number of the connection as $k = -\frac{1}{2}p_1(E)$. For large enough k clearly $N_0 > 0$, so we expect ASD connections to exist, but even if $N_0 > 0$, the connection may nevertheless be reducible to a connection with a smaller structure group. Following [44], we can derive the conditions for irreducibility of an ASD connection on K3 for the simple Lie groups ($n > 2$ for SU(n), $n > 6$ for SO(n), and $n > 2$ for Sp(n)):

$$
\begin{aligned}
&\text{SU(2)} : k \geq 2\,, && \text{SO(5)} : k \geq 5\,, && \\
&\text{SU}(n) : k \geq 2n\,, && \text{SO}(n) : k \geq n\,, && \text{Sp}(n) : k \geq 4n\,, \\
&E_6 : k \geq 9\,, && E_7 : k \geq 10\,, && E_8 : k \geq 10\,, \\
&F_4 : k \geq 9\,, && G_2 : k \geq 7\,. &&
\end{aligned}
\tag{4.7.46}
$$

Suppose we compactify the $E_8 \times E_8$ heterotic string on such a geometry. The $\mathbb{R}^{1,5}$ spacetime fermions arise from the massless modes of the ten-dimensional gravitino and dilatino, as well as the gauginos reduced on the non-trivial background geometry. In addition to the universal sector, i.e. the gravity multiplet and the single tensor multiplet (this contains the heterotic dilaton as a modulus), the gravity multiplet also leads to 20 neutral hypermultiplets that describe the geometry of the K3 and the choice of closed B-field. The massless gauginos are the fermionic components of hypermultiplets, and their number is determined by the index theorem as follows: decompose $\mathfrak{e}_8 \supset \mathfrak{g} \oplus \mathfrak{h}$, where $\mathfrak{h}$ is the commutant of $\mathfrak{g} \subset \mathfrak{e}_8$—this will be the spacetime gauge algebra. The ten-dimensional gauginos then decompose according to

$$
\mathfrak{e}_8 \supset \mathfrak{g} \oplus \mathfrak{h}
$$
$$
\mathbf{248} = \oplus_i (r_i, \mathbf{R}_i) \oplus (\mathbf{1}, \text{adj}\,\mathfrak{h})\,,
\tag{4.7.47}
$$

where $i = 0$ corresponds to (adj $\mathfrak{g}$, $\mathbf{1}$). The last term in the sum obviously describes the spacetime gauginos in vector multiplets, while the remaining terms yield a charged spectrum with hypermultiplets in the representation[49]

$$
\mathbf{R} = \mathbf{1}^{\oplus N_0} \oplus_i \mathbf{R}_i^{\oplus N_i}\,,
\tag{4.7.48}
$$

where

$$
N_i = -\frac{\ell(r_i)p_1(E)}{4} - \dim r_i = 12\ell(r_i) - \dim r_i\,.
\tag{4.7.49}
$$

---

[49]When $\mathbf{R}_i$ is real or complex, N_i is the number of hypermultiplets; when $\mathbf{R}_i$ is pseudoreal, then representation is actually $2N_i$ $\frac{1}{2}$-hypermultiplets, each in representation $\mathbf{R}_i$.

Recall that $\ell(r_i)$ is the Dynkin index of the representation r_i of $\mathfrak{g}$. In the second equality we used the anomaly cancelation condition that $p_1(E)/2 = -24$. Since $h_\mathfrak{g} = \frac{1}{2}\ell(\text{adj}(\mathfrak{g}))$, N_0 matches (4.7.45).

When $c_1(E) = 0$, there are no further constraints on the geometry, so that the massless spectrum has

$$N_H = 20 + \sum_i N_i \dim \boldsymbol{R}_i , \qquad N_V = \dim \mathfrak{h} + \dim \mathfrak{e}_8 . \qquad (4.7.50)$$

This seems quite rigid (i.e. independent of the geometry and the bundle moduli), but the rigidity is an illusion, even away from the singular locus of the SCFT. The expression for N_i is a consequence of an index theorem, which really counts, for each representation $\boldsymbol{R}_i$ of $\mathfrak{h}$, the number of massless hyperinos minus the number of massless gauginos. Special points in the moduli space can then exhibit the "un-Higgsing" mentioned above. It may also appear from the formula that the 20 geometric and B-field moduli have a canonical splitting from the other moduli. Metrically, this is not the case as a consequence of the non-trivial gauge-transformation properties of B; aspects of the quaternion-Kähler metric are discussed in [9].

It is a simple matter to work out the reduction for various subgroups of E_8 or $\text{Spin}(32)/\mathbb{Z}_2$. For instance, we might take the embedding

$$\mathfrak{e}_8 \supset \mathfrak{g}_2 \oplus \mathfrak{f}_4 ,$$

$$\boldsymbol{248} = (\boldsymbol{14}, \boldsymbol{1}) \oplus (\boldsymbol{7}, \boldsymbol{26}) \oplus (\boldsymbol{1}, \boldsymbol{52}) . \qquad (4.7.51)$$

G_2 instantons with $k = 12$ are irreducible, and looking back to (2.3.6), the massless hypermultiplet spectrum is then

$$\boldsymbol{1}^{\oplus 20} \oplus \boldsymbol{1}^{\oplus 82} \oplus \boldsymbol{26}^{\oplus 17} , \qquad (4.7.52)$$

and it is easy to check that $N_H - N_V = 244$ is satisfied.

Massless Spectrum in K3 Compactification: Worldsheet Perspective

As in the case of the non-linear sigma model on a Calabi–Yau 3-fold, it is an easy matter to obtain the massless spectrum in the large radius limit of a K3 compactification; moreover, the results obtained are much stronger due to the powerful non-renormalization theorems: away from the SCFT singular locus and various special "un-Higgs" loci, the large radius limit results describe the massless spectrum at a generic point in the moduli space. The topological heterotic ring is now a much simpler structure because the worldsheet fields are organized into multiplets of the $\widehat{\mathfrak{su}(2)}_1$ KM algebra. A heterotic perturbative compactification to $\mathbb{R}^{1,5}$ and eight spacetime supercharges requires the worldsheet theory with a small $N = 4$ superconformal algebra with $\bar{c} = 6$ [46, 165], so we set our description in that general context and work out the correspondence between spacetime massless fermions and the $\overline{R}$ ground states of the internal theory. For most of our discussion we will also have a $U(1)_L$ symmetry with integral charges, which will allow a definition of the topological heterotic rings.

We begin by quoting some basic facts about the little group in $\mathbb{R}^{1,5}$ and the representations for massless fermions.[50] Since much of this is a direct parallel of the story worked out above, we will not give all of the details. Working these out in detail is a nice exercise.

A Dirac spinor of $\mathfrak{so}(1,5)$ has eight complex components; these decompose into two Weyl multiplets $\mathbf{4}$ and $\mathbf{4'}$, and these are not conjugate representations (unlike, say, $\mathbf{4}$ and $\overline{\mathbf{4}}$ of $\mathfrak{so}(6)$). The eight real chiral supercharges then transform as $\mathbf{4}$. Denoting the little group's Lie algebra as $\mathfrak{su}(2)_- \oplus \mathfrak{su}(2)_+$, the supercharges decompose as $\mathbf{4} = (\mathbf{2}, \mathbf{1}) \oplus (\mathbf{1}, \mathbf{2})$, and the massless content of $(1,0)$ supersymmetry multiplets decomposes as

$$\tfrac{1}{2}\text{-hyper} = \underline{(\mathbf{2}, \mathbf{1})} \oplus (\mathbf{1}, \mathbf{1})^{\oplus 2},$$

$$\text{vector} = (\mathbf{2}, \mathbf{2}) \oplus \underline{(\mathbf{1}, \mathbf{2})}^{\oplus 2},$$

$$\text{tensor} = (\mathbf{3}, \mathbf{1}) \oplus \underline{(\mathbf{2}, \mathbf{1})}^{\oplus 2} \oplus (\mathbf{1}, \mathbf{1}),$$

$$\text{sugra} = (\mathbf{3}, \mathbf{3}) \oplus \underline{(\mathbf{2}, \mathbf{3})}^{\oplus 2} \oplus (\mathbf{1}, \mathbf{3}). \tag{4.7.53}$$

We underlined the fermion content of each multiplet. This decomposition reflects the real on-shell degrees of freedom; as a check note that the number of bosons matches that of the fermions.

To construct a heterotic compactification of the $E_8 \times E_8$ based on the $(0, 4)$ SCFT with $\overline{c} = 6$, $c = 4 + r$, and a rank r $U(1)_L$ symmetry, we add the $\mathbb{R}^{1,5}$ Minkowski sector, a copy of the left-moving hidden $\mathfrak{e}_8$ algebra, and $16 - 2r$ free fermions ξ^A. The right-moving fermions of the free $\mathbb{R}^{1,5}$ SCFT generate a level 1 $\mathfrak{su}(2)_- \oplus \mathfrak{su}(2)_+$ KM algebra, which is identified with the algebra of the spacetime little group. The $\overline{R}$ ground states of the Minkowski sector then decompose as

$$(\mathbf{1}, \mathbf{2}) \oplus (\mathbf{2}, \mathbf{1}) \equiv \mathcal{W}_+ \oplus \mathcal{W}_- , \tag{4.7.54}$$

where the subscript denotes the eigenvalue of $(-1)^{\overline{F}}$.

The $\overline{R}$ ground states of the internal theory are of the form $|\mathcal{O}_c^{\overline{q}}; \overline{R}\rangle_{\overline{q}-1}$, where $\mathcal{O}_c^{\overline{q}}$ is a right-chiral-primary operator, and unitarity restricts $\overline{q} \in \{0, 1, 2\}$. Spectral flow yields the isomorphism between the right-chiral primary operators with $\overline{q} = 0$ and those with $\overline{q} = 2$, and therefore the internal $\overline{R}$ ground states are organized into

---

[50]There is a nice presentation of these ideas in the Appendix of [327].

doublets and singlets of $\mathfrak{su}(2)_R$ according to

$$\begin{pmatrix} |\mathcal{O}_c; \overline{R}\rangle_{-1} \\ |\mathcal{O}'_c; \overline{R}\rangle_{+1} \end{pmatrix} \qquad |\mathcal{O}_c^{\overline{q}=1}; \overline{R}\rangle_0 , \qquad (4.7.55)$$

where $\mathcal{O}_c$ has $\overline{q} = 0$ and $\mathcal{O}'_c$ is its spectral flow partner with $\overline{q} = +2$. The GSO projection then pairs up the doublets with $\mathcal{W}_+$ and the singlets with $\mathcal{W}_-$.

Exercise 4.20 Every $(0, 4)$ $\overline{c} = 6$ SCFT with integral $U(1)_R$ charges has a unique right-chiral-primary operator Ω with $\overline{q} = 2$, which is the spectral flow of the identity operator. Consider the massless states in the NS,$\overline{R}$ sector and show that the identity and Ω, when tensored with the Minkowski bosonic excitations ∂X^I, $I = 1, \ldots, 4$ lead to exactly the fermion content of supergravity+tensor multiplet. Similarly, show that there are also universal states constructed by applying J_L and the currents $\xi^A \xi^B$, as well as the raising operators of the "hidden" $\mathfrak{e}_8$, that generate the $\mathfrak{u}(1)_L \oplus \mathfrak{so}(16 - r) \oplus \mathfrak{e}_8$ spacetime gauginos. These are the universal states that will arise in any compactification in this class. $\qquad\qquad\square$

Exercise 4.21 Combine the technologies presented in this and previous section and describe the massless spectrum for a rank $r = 3$ heterotic compactification, where the internal theory is a holomorphic rank $r = 3$ SU(3) structure bundle $\mathcal{E} \to M$ over a K3 with $c_1(E) = 0$ and $c_2(E) = -24$. In the NS,$\overline{R}$ sector you should find the following gauge-charged $\frac{1}{2}$-hypermultiplet states, organized in representations of the linearly realized $\mathfrak{so}(10) \oplus \mathfrak{u}(1)_L$ and counted by various Dolbeault cohomology groups:

$$\mathbf{10}_{-1} : H_{\overline{\partial}}^{0,1}(M, \mathcal{E}^*) , \qquad\qquad \mathbf{10}_{+1} : H_{\overline{\partial}}^{0,1}(M, \mathcal{E}) ,$$

$$\mathbf{1}_{+2} : H_{\overline{\partial}}^{0,1}(M, \wedge^2 \mathcal{E}) , \qquad\qquad \mathbf{1}_{-2} : H_{\overline{\partial}}^{0,1}(M, \wedge^2 \mathcal{E}^*) . \qquad (4.7.56)$$

The states in the R, $\overline{R}$ sector complete these (show it!) to the $\mathfrak{e}_6$-charged hypermultiplet spectrum $\mathbf{27}^{\oplus h^1(M, \mathcal{E})}$.

It is not hard to construct also the operators for the gauge-neutral states. Restricting to the lowest modes in the NS,$\overline{R}$ sector these are of the form

$$\left[\partial \phi^a \mu_{a\overline{a}} + \gamma_A \overline{\gamma}^B \lambda_{B\overline{a}}^A + (\rho_b - \mathcal{A}_{Bb}^A \overline{\gamma}_A \gamma^B) Z_{\overline{a}}^b \right] \overline{\psi}^{\overline{a}} |\overline{R}\rangle , \qquad (4.7.57)$$

and restricting to $\overline{Q}$ cohomology leads to $\frac{1}{2}$-hypermultiplets counted by

$$Z \in H_{\overline{\partial}}^{0,1}(M, T_M) , \qquad \alpha(Z) = 0 , \qquad \lambda \in H_{\overline{\partial}}^{0,1}(M, \text{End}\,\mathcal{E}) ,$$

$$\mu \in H_{\overline{\partial}}^{0,1}(M, T_M^*) . \qquad\qquad\qquad\qquad\qquad (4.7.58)$$

On K3 this structure is much simpler than on a three-fold. First, Kodaira–Serre duality implies $H_{\bar{\partial}}^{0,2}(M, \text{End}\,\mathcal{E}) \simeq H_{\bar{\partial}}^{0,0}(M, \text{End}\,\mathcal{E})$. But, for an irreducible SU(3) structure bundle the latter cohomology group is zero. Thus, the Atiyah class is trivially vanishing. Similarly, we also have $h^1(M, T_M) = h^1(M, T_M^*)$. Finally, we can compute the multiplicities by applying the Hirzebruch–Riemann–Roch index theorem. For any holomorphic bundle $\mathcal{E}$ over a compact complex manifold M we have

$$\chi(M, \mathcal{E}) = h^0(M, \mathcal{E}) - h^1(M, \mathcal{E}) + h^2(M, \mathcal{E}) - \ldots = \int_M \text{ch}(\mathcal{E})\text{Td}(M),$$

(4.7.59)

where the Todd class has the expansion [161]

$$\text{Td}(M) = 1 + \tfrac{1}{2}c_1(M) + \tfrac{1}{12}(c_2(M) + c_1(M)^2) + \tfrac{1}{24}c_1(M)c_2(M) + \ldots .$$

(4.7.60)

Specializing to the K3 manifold and a rank r bundle with $c_1(\mathcal{E}) = 0$, this leads to leads to

$$h^0(M, \mathcal{E}) - h^1(M, \mathcal{E}) + h^2(M, \mathcal{E}) = 2r - c_2(\mathcal{E}).$$

(4.7.61)

For $r < 12$ it follows that $h^1(M, \mathcal{E}) \neq 0$, and if the bundle is generic enough so that $h^0(M, \mathcal{E}) = h^2(M, \mathcal{E}) = 0$, we can conclude $h^1(M, \mathcal{E}) = c_2(\mathcal{E}) - 2r$.

Use these facts and $c_2(\mathcal{E} \otimes \mathcal{E}^*) = 2rc_2(\mathcal{E}) = 144$ to show that the generic hypermultiplet spectrum for the rank $r = 3$ bundle is given by[51]

$$\mathbf{1}^{\oplus 20} \oplus \mathbf{1}^{\oplus 64} \oplus \mathbf{27}^{\oplus 18}.$$

(4.7.62)

□

The result of the previous exercise is in perfect agreement with the spacetime description of an SU(3) instanton embedded in $\mathfrak{e}_8$. This will always be the case for the non-linear sigma models for $\mathcal{E} \to M$, where $\mathcal{E}$ is a rank r holomorphic bundle over K3. Not every spacetime gauge group can be obtained by such a simple construction that just generalizes the free fermion realizations of the ten-dimensional heterotic gauge groups. For instance, the example of the $\mathfrak{f}_4$ spacetime gauge algebra is a perfectly sensible six-dimensional (1,0) theory, but its worldsheet realization is far from simple. On the other hand, the spacetime point of view shows that there are perfectly good (0,4) SCFTs that realize an $\mathfrak{f}_4$ compactification; the

---

[51] Taking a look at the characteristic classes discussion above, it is easy to see that for a holomorphic bundle over any complex surface $\text{ch}(\mathcal{E} \otimes \mathcal{E}^*) = r^2 + r(c_1(\mathcal{E})^2 - 2c_2(\mathcal{E}))$. To obtain the counting above it is also important to remember that the worldsheet states count $\tfrac{1}{2}$-hypermultiplets.

methods of [147] can be used to provide explicit realizations as chiral WZW models fibered over a K3 geometry. Those constructions deserve more attention.

We can see this explicitly from basic facts about the Higgs mechanism in theories with 8 supercharges—see, for instance, the classic discussions of $\mathcal{N} = 2, d = 4$ gauge theories in [25, 26], or the more recent text [356] . Consider for instance, the $\mathfrak{e}_6$ theory we just discussed and the branching

$$\mathfrak{e}_6 \supset \mathfrak{f}_4 \,,$$

$$\mathbf{27} = \mathbf{26} \oplus \mathbf{1} \,,$$

$$\mathbf{78} = \mathbf{52} \oplus \mathbf{26} \,. \tag{4.7.63}$$

As far as representation theory goes, we see that it is possible to Higgs $\mathfrak{e}_6 \to \mathfrak{f}_4$ by appropriately giving expectation values to the $\mathbf{27}$ scalars. Is it possible to satisfy the constraints from the vanishing of the D-term potential? The answer is yes: as long as we give expectation values to at least two different hypermultiplets in $\mathbf{27}$, the D-term constraints can be satisfied. Our theory, with $\mathbf{27}^{\oplus 18}$, has plenty of charged matter, and the resulting massless hypermultiplet spectrum on the $\mathfrak{f}_4$ Higgs branch is therefore

$$\mathbf{1}^{\oplus 20} \oplus \mathbf{1}^{\oplus 64} \oplus \mathbf{1}^{\oplus 18} \oplus \mathbf{26}^{\oplus 17} \,. \tag{4.7.64}$$

This matches the spacetime result based on a $\mathfrak{g}_2$ instanton embedded in $\mathfrak{e}_8$, and we now understand that this seemingly exotic theory has a locus in the moduli space where the structure group is reduced as $G_2 \to SU(3)$, and there is a standard worldsheet construction.

Abelian Instantons on K3

In all of the geometric examples we discussed so far, the vector bundle for the left-moving fermions satisfied $c_1(E) = 0$. This is not required by anomaly cancellation, and simplicity is the only reason to make the restriction. On the other hand, there is a simple class of bundles on K3 with $c_1(E) \neq 0$, for which the conditions of anti-self-duality are easy to understand: these are sums of line bundles: $E = \oplus_{i=1}^r \mathcal{L}_i$, where each $\mathcal{L}_i$ is a rank 1 holomorphic bundle [231, 278, 301]. In this case anomaly cancellation requires

$$c_1(E) = \sum_i c_1(\mathcal{L}_i) \in H^2(M, 2\mathbb{Z}) \,, \quad p_1(E) = \sum_{i=1}^r c_1(\mathcal{L}_i)^2 = -48 \,. \tag{4.7.65}$$

The instanton equations are then equivalent to

$$\Omega \wedge \mathcal{F}_i = 0 \,, \qquad\qquad \omega \wedge \mathcal{F}_i = 0 \,. \tag{4.7.66}$$

Each $c_1(\mathcal{L}_i) = [i\mathcal{F}_i/2\pi]$ defines a class in $H^2(M, \mathbb{Z})$, while, on the other hand, the holomorphic (2,0) form Ω and the Kähler form ω define a Ricci-flat metric on M [31]. Thus, we can think of the conditions as restrictions on the metric on K3. The first equation concerns the Picard lattice of M. On K3 (4.5.38) simplifies because $H^1(M, \mathcal{O}) = 0$, so that $\text{Pic}(M) = H^2(M, \mathbb{Z}) \cap H^{1,1}(M)$. For a generic complex structure $\text{Pic}(M) = 0$, and there are no holomorphic line bundles on M. If the $c_1(\mathcal{L}_i)$ are linearly independent classes in $H^2(M, \mathbb{R})$, then the rank of $\text{Pic}(M)$ is at least r. The second equation is the statement of anti-self-duality.

While a generic choice of $\Omega, \omega \in H^2(M, \mathbb{R})$ compatible with SU(2) structure constraints

$$\Omega \wedge \omega = 0, \qquad \Omega \wedge \Omega = 0, \qquad \omega \wedge \omega = \frac{1}{2}\Omega \wedge \overline{\Omega} \qquad (4.7.67)$$

leads to a smooth metric on M, the metric will degenerate at singular points in the moduli space. These have a beautiful description: a choice of $\Omega, \omega \in H^2(M, \mathbb{R})$ compatible with SU(2) structure leads to a singular metric if and only if for some class $C \in \text{Pic}(M)$ with $C \cdot C = -2$ and $\int_C \omega = 0$ [31, 49].[52] Thus, in choosing the line bundles we must avoid $c_1(\mathcal{L}_i)^2 = -2$. For an anti-self-dual connection $c_1(\mathcal{L}_i)^2 < 0$, and, quite generally, for any $e \in H^2(M, \mathbb{Z})$, $e \cdot e \in 2\mathbb{Z}$. Thus, the Bianchi identity leads to a bound on $c_1(\mathcal{L}_i)^2$ for smooth compactifications:

$$-48 + 4(r - 1) \le c_1(\mathcal{L}_i)^2 \le -4. \qquad (4.7.68)$$

It also follows that $r \le 12$.

In all of the examples of irreducible bundles with $c_1(E) = 0$, we saw that the moduli of the Calabi–Yau metric on K3, together with a choice of closed B-field, led to the 20 gauge-neutral massless hypermultiplets in the spacetime theory. Clearly this is not the case here, since each linearly independent class $c_1(\mathcal{L}_i) \in H^2(M, \mathbb{Z})$ imposes 3 linear constraints on Ω and ω. The 3 does not sound quite "hyper" enough, and indeed there is more to the story: the B-field transforms under constant gauge transformations ϵ_i of the spacetime fields associated to the spacetime $\mathfrak{u}(1)^{\oplus r}$ gauge symmetry: $\delta B = -\frac{1}{2}\epsilon^i \mathcal{F}_i$. Hence, the B-field moduli are characterized by $H^2(M, \mathbb{R})/\{\text{span}(\mathcal{F}_1, \ldots, \mathcal{F}_r)\}$.

Thus, each linearly independent $c_1(\mathcal{L}_i)$ removes a hypermultiplet's worth of gauge-neutral scalars from the geometric moduli. Again, this sounds not "hyper" enough. How can we just lose a hypermultiplet and yet satisfy anomaly cancellation? There is a satisfying answer to this as well: the spacetime gauge algebra factor $\mathfrak{u}(1)^{\oplus r}$ is in general anomalous. When the $c_1(\mathcal{L}_i)$ are linearly independent, all of these are anomalous; in the spacetime interpretation, the B-field components proportional to the $\mathcal{F}_i$ of the previous paragraph act as Stückelberg fields for the

---

[52]We denote the intersection product $H^2(M, \mathbb{Z}) \times H^2(M, \mathbb{Z}) \to H^4(M, \mathbb{Z})$ by $\cdot$; the product is symmetric and has signature (3, 19).

anomalous gauge bosons. Such massive "anomalous" U(1)s are a classic of string compactification [141]; some recent studies are [8, 344]. In the context of K3 compactifications they have been discussed in, for instance,[231].

The fact that geometric moduli get "stabilized" may at first sight seem very exciting, since it suggests a route to finding SCFTs with a small number of marginal deformations. This is less exciting than it at first seems: as one makes the bundle E more complicated to reduce the geometric moduli, the number of gauge-neutral hypermultiplets usually increases. This is again a manifestation of the phenomenon we met before: splitting the marginal deformations of a (0,2) SCFT into classes such as "geometry" or "bundle" is not in general meaningful, and when one considers the full set of marginal deformations, any imagined split usually becomes illusory.

Exercise 4.22 Consider a compactification of the Spin(32)/$\mathbb{Z}_2$ heterotic string on a K3 M with r linearly independent line bundles $\mathcal{L}_i$ as above. Show that the spacetime gauge algebra is $\mathfrak{so}(k)$ with $k = 2(16 - r)$, and hypermultiplets in the representation

$$1^{\oplus n_0} \oplus k^{\oplus 2(12-r)} , \qquad n_0 = 20 - r + 2(24 - r)(r - 1) . \qquad (4.7.69)$$

Details can be found in [301].

Now use the spacetime perspective to verify that for $r > 1$ the same spectrum is obtained by using a bundle with structure group $G = SO(2r)$, and the gauge-neutral hypermultiplets are now counted by the index theorem as

$$n_0 = 20 + \underbrace{48(r - 1) - r(2r - 1)}_{\text{"bundle" defs}} . \qquad (4.7.70)$$

Note that irreducibility of the connection requires $r \leq 12$.

Even the $r = 1$ case is not really special, since its Higgs branch includes the other examples, and, conversely, it can be understood as a special point in the moduli space of the generic $r = 12$ theory with spacetime gauge algebra $\mathfrak{so}(8)$ and no gauge-charged matter. Note that $r = 12$ is the maximum value that leads to an irreducible $\mathfrak{so}(2r)$ connection with instanton number 12.   □

There are a few lessons from this example that are relevant for constructions based on sums of line bundles.

1. The class of models offers a large class of examples where the anti-self-duality constraint is easily understood.
2. The models very nicely illustrate the mixing of different "types" of moduli through the Green–Schwarz mechanism associated to massive U(1)s.
3. The examples can be best understood as special points in a larger moduli space of theories; this perspective is useful for dispelling any illusions about the class leading to "moduli stabilization" : the fixing of geometric parameters is only exciting if one forgets about the other marginal deformations.

Nothing prevents one from using sums of line bundles for more general (0,2) non-linear sigma models, and the approach does offer a tractable way to construct many examples of large radius limit compactifications [15]; the lessons also apply to those higher-dimensional constructions, but now there are additional complications due to non-trivial obstructions to marginality.

4.7.7   Remarks on (0,2) Sigma Models with Flux

The reader has perhaps felt a curious dichotomy in the preceding presentation of (0,2) geometry. On the one hand, we labored at some length to provide a geometric setup that goes beyond compactification on Kähler manifolds; on the other hand, whenever the matter came to discussing a putative SCFT, we retreated to theories with a large radius limit and vanishing torsion. The reason for this should be fairly clear. In theories without a large radius limit, the α' expansion is a formal procedure, since the corrections are in no way controlled.

This should not (and has not) prevented investigations of these geometries. That such compact heterotic solutions should exist and have an interpretation as non-Kähler compactification was pointed out in [125] and developed further in many works, including [59–62, 176, 199]. The best-understood examples preserve $\mathcal{N} = 2$ spacetime supersymmetry in $d = 4$; all such geometries must be principal T^2 fibrations over a K3 manifold [295, 301], and the heterotic flux then supports the non-trivial topology of the fibration. These string vacua are probably best understood as heterotic compactifications with an eight-dimensional "stop-over". More precisely, these theories have a marginal deformation that makes the base K3 geometry large. The de-compactification limit can then be understood as eight-dimensional heterotic supergravity. In that sense, the examples are closely related to constructions that first appeared in the context of type II/heterotic duality [258].

We can be reasonably confident that these non-linear sigma models, although formal, do correspond to SCFTs. The theories can be realized as IR limits of "torsional linear sigma models" introduced in [5], and this UV completion has been put to use to quantitatively study the SCFT in a series of elegant papers [19, 249, 250].

It is likely that there are also (0,2) SCFTs that correspond to $\mathcal{N} = 1$ $d = 4$ compactifications and can be formally described by a (0,2) non-linear sigma model with flux. However, given a particular (0,2) sigma model geometry that satisfies all of the criteria in Sect. 4.5.4, it is by no means clear that it defines a (0,2) SCFT. For instance, it is unknown which, if any, of the geometries in [164] leads to a heterotic vacuum. Any progress on these questions would constitute a big step in our understanding of (0,2) theories and their relation to geometry.

We finish our cursory discussion of this mysterious class of solutions with an observation on world-sheet instantons. If we suppose that it is sensible to treat the geometry as a useful starting point for describing the SCFT, then, as in the more conventional theories described above, we would expect non-trivial field configurations to play an important role, perhaps determining some class of protected couplings. As already indicated in Exercise 4.1, the presence of a topologically non-trivial $\mathcal{H}$ means that one must probably carry out an analytic continuation in field space to obtain saddle-points of the Euclidean action. Such continuations are known in quantum mechanics—see, for instance [170]—and have also been discussed in the context of Chern–Simons theory [220] and a variety of quantum field theories, e.g. [64, 106]. It should be interesting to study the interplay between the analytic continuation and the global geometry of the target space.

Chapter 5
Gauged Linear Sigma Models

Abstract In the previous chapter we gave a number of disparate constructions of (0,2) theories. The linear sigma models that we will discuss in this final chapter provide a surprisingly unified framework for most of these. We will now define these theories and describe a number of linear sigma model successes, as well as some of the key (0,2) puzzles that still remain but are greatly informed by the linear point of view. Along the way we will present a review of some toric geometry that is essential in linear sigma model exploration.

5.1 General Gauge Theory

Leaving, perhaps with a sigh of relief, all of the geometric intricacies of the non-linear sigma model, we return to the free (0,2) theory with N_Φ chiral multiplets Φ_a, which we assemble into a N_Φ-dimensional complex vector $\mathbf{\Phi}$, and N_Γ chiral fermi multiplets Γ^A, similarly assembled into an N_Γ-dimensional complex vector $\mathbf{\Gamma}$, and action

$$S_{\text{free}} = \frac{1}{2\pi} \int d^2z \mathcal{D}\overline{\mathcal{D}} \left\{ \tfrac{1}{4}(\mathbf{\Phi}^\dagger \partial \mathbf{\Phi} - (\partial \mathbf{\Phi}^\dagger)\mathbf{\Phi}) - \tfrac{1}{2}\mathbf{\Gamma}^\dagger \mathbf{\Gamma} \right\} . \tag{5.1.1}$$

The $\dagger$ notation is the Hermitian conjugate that also sends a chiral field to its anti-chiral conjugate field. The equations of motion are $\overline{\mathcal{D}}\mathbf{\Gamma}^\dagger = 0$ and $\overline{\mathcal{D}}\partial \mathbf{\Phi}^\dagger = 0$. S_{free} has a global symmetry $G_0 = \mathrm{U}(N_\Phi) \times \mathrm{U}(N_\Gamma)$ under which the $\mathbf{\Phi}$ transform in $(N_\Phi, \mathbf{1})$, while $\mathbf{\Gamma}$ transform in $(\mathbf{1}, N_\Gamma)$. The first step in building a gauged linear sigma model is to gauge a subgroup $G \subset G_0$.

© Springer Nature Switzerland AG 2019

I. V. Melnikov, *An Introduction to Two-Dimensional Quantum Field Theory with (0,2) Supersymmetry*, Lecture Notes in Physics 951,
https://doi.org/10.1007/978-3-030-05085-6_5

5.1.1 Supersymmetric Background Gauge Fields

To set the stage, we turn on a background gauge field for a general global symmetry in a (0,2) QFT. As we discussed in Sect. 3.2, a global symmetry current that commutes with (0,2) SUSY fits into the following multiplet structure:

1. $\mathcal{K}_1$ is an imaginary spin 1 and mass dimension 1 operator;
2. Ψ is chiral spin $-1/2$ and mass dimension $1/2$ operator, and its conjugate is the anti-chiral $\overline{\Psi}$;
3. The two are related via

$$\overline{\mathcal{D}}\mathcal{K}_1 = -2\partial\Psi, \qquad\qquad \mathcal{D}\mathcal{K}_1 = +2\partial\overline{\Psi}, \qquad\qquad (5.1.2)$$

which imply the conservation equation

$$\overline{\partial}\mathcal{K}_1 + \partial(\overline{\mathcal{D}\Psi} - \mathcal{D}\Psi) = 0. \qquad\qquad (5.1.3)$$

This multiplet can be naturally coupled to a background gauge field encoded in (0,2) superfields $\mathcal{V}_0$ and $\mathcal{V}_1'$: the first of these is a real bosonic multiplet, and the second is a chiral spin 1 multiplet. The linearized coupling is

$$\Delta S = \int d^2z \left\{ \mathcal{D}\overline{\mathcal{D}}(\mathcal{K}_1 \mathcal{V}_0) - \mathcal{D}(\Psi \mathcal{V}_1') - \overline{\mathcal{D}}(\overline{\Psi}\overline{\mathcal{V}}_1') \right\}. \qquad\qquad (5.1.4)$$

Exercise 5.1 Use (5.1.2) to show that ΔS is invariant under $\mathcal{V}_0 \to \mathcal{V}_0 - (\Lambda + \overline{\Lambda})/2$ and $\mathcal{V}_1' \to \mathcal{V}_1' - \partial\Lambda$ for any chiral multiplet Λ.   □

The structure simplifies in the free theory (or, indeed, in the general class of Yukawa models we discussed in Chap. 3) since $\Psi = \overline{\mathcal{D}}\mathcal{K}_0$, where $\mathcal{K}_0$ is a spin 0 real bosonic field, so that $\mathcal{K}_0$ and $\mathcal{K}_1$ are constrained through

$$\overline{\mathcal{D}}(\mathcal{K}_1 + 2\partial\mathcal{K}_0) = 0 \qquad\qquad \mathcal{D}(\mathcal{K}_1 - 2\partial\mathcal{K}_0) = 0, \qquad\qquad (5.1.5)$$

and the conservation equation is now

$$\overline{\partial}\mathcal{K}_1 + \partial[\mathcal{D}, \overline{\mathcal{D}}]\mathcal{K}_0 = 0, \qquad\qquad (5.1.6)$$

and the linearized coupling to a background gauge field with components $\mathcal{V}_0$ and $\mathcal{V}_1$ is

$$\Delta S = \int d^2z \mathcal{D}\overline{\mathcal{D}} \{\mathcal{K}_1 \mathcal{V}_0 - \mathcal{K}_0 \mathcal{V}_1\}. \qquad\qquad (5.1.7)$$

The action is real and supersymmetric if $\mathcal{V}_0$ is real and $\mathcal{V}_1$ is imaginary. Unlike $\mathcal{V}_1'$, $\mathcal{V}_1$ need not be chiral. It follows from (5.1.5) that ΔS is gauge-invariant under

$$\mathcal{V}_0 \to \mathcal{V}_0 - \tfrac{1}{2}(\Lambda + \overline{\Lambda}) , \qquad \mathcal{V}_1 \to \mathcal{V}_1 - \partial(\Lambda - \overline{\Lambda}) . \tag{5.1.8}$$

These multiplets have the right structure to introduce a two-dimensional spin one gauge field without introducing higher spin fields: the highest component of $\mathcal{V}_0$ will be proportional to the $\bar{z}$-component of the gauge field, while the lowest component of $\mathcal{V}_1$ will be proportional to the z-component of the gauge field.

It will also be convenient to define

$$\begin{aligned} \mathcal{U}_1 &= \partial \mathcal{V}_0 + \tfrac{1}{2}\mathcal{V}_1 , & \mathcal{U}_1 &\to \mathcal{U}_1 - \partial\Lambda , \\ \overline{\mathcal{U}}_1 &= \partial \mathcal{V}_0 - \tfrac{1}{2}\mathcal{V}_1 , & \overline{\mathcal{U}}_1 &\to \overline{\mathcal{U}}_1 - \partial\overline{\Lambda} . \end{aligned} \tag{5.1.9}$$

5.1.2 Gauge-Invariant Kinetic Terms

Coming back to the free action (5.1.1), we now wish to promote the global invariance under G to a local invariance. The procedure is standard and parallel to, say, the $\mathcal{N} = 1$, $d = 4$ case that is perhaps familiar to the reader [24]. Any global G-transformation connected to the identity has the form

$$\begin{aligned} \Phi &\to \Phi_\lambda = \exp\left[\lambda_\alpha T_\Phi^\alpha\right] \Phi , & \Gamma &\to \Gamma_\lambda = \exp\left[\lambda_\alpha T_\Gamma^\alpha\right] \Gamma , \\ \Phi^\dagger &\to \Phi_\lambda^\dagger = \Phi^\dagger \exp\left[\overline{\lambda}_\alpha T_\Phi^\alpha\right] , & \Gamma &\to \Gamma_\lambda = \Gamma^\dagger \exp\left[\overline{\lambda}_\alpha T_\Gamma^\alpha\right] . \end{aligned} \tag{5.1.10}$$

The T^α are Hermitian generators of the Lie algebra $\mathfrak{g} \subset \mathfrak{u}(N_\Phi) \oplus \mathfrak{u}(N_\Gamma)$ of G, and the λ_α, $\alpha = 1, \ldots, \dim\mathfrak{g}$ are constant imaginary parameters.

To promote this to a local invariance, we must make the λ parameters worldsheet dependent, but we want to do so while preserving the chirality conditions on the fields. We can do this by promoting $\lambda_\alpha \to \Lambda_\alpha$, a chiral superfield and $\overline{\lambda}_\alpha \to \overline{\Lambda}_\alpha$, the conjugate anti-chiral superfield. In the process, we inevitably enlarge the gauge group G to its complexification: $G \to G_\mathbb{C}$. The next step is to minimally couple S_{free} to the $(\mathcal{V}_0, \mathcal{V}_1)$ gauge fields so that the resulting action S_{min} is $G_\mathbb{C}$-invariant. There are two issues: we need to introduce factors that absorb the homogeneous $G_\mathbb{C}$-dependence, and we need to replace the explicit derivative ∂ by its covariant form ; the first issue is addressed by dressing the kinetic terms with $\mathcal{V}_0$, while the second uses a non-abelian generalization of $\mathcal{U}_1$.

For any gauge transformation connected to the identity, we define $\mathcal{V}_{0\alpha}^\lambda$ as the unique solution to

$$\exp\left[2\mathcal{V}_0^\lambda \cdot \boldsymbol{T}\right] = \exp\left[-\overline{\Lambda} \cdot \boldsymbol{T}\right] \exp\left[2\mathcal{V}_0 \cdot \boldsymbol{T}\right] \exp\left[-\Lambda \cdot \boldsymbol{T}\right] , \tag{5.1.11}$$

where the T^α are the generators of $\mathfrak{g}_{\mathbb{C}}$ in the adjoint representation, and $\mathcal{V}_0 \cdot T = \mathcal{V}_{0\alpha} T^\alpha$, etc. Now if we set the gauge transformation of $\mathcal{V}_0$ to be

$$\mathcal{V}_0 \rightarrow \mathcal{V}_0^\lambda , \tag{5.1.12}$$

then $\Gamma^\dagger \exp\left[2\mathcal{V}_{0\alpha} T_\Gamma^\alpha\right] \Gamma$ is $G_{\mathbb{C}}$-invariant. In the abelian case this leads to the transformation of $\mathcal{V}_0$ in (5.1.8).

We now take $\mathcal{U}_1$ to be a connection valued in the $(N_\Phi, \mathbf{1})$ representation of $\mathfrak{g}_{\mathbb{C}}$ with

$$\mathcal{U}_1^\lambda = \exp[\Lambda \cdot T_\Phi]\mathcal{U}_1 \exp[-\Lambda \cdot T_\Phi] - (\partial \exp[\Lambda \cdot T_\Phi]) \exp[-\Lambda \cdot T_\Phi] . \tag{5.1.13}$$

In the abelian case this too reduces to the simple transformation law of $\mathcal{U}_1$. With this definition

$$(\partial + \mathcal{U}_1^\lambda)\Phi_\lambda = \exp[\Lambda \cdot T_\Phi] (\partial + \mathcal{U}_1)\Phi , \tag{5.1.14}$$

and therefore we have a simple form for the minimal supersymmetric coupling:

$$S_{\min} = \frac{1}{2\pi} \int d^2z \mathcal{D}\overline{\mathcal{D}} \left\{ \frac{1}{4} \left[\Phi^\dagger e^{2\mathcal{V}_{0\alpha} T_\Phi^\alpha}(\partial + \mathcal{U}_1)\Phi - \Phi^\dagger(\overleftarrow{\partial} + \overline{\mathcal{U}}_1)e^{2\mathcal{V}_{0\alpha} T_\Phi^\alpha}\Phi \right] \right.$$
$$\left. - \frac{1}{2}\Gamma^\dagger e^{2\mathcal{V}_{0\alpha} T_\Gamma^\alpha}\Gamma \right\} . \tag{5.1.15}$$

Since the non-homogeneous term in $\mathcal{U}_1^\lambda$ is chiral, $\Upsilon = \overline{\mathcal{D}}\mathcal{U}_1$ transforms homogeneously, and therefore is a natural candidate for the gauge field strength multiplet. Υ is a fermi multiplet, and its lowest component is the left-moving gaugino. This is again, much like the perhaps more familiar $\mathcal{N} = 1$ d=4 case. Thus, there is a natural term we can add to the action:

$$S_{\mathrm{YM}} \propto \frac{1}{g_{\mathrm{YM}}^2} \int d^2z \, \mathcal{D}\overline{\mathcal{D}} \, \mathrm{Tr}\{\Upsilon^\dagger \Upsilon\} . \tag{5.1.16}$$

Exercise 5.2 Linearize $S_{\min}$ in the $\mathcal{V}_0$ and $\mathcal{U}_1$ gauge fields and show that it reproduces (5.1.7) with the expected linearized gauge invariance. You should encounter the currents

$$\mathcal{K}_{1\alpha} = \frac{1}{2}(\Phi^\dagger T_\Phi^\alpha \partial \Phi - \partial \Phi^\dagger T_\Phi^\alpha \Phi) - \Gamma^\dagger T_\Gamma^\alpha \Gamma , \quad \mathcal{K}_{0\alpha} = -\frac{1}{4}\Phi^\dagger T_\Phi^\alpha \Phi . \tag{5.1.17}$$

Use the free equations of motion to show that these satisfy (5.1.5). $\square$

5.1.3 Remarks on the General Structure

We now have all of the tools to build the general classical $(0,2)$ non-abelian gauge theory. Having fixed the gauge group G and identified the representation R_Γ in which the fermi multiplets transform, we construct the most general gauge-covariant superpotentials $E(\Phi) \in R_\Gamma$ and $J(\Phi) \in R_\Gamma^*$ such that $J(E) = 0$ and set

$$\overline{\mathcal{D}}\Gamma = E , \qquad\qquad W = \int d^2z\, \mathcal{D}[J\Gamma] . \qquad (5.1.18)$$

In addition, we can specify Fayet-Iliopoulos parameters for each abelian factor in the gauge theory, as well as continuous (for abelian gauge group) and discrete [378] (for non-simply connected gauge group components) theta-angles.

We would now embark on various standard investigations, such as the space of classical vacua X. On general grounds, this will be the zero set of E and J in

$$V = \{\phi \in \mathbb{C}^{N_\Phi} \mid D^\alpha(\phi, \overline{\phi}) = 0\}/G . \qquad (5.1.19)$$

The D^α are the D-term potentials that arise from the gauge couplings; D^α is a scalar field in the top component of Υ, and we will describe it in further detail below.

What is the physical interpretation of X? In a sensible gauge theory the non-linear sigma model on X may approximate the low energy excitations. This is because in two dimensions g_{YM} has units of mass, so we can formally take the $g_{YM} \to \infty$ in the classical action to get a (possibly very crude) approximation to the low energy dynamics of the gauge theory. In this limit the gauge fields act as Lagrange multipliers, and integrating them out amounts to imposing non-linear constraints on the matter fields. As a result, we obtain a $(0,2)$ non-linear sigma model studied in the previous chapter: there is a complex target space X and a holomorphic bundle $\mathcal{E} \to X$ describes the coupling of the left-moving fermions.

If X is smooth and compact, then, at least classically, we can think of the $(0,2)$ gauged linear sigma model as providing a UV completion for the non-linear sigma model. When M does not meet these requirements, there need not be any pathology in the gauged linear sigma model itself: the statement is simply that the IR dynamics of the gauge theory unavoidably involve strongly coupled (because g_{YM} has units of mass) dynamical gauge fields.

5.1.4 Local Gauge Anomalies

The linear sigma model may not be a sensible theory, since the gauge theory may be anomalous. A necessary condition for sensibility is the vanishing of the local gauge anomalies. Decompose the Lie algebra $\mathfrak{g}$ of G as

$$\mathfrak{g} = \oplus_i \mathfrak{h}_i \oplus \mathfrak{u}(1)^{\oplus r} , \qquad (5.1.20)$$

where each $\mathfrak{h}_i$ is a simple Lie algebra. If R_Γ and R_Φ denote the representations of the fermi and chiral multiplets, then the local non-abelian anomalies will vanish if for each $\mathfrak{h}_i$ factor the Dynkin indices match:

$$\ell_{\mathfrak{h}_i}(R_\Gamma) = \ell_{\mathfrak{h}_i}(R_\Phi) . \tag{5.1.21}$$

Let Q_A^p and Q_a^p denote the abelian gauge charges of Γ^A and Φ_a under the p-th $\mathfrak{u}(1)$ factor. The abelian local anomalies will vanish if for all $p_1, p_2 \in \{1, \dots, r\}$

$$\sum_A Q_A^{p_1} Q_A^{p_2} = \sum_a Q_a^{p_1} Q_a^{p_2} . \tag{5.1.22}$$

While these conditions are sufficient to eliminate the local gauge anomaly, they are not necessary. In some cases abelian gauge anomalies can be canceled by the two-dimensional analogue of the Green-Schwarz mechanism: this requires additional degrees of freedom encoded in chiral (or more precisely linear) matter multiplets that transform by linear shifts under the gauge symmetry, such that the quantum transformation of the measure is canceled by a local non-invariant term in the action that couples the gauge field strength to the additional scalars [5]. This cancelation mechanism can be used to produce linear sigma models related to non-linear sigma models with torsion [4, 300, 329, 330]. Two-dimensional theories, much like six-dimensional ones, can have chiral bosonic degrees of freedom; while these are typically non-Lagrangian, if one is willing to introduce such degree of freedom, it is also possible to devise more general mechanisms to cancel the standard local gauge anomaly.[1]

There may also be global gauge anomalies. Their vanishing requires a conspiracy between the homotopy groups of G and the matter representations. We refer the reader to the elegant [187] and references therein for a very readable and modern presentation.

General (0,2) gauge theories have not been extensively studied, mainly because of the challenges presented by the non-abelian case. Nevertheless there have been some important developments. Non-abelian (2,2) theories have been examined in the context of mirror symmetry starting with [235] and, more recently in [196, 215, 232, 234], and, even earlier, in the context of quantum cohomology [392]. More recently, localization techniques have also been applied to give general formulas for partition functions of (2,2) theories on spheres [68, 156], half-twisted correlation functions [116], as well as general formulas for elliptic genera of (0,2) Lagrangian theories [69, 70, 180]. A relation between (0,2) gauge theories and geometry of 4-manifolds was proposed in [182], and new dualities and IR fixed points were developed and described in [183, 184].

---

[1] Such degrees of freedom occur when one reduces a six-dimensional (1,0) theory on a Kähler manifold; that construction makes it clear that there are many possible consistent choices for these degrees of freedom in two dimensions [23].

5.2 Abelian Gauge Theory

In our study of (0,2) quantum field theory, we have often retreated from the general framework to some special cases. We will do so again and restrict attention to abelian gauge theory. We bid the reader not to despair. To quote one poet on a better one, "here is God's plenty" [157]. The reader can take comfort that these theories provide the UV completions for non-linear sigma models in the largest class of Calabi-Yau manifolds currently known—the smooth complete intersections in toric varieties. Every (0,2) Landau-Ginzburg theory is also trivially included in this class. A consequence is the Calabi-Yau—Landau-Ginzburg correspondence and its generalizations that we will discuss below.

Given the author list, it should not be surprising that the seminal paper on the subject [388] remains a must-read for any serious student and anticipated most developments that followed it

5.2.1 Supersymmetric Lagrangian

We now fix our conventions and assumptions for abelian linear sigma models. The gauge group will most often be $G = U(1)^{n_G}$, but occasionally it will be useful to include an additional abelian discrete component in G. Denoting the imaginary circle-valued gauge parameters by λ_α, $\alpha = 1, \ldots, n_G$, the gauge action on the matter fields is specified by the charges Q_a^α and Q_A^α:

$$\Phi^a \to \left(\prod_\alpha e^{\lambda_\alpha Q_a^\alpha} \right) \Phi^a , \qquad \Gamma^A \to \left(\prod_\alpha e^{\lambda_\alpha Q_\alpha^A} \right) \Gamma^A . \qquad (5.2.1)$$

The matter fields are indexed by $a \in \{1, \ldots, n_\Phi\}$ and $A \in \{1, \ldots, n_\Gamma\}$. We do not consider theories with non-homogeneous transformations for the bosons, but some of the fields can be gauge-neutral.

It will be useful to combine the charges into matrices Q_Φ and Q_Γ, where the rows index the gauge group factor, and the columns the matter representation. For instance, the absence of a local gauge anomaly is expressed by

$$Q_\Gamma (Q_\Gamma)^T = Q_\Phi (Q_\Phi)^T . \qquad (5.2.2)$$

The charges are quantized: the entries in Q_Φ and Q_Γ are integers. The charges Q_Φ and Q_Γ also describe the action of the complexified gauge group $G_\mathbb{C} = (\mathbb{C}^*)^{n_G}$, where $\mathbb{C}^* = \mathbb{C} \setminus \{0\}$. In analogy with $U(1)^{n_G} = (S^1)^{n_G}$, $(\mathbb{C}^*)^{n_G}$ is known as the algebraic torus. If we let $t = (t_1, \ldots, t_{n_G}) \in (\mathbb{C}^*)^{n_G}$, then the $G_\mathbb{C}$ action on the fields is

$$t \cdot \Phi^a = \left(\prod_\alpha t_\alpha^{Q_a^\alpha} \right) \Phi^a , \qquad t \cdot \Gamma^A \to \left(\prod_\alpha t_\alpha^{Q_A^\alpha} \right) \Gamma^A . \tag{5.2.3}$$

The charges define the torus action on the fields in a particular basis. Two such actions are equivalent, i.e. define the same gauge theory, if and only if they are related by a $GL(r, \mathbb{Z})$ transformation M:

$$Q'_\Phi = M Q_\Phi , \qquad\qquad Q'_\Gamma = M Q_\Gamma . \tag{5.2.4}$$

It is now easy to specialize the general Yang-Mills Lagrangian to the abelian case. We set

$$(T_\Phi^\alpha)_a^b = Q_a^\alpha \delta_a^b , \qquad\qquad (T_\Gamma^\alpha)_A^B = Q_A^\alpha \delta_A^B , \tag{5.2.5}$$

and

$$(\mathcal{U}_1)_a^b = \mathcal{U}_{1\alpha} Q_a^\alpha \delta_a^b . \tag{5.2.6}$$

The gauge-invariant supersymmetric two-derivative terms are therefore

$$S_{\min} = \frac{1}{2\pi} \int d^2 z \mathcal{D} \overline{\mathcal{D}} \left\{ \frac{1}{4} \sum_{a=1}^{n_\Phi} e^{2 V_{0\alpha} Q_a^\alpha} \left[\overline{\Phi}^a \partial \Phi^a - \Phi^a \partial \overline{\Phi}^a + V_{1\alpha} \overline{\Phi}^a Q_a^\alpha \Phi^a \right] \right.$$

$$\left. - \frac{1}{2} \sum_{A=1}^{n_\Gamma} \overline{\Gamma}^A e^{2 V_{0\alpha} Q_A^\alpha} \Gamma^A \right\} . \tag{5.2.7}$$

Note that we leave the sum on the gauge index implicit.

Canonically Normalized Gauge Fields and Wess-Zumino Gauge

We have so far avoided any explicit component expansion of the gauge superfields. To fix a canonical normalization, we will need these details. In addition, we swept one technical point under the rug. In the construction of the general action we ensured that the constraints $\overline{\mathcal{D}} \Phi = 0$ and $\overline{\mathcal{D}} \Gamma = E(\Phi)$ are sensible by promoting the gauge transformations to chiral superfields. This is enough to construct a supersymmetric and gauge-invariant action, but the component action complicated, since it contains artifacts of the enlarged gauge invariance.

As in four dimensions, we can introduce the Wess-Zumino gauge to simplify these aspects of the theory. This is a partial gauge-fixing that eliminates the extra gauge freedom introduced in constructing the off-shell supersymmetric gauge theory couplings. To motivate the construction, consider the general gauge transformation of some superfield X in a U(1) theory. Assuming that X transforms homogeneously, from the setup above, we arrive at

$$X \to e^{m\Lambda + m'\overline{\Lambda}} X . \tag{5.2.8}$$

This seems strange: which coefficient here is "the" charge of the field X? Fortunately, there is a nice answer: the field $X_v = e^{(m+m')V_0} X$ transforms as

$$X_v \to e^{(m-m')(\Lambda - \overline{\Lambda})/2} X_v . \tag{5.2.9}$$

This is recognizable: when we set the θ components of the Λ and $\overline{\Lambda}$ superfields to zero, we obtain a standard gauge transformation with parameter proportional to the imaginary part of $\Lambda|_{\theta=\overline{\theta}=0}$ for a field of charge $Q = m - m'$. By suitably dressing the fields, we can therefore arrange them in Q-eigenspaces, with gauge transformations $X_v \to e^{Q(\Lambda - \overline{\Lambda})/2} X_v$ for all X_v in the Q-eigenspace.

Next, define the twisted superspace derivatives $\mathcal{D}_v$ and $\overline{\mathcal{D}}_v$ in the eigenspace of charge Q operators by

$$\mathcal{D}_v = e^{-QV_0} D e^{QV_0} , \qquad \overline{\mathcal{D}}_v = e^{QV_0} \overline{D} e^{-QV_0} . \tag{5.2.10}$$

It is not hard to check $\mathcal{D}_v^2 = 0, \overline{\mathcal{D}}_v^2 = 0$, and

$$\{\mathcal{D}_v, \overline{\mathcal{D}}_v\} = 2(\overline{\partial} + Q\tfrac{1}{2}[\overline{D}, D]V_0) . \tag{5.2.11}$$

Exercise 5.3 Show that $\tfrac{1}{2}[\overline{D}, D]V_0$ transforms like the $\overline{z}$ component of a connection:

$$\tfrac{1}{2}[\overline{D}, D]V_0 \to \tfrac{1}{2}[\overline{D}, D]V_0 - \tfrac{1}{2}\overline{\partial}(\Lambda - \overline{\Lambda}) . \tag{5.2.12}$$

Also check that in a Q-eigenspace $\mathcal{D}_v X$ and $\overline{\mathcal{D}}_v X$, given by

$$\mathcal{D}_v X = DX + Q(DV_0)X , \qquad \overline{\mathcal{D}}_v X = \overline{D}X - Q(\overline{D}V_0)X , \tag{5.2.13}$$

transform covariantly. □

The exercise shows that $\mathcal{D}_v$ and $\overline{\mathcal{D}}_v$ preserve the Q-eigenspace. Moreover, the space of $\overline{D}$-chiral operators is isomorphic to the space of $\overline{\mathcal{D}}_v$-chiral operators. We also have the left-moving covariant derivative, which in a Q-eigenspace takes the form

$$(\partial + \tfrac{1}{2}QV_1) , \tag{5.2.14}$$

and it is useful to study its commutators with $\mathcal{D}_v$ and $\overline{\mathcal{D}}_v$. The results are

$$[\mathcal{D}_v, (\partial + \tfrac{1}{2}Q\mathcal{V}_1)] = -\tfrac{1}{\sqrt{2}}Q\overline{\Upsilon}, \qquad [\overline{\mathcal{D}}_v, (\partial + \tfrac{1}{2}Q\mathcal{V}_1)] = \tfrac{1}{\sqrt{2}}Q\Upsilon, \qquad (5.2.15)$$

where Υ is the chiral gauge-invariant superfield

$$\Upsilon = \sqrt{2}\,\overline{\mathcal{D}}(\partial\mathcal{V}_0 + \tfrac{1}{2}\mathcal{V}_1). \qquad (5.2.16)$$

Since Υ is gauge-invariant, it is chiral with respect to either $\overline{\mathcal{D}}$ or $\overline{\mathcal{D}}_v$.

The real superfield $\mathcal{V}_0$ and the imaginary superfield $\mathcal{V}_1$ have θ expansions (the numeric coefficients are chosen for later convenience)

$$\mathcal{V}_0 = a + \theta\zeta - \overline{\theta\zeta} + i\theta\overline{\theta}\overline{A_{\overline{z}}}, \qquad \mathcal{V}_1 = 2i\,A_z + \sqrt{2}(\overline{\theta}\upsilon - \theta\overline{\upsilon}) + \theta\overline{\theta}D, \qquad (5.2.17)$$

with a A_z, $\overline{A_{\overline{z}}}$, and D real fields (with respect to charge conjugation $\mathcal{C}$).[2]

Under a gauge transformation with

$$\Lambda = \lambda + 2\theta\eta + \theta\overline{\theta}\partial\lambda, \qquad (5.2.18)$$

we see that

$$\mathcal{V}_0 \rightarrow \left[a - \tfrac{1}{2}(\lambda + \overline{\lambda})\right] + \theta(\zeta - \eta) - \overline{\theta}(\overline{\zeta} - \overline{\eta}) + i\theta\overline{\theta}\left[\overline{A_{\overline{z}}} - \partial\tfrac{1}{2i}(\lambda - \overline{\lambda})\right]. \qquad (5.2.19)$$

Thus, $\overline{A_{\overline{z}}}$ transforms like the $\overline{z}$ component of a connection, and the same is true of A_z in the $\mathcal{V}_1$ multiplet. We define the gauge-covariant derivatives

$$\overline{\nabla}X = (\overline{\partial} + iQ\overline{A_{\overline{z}}})X, \qquad \nabla X = (\partial + iQA_z)X. \qquad (5.2.20)$$

This implies $\{\overline{\mathcal{D}}_v, \mathcal{D}_v\} = 2\overline{\nabla} + O(\theta, \overline{\theta})$. We also see that all but the top component of the $\mathcal{V}_0$ superfield can be eliminated algebraically by a super-gauge transformation. Wess-Zumino gauge is the partial gauge fixing to

$$\mathcal{V}_0 = i\theta\overline{\theta}\,\overline{A_{\overline{z}}}. \qquad (5.2.21)$$

---

[2]Recall our $\mathcal{C}$ conventions: $\mathcal{C}^2 = 1$, $\mathcal{C}(\theta) = -i\overline{\theta}$, $\mathcal{C}(\overline{\theta}) = -i\theta$; for a right-moving Weyl fermion $\mathcal{C}(\psi) = +i\overline{\psi}$, $\mathcal{C}(\overline{\psi}) = +i\psi$, while for a left-moving fermion $\mathcal{C}(\gamma) = -i\overline{\gamma}$, and $\mathcal{C}(\overline{\gamma}) = -i\gamma$; on scalar bosonic fields $\mathcal{C}$ acts by complex conjugation. Since in our definition $\mathcal{C}$ does not act on the worldsheet derivatives ∂ and $\overline{\partial}$, we define the conjugation on the gauge fields as $\mathcal{C}(\overline{A_{\overline{z}}}) = \overline{A_{\overline{z}}}$, and similarly $\mathcal{C}(A_z) = A_z$. As before, the point of these definitions is that $\mathcal{C}$ is the Euclidean manifestation of usual complex conjugation in Minkowski space.

The residual gauge symmetry is simply the ordinary gauge symmetry with $\Lambda = \lambda(z, \bar{z})$ for a pure imaginary field λ, and the twisted superspace derivatives are

$$\mathcal{D}_v = \partial_\theta + \bar{\theta} \,\overline{\nabla} \,, \qquad\qquad \overline{\mathcal{D}}_v = \partial_{\bar{\theta}} + \theta \overline{\nabla} \,. \qquad (5.2.22)$$

The superfields Υ and $\overline{\Upsilon}$ simplify to

$$\Upsilon = v + \frac{1}{\sqrt{2}}\theta(F_{12} - D) + \theta\bar{\theta}\bar{\partial}v \qquad \overline{\Upsilon} = \bar{v} - \frac{1}{\sqrt{2}}\bar{\theta}(F_{12} + D) - \theta\bar{\theta}\bar{\partial}\bar{v} \,, \qquad (5.2.23)$$

where the gauge field strength is (compare to (2.5.8))

$$F_{12} = \partial_1 A_2 - \partial_2 A_1 = 2i(\bar{\partial}A_z - \partial\overline{A_{\bar{z}}}) \,. \qquad (5.2.24)$$

The fermi multiplet Υ is often called the chiral gauge-field strength multiplet. D has mass dimension 2, while the gaugino v has mass dimension $3/2$. We can therefore write the kinetic term for the gauge fields as

$$S_{\text{gauge}} = \frac{1}{2\pi e^2} \int d^2z \mathcal{D}\overline{\mathcal{D}} \left\{ -\tfrac{1}{2}\overline{\Upsilon}\Upsilon \right\} = \frac{1}{2\pi e^2} \int d^2z \left\{ \bar{v}\partial v + \frac{1}{4}\left(F_{12}^2 - D^2\right) \right\} \,, \qquad (5.2.25)$$

where the gauge coupling e has dimension of mass.

Similarly, in Wess-Zumino gauge, the $\overline{\mathcal{D}}_v$-chiral superfields and their anti-chiral conjugates take the form

$$\Phi_v = e^{Q V_0}\Phi = \phi + \sqrt{2}\theta\psi + \theta\bar{\theta}\,\overline{\nabla}\phi \,, \quad \overline{\Phi}_v = e^{Q V_0}\overline{\Phi} = \bar{\phi} - \sqrt{2}\theta\bar{\psi} - \theta\bar{\theta}\,\overline{\nabla}\bar{\phi} \,. \qquad (5.2.26)$$

The fermi multiplets, which satisfy $\overline{\mathcal{D}}_v\Gamma_v = \sqrt{2}E(\Phi_v)$, and their conjugates, have the component expansions

$$\Gamma_v = \gamma + \sqrt{2}\theta G + \theta\bar{\theta}\,\overline{\nabla}\gamma + \sqrt{2}\bar{\theta}E(\Phi_v) \,,$$
$$\overline{\Gamma}_v = \bar{\gamma} + \sqrt{2}\theta G - \theta\bar{\theta}\,\overline{\nabla}\,\bar{\gamma} + \sqrt{2}\theta\overline{E}(\overline{\Phi}_v) \,. \qquad (5.2.27)$$

Supersymmetry Transformations

The supersymmetry variation of the V_0 vector field

$$\delta_\xi V_0 = \frac{1}{\sqrt{2}}\left(\xi\mathcal{Q} - \bar{\xi}\overline{\mathcal{Q}}\right)V_0 = -\theta\left(\frac{i}{\sqrt{2}}\xi\overline{A_{\bar{z}}}\right) - \bar{\theta}\left(\frac{i}{\sqrt{2}}\xi\overline{A_z}\right) \qquad (5.2.28)$$

does not preserve Wess-Zumino gauge. From (5.2.19) we see that Wess-Zumino gauge can be restored by a compensating gauge transformation with gauge parameter

$$\Lambda_\xi = 2\theta \left(-\frac{i}{\sqrt{2}}\xi \overline{A}_{\bar z}\right) , \qquad \overline{\Lambda}_\xi = 2\overline{\theta}\left(-\frac{i}{\sqrt{2}}\overline{\xi}A_{\bar z}\right) . \qquad (5.2.29)$$

This motivates a modified supersymmetry transformation that combines the usual supersymmetry variation with the compensating gauge transformation:

$$\delta_\xi^{\mathrm{mod}} = \delta_\xi + \delta_{\Lambda_\xi}^{\mathrm{gauge}} . \qquad (5.2.30)$$

The modified transformation is a supersymmetry of the action, and it allows us to retain the virtues of Wess-Zumino gauge.[3]

The price to pay for this is that the algebra realized on the fundamental fields no longer closes onto translations; since we are ultimately interested in gauge-invariant operators, this is a small price to pay. On the space of gauge invariant operators we will obtain the usual supersymmetry algebra.

Exercise 5.4 Show that on a Q-eigenspace, where operators transform as $X \to e^{Q(\Lambda - \overline{\Lambda})/2}X$, the supersymmetry transformations are implemented by

$$\delta_\xi^{\mathrm{mod}} = \frac{1}{\sqrt{2}}\left\{\xi Q^{\mathrm{mod}} - \overline{\xi}\overline{Q}^{\mathrm{mod}}\right\} , \qquad Q^{\mathrm{mod}} = \partial_\theta - \overline{\theta}\,\overline{\nabla} , \qquad \overline{Q}^{\mathrm{mod}} = \partial_{\overline{\theta}} - \theta\overline{\nabla} . \qquad (5.2.31)$$

□

Since Q^{mod} and $\overline{Q}^{\mathrm{mod}}$ anticommute with $\mathcal{D}_v$ and $\overline{\mathcal{D}}_v$, it becomes an easy matter to obtain the supersymmetry variations of the fields in exact parallel to (1.8.18):

$$
\begin{array}{llll}
\overline{Q}\cdot\phi = 0, & Q\cdot\phi = -\psi, & \overline{Q}\cdot\gamma = -E(\phi), & Q\cdot\gamma = -G, \\
\overline{Q}\cdot\overline{\phi} = \overline{\psi} , & Q\cdot\overline{\phi} = 0, & \overline{Q}\cdot\overline{\gamma} = -\overline{G}, & Q\cdot\overline{\gamma} = -\overline{E}(\overline{\phi}), \\
\overline{Q}\cdot\psi = -\overline{\nabla}\phi , & Q\cdot\psi = 0, & \overline{Q}\cdot G = E'(\phi)\psi - \overline{\nabla}\gamma, & Q\cdot G = 0, \\
\overline{Q}\cdot\overline{\psi} = 0, & Q\cdot\overline{\psi} = \overline{\nabla}\phi, & \overline{Q}\cdot\overline{G} = 0 , & Q\cdot\overline{G} = -\overline{E}'(\overline{\phi})
\end{array}
$$

$$\overline{\psi} - \overline{\nabla}\overline{\gamma}. \qquad (5.2.32)$$

---

[3]This clearly works on any open patch of the world-sheet; on a curved world-sheet the construction is complicated by the non-trivial patching of the gauge field, but that complication is overshadowed by the fact that a non-trivial metric typically breaks the (0,2) supersymmetry. These complications can be addressed in half-twisted theories: the scalar supersymmetry operator $\overline{Q}$ exists, and it is possible to consistently work in Wess-Zumino gauge.

For the vector superfield components we find

$$\overline{Q} \cdot \overline{A}_{\bar{z}} = 0 \,, \qquad\qquad Q \cdot \overline{A}_{\bar{z}} = 0 \,,$$

$$\overline{Q} \cdot A_z = \tfrac{i}{2} \upsilon \,, \qquad\qquad Q \cdot A_z = -\tfrac{i}{2} \overline{\upsilon} \,,$$

$$\overline{Q} \cdot \upsilon = 0 \,, \qquad\qquad Q \cdot \upsilon = -\tfrac{1}{2}(F_{12} - D) \,,$$

$$\overline{Q} \cdot \overline{\upsilon} = \tfrac{1}{2}(F_{12} + D) \,, \qquad\qquad Q \cdot \overline{\upsilon} = 0 \,,$$

$$\overline{Q} \cdot D = \bar{\partial} \upsilon \,, \qquad\qquad Q \cdot D = \bar{\partial} \overline{\upsilon} \,. \tag{5.2.33}$$

Linear Sigma Model Lagrangian

In this section we use the previous developments to present the action of the class of linear sigma models that we will examine.

The gauge-kinetic terms for the $U(1)^r$ theory are

$$S_{\text{gauge}} = \frac{1}{2\pi e^2} \int d^2z D\overline{D} \left\{ -\tfrac{1}{2} \overline{\Upsilon}_\alpha \Upsilon_\alpha \right\}$$

$$= \frac{1}{2\pi e^2} \int d^2z \left\{ \overline{\upsilon}_\alpha \bar{\partial} \upsilon_\alpha + \frac{1}{4} \left(F_{12\alpha}^2 - D_\alpha^2 \right) \right\} \,. \tag{5.2.34}$$

Nothing prevents us from taking different gauge couplings e_α for the different multiplets, but since in all of our applications we will be interested in e-independent quantities, we use the same coupling for all $\alpha = 1, \ldots, r$.

The matter kinetic terms are best described by introducing the Φ_υ and Γ_υ multiplets of the previous section, so that

$$S_{\text{min}} = \frac{1}{2\pi} \int d^2z D\overline{D} \left\{ \tfrac{1}{4} \overline{\Phi}_\upsilon^a (\partial + \tfrac{1}{2}(\mathcal{V}_1 Q_\Phi)_a) \Phi_\upsilon^a - \tfrac{1}{4} \Phi_\upsilon^a (\partial \right.$$

$$\left. -\tfrac{1}{2}(\mathcal{V}_1 Q_\Phi)_a) \overline{\Phi}_\upsilon^a - \tfrac{1}{2} \overline{\Gamma}_\upsilon^A \Gamma_\upsilon^A \right\} \,. \tag{5.2.35}$$

The component expansion can either be found by working out the θ expansion explicitly, or by observing that, since the Lagrangian is gauge-invariant, the derivatives D and $\overline{D}$ can be replaced with the covariant D_υ and $\overline{D}_\upsilon$. A steady application of the commutation relations then leads to

$$S_{\text{min}} = \frac{1}{2\pi} \int d^2z \left\{ \tfrac{1}{2} (\overline{\nabla\phi}^a \nabla\phi^a + \nabla\overline{\phi}^a \overline{\nabla}\phi^a) + \overline{\psi}^a \nabla\psi^a \right.$$

$$-\tfrac{1}{2} \upsilon_\alpha (Q_\Phi)_\alpha^a \overline{\phi}^a \psi^a - \tfrac{1}{2} \overline{\psi}^a \phi^a (Q_\Phi)_\alpha^a \overline{\upsilon}_\alpha - \tfrac{1}{4} D_\alpha (Q_\Phi)_\alpha^a \overline{\phi}^a \phi^a$$

$$\left. + \overline{\gamma}^A \overline{\nabla}\gamma^A - \overline{G}^A G^A + \overline{E}^A E^A - \overline{E}_{,a}^A \overline{\psi}^a \gamma^A - \overline{\gamma}^A E_{,a}^A \psi^a \right\} \,. \tag{5.2.36}$$

We can add a superpotential interaction just as for the Yukawa theories we discussed in Chap. 1: $J_A(\Phi)$ are holomorphic functions, with charge $(-Q_\Gamma)_A^\alpha$, subject to the supersymmetry constraint

$$\sum_A E^A J_A = 0 \,, \tag{5.2.37}$$

and the interaction terms are

$$
\begin{aligned}
S_J &= \frac{1}{2\pi\sqrt{2}} \int d^2z \left\{ \mathcal{D}(\Gamma^A J_A) + \overline{\mathcal{D}}(\overline{\Gamma}^A \overline{J}_A) \right\} \\
&= \frac{1}{2\pi} \int d^2z \left\{ G^A J_A + \overline{G}^A \overline{J}_A - \gamma^A J^A_{,a} \psi^a - \overline{\psi}^a \overline{J}^A_{,a} \overline{\gamma}^A \right\} \,. \tag{5.2.38}
\end{aligned}
$$

At first blush, since Υ is a gauge-invariant chiral fermi multiplet, it is natural to contemplate another potential term

$$S_T = \frac{i}{2\sqrt{2}} \int d^2z \left\{ \mathcal{D}(\Upsilon_\alpha T^\alpha(\Phi)) - \overline{\mathcal{D}}(\overline{\Upsilon}_\alpha T^\alpha(\overline{\Phi})) \right\} \,, \tag{5.2.39}$$

where $T^\alpha(\Phi)$ are gauge-invariant and holomorphic functions. As we will see, such terms arise naturally in effective descriptions, where some of the degrees of freedom of the gauge theory are integrated out, but the general theory of such couplings in a UV description remains to be worked out. In particular, criteria for compactness and aspects of strongly coupled gauge dynamics remain to be understood. These questions are not well-addressed even for (2,2) theories, where they lead to a wide class of generalized Kähler "geometries." The quotes again refer to the singularity problem: it is not so easy to construct examples that are smooth. The reader interested in these issues may find [4, 233, 300] useful. The classic story of the bosonized Schwinger model illustrates some of the simpler aspects of such couplings. We illustrate these in an example.

Exercise 5.5 Consider the two-dimensional Minkowski (signature $(+,-)$) Lagrangian for a gauge field A_μ and a real scalar Θ:

$$\mathcal{L}_0 = -\frac{1}{4e^2} F_{\mu\nu} F^{\mu\nu} + \tfrac{1}{2} \partial_\mu \Theta \partial^\mu \Theta \,, \tag{5.2.40}$$

and perturb it by

$$\mathcal{L}_{\text{int}} = b A^\mu \partial_\mu X + \tfrac{m}{2} A^\mu A_\mu + a \epsilon^{\mu\nu} \Theta F_{\mu\nu} \,. \tag{5.2.41}$$

Show that there are two ways to choose the parameters b, m, a so as to obtain a gauge-invariant action under $\delta_\epsilon \Theta = q\epsilon$ and $\delta_\epsilon A = -d\epsilon$:

i. $q = b$, $m = b^2$ and $a = 0$;
ii. $q = b = 0$.

Now show that cases (i) with $q \neq 0$ is equivalent to case (ii) with $a \neq 0$, provided we relate the parameters via $q = 2a$, and in each case the theory has a single massive vector degree of freedom. Case (i) is rather trivial, but with case (ii) the situation is not so obvious; note that this is exactly the case of a dynamical theta angle. To establish the equivalence we can use a trick familiar from T-duality: introduce an auxiliary field V_μ and replace

$$\tfrac{1}{2} \partial_\mu \Theta \partial^\mu \Theta \to \tfrac{1}{2} V_\mu V^\mu + \partial_\mu \Theta \epsilon^{\mu\rho} V_\rho \,. \tag{5.2.42}$$

Upon integrating out V_μ one recovers the original theory; integrating out Θ leads to the claimed equivalence.

The $(0,2)$ linear sigma models with non-constant $T(\Phi)$ couplings generalize this phenomenon. □

At any rate, we will once again retreat from generality by setting

$$T^\alpha \to \tau^\alpha = \frac{i r^\alpha}{2\pi} + \frac{\vartheta^\alpha}{2\pi} = \text{constant} \,, \tag{5.2.43}$$

with the result

$$S_T \to S_{\text{Fayet-Iliopoulos}} = \frac{i}{2\sqrt{2}} \int d^2z \left\{ \tau^\alpha \mathcal{D} \Upsilon_\alpha - \overline{\tau}^\alpha \overline{\mathcal{D}} \overline{\Upsilon}_\alpha \right\}$$

$$= \frac{1}{2\pi} \int d^2z \left\{ \tfrac{1}{2} D_\alpha r^\alpha \right\} + i \frac{\theta^\alpha}{2\pi} \int F_\alpha \,. \tag{5.2.44}$$

This is a very beautiful two-dimensional coupling: the Fayet-Iliopoulos parameter r^α is complexified by the ϑ-angle ϑ^α. Since for our Euclidean worldsheet $\int F_\alpha \in 2\pi \mathbb{Z}$, $\vartheta^\alpha \sim \vartheta^\alpha + 2\pi$. While the ϑ^α angle couplings do not affect the perturbative physics, the r^α do directly contribute to the bosonic potential. After integrating out the auxiliary fields G^A and D_α, we obtain

$$U_{\text{bos}} = \sum_{A=1}^{n_\Gamma} \left(E^A \overline{E}^A + J_A \overline{J}_A \right) + \frac{1}{4e^2} \sum_{\alpha=1}^{n_G} D_\alpha D_\alpha \,, \tag{5.2.45}$$

where

$$D_\alpha = -\frac{e^2}{2} \left((Q_\Phi)^\alpha_a \overline{\phi}^a \phi^a - 2r^\alpha \right) \,. \tag{5.2.46}$$

As in the Yukawa theories considered in Chap. 3, it is possible to exchange the J_A and E^A potentials by exchanging $\Gamma \leftrightarrow \overline{\Gamma}$ and sending $Q_\Gamma \to -Q_\Gamma$.

Exercise 5.6 All of the terms in the action are $\overline{Q}$-closed, since the action is supersymmetric. Most of the terms in the action are $\overline{Q}$-exact. In this exercise we work this out explicitly and verify that the only closed but not exact terms are the holomorphic superpotential couplings. This will be important when we study the half-twisted theory and topological heterotic rings.

Show that

$$S_{\text{gauge}} = \overline{Q} \cdot V_{\text{gauge}} , \qquad\qquad S_{\text{min}} = \overline{Q} \cdot V_{\text{min}} , \qquad (5.2.47)$$

with

$$V_{\text{gauge}} = \frac{1}{4\pi e^2} \int d^2z \, \overline{v}_\alpha (F_{12} - D)_\alpha ,$$

$$V_{\text{min}} = \frac{1}{2\pi} \int d^2z \left[-\psi^a \nabla \overline{\phi}^a - \tfrac{1}{2} Q_a^\alpha \overline{v}_\alpha \overline{\phi}^a \phi^a + \overline{\gamma}^A G^A - \gamma^A \overline{E}^A \right] , \qquad (5.2.48)$$

while

$$S_J = \frac{1}{2\pi} \int d^2z \left[G^A J_A - \gamma^A J_{,a}^A \psi^a \right] + \overline{Q} \cdot \frac{1}{2\pi} \int d^2z \left[-\overline{\gamma}^A \overline{J}_A \right] , \qquad (5.2.49)$$

and

$$S_{\text{Fayet-Iliopoulos}} = \frac{i}{4} \int d^2z \, \tau^\alpha (F_{12} - D)_\alpha + \overline{Q} \cdot \frac{i}{2} \int d^2z \, \overline{\tau}^\alpha \overline{v}_\alpha . \qquad (5.2.50)$$

From this we immediately have the formal statement that the chiral algebra of the linear sigma model will depend holomorphically on the τ^α and the coefficients in the J_A potential.   □

Exercise 5.7 The action satisfies a Bogomolnyi bound [388]. Show that after integrating out the auxiliary fields the bosonic action takes the form

$$S_{\text{bos}} = \frac{1}{2\pi} \int d^2z \left[|\nabla \phi^a|^2 + \tfrac{1}{4e^2}(F_{12\alpha} + D_\alpha)^2 + \left| E^A \right|^2 + |J_A|^2 \right] - 2\pi i \tau^\alpha N_\alpha , \qquad (5.2.51)$$

where the instanton numbers $N_\alpha \in \mathbb{Z}$ are defined to be

$$N_\alpha = -\frac{1}{2\pi} \int F_\alpha . \qquad (5.2.52)$$

Check that on bosonic field configurations annihilated by $\overline{Q}$ the action therefore satisfies

$$e^{-S} = \exp[2\pi i \textstyle\sum_{\alpha=1}^{n_G} \tau^\alpha N_\alpha] = \prod_{\alpha=1}^{n_G} q_\alpha^{N_\alpha} , \qquad (5.2.53)$$

where the q_α are defined by

$$q_\alpha = e^{2\pi i \tau_\alpha} . \tag{5.2.54}$$

It is useful to compare this to the non-linear sigma model discussion in Sect. 4.6.1.

□

5.2.2 Some Preliminary Observations

Our task in the remainder of the text is to study the linear sigma model just defined. Throughout, our interest will be on the low energy physics, where the gauge theory becomes strongly coupled. We will be therefore content to make some statements about the vacuum, the topological heterotic rings, and relations to other constructions of non-trivial (0,2) SCFTs. There is a host of interesting quantum field theory questions beyond the chiral algebra and the IR limit. That realm is largely unexplored, though see [183].

In order to make some general progress without a complete retreat to very specific examples, we will need a bit more geometric machinery, namely toric geometry. We will turn to those developments in the next section. Before we make that jump, we will spend a little time on some simple aspects that can be understood without the toric toolbox.

The Classical $e \to \infty$ Limit

Consider the theory just described with $E = J = 0$ (these couplings can be excluded by demanding certain symmetries of the action). Our theory has a dimensionful parameter—the gauge coupling, and we obtain a classical approximation to the IR physics by sending $e \to \infty$. Let us consider a particularly simple case to get an idea of how the resulting limit works. We set $n_G = 1$, $n_\Phi = 2$, and $n_\Gamma = 1$; furthermore we will take $Q^a_\Phi = 1$, and leave Q_Γ as a parameter. Up to boundary terms, the $e \to \infty$ action is then

$$S_{e\to\infty} = \frac{1}{2\pi} \int d^2z \mathcal{D}\overline{\mathcal{D}} \left[\tfrac{1}{4} e^{2V_0} (\overline{\Phi}^a \partial \Phi^a - \Phi^a \partial \overline{\Phi}^a + V_1 (\overline{\Phi}^a \Phi^a)) \right.$$
$$\left. - \tfrac{1}{2} e^{2Q_\Gamma V_0} \overline{\Gamma} \Gamma - \tfrac{r}{2} V_1 \right] . \tag{5.2.55}$$

To obtain the last term we used (5.2.16). It may seem that we have undone much of our good work of fixing Wess-Zumino gauge. That is indeed the case, but it turns out to be useful to work with the complexified gauge invariance. This is an example of the equivalence between symplectic and holomorphic quotient constructions we will meet in the next section.

The $\mathcal{V}_1$ vector field acts as a Lagrange multiplier that imposes the superfield constraint

$$\tfrac{1}{4}e^{2\mathcal{V}_0}\overline{\Phi}^a\Phi^a - \tfrac{r}{2} = 0 . \tag{5.2.56}$$

Evidently, if $r < 0$, there are no solutions possible, and classical supersymmetry is broken. When $r > 0$ then at least one of the Φ^a fields has a non-vanishing expectation value. Supposing $\Phi^2 \neq 0$, we fix the $G_{\mathbb{C}}$ gauge symmetry by setting

$$\Phi^2 = 1 . \tag{5.2.57}$$

We now solve for $e^{2\mathcal{V}_0}$ in terms of the gauge-invariant ratio $Z = \Phi^1/\Phi^2$:

$$e^{2\mathcal{V}_0} = \frac{2r}{1+|Z|^2} . \tag{5.2.58}$$

Plugging the result back into the action we then obtain

$$S_{e\to\infty} = \frac{1}{2\pi}\int d^2z \mathcal{D}\overline{\mathcal{D}}\left\{ \frac{r}{2(1+|Z|^2)}(\overline{Z}\partial Z - Z\partial\overline{Z}) - \tfrac{1}{2}\left(\frac{2r}{1+|Z|^2}\right)^{\varrho_\Gamma}\overline{\Gamma}\Gamma \right\} . \tag{5.2.59}$$

Comparing to the general (0,2) non-linear sigma model action (4.7.14), we recognize this: the (0,2) 1-form potential is given by

$$\mathcal{K} = \frac{r\overline{Z}dZ}{1+|Z|^2} = r\partial \log(1+|Z|^2) , \tag{5.2.60}$$

and the Hermitian metric on the rank 1 bundle is

$$h = \tfrac{1}{2}\left(\frac{2r}{1+|Z|^2}\right)^{\varrho_\Gamma} . \tag{5.2.61}$$

All of this could also be done in the patch with $\Phi^1 \neq 0$, and we obtain the same results but written in terms of a coordinate $Z' = \Phi^2/\Phi^1$; on the overlap, where Z and Z' are both non-zero the two presentations are gauge equivalent (or, in the non-linear sigma model language, diffeomorphic). At any rate, the (0,2) geometry obtained in this fashion is the simplest non-trivial case: it is a line bundle $\mathcal{O}(\varrho_\Gamma) \to \mathbb{P}^1$ (compare with Exercise B.23), and r is the radius of the sphere.

It should be clear that this construction has a huge generalization. By choosing $(\varrho_\Phi)^\alpha_a$ appropriately, the generalization of the above solution, carried out patch by patch, will lead to a complex manifold V of dimension $n_\Phi - n_G$. If the charges allow it, the construction can then be extended by introducing the E^A and J_A potentials. These will naturally be sections of various line bundles over V, and the low energy excitations will be confined by the bosonic potential terms to the locus

$$\{E^A = 0, \quad J_A = 0, A = 1, \ldots, n_\Gamma\} \in V. \tag{5.2.62}$$

By making further combinatorial assumptions it may be possible to arrange matters so that this zero locus is a smooth compact submanifold $X \subset V$. In general the resulting theory will be a hybrid model [82, 84, 388]—a theory with low energy excitations that can be described by a Landau-Ginzburg theory with potential that depends on the base manifold X. It is possible to arrange matters so that the Landau-Ginzburg fiber is trivial, i.e. the classical low energy limit is a (0,2) non-linear sigma model with bundle $\mathcal{E} \to X$.

The combinatorial conditions that ensure various levels of simplification in the previous paragraph are not well-understood. Indeed, even in the case of (2,2) linear sigma models that should flow to IR SCFTs with $c = \bar{c} = 9$ and integral R-charges, as of this writing, only sufficient conditions have been established [35]. It is perhaps not too surprising that the step that is understood the best is the construction of V— this is the realm of toric geometry, and those details, along with various observations on the so-called "V-models" will occupy us in what follows.

$U(1)_L \times U(1)_R$ and Gauge Anomalies

The $U(1)_L \times U(1)_R$ symmetry has been an important guide to IR physics in our discussion of Landau-Ginzburg and non-linear models. The same will be true for the linear models.

Suppose that there is a classical symmetry

	θ	Φ^a	Γ^A	J_A	E^A	Υ_α
$U(1)_L$	0	q_Φ^a	q_Γ^A	$-q_\Gamma^A$	q_Γ^A	0
$U(1)_R$	1	$\bar{q}_\Phi^a$	$\bar{q}_\Gamma^A$	$1 - \bar{q}_\Gamma^A$	$1 + \bar{q}_\Gamma^A$	1

$$\tag{5.2.63}$$

To have a gauge theory with these symmetries we need (5.2.2) for gauge anomaly cancelation, as well as anomaly cancellation conditions for the global symmetries:

$$U(1)_L: \quad Q_\Phi q_\Phi = Q_\Gamma q_\Gamma, \quad U(1)_R: \quad Q_\Phi(\bar{q}_\Phi - 1) = Q_\Gamma \bar{q}_\Gamma. \tag{5.2.64}$$

We arranged the charges q_Φ, $\bar{q}_\Phi$, q_Γ and $\bar{q}_\Gamma$ into column vectors, and $(\bar{q}_\Phi - 1)^a = \bar{q}_\Phi^a - 1$. The charges of Υ_α are fixed by $S_{\text{Fayet-Iliopoulos}}$. Unless the E^A vanish or the J_A vanish it is not possible to take $\bar{q}_\Phi$ to be proportional to q_Φ: therefore the potentials E^A and J_A must admit at least two linearly independent quasi-homogeneity conditions.

The $U(1)_L$ or the $U(1)_R$ charges of the fundamental fields are only determined up to shifts by the gauge charges:

$$q_\Phi \to q_\Phi + s Q_\Phi \, , \qquad q_\Gamma \to q_\Gamma + s Q_\Gamma \, , \qquad \bar{q}_\Phi \to \bar{q}_\Phi + \bar{s} Q_\Phi \, ,$$

$$\bar{q}_\Gamma \to \bar{q}_\Gamma + \bar{s} Q_\Gamma \tag{5.2.65}$$

preserve the conditions for any choice of $s, \bar{s} \in \mathbb{R}^r$. This ambiguity does not affect the $U(1)_L \times U(1)_R$ charges of gauge-invariant operators.

With optimistic assumptions about compactness and lack of accidents, just as in Chap. 3, c-extremization can again be applied to fix the charges of the IR R-symmetry. When the gauge symmetry and $U(1)_L \times U(1)_R$ are the only continuous symmetries of the UV Lagrangian, then the extremized charges satisfy [145]

$$q_\Phi^T (\bar{q}_\Phi - 1) = q_\Gamma^T \bar{q}_\Gamma \, , \tag{5.2.66}$$

and the putative IR central charges are given by

$$\frac{\bar{c}}{3} = (\bar{q}_\Phi - 1)^T (\bar{q}_\Phi - 1) - \bar{q}_\Gamma^T \bar{q}_\Gamma - n_G \, , \quad c - \bar{c} = n_G + n_\Gamma - n_\Phi \, . \tag{5.2.67}$$

The second equation is just the gravitational anomaly, and the first is obtained by computing the level of the putative IR $U(1)_R$ symmetry. Similarly, the level of the $U(1)_L$ current is

$$r_L = q_\Gamma^T q_\Gamma - q_\Phi^T q_\Phi \, . \tag{5.2.68}$$

Equation (5.2.66) is simply the requirement that there is no mixed anomaly between the $\mathcal{J}_L$ and $\mathcal{J}_R$ currents.

Exercise 5.8 Check that (5.2.66) and (5.2.67) are not affected by the ambiguity (5.2.65). □

To explore these theories and to be able to assess the validity of various optimistic assumptions necessary for these expressions to hold, we now turn to the development of a toric geometry toolbox.

5.3 Toric Geometry

There are many sources on toric geometry. There is now a quite comprehensive and modern treatment [122]. The classic introductory treatment of [178] is very readable. The toric geometry chapters of [120] are excellent. Our survey, mainly based on those references, will not be comprehensive, nor is it in any way original; the goal is to highlight and summarize the results most relevant for linear sigma model exploration and to give the reader some guidance to useful literature for further information.

5.3.1 Algebraic Generalities

The discussion of toric geometry is best carried out in the language of algebraic geometry and therefore algebraic varieties. We give a brief discussion, essentially to set up some definitions, following [121, 162, 210].

All of our discussion will be over the complex numbers. An algebraic variety is a zero set of a collection of polynomials. We distinguish two types: affine algebraic varieties, where the polynomials can be viewed as functions on $\mathbb{C}^n$, and projective algebraic varieties, where we consider the simultaneous vanishing of homogenous polynomials in $\mathbb{P}^n$. A central tenet of algebraic geometry is the relationship between properties of the zero set and the ideal generated by the polynomials:

$$V = \{z \in \mathbb{C}^n | \, p_1(z) = 0 \, , p_2(z) = 0, \, , \ldots, p_k(z) = 0\}$$
$$I = \langle p_1, p_2, \ldots, p_k \rangle \in \mathbb{C}[z_1, \ldots, z_n] \, . \tag{5.3.1}$$

This correspondence can be read both ways, i.e. we have assignments $I(Z)$ and $V(I)$, with some standard relations that the reader may ponder:

$$I \subseteq I' \implies V(I) \supseteq V(I')$$
$$Z \subseteq Z' \implies I(Z) \supset I(Z')$$
$$V(II') = V(I) \cup V(I')$$
$$V(I + I') = V(I) \cap V(I') \, . \tag{5.3.2}$$

The algebraic torus $\mathbb{C}^*$ is a particular example of an affine variety. If $t \in \mathbb{C}^*$, then a general "polynomial" on $\mathbb{C}^*$ takes the form

$$f = a_{-m}t^{-m} + a_{1-m}t^{1-m} + \cdots + a_{n-1}t^{n-1} + a_n t^n \, . \tag{5.3.3}$$

This is a finite Laurent polynomial, and it is sensible because $t \neq 0$. We can write the corresponding polynomial ring as

$$\mathbb{C}[t, t^{-1}] = \mathbb{C}[s, t]/(ts - 1) \, . \tag{5.3.4}$$

We will typically be interested in irreducible varieties: an example of a reducible variety is

$$\{(z_1, z_2) \mid z_1 z_2 = 0\} \, . \tag{5.3.5}$$

The two reducible components are $z_1 = 0$ and $z_2 = 0$.

The quotient $R = \mathbb{C}[x_1, \ldots, x_n]/I$ is the ring of holomorphic functions on $V(I)$, and we set

$$\text{Spec}(R) = \{\text{the set of prime ideals in } R\}. \tag{5.3.6}$$

Every element $v \in \text{Spec}(R)$ is an irreducible subvariety of $V(I)$, and $\text{Spec}(R)$ can be endowed with a natural topology, known as the Zariski topology. A maximal ideal $p \in \text{Spec}(R)$ corresponds to a point in $V(I)$, so that we can think of $\text{Spec}(R)$ as containing the points of $V(I)$ but also keeping track of the other irreducible subvarieties.

The algebraic torus $\mathbb{C}^*$ has a natural structure of a multiplicative group, since for $(s, t) \in \mathbb{C}^* \times \mathbb{C}^*$ the product $st \in \mathbb{C}^*$. An algebraic torus of dimension k, $(\mathbb{C}^*)^k$, has the obvious generalization of this product, where the multiplication works coordinate by coordinate:

$$(s_1, \ldots, s_k) \cdot (t_1, \ldots, t_k) = (s_1 t_1, \ldots, s_k t_k). \tag{5.3.7}$$

A toric variety V is, by definition, a (normal) variety that satisfies two properties:

i. V has $(\mathbb{C}^*)^k$ as a dense open subset;
ii. there is an action $(\mathbb{C}^*)^k : V \to V$ that is the natural extension of (5.3.7).

In other words, V is a (perhaps partial) compactification of an algebraic torus. Examples include:

$$(\mathbb{C}^*)^k,$$
$$\mathbb{C} \simeq \mathbb{C}^* \cup \{0\},$$
$$\mathbb{P}^1 \simeq \mathbb{C}^* \cup \{0\} \cup \{\infty\}, \tag{5.3.8}$$

and many other spaces that we will meet soon. Of course many varieties, such as compact Calabi-Yau manifolds or Riemann surfaces of genus ≥ 1 are not toric.

Toric varieties are constructed by gluing together affine varieties constructed in a particular way. For instance, the simplest compact toric variety is our old friend $\mathbb{P}^1$, and we can think of it as follows:

$$
\begin{array}{ccccc}
U_1 \simeq \mathbb{C} & \longleftarrow & U_1 \cap U_2 \simeq \mathbb{C}^* & \longrightarrow & U_2 \simeq \mathbb{C} \\
z & & zw = 1 & & w \\
\mathbb{C}[z] & \longrightarrow & \mathbb{C}[z, w]/(zw - 1) & \longleftarrow & C[w]
\end{array}
\tag{5.3.9}
$$

Every function on U_i restricts in an obvious fashion to a function on the overlap. The most straightforward construction of toric varieties generalizes this [178]: the affine charts and the gluing maps are constructed from two types of combinatorial objects: cones and fans; the cones correspond to affine charts, and the fans, which describe how cones fit together, provide the gluing maps.

There are two more standard constructions of $\mathbb{P}^1$. First, we may view the space as the holomorphic quotient

$$\mathbb{P}^1 \simeq \frac{\mathbb{C}^2 \setminus \{0\}}{\mathbb{C}^*}, \tag{5.3.10}$$

where $\mathbb{C}^*$ acts on $(Z_1, Z_2) \in \mathbb{C}^2$ by $t \cdot (Z_1, Z_2) = (tZ_1, tZ_2)$. Second, we can also build $\mathbb{P}^1 \simeq S^2$ by a symplectic quotient: the map $h : \mathbb{C}^2 \to \mathbb{R}^3$,

$$h(z_1, z_2) = (|z_1|^2 - |z_2|^2, z_1\bar{z}_2 + \bar{z}_1 z_2, i(\bar{z}_1 z_2 - z_1\bar{z}_2)) \tag{5.3.11}$$

is smooth when restricted to $S^3 \subset \mathbb{C}^2$ parameterized by $|z_1|^2 + |z_2|^2 = 1$, and its image is S^2. This defines the Hopf fibration

$$\begin{array}{c} S^1 \longrightarrow S^3 \\ \downarrow{h} \\ S^2 \end{array} , \tag{5.3.12}$$

and there is a natural $U(1)$ action on the S^1 fiber. S^2 can be constructed in two steps: restrict to S^3; take the quotient by the $U(1)$ action. This is the symplectic (in fact Kähler) quotient construction. A generalization leads to $\mathbb{P}^n$:

$$\begin{array}{c} S^1 \longrightarrow S^{2n+1} \\ \downarrow{h} \\ \mathbb{P}^n \end{array} , \tag{5.3.13}$$

but there is a wider generalization that applies to a large class of toric varieties.

5.3.2 Cones and Fans

Fix a lattice $N \simeq \mathbb{Z}^d$ and set $N_{\mathbb{R}} \simeq N \otimes_{\mathbb{Z}} \mathbb{R} \simeq \mathbb{R}^d$. The "N" lattice has a dual "M" lattice

$$M = \mathrm{Hom}(N, \mathbb{Z}), \tag{5.3.14}$$

which is the space of $\mathbb{Z}$-linear functionals on N. We will denote the linear pairing for $m \in M$ and $n \in N$ by $\langle m, n \rangle$.

The cones reside in $N_\mathbb{R}$, and we define a strongly convex rational polyhedral cone $\sigma \in N_\mathbb{R}$ to be a cone with:

i. an apex at the origin (strongly convex);
ii. generators that lie in the lattice N (rational),
iii. and a finite number of generators (polyhedral).

It is useful to relax the "strongly convex" condition: this allows σ to contain a hyperplane that passes through the origin. In either case every $v \in \sigma$ has a presentation as $v = \sum_i a_i v_i$, where the v_i are the generators, and a_i are non-negative real coefficients. For any cone $\sigma \in N_\mathbb{R}$, the dual cone $\check{\sigma} \subseteq M_\mathbb{R}$ is the set of non-negative linear functionals on σ:

$$\check{\sigma} = \{u \in M_\mathbb{R} \mid \langle u, v \rangle \geq 0 \quad \text{for all } v \in \sigma\} . \tag{5.3.15}$$

For example, we have

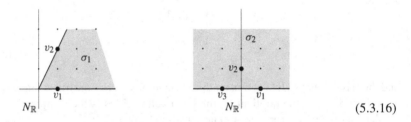

$$\tag{5.3.16}$$

The cone on the right is not strongly convex. The dual cones are given by

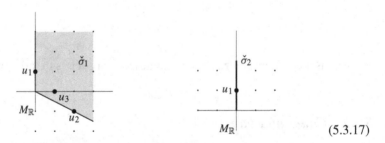

$$\tag{5.3.17}$$

Exercise 5.9 Verify that the dual cones $\check{\sigma}$ take the claimed form.   □

We can now define the affine toric variety U_σ encoded by this structure. $\check{\sigma} \cap M$ has the structure of an additive semigroup since $0 \in \check{\sigma} \cap M$ and for all $u, w \in \check{\sigma} \cap M$ $u + w \in \check{\sigma} \cap M$. There is a corresponding multiplicative group S_σ, where to each $u \in \check{\sigma} \cap M$ there corresponds a monomial χ^u:

$$\check{\sigma} \cap M \leftrightarrow S_\sigma$$

$$u \leftrightarrow \chi^u$$

$$0 \leftrightarrow 1 , \tag{5.3.18}$$

and the commutative multiplication is defined by $\chi^u \cdot \chi^v = \chi^v \cdot \chi^u = \chi^{u+v}$. S_σ is a commutative, finitely generated semigroup without nilpotents, and it is a theorem that $\mathrm{Spec}(\mathbb{C}[S_\sigma])$ is an affine variety: this is the definition of U_σ.

The points in U_σ, i.e. the maximal ideals in $\mathrm{Spec}(\mathbb{C}[S_\sigma])$, have a nice presentation: a point $p \in U_\sigma$ is a consistent assignment of complex values to the lattice points in $\check{\sigma} \cap M$, i.e. a map $p : S_\sigma \to \mathbb{C}$ satisfying $p(\chi^u \chi^v) = p(\chi^u)p(\chi^v)$ for all $u, v \in \check{\sigma} \cap M$.

Examples

To make this more concrete, consider the example of two-dimensional cone σ_1 from above. $\check{\sigma} \cap M$ is generated by the three lattice points u_1, u_2 and u_3 in M subject to the relation $u_1 + u_2 = 2u_3$.[4] To each generator u_i we assign a monomial $\chi_i \in S_\sigma$, so that for any $u = au_1 + bu_2 + cu_3$ we have $\chi^u = \chi_1^a \chi_2^b \chi_3^c$, and the relation implies $\chi_1 \chi_2 = \chi_3^2$; therefore

$$\mathbb{C}[S_\sigma] = \mathbb{C}[\chi_1, \chi_2, \chi_3]/(\chi_1 \chi_2 - \chi_3^2) . \tag{5.3.19}$$

This is a famous singular affine variety $\mathbb{C}^2/\mathbb{Z}_2$ presented as a quadric in $\mathbb{C}^3$.

The example also illustrates the definition of a point in U_σ. The multiplicative structure on S_σ implies that we associate the value 1 to the origin, and then complex values z_1, z_2, z_3 to u_1, u_2, u_3, but these are constrained by $z_1 z_2 - z_3^2 = 0$, which is the defining equation for the variety in $\mathbb{C}^3$.

As another example, we can consider the cone $\sigma = \{0\} \in \mathbb{R}^d$. In this case $\check{\sigma} = M$ is generated by the standard unit vectors e_i and their negatives $-e_i$, and

$$\mathbb{C}[S_{\{0\}}] = \mathbb{C}[\chi^{e_1}, \chi^{-e_1}, \ldots, \chi^{e_d}, \chi^{-e_d}]/(\chi^{e_1}\chi^{-e_1} - 1, \ldots, \chi^{e_d}\chi^{-e_d} - 1),$$
$$\tag{5.3.20}$$

so that $U_{\{0\}} = (\mathbb{C}^*)^d$. This is the dense open algebraic torus that will be present in any toric variety.

Exercise 5.10 Consider the two-dimensional cone σ generated by the vectors $\binom{2}{1}$ and $\binom{-1}{3}$. Show that U_σ is a non-complete intersection in $\mathbb{C}^5$ with defining equations

$$z_2^3 = z_1 z_3 , \qquad z_2 z_5 = z_3 z_4 , \qquad z_2^2 z_3 = z_1 z_4 , \qquad z_1 z_5 = z_4 z_2^2 . \tag{5.3.21}$$

Note that each relation is of the form monomial=monomial. This is a general feature of affine toric varieties. □

---

[4] The cone $\check{\sigma}_1 \in N_{\mathbb{R}}$ is generated by u_1 and u_2, but the semigroup $\check{\sigma}_1 \cap M$ requires three generators.

Cone Properties

Convex rational polyhedral cones have beautiful combinatorial properties that have a direct application to the algebraic geometry of toric varieties. We leave most of these to the references and just quote a few basic ones. First, every σ that spans $N_{\mathbb{R}}$ is an intersection of half-spaces. We define a face τ of σ, written as $\tau \prec \sigma$, as the intersection of σ with one of the supporting hyperplanes: every $\tau \prec \sigma$ takes the form $\sigma \cap u^{\perp}$, where $u \in \check{\sigma} \cap M$, and $u^{\perp} = \{v \in N_{\mathbb{R}} \,|\, \langle u, v \rangle = 0\}$. Every face $\tau \prec \sigma$ is itself also a rational convex polyhedral cone and therefore has its own S_{τ} and U_{τ}. The boundary of σ is then the union of its proper faces, and for any face $\tau \prec \sigma$ with $\tau = \sigma \cap u^{\perp}$

$$\check{\tau} = \check{\sigma} + \mathbb{R}_{\geq 0}(-u) , \tag{5.3.22}$$

and therefore

$$S_{\tau} = S_{\sigma} + \mathbb{Z}_{\geq 0}(-u) . \tag{5.3.23}$$

For instance, in (5.3.15), $v_1 \in \sigma_1$ generates a face $\tau \prec \sigma$, with $\tau = \sigma \cap u_1^{\perp}$, and $\check{\tau}$ consists of the right-half-space in $M_{\mathbb{R}}$. $\check{\tau} \cap M$ is generated by u_1, u_3 and $-u_1$. Thus, $U_{\tau} = \mathbb{C} \times \mathbb{C}^*$.

Gluing via Fans

For any face $\tau \prec \sigma$ $U_{\tau} \subset U_{\sigma}$ embeds U_{τ} as an open subset of U_{σ}. This follows easily from the definition of a point p in U_{σ} and the observation $\check{\sigma} \subset \check{\tau}$. Since p is a consistent assignment of complex values to the lattice points in $\check{\sigma} \cap M$, i.e. a map $p : S_{\sigma} \to \mathbb{C}$ satisfying $p(\chi^u \chi^v) = p(\chi^u) p(\chi^v)$ for all $u, v \in \check{\sigma} \cap M$, for any point $p \in U_{\tau}$ we obtain $p \in U_{\sigma}$ by just restricting the map $p : S_{\tau} \to \mathbb{C}$ to a map $p : S_{\sigma} \to \mathbb{C}$.

If σ and σ' are two convex polyhedral cones with a common face τ that is also their intersection $\tau = \sigma \cap \sigma'$, $\tau \prec \sigma$, and $\tau \prec \sigma'$, then

$$S_{\tau} = S_{\sigma} + S_{\sigma'} . \tag{5.3.24}$$

This motivates the definition of a fan. A fan $\Sigma \subseteq N_{\mathbb{R}}$ is a collection of rational strongly convex polyhedral cones $\sigma \in N_{\mathbb{R}}$ such that:

i. each face of a cone in Σ is also a cone in Σ;
ii. the intersection of any two cones in Σ is also a face in each.

Each $\sigma \in \Sigma$ defines an affine toric variety U_{σ}, and the gluing is then accomplished by

$$U_{\sigma_1} \cap U_{\sigma_2} = U_{\sigma_1 \cap \sigma_2} . \tag{5.3.25}$$

It is best to see how this works in practice, so we turn to a few classic examples and exercises.

Our first example is $\mathbb{P}^1$, now described from our enlightened toric perspective. The $N_\mathbb{R}$ is one-dimensional with basis vector e, and the fan has two cones: $\sigma_1 = \mathbb{R}_{\geq 0}(e)$, and $\sigma_2 = \mathbb{R}_{\geq 0}(-e)$. The duals of these in $M_\mathbb{R}$ are then

$$\overline{\qquad\qquad\underset{\check{\sigma}_2}{\qquad}\overset{0}{\bullet}\underset{\check{\sigma}_1}{\qquad}\qquad\qquad} \tag{5.3.26}$$

In this case $U_{\sigma_1} \simeq \mathbb{C}$ and $U_{\sigma_2} \simeq \mathbb{C}$: a point in U_{σ_1} is determined by the value assigned to $\chi^{(1,0)}$, say z, and a point in U_{σ_2} is determined by the value of $\chi^{(-1,0)}$, say w. Thus, we can decorate the picture in $M_\mathbb{R}$ as follows:

$$\overline{\quad \overset{w^3\quad w^2\quad w^1}{\underset{\check{\sigma}_2}{\qquad}}\ \overset{1}{\underset{0}{\bullet}}\ \overset{z^1\quad z^2\quad z^3}{\underset{\check{\sigma}_1}{\qquad}}\quad} \tag{5.3.27}$$

The gluing is then made on $U_{\sigma_1 \cap \sigma_2} \simeq \mathbb{C}^*$, where points are described by $wz = 1$. The result is that two affine varieties, each a copy of $\mathbb{C}$, are glued together to form $\mathbb{P}^1$.

Exercise 5.11 Blow up of $\mathbb{C}^2$. In this case the fan is

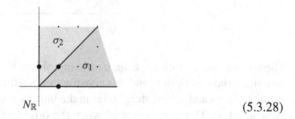

$$\tag{5.3.28}$$

Show that the dual cones are as shown (note that they do not form a fan!):

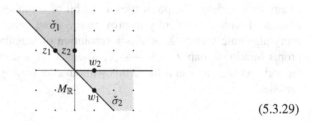

$$\tag{5.3.29}$$

As in our $\mathbb{P}^1$ example, we decorated the generators of $\check{\sigma}_1$ and $\check{\sigma}_2$ by complex numbers. Convince yourself that U_{σ_1} and U_{σ_2} are each isomorphic to $\mathbb{C}^2$, with coordinates, respectively (z_1, z_2) and (w_1, w_2). Show that the overlap is $U_{\sigma_1 \cap \sigma_2} \simeq$

$\mathbb{C} \times \mathbb{C}^*$, and on the overlap

$$z_1 w_1 = 1 \,, \qquad\qquad z_2 w_1 = w_2 \,. \qquad\qquad (5.3.30)$$

$\square$

The space just constructed can be given a concrete form by writing z_1 and w_1 in terms of projective coordinates $[Z_0 : Z_1]$ with $z_1 = Z_1/Z_0$ and $w_1 = Z_0/Z_1$. This leaves us with the second relation, which we can recast as

$$Z_0 z_2 = Z_1 w_2 \,. \qquad\qquad (5.3.31)$$

Thus, the identification is constructing the variety as the vanishing set of $Z_0 z_2 - Z_1 w_2$ in $\mathbb{P}^1 \times \mathbb{C}^2$. Away from the origin $z_2 = w_2 = 0$, the equation just fixes a point on $\mathbb{P}^1$, but at the origin itself the variety consists of a whole $\mathbb{P}^1$! The resulting variety V is the "blow-up of $\mathbb{C}^2$ at the origin." The fan (5.3.28) is said to be a refinement of the fan for $\mathbb{C}^2$:

$$(5.3.32)$$

There is a natural birational map, a toric morphism $V \to \mathbb{C}^2$ that has a completely combinatorial description, and this construction readily generalizes: by introducing additional rays and subdividing cones in the fan Σ for a variety V, we obtain a toric blow-up of V. This is most useful when the original toric variety is singular, and one wants to construct a blow-up to resolve the singularities. It can be shown that a smooth blow-up geometry can be obtained by successively subdividing the cones of any toric variety. The proof uses a fair bit of combinatorics, but this should be contrasted with the difficulty in proving the more general theorem of Hironaka: every algebraic variety X admits a resolution of singularities, i.e. there exists a proper birational map $f : X' \to X$, where X' is a smooth algebraic variety. In the toric setting one can achieve this without ever leaving the tame world of toric varieties.

Exercise 5.12 As a final example, work out the transition functions for the Hirzebruch surfaces $\mathbb{F}_p$, p a non-negative integer, with fan given by

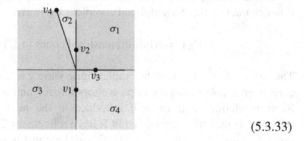

$$(5.3.33)$$

The generators are

$$v_1 = \begin{pmatrix} 0 \\ -1 \end{pmatrix}, \qquad v_2 = \begin{pmatrix} 0 \\ 1 \end{pmatrix}, \qquad v_3 = \begin{pmatrix} 1 \\ 0 \end{pmatrix}, \qquad v_4 = \begin{pmatrix} -1 \\ p \end{pmatrix}. \qquad (5.3.34)$$

Convince yourself that the space is smooth (because every chart is isomorphic to $\mathbb{C}^2$) and can be thought of as a $\mathbb{P}^1$ bundle over $\mathbb{P}^1$. □

5.3.3 Basic Properties of Compact Toric Varieties

In this short section we list some useful relations between the combinatorics of the fan Σ and the geometry and topology of the corresponding toric variety V_Σ. The proofs and further discussion are in [122].

A full-dimensional cone $\sigma \in N_\mathbb{R}$ is said to be simplicial if it is generated by $\dim N_\mathbb{R} = d$ vectors; if $d \leq 2$ this condition is vacuous, but it becomes non-trivial for $d \geq 3$. A full-dimensional cone is said to be unimodular if its generators yield a basis for N. Concretely, if $v_1, \ldots, v_k$ are the generators of a simplicial cone σ, then σ is unimodular if and only if $|\det(v_1, v_2, \ldots, v_d)| = 1$.

With this result in hand, we can state the first property: the affine variety U_σ for a full-dimensional cone σ is smooth if and only if σ is simplicial and unimodular, in which case $U_\sigma \simeq \mathbb{C}^l \times (\mathbb{C}^*)^{d-l}$; V_Σ is smooth if and only if every full-dimensional cone in Σ is simplicial and unimodular. We say V_Σ is simplicial if every cone $\sigma \in \Sigma$ is simplicial.

The singularities of a simplicial toric variety are mild: at worst they are orbifold singularities invariant under the $T_N = (\mathbb{C}^*)^d$ action. The origin in the $\mathbb{C}/\mathbb{Z}_2$ example above is the simplest example of such a toric singularity.

Next, we have the result on completeness, a generalization of compactness appropriate for algebraic varieties. If the variety is smooth, then completeness is equivalent to compactness. A toric variety V_Σ is complete if and only if the fan Σ covers $N_\mathbb{R}$.

Every toric variety whose fan contains a full-dimensional cone is simply connected: $\pi_1(V_\Sigma) = 0$. In what follows, we will only work with fans that contain a full-dimensional cone, so all of our toric varieties will be simply connected. The Euler characteristic of a simplicial complete V_Σ is given by

$$\chi(V_\Sigma) = \#(\text{full-dimensional cones in } \Sigma) . \tag{5.3.35}$$

The cohomology of a smooth complete toric variety is zero in odd degree, while the even degree cohomology groups are torsion-free and determined by the Stanley-Reisner relations that we will introduce in the next section. One final useful statement is that every complete fan Σ has a refinement Σ' such that $V_{\Sigma'}$ is a smooth projective toric variety; such a fan Σ' can be constructed from a polytope, as we discuss below.

Most of these results are obtained by using the $T_N = (\mathbb{C}^*)^d$ action on the toric variety to stratify the space into simpler subsets, so we will leave this section with a few remarks on that structure. At its most basic level, whenever a group (in this case T_N) acts on a set (in this case the points in V_Σ), the set is stratified into orbits. In a toric variety the orbits are related to the cones through a dimension-reversing correspondence.

For starters, every toric variety includes the cone $\{0\} \in \Sigma$; its dual is $M_\mathbb{R}$, and $U_{(0)} = (\mathbb{C}^*)^d$. A point p in $U_{\{0\}}$ is specified by the map $p(\chi^{\pm e_i}) = s_i^{\pm 1}$, where $i = 1, \ldots, d$, the e_i are a basis for M, and $s_i \in \mathbb{C}^*$. Such points admit a natural fixed-point free action of T_N: if $t = (t_1, \ldots, t_d) \in T_N$, then $(t \cdot p)_i = t_i s_i$. Thus $U_{\{0\}}$ is a d-dimensional orbit of the T_N action.

The T_N action extends to any U_σ: for any $u \in \check\sigma \cap M$ and a point $p : \chi^u \to \mathbb{C}$, we set $t \cdot p = t_u p(\chi^u)$, where $t_u = \prod_i t_i^{u^i}$, and $u = u^i e_i$. Since $\sigma^\perp \prec \check\sigma$, we also have a distinguished point $p_\sigma \in U_\sigma$: for any $u \in \check\sigma \cap M$ set

$$p_\sigma(\chi^u) = \begin{cases} 1 & \text{if } u \in \sigma^\perp , \\ 0 & \text{otherwise} \end{cases} . \tag{5.3.36}$$

If σ is full-dimensional, p_σ is the unique point in U_σ fixed by the T_N action, and defines the orbit $O_\sigma = p_\sigma$. If σ is not full-dimensional, then we define O_σ to be the T_N orbit of p_σ. If σ has dimension k then $O_\sigma = (\mathbb{C}^*)^{d-k}$. V_Σ is then the disjoint union of the orbits O_σ. Each orbit has a closure $\overline{O_\tau}$ that is a closed toric subvariety of V_Σ, and thus we obtain an interesting class of T_N-invariant subvarieties of V_Σ. A particularly important class of these arise from $\rho \in \Sigma(1)$, the set of one-dimensional cones in the fan: $\overline{O_\rho}$ have dimension $d - 1$ and correspond to T_N-invariant divisors on V_Σ. We will have more to say about them below, but first we discuss an alternative presentation of simplicial toric varieties.

In what follows we will often drop the Σ from V_Σ.

5.3.4 The Audin-Cox Homogeneous Coordinate Ring

The holomorphic quotient construction of $\mathbb{P}^n$ has three basic ingredients[5]:

1. an affine variety $Y = \mathbb{C}^{n+1}$;
2. a $G_{\mathbb{C}} = \mathbb{C}^*$ action on Y;
3. the exceptional set F, where the orbit of $\mathbb{C}^*$ fails to be one-dimensional—the origin $\{0\} \in Y$.

The quotient $Y/G_{\mathbb{C}}$ is not well-behaved—the quotient topology fails to separate points. However, by removing the exceptional set, we get a nice quotient:

$$\mathbb{P}^n = \frac{Y \setminus F}{G_{\mathbb{C}}} . \qquad (5.3.37)$$

This is made explicit by giving homogeneous coordinates $[Z_0, Z_1, \ldots, Z_n]$, with the $\mathbb{C}^*$ action

$$t \cdot [Z_0, Z_1, \ldots, Z_n] = [t Z_0, t Z_1, \ldots, t Z_n] . \qquad (5.3.38)$$

Suppose that V_Σ is a simplicial toric variety. Let $\Sigma(1)$ denote the one-dimensional cones in the fan (these are often called the rays); suppose there are n of them, denoted by $\rho \in \Sigma(1)$. We will often conflate the label ρ with the vector that generates the ray. We can then define homogeneous coordinates and a holomorphic quotient with a geometric interpretation by the same three steps as in the construction of $\mathbb{P}^n$.

1. To each $\rho \in \Sigma(1)$ we associate a coordinate Z_ρ on $\mathbb{C}^n$ and a generator of the polynomial ring $S = \mathbb{C}[Z_{\rho_1}, \ldots, Z_{\rho_n}]$. We set $Y = \mathbb{C}^n$. Y has a natural action of a "big torus" $(\mathbb{C}^*)^n$:

$$(Z_{\rho^1}, Z_{\rho^2}, \ldots, Z_{\rho^n}) \rightarrow (t_1 Z_{\rho^1}, t_2 Z_{\rho^2}, \ldots, t_n Z_{\rho^n}) . \qquad (5.3.39)$$

This harkens back to large global symmetry group of the free theory with which we started in our quest for the linear sigma model.

2. The group $G_{\mathbb{C}}$ is a subgroup of the big torus and fits into the short exact sequence

$$1 \longrightarrow G_{\mathbb{C}} \longrightarrow (\mathbb{C}^*)^n \xrightarrow{\ R\ } T_N \longrightarrow 1 . \qquad (5.3.40)$$

---

[5]The terminology "holomorphic quotient" is frequently used in the physics literature to distinguish it from the Kähler or symplectic quotient construction; in the mathematics literature, the same construction is known as the categorical quotient of GIT; when the quotient is nice—meaning that, as in the $\mathbb{P}^d$ example, it has an obvious geometric interpretation, and the quotient topology separates points—the notion coincides with the simpler "geometric quotient."

If we fix a basis $\{e^1, \ldots, e^d\}$ for N, so that $\rho^a = \rho_i^a e^i$, then the map R is given by

$$R \cdot (t_1, t_2, \ldots, t_n) \rightarrow \left(\prod_a t_a^{\rho_1^a}, \prod_a t_a^{\rho_2^a}, \ldots, \prod_a t_a^{\rho_d^a} \right). \qquad (5.3.41)$$

The group $G_{\mathbb{C}}$ is then simply $G_{\mathbb{C}} = \ker R$. Since $N_{\mathbb{R}} = \mathrm{span}\{\Sigma(1)\}$, V_Σ,

$$G_{\mathbb{C}} \simeq (\mathbb{C}^*)^{n-d} \times H, \qquad (5.3.42)$$

where H is a finite abelian group. The continuous part of $G_{\mathbb{C}}$ arises from integer solutions to

$$\sum_{a=1}^{n} v_a \rho^a = 0. \qquad (5.3.43)$$

We can choose an explicit basis for such $v \in \mathbb{Z}^n$: any v annihilated by R takes the form

$$v = v_\alpha Q_a^\alpha \qquad (5.3.44)$$

with $n - d$ generators $Q^\alpha \in \mathbb{Z}^n$ that satisfy $Q_a^\alpha \rho^a = 0$ for all α. In the gauge theory these will be the gauge charges. We will often denote them by Q_ρ^α, with the understanding that ρ is an index for the the cones in $\Sigma(1)$.

The discrete part H is also easy to understand: it arises simply because the equation $t^k = 1$ for integer k is solved by setting t to be a k-th root of unity, and in the gauge theory it has the interpretation of a discrete gauge symmetry: an orbifold of the theory by H.

The action of $G_{\mathbb{C}}$ on the coordinates Z_ρ leads to a grading of the polynomial ring $S = \mathbb{C}[Z_{\rho^1}, \ldots, Z_{\rho^n}]$. In particular each monomial $\mathsf{M} \in S$ has a definite charge with respect to the $(\mathbb{C}^*)^{n-d}$ action, which we will denote by $\deg(\mathsf{M})$.

3. The exceptional set $F \subset Y$ is the vanishing locus of the "irrelevant ideal" $B(\Sigma) \subset S$:

$$B(\Sigma) = \left(\prod_{\rho \notin \sigma} Z_\rho \,|\, \sigma \in \Sigma(d) \right). \qquad (5.3.45)$$

The irrelevant ideal is generated by monomials; there is one monomial for each full-dimensional cone $\sigma \in \Sigma(d)$, and it is a product of all coordinates Z_ρ for $\rho \not\subset \sigma$.

Equivalently, a collection of rays $\{\rho^a\}_{a \in I}$ that does not belong to a full-dimensional cone in Σ is called a primitive collection. F is a union of intersections of coordinate hyperplanes $\bigcap_{a \in I} \{Z_{\rho^a} = 0\}$ for each primitive collection.

The exceptional set does not have an immediately obvious counter-part in the gauge theory, and this is for a good reason: its form depends on the Fayet-Iliopoulos parameters. We will explore this more in detail in what follows.

Finally, with all of these ingredients, we set

$$V_\Sigma = \frac{Y \setminus F}{G_\mathbb{C}} . \tag{5.3.46}$$

Exercise 5.13 Consider the grading of the polynomial ring S in a bit more detail. Prove that two Laurent monomials

$$M_a = \prod_\rho Z_\rho^{a_\rho} \qquad\qquad M_b = \prod_\rho Z_\rho^{b_\rho} \tag{5.3.47}$$

have the same degrees if and only if there exists a lattice point $m \in M$ such that

$$a_\rho = b_\rho + \langle m, \rho \rangle . \tag{5.3.48}$$

$\square$

Exercise 5.14 Apply the construction to all of the examples given above. In particular, verify that for the Hirzebruch surfaces $\mathbb{F}_p$ we have

$$\mathbb{F}_p = \frac{\mathbb{C}^4 \setminus F}{(\mathbb{C}^*)^2} , \tag{5.3.49}$$

with

$$B(\Sigma) = (Z_1 Z_4, Z_1 Z_3, Z_2 Z_3, Z_2 Z_4) ,$$
$$F = \{Z_1 = Z_2 = 0\} \cup \{Z_3 = Z_4 = 0\} , \tag{5.3.50}$$

and a charge matrix

$$Q = \begin{pmatrix} p & 0 & 1 & 1 \\ 1 & 1 & 0 & 0 \end{pmatrix} . \tag{5.3.51}$$

This form makes it clear that this manifold is a $\mathbb{P}^1$ bundle—with fiber parameterized by the Z_3, Z_4 coordinates—over $\mathbb{P}^1$ (parameterized by Z_1, Z_2). Note also that $\mathbb{F}_0$ is simply the product $\mathbb{P}^1 \times \mathbb{P}^1$.

You should also see that in the $\mathbb{C}^2/\mathbb{Z}_2$ example the quotient by $\mathbb{Z}_2$ arises as the finite group H.

$\square$

The reader might be pondering the reason for the assumption that Σ is simplicial: we do not need the assumption to construct the quotient. However, the assumption is indispensable if we want to think of the quotient in geometric terms. When Σ is simplicial, this is the case: the homogeneous coordinates yield a good description of points in V_Σ: two points Z and Z' in $Y \setminus F$ define the same point in V_Σ if and only if $Z' = g \cdot Z$ for some $g \in G$. To see what goes wrong, we take a look at a famous non-simplicial example

Exercise 5.15 The conifold. In this exercise we consider the famous conifold, our first 3-dimensional toric variety. The fan consists of a single cone σ with generators

$$v_1 = \begin{pmatrix} 0 \\ 0 \\ 1 \end{pmatrix}, \qquad v_2 = \begin{pmatrix} 1 \\ 0 \\ 1 \end{pmatrix}, \qquad v_3 = \begin{pmatrix} 1 \\ 1 \\ 1 \end{pmatrix}, \qquad v_4 = \begin{pmatrix} 0 \\ 1 \\ 1 \end{pmatrix}. \qquad (5.3.52)$$

Since all the points lie in a hyperplane, a nice way to visualize the fan is to intersect it with the $z = 1$ hyperplane in $N_\mathbb{R}$; the result is a polytope:

$$
\begin{array}{cc}
v_4 & v_3 \\
\bullet & \bullet \\
\square & \\
\bullet & \bullet \\
v_1 & v_2
\end{array}
\qquad (5.3.53)
$$

Show that the dual cone $\check{\sigma}$ has generators

$$u_1 = (0, 1, 0), \qquad u_2 = (1, 0, 0), \qquad u_3 = (-1, 0, 1), \qquad u_4 = (0, -1, 1) \qquad (5.3.54)$$

and therefore U_σ is the variety $z_1 z_4 - z_2 z_3 = 0$ in $\mathbb{C}^4$. This is the singular conifold geometry, and now we see explicitly that it is toric.

Next, describe the quotient construction and show that

$$z_1 = Z_1 Z_2, \qquad z_2 = Z_1 Z_4, \qquad z_3 = Z_2 Z_3, \qquad z_4 = Z_3 Z_4 \qquad (5.3.55)$$

are G-invariant combinations of the homogeneous "coordinates." However, show that the origin $z_1 = z_2 = z_3 = z_4 = 0$ does not correspond to a unique $G_\mathbb{C}$-orbit; in fact, there are even orbits that have different stabilizer subgroups in $G_\mathbb{C}$. This is the classic example of the pathology in the quotient construction for a non-simplicial toric variety. As we discuss below, by including stability conditions we can remove the offending orbits and produce a separated albeit still singular space that matches the hypersurface U_σ.

There are two classic ways to fix the troubles of the conifold singularity. There is the non-toric deformation of the defining equation: $z_1 z_4 - z_2 z_3 = \epsilon$, and it is not hard to argue that the resulting variety is smooth for $\epsilon \neq 0$ but not toric. There is also a toric (small) resolution obtained by subdividing the fan:

$$F = \{Z_2 = Z_4 = 0\} \qquad \text{or} \qquad F = \{Z_1 = Z_3 = 0\} . \tag{5.3.57}$$

Argue that in each case the manifold can be thought of as a bundle $\mathcal{O}(1) \oplus \mathcal{O}(1) \to \mathbb{P}^1$.

The two resolutions are related by what is known as the "flop transition." As we will discuss below, the linear sigma model provides a way of smoothly relating the flopped geometries. □

Orbits and Automorphisms

The homogeneous coordinate ring yields a nice description of two important geometric features of a simplicial toric variety V_Σ.[6] First, we already described the T_N orbit orbit closures that assign to each cone $\tau \in \Sigma$ a toric subvariety $\overline{O_\tau} \subset V_\Sigma$. These have a simple presentation in terms of the homogeneous coordinates:

$$\overline{O_\tau} = \{Z \in Y \setminus F \mid Z_\rho = 0 \qquad \text{for all} \quad \rho \prec \tau\}/G_{\mathbb{C}} . \tag{5.3.58}$$

That is $\overline{O_\tau}$ is described "upstairs" by setting $Z_\rho = 0$ for all one-dimensional cones ρ contained in τ. The toric divisors,

$$D_\rho = \overline{O_\rho} , \tag{5.3.59}$$

which correspond to just setting $Z_\rho = 0$ play a particularly important role in what follows.

The second feature we discuss here is the automorphism group of V_Σ. For any algebraic variety we denote by $\text{Aut}(V)$ the group of invertible holomorphic maps from V to itself. When V is a smooth complex manifold these are just the diffeomorphisms that preserve the complex structure on V. The automorphism group of a toric variety at least contains T_N, the dense algebraic torus that acts

---

[6]We follow here the discussion in [120].

on V, but in general $\mathrm{Aut}(V)$ may be larger. In fact, it fits into a short exact sequence

$$1 \longrightarrow G_{\mathbb{C}} \longrightarrow \widetilde{\mathrm{Aut}}(V) \longrightarrow \mathrm{Aut}(V) \longrightarrow 1 \ , \tag{5.3.60}$$

where $\widetilde{\mathrm{Aut}}$ is a group that acts on the homogeneous coordinates. The group $\mathrm{Aut}(V)$ is generated by three types of elements: (i) the generators of T_N, (ii) additional one-parameter coordinate transformations of the homogeneous coordinates, and (iii) discrete symmetries that arise from symmetries of the fan.

The connected component of $\mathrm{Aut}(V)$, denoted by $\mathrm{Aut}_0(V)$, is generated by type (i) and (ii) elements. For us it will be more convenient to lift that action to the homogeneous coordinates and discuss $\widetilde{\mathrm{Aut}}_0(V)$, which is also generated by two types of elements. The first of these is simply a rescaling of the coordinates by an action of the "big" torus of (5.3.39). To describe the second class of generators, let

$$Y_\rho = \{\text{monomial } \mathsf{M} \in S, \mathsf{M} \neq Z_\rho \mid \deg[\mathsf{M}] = \deg[Z_\rho]\} \ . \tag{5.3.61}$$

Each $\mathsf{M} \in Y_\rho$ can be used to construct a coordinate transformation

$$Z_\rho \to Z_\rho + \epsilon \mathsf{M} \ , \tag{5.3.62}$$

and the set of all such transformations is precisely the set of type (ii) generators of $\widetilde{\mathrm{Aut}}_0(V)$.

5.3.5 Divisors, Line Bundles, and Cohomology

For any variety X there is a close relationship between line bundles and complex co-dimension one subvarieties. We explain a few of the relevant definitions in a general setting and then discuss their realization in the toric setting of most interest to us—complete simplicial toric varieties V. It may help to take a look at the short sheaf discussion in Sect. B.3.2 before tackling these mysteries.

We start by stating some general terminology applicable to the wide setting of normal irreducible algebraic varieties and then quickly specialize and illustrate these notions in the toric setting. A very readable reference on divisors and line bundles in general is [282].

1. A Weil divisor on X is a formal sum over co-dimension one subvarieties with integer coefficients. A Weil divisor is effective if the coefficients are non-negative.
2. For any open set $U \subset X$ we define the sheaf of non-zero holomorphic functions as $\mathcal{O}^*(U)$; similarly, $\mathcal{M}^*(U)$ is the sheaf of non-zero meromorphic functions, where every element is of the form f/g for some $f, g \in \mathcal{O}^*(U)$. A Cartier divisor D is then a global section of the sheaf $\mathcal{M}^*/\mathcal{O}^*$. This means that we can choose a cover for X, such that on every U_α there is a meromorphic function

f_α / g_α, and on overlaps $U_{\alpha\beta}$ $f_\alpha = \lambda_{\alpha\beta} f_\beta$ and $g_\alpha = \lambda_{\alpha\beta} g_\beta$ for some $\lambda_{\alpha\beta} \in \mathcal{O}^*(U_{\alpha\beta})$. The divisor then keeps track of the order of vanishing of f (which makes positive contributions to the formal sum) and the order of vanishing of g (which makes negative contributions to the formal sum).

We set $\mathrm{Div}(X) = \mathcal{O}(X, \mathcal{M}^*/\mathcal{O}^*)$ The sheaves in the construction have a natural multiplicative structure, but the corresponding operation on the divisors is written additively: $D + D'$; either way $\mathrm{Div}(X)$ has the structure of an abelian group.

3. A global section $f \in \mathcal{O}(X, \mathcal{M}^*)$, also known as a meromorphic or rational function on X, defines a principal divisor, denoted by $\mathrm{div}(f)$. So, Cartier divisors are defined by local meromorphic functions, while principal divisors are defined by global meromorphic functions. Principal divisors obey $\mathrm{div}(fg) = \mathrm{div}(f) + \mathrm{div}(g)$ and form an additive subgroup $\mathrm{Div}_0(X) \subset \mathrm{Div}(X)$.

4. Two Cartier divisors are said to be linearly equivalent, written $D \sim D'$ if and only if they differ by a principal divisor $\mathrm{div}(f)$ for some global function.

5. Every Cartier divisor defines a Weil divisor, and a Weil divisor D is said to be $\mathbb{Q}$-Cartier if pD is Cartier for some integer $p > 1$.

6. The Chow group $A_{d-1}(X)$ is the additive group of Weil divisors modulo linear equivalence; d is the dimension of X.

7. Isomorphism classes of line bundles on X are classified by the sheaf cohomology group $H^1(X, \mathcal{O}^*) = \mathrm{Pic}(X)$; the additive structure in $\mathrm{Pic}(X)$ corresponds to the natural multiplicative structure on $H^1(X, \mathcal{O}^*)$, which is reflected by the line bundle tensor product $\mathcal{L}_1 \otimes \mathcal{L}_2$.

8. If a line bundle $\mathcal{L} \to X$ has a global (not point-wise zero) section s, then its zero set defines an effective divisor. More generally, a rational section of $\mathcal{L} \to X$ defines a divisor that is not necessarily effective.

9. The defining data of a Cartier divisor D also yields transition functions for a line bundle denoted by $\mathcal{O}(D)$. $\mathcal{O}(D)$ is isomorphic to $\mathcal{O}(D')$ if and only if $D \sim D'$.

10. When X is irreducible, every line bundle is of the form $\mathcal{O}(D)$ for some Cartier divisor; D is linearly equivalent to the divisor constructed from any rational section of $\mathcal{O}(D)$.

Let us see how these concepts are realized in the toric setting. We begin with an easy example familiar from above: $V = \mathbb{C}^2/\mathbb{Z}_2$ with fan

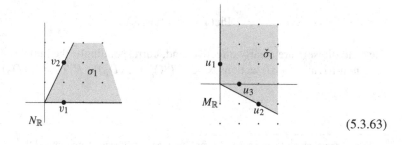

$$(5.3.63)$$

In the previous section we described the toric divisors D_ρ, and using these we construct a toric Weil divisor

$$D = a_1 D_{\rho_1} + a_2 D_{\rho_2} \, . \tag{5.3.64}$$

We also have a natural class of toric rational functions: for any point $m \in M$ the monomial χ^m defines a rational function, and the corresponding divisor is

$$\mathrm{div}\chi^m = \sum_\rho \langle m, \rho \rangle D_\rho \, . \tag{5.3.65}$$

In our example we have $m = (m_1, m_2)$ and

$$\mathrm{div}\chi^m = m_1 D_{\rho_1} + (m_1 + 2m_2) D_{\rho_2} \, . \tag{5.3.66}$$

Evidently, while $D_{\rho_1} + D_{\rho_2}$, $2D_{\rho_1}$, and $2D_{\rho_2}$ are Cartier (and in fact principal), there is no toric rational function that yields either D_{ρ_1} or D_{ρ_2}; these are, in fact, Weil divisors that are not Cartier; however, they are both $\mathbb{Q}$-Cartier.

If V is simplicial, then every Weil divisor is $\mathbb{Q}$-Cartier, and $A_{d-1}(V)$ is generated by the toric divisors D_ρ; there is a short exact sequence

$$1 \longrightarrow M \longrightarrow \mathbb{Z}^{\Sigma(1)} \longrightarrow A_{d-1}(V) \longrightarrow 1 \, , \tag{5.3.67}$$

where the second map is $m \to (\langle m, \rho_1 \rangle, \ldots, \langle m, \rho_n \rangle)$, and the third map sends $a \in \mathbb{Z}^{\Sigma(1)}$ to $a \to \sum_\rho a_\rho D_\rho$ [122]. So, for instance, applying this to $V = \mathbb{C}^2/\mathbb{Z}_2$, we find that the Chow group $A_1(V) \simeq \mathbb{Z}_2$ is pure torsion and matches the torsion of the group G. This is not a coincidence: comparing to (5.3.40), $G_{\mathbb{C}} = \mathrm{Hom}_{\mathbb{Z}}(A_{d-1}(V), \mathbb{C}^*)$. The definition of the Hom uses the following fact: $\mathrm{Hom}_{\mathbb{Z}}(M, \mathbb{C}^*) = M \otimes \mathbb{C}^* = T_M$; a point $(t_1, \ldots, t_d) \in T_M$ yields the homomorphism $t : M \to \mathbb{C}^*$ with $t(m) = \prod_i t_i^{m_i}$.

The toric divisors D_ρ generate the Chow group $A_{d-1}(V)$; the group of Cartier divisors modulo linear equivalence is isomorphic to $\mathrm{Pic}(V)$, so we can usefully think of $\mathrm{Pic}(V)$ as a subgroup of $A_{d-1}(V)$, and the two differ modulo torsion. For a complete simplicial toric variety the sheaf cohomology groups of the structure sheaf $\mathcal{O}_V$ satisfy $H^i(V, \mathcal{O}_V) = 0$ for $i > 0$[7]; this means that (4.5.38) simplifies to

$$\mathrm{Pic}(V) = H^2(V, \mathbb{Z}) \, . \tag{5.3.68}$$

The toric divisors are clearly effective, and, correspondingly, the bundles $\mathcal{O}_V(D_\rho)$ have non-trivial global sections $s \in H^0(V, \mathcal{O}_V(D_\rho))$. $H^0(V, \mathcal{O}_V(D_\rho))$ has a

---

[7]This is a special case of a more general vanishing theorem due to Demazure [122].

simple basis in terms of the homogeneous coordinates: it is the vector space of all monomials $M \in S$ with $\deg(M) = \deg(Z_\rho)$—compare to (5.3.61).

When V is smooth and complete, $\text{Pic}(V) = A_{d-1}(V)$, and Poincaré duality for singular homology of a closed oriented manifold, $H_k(V, \mathbb{Z}) \simeq H^{d-k}(V, \mathbb{Z})$, implies that $H^2(V, \mathbb{Z})$ is generated by the classes isomorphic to the toric divisors D_ρ. Let $\xi_\rho \in H^2(V, \mathbb{Z})$ be the class that corresponds to the divisor D_ρ. Tracing through (5.3.67) and our definition of the gauge group G, it follows that

$$H^2(V, \mathbb{Z}) \simeq \mathbb{Z}^{\oplus(n-d)} , \tag{5.3.69}$$

and we can pick a basis $\{\eta_\alpha\}_{\alpha=1,\dots,n-d}$ for $H^2(V, \mathbb{Z})$ such that

$$\xi_\rho = Q_\rho^\alpha \eta_\alpha . \tag{5.3.70}$$

The ξ_ρ are related to the line bundles $\mathcal{O}(D_\rho)$ by

$$c_1(\mathcal{O}(D_\rho)) = \xi_\rho . \tag{5.3.71}$$

In fact, the full integral cohomology $H^\bullet(V, \mathbb{Z})$, together with the product (we can think of this abstractly as the cup product on cohomology or in terms of the wedge product of representative classes in de Rham cohomology) is generated by the ξ_ρ modulo a set of relations. These so-called Stanley-Reisner relations are determined by the exceptional set F of the holomorphic quotient construction:

$$H^\bullet(V, \mathbb{Z}) = \mathbb{Z}[\eta_1, \dots, \eta_{n-d}]/SR(\Sigma) . \tag{5.3.72}$$

The Stanley-Reisner ideal $SR(\Sigma)$ is determined as follows: for each primitive collection $\{\rho_i\}_{i \in I}$, $SR(\Sigma)$ includes the generator $\prod_{i \in I} \xi_{\rho_i}$.

The origin of these relations is easily understood in terms of the toric divisors: if two divisors D_ρ and $D_{\rho'}$ do not intersect in V, then, correspondingly, $\xi_\rho \xi_{\rho'} = 0$ in $H^4(V, \mathbb{Z})$. The toric divisors fail to intersect if and only if $\{Z_\rho = Z_{\rho'} = 0\}$ is contained in the exceptional set. This generalizes to intersections of three or more divisors. Thus, the non-trivial part of (5.3.72) is to show that all of $H^\bullet(V, \mathbb{Z})$ is generated in this fashion. The details can be found in [122].

The intersection of d divisors $D_{\rho_1}, D_{\rho_2}, \dots, D_{\rho_d}$ is a number of points, and when counted with orientation, they yield the intersection number $\#(D_{\rho_1} \cdot D_{\rho_2} \cdots D_{\rho_d}) \in \mathbb{Z}$, a topological invariant of V. From the previous paragraph it follows that for a smooth complete variety this number is 1 if the collection $\rho_1, \dots, \rho_d$ belongs to a full-dimensional cone and zero otherwise. Equivalently, we can think of this number as the integral of the top form

$$\#(D_{\rho_1} \cdot D_{\rho_2} \cdots D_{\rho_d}) = \int_V \xi_{\rho_1} \wedge \xi_{\rho_2} \wedge \cdots \wedge \xi_{\rho_d} . \tag{5.3.73}$$

We will frequently conflate notation and write $\#(\xi_{\rho_1}\xi_{\rho_2}\cdots\xi_{\rho_d})$ for this intersection number.

These results have a generalization to complete simplicial toric varieties: $A_{n-1}(V)$ and $H^2(V,\mathbb{Z})$ are isomorphic modulo torsion, and the ξ_ρ generate $H^\bullet(V,\mathbb{R})$ in precisely the analogous fashion. The intersection number of d toric divisors satisfies

$$\#(D_{\rho_1}\cdot D_{\rho_2}\cdots D_{\rho_d}) = \begin{cases} |\det(\rho_1\rho_2\cdots\rho_d)|^{-1} & \text{if } \rho_1,\ldots,\rho_d \text{ belong to a full} \\ & \text{dimensional cone} \\ 0 & \text{otherwise .} \end{cases}$$

(5.3.74)

The rational numbers that show up here are indicative of the orbifold singularities in V and reflect the fact that when V is simplicial but singular the toric divisors are $\mathbb{Q}$-Cartier.

Exercise 5.16 Apply the machinery to the Hirzebruch surfaces $\mathbb{F}_p$ and show that the intersection form $H^2(\mathbb{F}_p,\mathbb{Z}) \times H^2(\mathbb{F}_p,\mathbb{Z}) \to H^4(\mathbb{F}_p,\mathbb{Z})$ is given by (in the η_1,η_2 basis)

$$\mathcal{I}_p = \begin{pmatrix} 0 & 1 \\ 1 & -p \end{pmatrix}.$$

(5.3.75)

For extra entertainment show that $\mathcal{I}_p$ and $\mathcal{I}_k$ are related by an $\mathrm{SL}(2,\mathbb{Z})$ similarity transformation if and only if $p = k \mod 2$. This is as it should be, since $\mathbb{F}_p$ is diffeomorphic to $\mathbb{F}_k$ if and only if $p = k \mod 2$ [226].   □

Much more is known about divisors and line bundles (or more generally rank 1 sheaves) on toric varieties, and many problems are eminently computable. For example the evaluation of the sheaf cohomology groups $H^\bullet(V,\mathcal{O}_V(D))$, where $D = \sum_\rho a_\rho D_\rho$ is reduced to a computational combinatorics problem. The reader may not want to carry this out by hand, but there are computational algebra packages that greatly facilitate the computations.[8]

We end our discussion with one final ingredient that we will need for our linear sigma model purposes: the presentation of the tangent bundle and the canonical bundle of a complete smooth toric variety. Again, the motivation for the general construction comes from some familiar results on $\mathbb{P}^n$. The cohomology of $\mathbb{P}^n$ is generated by a single class $\xi \in H^2(\mathbb{P}^n,\mathbb{Z})$ subject to $\xi^{n+1} = 0$, and the $n+1$ toric divisors described as the hypersurfaces $D_i = \{Z_i = 0\}$ are all linearly

---

[8]Macaulay2 has the packages ToricVectorBundles, authored by Birkner, Ilten, and Petersen, and NormalToricVariety authored by Smith. A discussion of efficient algorithms can be found in [89, 257].

equivalent; the hyperplane class H then generates $H_{n-2}(\mathbb{P}^n, \mathbb{Z})$ and is dual to ξ. The homogeneous coordinates are sections of the line bundle $\mathcal{O}(H)$.

A holomorphic section of $T_{\mathbb{P}^n}$ can be presented in homogeneous coordinates as follows:

$$v = \sum_{i=0}^{n} Z_i A_{ij} \frac{\partial}{\partial Z_j} , \tag{5.3.76}$$

where A_{ij} is a constant matrix. Note that s is well-defined with respect to rescaling the homogeneous coordinates, and the coefficients of the $\frac{\partial}{\partial Z_i}$ can evidently be thought of as a section of $\mathcal{O}(H)^{\oplus(n+1)}$.

Now clearly this is a bit much—after all, the tangent bundle has rank n, and not $n+1$. However, we understand the point: consider, for definiteness, a patch U of $\mathbb{P}^n$ with $Z_0 \neq 0$. The holomorphic functions on U are of the form $Z_0^{-d} f(Z_0, \ldots, Z_n)$, where d is the degree of the homogeneous polynomial f, and therefore they are annihilated by

$$v_0 = \sum_{i=0}^{n} Z_i \frac{\partial}{\partial Z_i} , \tag{5.3.77}$$

since $V_0(f) = \deg(f)f$ for any holomorphic homogeneous f. Thus, to obtain holomorphic vectors on $\mathbb{P}^n$, we should quotient the space spanned by the v by v_0. This leads to the result that the tangent bundle of $\mathbb{P}^n$ is encoded in the short exact sequence (known as the Euler sequence)

$$0 \longrightarrow \mathcal{O}_{\mathbb{P}^n} \xrightarrow{E_{22}} \mathcal{O}_{\mathbb{P}^n}(H)^{\oplus(n+1)} \longrightarrow T_{\mathbb{P}^n} \longrightarrow 0 , \tag{5.3.78}$$

where the map E_{22} acts by $E_{22}(\lambda) = \lambda E_{22}$, where the matrix E is

$$E_{22} = (Z_0, Z_1, Z_2, \ldots, Z_n) . \tag{5.3.79}$$

This construction generalizes to the tangent sheaf of any projective toric variety V, which is equivalent to the following exact sequence of sheaves:

$$0 \longrightarrow H_2(V, \mathbb{Z}) \otimes \mathcal{O}_V \xrightarrow{E_{22}} \oplus_\rho \mathcal{O}_V(D_\rho) \longrightarrow T_V \longrightarrow 0 . \tag{5.3.80}$$

The map E_{22} is given by

$$E_{22} : H_2(V, \mathbb{Z}) \otimes \mathcal{O}_V \rightarrow \oplus_\rho \mathcal{O}_V(D_\rho)$$

$$E_{22} : (\lambda_1, \lambda_2, \ldots, \lambda_{n-d}) \mapsto \oplus_\rho \lambda_\alpha Q_\rho^\alpha Z_\rho . \tag{5.3.81}$$

We will have many uses for this presentation. For instance, we compute $c_1(T_V)$:

$$c_1(T_V) = \sum_\rho c_1(\mathcal{O}_V(D_\rho)) \, . \tag{5.3.82}$$

We use here a fundamental property of the Chern class: given an exact sequence of vector bundles $0 \longrightarrow \mathcal{E}_1 \longrightarrow \mathcal{E}_2 \longrightarrow \mathcal{E}_3 \longrightarrow 0$ the Chern classes obey $c(\mathcal{E}_2) = c(\mathcal{E}_1)c(\mathcal{E}_3)$. This follows from the facts that the Chern classes are topological invariants, the bundles in the exact sequence obey $\mathcal{E}_2 = \mathcal{E}_1 \oplus \mathcal{E}_3$ as C^∞ bundles, and we have the Whitney product formula for the Chern class $\mathcal{E}_1 \oplus \mathcal{E}_3$ [94]. For our application we therefore have $c(T_V) = c(\oplus_\rho \mathcal{O}_V(D_\rho))$.[9]

Using (5.3.71) we obtain

$$c_1(T_V) = \sum_\rho \xi_\rho \, . \tag{5.3.83}$$

Equivalently, the canonical divisor K_V, which corresponds to the canonical line bundle $\wedge^d T_V^*$, is

$$K_V = -\sum_\rho D_\rho \, . \tag{5.3.84}$$

Another simple application is that the sequence shows that T_V can be deformed to a more general bundle $\mathcal{E} \to V$ by the simple expedient of deforming the map in the sequence $E_{22} \to E$ that is consistent with the degree assignment. While T_V may have deformations that cannot be described by deforming the map—see, for instance, [152]—these, so-called "monadic" deformations are particularly nice: they are obviously unobstructed, since we can just tune parameters at will, and they have a natural interpretation as deformations of the (0,2) linear sigma model.

5.3.6 Symplectic Quotient and the Secondary Fan

We turn to yet another illuminating way to describe toric varieties: the symplectic quotient. The constructions we have given so far have just described the complex geometry of toric varieties, but for our physical applications we need additional metric data. This turns out to be nicely encoded in the symplectic quotient construction. Our discussion here follows [311].

Suppose V is a smooth toric variety, so that its holomorphic quotient construction takes the form $(Y \setminus F)/G_\mathbb{C}$, with $G_\mathbb{C} = (\mathbb{C}^*)^{n-d}$. $Y = \mathbb{C}^n$ is a Kähler manifold, and the Kähler form is clearly preserved by the action of $G = U(1)^{n-d}$, the maximal

---

[9]There is a very useful generalization of this relation to free resolutions of coherent sheaves [222].

compact subgroup of $G_{\mathbb{C}}$. This means that the diffeomorphisms g_α that generate the G action via $g_\alpha : Z_\rho \rightarrow e^{i\theta Q_\rho^\alpha} Z_\rho$ are symplectomorphisms. In fact, they are Hamiltonian, meaning that the vector field corresponding to an infinitesimal g_α transformation

$$v^\alpha = i \sum_\rho Q_\rho^\alpha \left(\frac{\partial}{\partial Z_\rho} - \frac{\partial}{\partial \overline{Z}_\rho} \right) \tag{5.3.85}$$

can be derived from a G-invariant function[10]

$$\mu^\alpha = \sum_\rho Q_\rho^\alpha |Z_\rho|^2 - 2r^\alpha . \tag{5.3.86}$$

The r^α are real constants valued in the dual of the Lie algebra of G, and tracing through the relations between $G_{\mathbb{C}}$ and $H^2(V, \mathbb{R})$ described above, a choice of the r^α is equivalent to a choice of a class $\omega = r^\alpha \eta_\alpha \in H^2(V, \mathbb{R})$. The symplectic quotient is then defined as

$$S(r) = \mu^{-1}(0)/G . \tag{5.3.87}$$

When G acts freely and properly on $\mu^{-1}(0)$, and $\mu^{-1}(0)$ is a smooth submanifold of Y, then $S(r)$ defines the symplectic quotient, in this case a Kähler quotient with Kähler class ω. If the defining data arises from a smooth projective toric variety X, then there exists a choice of r^α such that $S(r)$ is diffeomorphic to V [218]; if X is projective and simplicial, then the same result holds, except that $S(r)$ and V are isomorphic as orbifolds. Now we have V realized explicitly as a Kähler manifold with Kähler class ω. In fact, the ω that lead to $S(r) \simeq_{\text{diffeo}} V$ span the Kähler cone of V, $K(V) \subset H^2(V, \mathbb{R})$. In the smooth case the points in $K(V)$ are in 1:1 correspondence with Kähler classes realized by smooth Kähler metrics on V in the fixed complex structure.

Exercise 5.17 This exercise shows that the condition for X to be projective in order to obtain isomorphic symplectic and holomorphic quotient constructions is not necessary, nor is it trivial. To show that it is not necessary, consider the simplicial fan

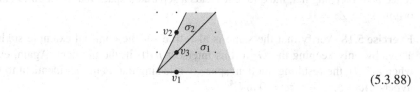

$$\tag{5.3.88}$$

---

[10]More details on the symplectic quotient construction can be found in [102, 290].

Use the holomorphic quotient construction to show that this defines the toric variety which is the total space of the line bundle $\mathcal{O}(-2) \rightarrow \mathbb{P}^1$. Now verify that the symplectic quotient leads to the same space for an appropriately chosen sign of r. When you change the sign of r you should find the symplectic quotient construction of $\mathbb{C}^2/\mathbb{Z}_2$—this is a first glimpse of the secondary fan structure discussed below.

To show that the condition is non-trivial, consider the simplicial but incomplete fan

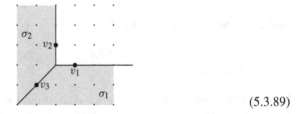

$$(5.3.89)$$

Use the holomorphic quotient construction to show that the toric variety is $\mathbb{P}^2$ with one point removed; equivalently, it is the total space of $\mathcal{O}(1) \rightarrow \mathbb{P}^1$. Now verify that the symplectic quotient yields $\mathbb{P}^2$ for one sign of r and is empty for the opposite sign. Thus, it adds the extra point back into the space!   □

A Glimpse of GIT

Since Y is stratified into $G_\mathbb{C}$ orbits, we can ask which orbits contribute to points in $S(r)$, or, equivalently, which $G_\mathbb{C}$ orbits contribute G-orbits to $\mu^{-1}(0)$. This is a basic ingredient in the Geometric Invariant Theory approach to quotients like $Y/G_\mathbb{C}$: the choice of the moment map, i.e. the parameters r^α, is a "stability condition" that can be used to eliminate bad orbits. As a simple example, consider $Y = \mathbb{C}^2$, and a $\mathbb{C}^*$ action with charges $+1$ and -1 for Z_1 and Z_2, respectively. We then have

$$\mu = |Z_1|^2 - |Z_2|^2 - 2r . \qquad (5.3.90)$$

We can now impose a condition on the orbits of the $\mathbb{C}^*$ action: we keep the orbits that intersect the zero set $\mu(0)$. In this fashion, even when $r = 0$, so that the resulting space is in fact singular, the quotient yields a separated space that is singular at the origin $Z_1 = Z_2 = 0$.

Exercise 5.18 Verify that the same is also true for the conifold example studied above by only keeping the $G_\mathbb{C}$ orbits intersect $\mu(0)$ in the quotient. Again, even when $r = 0$, the resulting quotient space, while singular, is in fact identical to the hypersurface $z_1 z_2 - z_3 z_4 = 0$ in $\mathbb{C}^4$.   □

The Secondary Fan and Phases

We now have a natural question before us: when ω is in the Kähler cone K$(V) \subset$ $H^2(V, \mathbb{R})$, the symplectic quotient is diffeomorphic to the original toric variety, but what if ω is not in K(V)? The answer is described by the secondary fan associated to the toric variety X.

First, we note that in general $S(r)$ may be empty. Reversing the logic of the moment map (5.3.86), it is clear that the set of r^α where $S(r) \neq \emptyset$ is a full-dimensional cone in $H^2(V, \mathbb{R})$ generated by the charge vectors Q_ρ and denoted by $K_{cl}(V)$. A pretty argument [311] shows that $K_{cl}(V)$ is a strongly convex cone whenever V is compact. If the Q_ρ generate all of $H^2(V, \mathbb{R})$, then there is a choice of non-negative integers m^ρ, not all simultaneously zero, such that $\sum_\rho m^\rho Q_\rho = 0$. This means $\prod_\rho Z_\rho^{m^\rho}$ is a $G_{\mathbb{C}}$-invariant non-constant function that descends to a non-constant holomorphic function on V, and therefore V cannot be compact.

The cone $K_{cl}(V)$ has a further decomposition into a fan (this is the GKZ decomposition, and the resulting fan is the secondary fan). One of the cones in $K_{cl}(V)$ is the (closure of) the Kähler cone $K(V)$, while the remaining cones are associated to other toric varieties obtained by either taking a different triangulation of the points in $\Sigma(1)$—the flop of the resolved conifold is the simplest example—or by removing some of the rays in $\Sigma(1)$. There is a nice physics terminology for these different varieties—they are the "phases" of the gauge theory. The terminology is apt, since at least in classical geometry the boundaries between phases are singular geometries.

We illustrate this in a simple example, and let the reader tackle a more entertaining one in the exercise below. Our simple example is based on $\mathbb{F}_1$, which is also isomorphic to the first del Pezzo surface dP$_1$—a blow up of $\mathbb{P}^2$ at one point. The complexified gauge group is $(\mathbb{C}^*)^2$ and has two generators g_1 and g_2 with charges

$$Q = \begin{pmatrix} 1 & 0 & 1 & 1 \\ 1 & 1 & 0 & 0 \end{pmatrix} \tag{5.3.91}$$

we see that setting $\mu(r) = 0$ leads to

$$|Z_1|^2 + |Z_3|^2 + |Z_4|^2 = 2r^1, \qquad |Z_1|^2 + |Z_2|^2 = 2r^2, \tag{5.3.92}$$

and their difference yields

$$|Z_3|^2 + |Z_4|^2 - |Z_2|^2 = 2r^1 - 2r^2. \tag{5.3.93}$$

We see that K_{cl} is the positive orthant in (r^1, r^2). From the moment map we see that for any $r \in K_{cl}$, the exceptional set includes the loci $Z_1 = Z_3 = Z_4 = 0$ and $Z_1 = Z_2 = 0$. We abbreviate this as $F \supset \{134\}$ and $F \supset \{12\}$. From (5.3.93) we see that if $r^1 > r^2$ then $F \supset \{34\}$, while if $r^1 < r^2$, then $F \supset \{2\}$. Thus, in the former case, we have $F = \{12\} \cup \{34\}$, and therefore the quotient yields $\mathbb{F}_1$. In the latter case the exceptional set is $\{134\} \cup \{2\}$, and this too has a nice interpretation:

$g_2 \subset G_{\mathbb{C}}$ is fixed by setting $Z_2 = 1$, and we are left with $X = \{\mathbb{C}^3 \setminus \{0\}\}/\mathbb{C}^*$. This is exactly the "blow-down" space $\mathbb{P}^2$ obtained by shrinking the "exceptional divisor" $\{Z_2 = 0\}$ of $\mathbb{F}_1$ to zero size and dropping the corresponding generator of $\Sigma(1)$. Thus, the secondary fan takes the form

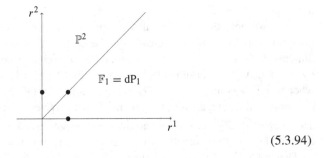

$$(5.3.94)$$

It is also interesting to consider the "phase" boundaries. When $r^1 = 0$, the Z_ρ are restricted to the locus $Z_1 = Z_3 = Z_4 = 0$, and the action of g_1 preserves the locus. In standard physics terminology we say that g_1 generates an unbroken $\mathbb{C}^* \subset G_{\mathbb{C}}$ subgroup. Similarly, g_2 generates an unbroken $\mathbb{C}^* \subset G_{\mathbb{C}}$ when $r^2 = 0$. The inner phase boundary is slightly different: when $r^1 = r^2$ there is a point $Z_3 = Z_4 = Z_2 = 0$ with an unbroken subgroup is generated by $g_1 g_2^{-1}$, but note that there are other points in $\mu^{-1}(0)$ with a trivial stabilizer in $G_{\mathbb{C}}$.

A convenient way to depict the different phases is by marking the lattice points for the dropped rays by $\times$ as follows (anticipating some developments we also indicate a polytope that supports the fan by a dashed line):

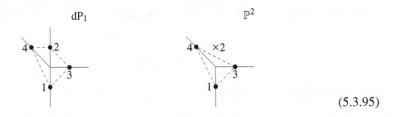

$$(5.3.95)$$

It is easy to carry out the same procedure for $\mathbb{F}_p$ with $p > 1$. The phases are now separated by the line $r^1 - p r^2 = 0$, so that the secondary fan takes the form

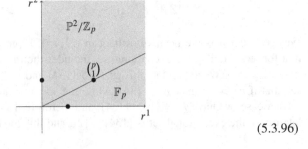

$$(5.3.96)$$

The reader can check that $\mathbb{P}^2/\mathbb{Z}_p$ is the quotient of $\mathbb{P}^2$ by the $\mathbb{Z}_p$ action

$$[Z_1, Z_3, Z_4] \simeq [Z_1, gZ_3, gZ_4], \tag{5.3.97}$$

where g is a p-th root of unity. The point $[1, 0, 0]$ is a $\mathbb{Z}_p$ orbifold singularity.

Exercise 5.19 The reader who really wants to appreciate the wonders of the secondary fan is encouraged to carry out a similar study for dP2—a blow up $\mathbb{P}^2$ at two points.[11] The fan for dP2 is

$$(5.3.98)$$

Show that a basis for the charges of the $G_\mathbb{C}$ action is

$$Q = \begin{pmatrix} 1 & 0 & 1 & 0 & 1 \\ 1 & 0 & 0 & 1 & 0 \\ 0 & 1 & 0 & 0 & 1 \end{pmatrix}. \tag{5.3.99}$$

Setting the moment map to zero leads to

$$|Z_1|^2 + |Z_3|^2 + |Z_5|^2 = 2r^1, \quad |Z_1|^2 + |Z_4|^2 = 2r^2, \quad |Z_2|^2 + |Z_5|^2 = 2r^3. \tag{5.3.100}$$

This immediately shows that K_{cl} is again the positive orthant and the exceptional set contains

$$F \supset \{135\} \cup \{14\} \cup \{25\}. \tag{5.3.101}$$

To study the phase boundaries more systematically, we can look for subsets $A \subset \{1, 2, \ldots, 5\}$ such that if we set $Z_a = 0$ for all $a \in A$ the stabilizer of the G action is at least a $U(1)$. The remaining Z_i then generate a subcone in K_{cl} that is a phase boundary. Show that the non-trivial phase boundaries (i.e. ones that are not the faces of K_{cl}) are:

$$A_1 = \{3, 4, 5\} \implies r^1 = r^2, \qquad A_2 = \{1, 2, 3\} \implies r^1 = r^3,$$

$$A_3 = \{2, 3, 4\} \implies r^2 = r^2 + r^3. \tag{5.3.102}$$

---

[11] There is just one more toric del Pezzo surface, dP3.

Now show that this leads to five phases:

I	(dP$_2$) $F = \{14\} \cup \{25\} \cup \{24\} \cup \{13\} \cup \{35\}$	
II	(dP$_1$) $F = \{14\} \cup \{35\} \cup \{2\}$	
III	($\mathbb{P}^1 \times \mathbb{P}^1$) $F = \{14\} \cup \{25\} \cup \{3\}$	
IV	(dP$_1$) $F = \{13\} \cup \{25\} \cup \{4\}$	
V	($\mathbb{P}^2$) $F = \{135\}$	

$$(5.3.103)$$

Finally, to describe the combinatorics of the secondary fan intersect it with the hypersurface $r^1 + r^2 + r^3 = 1$, and show that the phase diagram is

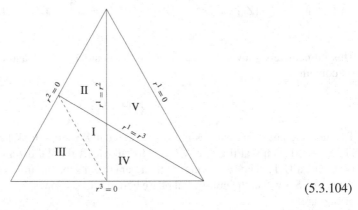

$$(5.3.104)$$

The dashed line indicates the intersection with the hypersurface $r^1 = r^2 + r^3$.   □

Ample Line Bundles and Fano Varieties

As we discuss in the geometry appendix, a compact Kähler manifold M is not necessarily projective. The Kodaira embedding theorem states that M is projective if and only if it admits a rational Kähler class. This means that there exists a line bundle $\mathcal{L} \to M$ with $c_1(\mathcal{L})$ an integer multiple of the rational Kähler class, and this guarantees that a tensor power $\mathcal{L}^{\otimes k} \to M$ has "enough" global sections to construct an explicit embedding $M \hookrightarrow \mathbb{P}^n$. In this case we say that $\mathcal{L}$ is an ample line bundle and $\mathcal{L}^{\otimes k}$ is a very ample line bundle.

We say that M is a Fano manifold if and only if the anti-canonical bundle is ample (by the line bundle—divisor correspondence, we can also speak of an ample divisor). Fano manifolds are very tightly constrained: for example, $H^k(M, \mathcal{O}_M) = 0$ for $k > 0$, and $\mathrm{Pic}(M) = H^2(B, \mathbb{Z})$; in addition, they are classified in dimension $d \le 3$ and admit powerful criteria for evaluating positivity properties of vector bundles [282]. A smooth Fano surface (i.e. a manifold of complex dimension 2) is isomorphic to one of the following: $\mathbb{P}^2$, $dP_0 = \mathbb{P}^1 \times \mathbb{P}^1$, or a del Pezzo surface dP_n with $1 \le n \le 8$—a blow up of $\mathbb{P}^2$ at n generic points.

There is a concrete characterization of Fano smooth toric varieties: a complete smooth toric variety X is Fano if and only if the anti-canonical class is ample, or, equivalently, the anti-canonical divisor is Cartier, and the corresponding class in $H^2(X, \mathbb{R})$, which we denote by $R = \sum_\rho \xi_\rho$, is strictly contained in the Kähler cone of X. This definition can be extended to the more general setting of simplicial complete toric varieties.

Exercise 5.20 Check that $\mathbb{F}_p$ is Fano if and only if $p < 2$. Check that for $\mathbb{F}_2$ the canonical class is contained in the boundary of the Kähler cone. This is an example of a slightly weaker notion of a NEF Fano variety. □

5.3.7 Polytopes

We met polytopes several times in our (0,2) explorations: in our analysis of (0,2) Landau-Ginzburg theories they arise naturally as Newton polytopes that describe the monomials in the J_A potentials; other polytopes can be obtained by intersecting a polyhedral cone with a hyperplane, and we also saw that certain toric fans are naturally associated to polytopes. We generalize these notions in two ways: the polytopes we discuss will encode both superpotential couplings and toric geometry, and there will be a rather precise duality between the two—this is the origin of Batyrev mirror symmetry. The geometry of polytopes is discussed in the classic work [214].

A polytope is a closed convex subset of $\mathbb{R}^d$ that is the convex hull of a finite collection of points—the vertices of the polytope. We will always take the polytope to be full-dimensional. For example, let Δ be the polytope

(5.3.105)

It is the convex hull of the three vertices, meaning

$$\Delta = \{a_i v_i \mid a_i \geq 0, \ a_1 + a_2 + a_3 = 1\} \subset \mathbb{R}^2 . \tag{5.3.106}$$

We say that Δ is a lattice polytope if all of its vertices belong to a lattice. Every polytope can be obtained as the intersection of a collection of half-spaces, and the bounding hyperplanes are known as the supporting hyperplanes. A face of a polytope is the intersection of Δ with a collection of its supporting hyperplanes. Every face of a polytope is itself a polytope.

A lattice polytope $\Delta \subset M_{\mathbb{R}}$ defines a family of polynomials in $(\mathbb{C}^*)^d$:

$$P_\Delta = \sum_{m \in \Delta \cap M} \alpha_m \prod_i t_i^{m_i} . \tag{5.3.107}$$

The α_m are complex coefficients that parameterize the family; their number is the number of lattice points in Δ, which we denote by $|\Delta \cap M|$. There is a corresponding (possibly singular) subvariety $X' = \{P_\Delta = 0\} \subset (\mathbb{C}^*)^d$.

Polytopes obey a combinatorial duality $\Delta \leftrightarrow \Delta^\circ$ with a dimension-reversing correspondence between the faces: that is, for every face $F \subset \Delta$ there is a face $F^\circ \subset \Delta^\circ$ with $\dim(F) + \dim(F^\circ) = d$; moreover, if $F_1 \subset F_2 \subset \Delta$, then $F_2^\circ \subset F_1^\circ \subset \Delta^\circ$. The dual polytope is defined in terms of the natural pairing between $N_{\mathbb{R}}$ and $M_{\mathbb{R}}$:

$$\Delta^\circ = \{x \in N_{\mathbb{R}} \mid \langle m, x \rangle \geq -1 \quad \text{for all} \quad m \in \Delta\} . \tag{5.3.108}$$

The facets, i.e. co-dimension 1 faces, of Δ° are dual to the vertices of Δ, so for every vertex $v \in \Delta$ we obtain a supporting hyperplane defined by the equation $\langle v, x \rangle = -1$. Clearly $(\Delta^\circ)^\circ = \Delta$, but Δ° need not be a lattice polytope.

Here is an example of a dual pair in $d = 2$ where both Δ and Δ° are lattice polytopes.

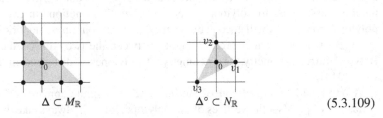

$$\Delta \subset M_{\mathbb{R}} \qquad\qquad \Delta^\circ \subset N_{\mathbb{R}} \tag{5.3.109}$$

Exercise 5.21 Construct the dual for Δ

$$\tag{5.3.110}$$

Since Δ has three vertices, $\Delta°$ will have three facets, and finding these is the most efficient way to build $\Delta°$. You should discover that $\Delta°$ is not an integral polytope. □

Our interest in this duality arises from the following observation: quite generally, Δ defines a complete toric variety V with fan $\Sigma \subset N_\mathbb{R}$. This variety is projective, and there is a compactification $\overline{\{P_\Delta = 0\}}$ of the subvariety defined above to a hypersurface $X \subset V$. There is a more intrinsic point of view on this that we will not explore in detail: the polytope Δ defines not only the projective variety V but also an ample divisor that yields the Kodaira embedding; the polynomial P_Δ is then a generic section of the corresponding line bundle, and its vanishing defines the hypersurface X. Every projective toric variety arises from a suitably chosen polytope. Details are given in the references.

When Δ contains the origin, the fan Σ is constructed by taking cones over the faces of $\Delta°$. In particular, the one-dimensional cones in $\Sigma(1)$ are generated by the vertices of $\Delta°$. When the V so constructed is simplicial, so that we can describe it through the holomorphic quotient and homogeneous coordinate ring as above, the global form for the hypersurface is

$$P_\Delta = \sum_{m \in \Delta \cap M} \alpha_m \prod_{\rho \in \Sigma(1)} Z_\rho^{\langle m, \rho \rangle + 1} . \tag{5.3.111}$$

As the following exercise shows, P_Δ is $G_\mathbb{C}$-equivariant, so that $P_\Delta = 0$ defines a hypersurface in V.

Exercise 5.22 Prove that P_Δ is $G_\mathbb{C}$ equivariant. That is, if $g \in G_\mathbb{C}$ acts as $g \cdot \prod_\rho Z_\rho \mapsto \tau \prod_\rho Z_\rho$ for some $\tau \in \mathbb{C}^*$, then $P(g \cdot Z) = \tau P(Z)$. □

The Torus in $\mathbb{P}^2$

Returning to the example in (5.3.109), we see that it describes the vanishing of a generic cubic in $\mathbb{P}^2$. This is a famous manifold: T^2. It is worthwhile to discuss it a little bit more explicitly. We will not work with the most generic form of P (which would have 10 monomials), but rather just keep the monomials corresponding to the origin and the vertices of Δ:

$$P = \alpha_1 Z_1^3 + \alpha_2 Z_2^3 + \alpha_3 Z_3^3 - 3\alpha_4 Z_1 Z_2 Z_3 . \tag{5.3.112}$$

The hypersurface $\{P = 0\} \subset \mathbb{P}^2$ will be singular if there is a point $p \in \mathbb{P}^2$ where P and dP vanish simultaneously in $\mathbb{P}^2$. That in turn can happen if and only if $Z_i \frac{\partial P}{\partial Z_i} = 0$ for all i for some p, which implies that the parameters lie on the locus singular locus $\mathcal{Z}_B$ defined as the vanishing locus

$$\alpha_1 \alpha_2 \alpha_3 - \alpha_4^3 = 0 . \tag{5.3.113}$$

The reader can compare this to the singular locus in (0,2) LG theory that we examined in Exercise 3.8—it is a very similar structure. As there, by changing the coefficients of the defining polynomial we can obtain a variety with a different complex structure.

How do we know that this is the torus? There are quite a few classic ways [210] to this conclusions, but let us discuss a perspective that will readily generalize to the more elaborate examples we will study below.

First, just like the tangent bundle of the toric variety T_V in (5.3.80), the tangent bundle T_X of the hypersurface can also be described by a complex of sheaves:

$$0 \longrightarrow \mathcal{O}_X \xrightarrow{\ E_{22}\ } \mathcal{O}_{\mathbb{P}^2}(H)^{\oplus 3}\big|_X \xrightarrow{\ J_{22}\ } \mathcal{O}_{\mathbb{P}^2}(3H)\big|_X \longrightarrow 0 \ , \tag{5.3.114}$$

where H is the hyperplane class, i.e. a toric divisor, E is as in (5.3.80), and the map J is

$$J : (S_1, S_2, S_3) \mapsto S_1 \frac{\partial P}{\partial Z_1} + S_2 \frac{\partial P}{\partial Z_2} + S_3 \frac{\partial P}{\partial Z_3} \ . \tag{5.3.115}$$

This is a complex since

$$J_{22} E_{22}(\lambda) = 3\lambda P \ , \tag{5.3.116}$$

and since we restrict the sheaves to X, the right-hand-side is zero. The tangent sheaf T_X is the cohomology of the complex: its sections are described as $\ker J_{22}/ \operatorname{im} E_{22}$. This means we have a nice way to compute the Chern classes of T_X: if $\iota : X \hookrightarrow V$ is the embedding, then

$$c(T_X) = \iota^* \frac{c(\mathcal{O}(H))^3}{c(\mathcal{O}(3H))} = 1 \qquad \Longrightarrow \quad c_1(T_X) = 0 \ . \tag{5.3.117}$$

When smooth, X is therefore a compact Riemann surface with $c_1(T_X) = 0$—it must be a torus.

Exercise 5.23 Generalize the discussion of the Euler sequence above to derive the formula for $c(T_X)$ by constructing a short exact sequence that has T_X as one of the terms and the subsheaf defined by $\ker J_{22}$ as another term.   □

The simple result for the torus generalizes in a straightforward way to the tangent bundle T_X for any smooth hypersurface $X = \{P = 0\} \subset V$. Let $[X]$ denote the divisor class. Then we have the complex

$$0 \longrightarrow H_2(V, \mathbb{Z}) \otimes \mathcal{O}_X \xrightarrow{\ E_{22}\ } \oplus_\rho \mathcal{O}_V(D_\rho)|_X \xrightarrow{\ J_{22}\ } \mathcal{O}_V([X])|_X \longrightarrow 0 \ , \tag{5.3.118}$$

where E_{22} is the map we met in the construction of T_V above (5.3.81), and

$$J_{22\rho} = \frac{\partial P}{\partial Z^\rho} \ . \tag{5.3.119}$$

As in the case of the torus, this is a complex because P is $G_\mathbb{C}$-equivariant, so that $J_{22}E_{22}$ is proportional to P and vanishes upon restricting to X. Its cohomology, which is concentrated in the middle term then describes the tangent bundle T_X: the space of sections of T_X is isomorphic to $\ker J_{22}/\operatorname{im} E_{22}$.

Generalizing the torus example, we compute

$$c(T_X) = \imath^* \frac{\prod_\rho c(\mathcal{O}(D_\rho))}{c(\mathcal{O}([X]))} = 1 + \oplus_\rho c_1(\mathcal{O}(D_\rho)) - c_1(\mathcal{O}([X])) + \dots,$$

(5.3.120)

so that if we choose $[X] = -K_V = \sum_\rho D_\rho$, it follows that $c_1(T_X) = 0$.

Just as deformations of the Euler sequence provide a simple way to deform the tangent bundle of a toric variety, so here, the deformations of the maps in the complex describe a class of deformations of T_X. These are again obviously unobstructed, and we will refer to them as the "monadic" deformations of T_X. There is one important difference between deforming the Euler sequence and the complex: in the latter case the deformed maps must satisfy the non-trivial constraint $JE|_X = 0$. All of these features will be reflected in the (0,2) linear sigma model.

Calabi-Yau Hypersurfaces and Batyrev Mirror Pairs

We now review the Batyrev construction of Calabi-Yau hypersurfaces in Fano toric varieties, mostly following [120]. To set the stage, we observe that the cubic in $\mathbb{P}^2$ is the zero locus of a generic section of the anti-canonical bundle. Suppose we take, more generally, a hypersurface defined by the vanishing of a global section of $\mathcal{O}(-K_V)$ for some smooth projective toric variety. If the result is a smooth hypersurface, then by the adjunction formula [210] it is guaranteed to be a Kähler manifold with trivial canonical bundle.[12]

In order for the idea to have a chance, the anti-canonical bundle on V should have holomorphic sections. A sufficient condition for that is to require V to be a Fano toric variety. Such V must arise from a polytope Δ, and Batyrev showed that this will be so if and only if Δ is a reflexive polytope: a lattice polytope is reflexive if and only if it contains the origin, and the dual polytope Δ° is a lattice polytope. It follows that reflexive polytopes come in pairs.

---

[12]The adjunction formula states that for a hypersurface in a complex manifold $X \subset V$, the canonical bundle on X is the pull-back of $\mathcal{O}(K_V) \otimes \mathcal{O}([V])$, where $[V]$ is the divisor class of the hypersurface.

A Few Remarks on Reflexive Polytopes

A reflexive polytope can also be characterized as follows: a lattice polytope is reflexive if and only if it has a unique interior lattice point that lies in an adjacent hyperplane to any facet. There are $1, 16, 4319$, and $473, 800, 776$ reflexive polytopes of dimensions, respectively $d = 1, 2, 3, 4$. More generally, the number of reflexive polytopes is finite for any fixed dimension d; on the other hand, it can also be shown that any lattice polytope is lattice equivalent to a face of some reflexive polytope [219]. Most reflexive polytopes do not lead to smooth Fano varieties: a reflexive polytope Δ corresponds to a smooth Fano variety if and only if the vertices in each facet of Δ° form a basis for the lattice N. These form a much more restricted set: there are $1, 5, 18, 124, 866$ smooth toric Fano manifolds in dimensions $d = 1, 2, 3, 4, 5$; see [273] and references therein.

There are excellent computational resources for examining the properties of reflexive polytopes. The classic is the PALP stand-alone package [99, 275]; it has since been incorporated as part of the SageMath mathematics software system [362].

Exercise 5.24 Construct the reflexive polytope pairs for each of the smooth toric Fano surfaces $\mathbb{P}^2$, $\mathbb{P}^1 \times \mathbb{P}^1$, $dP_{1,2,3}$.   □

Resolutions and Mirror Pairs

The normal fan of a generic reflexive polytope does not lead to a smooth variety. To see the trouble, we take a look at example (5.3.109), but now reversing the role of the two polytopes:

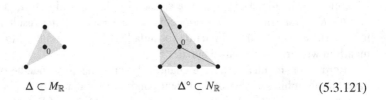

$$\Delta \subset M_\mathbb{R} \qquad\qquad \Delta^\circ \subset N_\mathbb{R} \qquad\qquad (5.3.121)$$

We included the one-dimensional rays of the normal fan Σ in Δ°, and it should be evident that V is singular: indeed, the reader can show without too much trouble that it is an orbifold $\mathbb{P}^2/\mathbb{Z}_3$ with $\mathbb{C}^* \times \mathbb{Z}_3$ action

$$g(t, \zeta_3) : [Z_1, Z_2, Z_3] \mapsto [\zeta_3 t Z_1, \zeta_3^2 t Z_2, t Z_3] . \qquad\qquad (5.3.122)$$

where ζ_3 is a third root of unity. The points $[1, 0, 0]$, $[0, 1, 0]$, and $[0, 0, 1]$ are all $\mathbb{Z}_3$ orbifold singularities. The singularities are mild: they are examples of Gorenstein canonical singularities. The toric diagram suggests a simple way to fix the trouble: we can refine the fan to Σ' by introducing 1-dimensional rays for all non-zero lattice points in Δ°: this is a toric resolution of singularities $f : V' \to V$, and the key point is that it preserves the canonical class of the variety: $K_{V'} = f^*(K_V)$. Such a resolution is called "crepant."[13] The resulting smooth variety is NEF Fano, and the beauty is that the anti-canonical class has zero intersection with the exceptional divisors corresponding to the added rays: we can introduce them so that V' is smooth, but the introduction does not affect the properties of the hypersurface $\{P_\Delta = 0\}$. This can be said another way: the variety V has singularities at three points, but a zero locus of a generic P_Δ will miss singularities, and the resulting space will be smooth. The fact that singularities have a crepant resolution then implies that the hypersurface will be Calabi-Yau whether we work in V or V'.

The two-dimensional example is misleading in two ways as guide to what happens in $d > 2$. First, for $d > 2$ the normal fan will in general not be simplicial: we must choose a triangulation to obtain a simplicial fan. This leads to the definition of a projective subdivision: this is a fan Σ that refines the normal fan, has $\Sigma(1) \subset \Delta^\circ \cap N \backslash 0$, and is chosen so that V_Σ is projective and simplicial. Just as in our simple example, a projective subdivision leads to a crepant resolution of singularities of the original variety. A projective-subdivision is very much non-unique—their number grows exponentially in the number of lattice points of Δ° [85], and even an enumeration is out of the question. Fortunately, they have a nice interpretation as phases in a secondary fan and the associated gauge theory; we will have more to say on this below.

A projective subdivision is maximal if $\Sigma(1)$ consists of all non-zero lattice points in Δ°. Following [276], we also define a minimal projective subdivision: in this case $\Sigma(1)$ consists of all non-zero lattice points of Δ° that are not contained in the interior of a facet. A maximal projective subdivision is a refinement of a minimal projective subdivision, and, correspondingly, there is a crepant resolution $f : V_{\max} \to V_{\min}$. The two only differ in the singular points corresponding to the non-empty facets of Δ°. As in the example, a generic hypersurface in $V_{\min}$ will miss these singular points, and the corresponding exceptional divisors in V' have zero intersection with the anti-canonical divisor $-K_{V'}$.

The second way in which the $d = 2$ example is misleading, is that in general there will not exist a maximal projective subdivision that leads to a smooth toric variety. Nevertheless, it is possible to choose a maximal projective subdivision so that the toric variety has terminal Gorenstein orbifold singularities. Many results that hold for smooth toric varieties continue to hold in this case as well [120]. One particularly nice property is that the singular locus of such a singularity has codimension at least 4. Thus, for generic choice of defining coefficients of the defining polynomial P_Δ, the Batyrev construction leads to smooth Calabi-Yau hypersurfaces in $d = 4$ Fano varieties.

---

[13] By the general toric results described above, every toric manifold has a toric resolution of singularities, but generally, the resolution introduces a discrepancy: $K_{V'} \neq f^*(K_V)$. For more information on singularity terminology the reader can consult [332].

The Total Space of the Canonical Bundle

Given a toric variety V, any vector bundle $\mathcal{E} \rightarrow V$ where $\mathcal{E}$ is a sum of toric line bundles is itself a toric variety. In particular, we can set $V_+ = \text{tot}\{\mathcal{O}(K_V) \rightarrow V\}$, the total space of the canonical bundle. V_+ will arise in a natural way in the linear sigma models we will study below. When V is Fano with a fan Σ given by a projective subdivision based on the lattice points in $\Delta^\circ \cap M$, it is easy to obtain the fan for V_+. Let $\Delta_+^\circ \subset N_{+\mathbb{R}} = N_\mathbb{R} \times \mathbb{R}$ be the polytope $\Delta_\circ^+ = \{\Delta\} \times \{1\}$, i.e. it is simply the original polytope lifted to "height one" in a space of one dimension higher. The one-dimensional cones for the fan Σ_+ are now $\{\Sigma(1)\} \times \{1\} \cup \{0\} \times \{1\}$. More generally, for any k-dimensional cone $\tau \prec \Sigma$, Σ_+ has a $k+1$-dimensional cone $\tau_+ \prec \Sigma_+$ generated by $\tau \times \{1\}$ and $\{0\} \times \{1\}$. This is surely much more confusing in words than in pictures, so we give the construction for $V = \mathbb{P}^2$ and $V_+ = \text{tot}\{\mathcal{O}(-2) \rightarrow \mathbb{P}^2\}$. The intersection of Σ_+ with the height 1 hyperplane is the polytope Δ°, which in this case is

$$w_0 = \begin{pmatrix} 0 \\ 0 \\ 0 \\ 1 \end{pmatrix}, \quad w_1 = \begin{pmatrix} 1 \\ 0 \\ 0 \\ 1 \end{pmatrix}, \quad w_2 = \begin{pmatrix} 0 \\ 1 \\ 0 \\ 1 \end{pmatrix}, \quad w_3 = \begin{pmatrix} -1 \\ -1 \\ 0 \\ 1 \end{pmatrix}.$$

$$(5.3.123)$$

Exercise 5.25 Consider the holomorphic quotient construction for the V_+ variety for any Gorenstein Fano variety V. Let Z_0 be the homogeneous coordinate corresponding to $\{0\} \subset \Delta^\circ$. Let ρ denote the one-dimensional cones in the fan Σ for V, and, by abuse of notation, all but one of the one-dimensional cones of Σ_+. Show that the exceptional set of V_+ is identical to that of V, as is the group $G_\mathbb{C}$. Show that the $G_\mathbb{C}$ action on Z_0 is inverse to that of the $G_\mathbb{C}$ action on the monomial $\prod_\rho Z_\rho$. $\square$

Having climbed this wonderful geometric edifice, we can finally appreciate the dual aspect of the construction: we can run the reflexive polytope both ways to produce a pair of Calabi-Yau manifolds.

1. X is the hypersurface $\{P_\Delta = 0\} \subset V$, where V is a maximal projective subdivision supported on Δ°; the Kähler class on X is pulled back from the Kähler class on V, which is linear in the r^α parameters of the moment map, and by varying the coefficients in P_Δ, we can change the complex structure on X.
2. X° is the hypersurface $\{P_{\Delta^\circ} = 0\} \subset V^\circ$, where V° is a maximal projective subdivision supported on Δ. Now the Kähler class on X° is pulled back from V°, and the coefficients of P_{Δ° encode variations of the complex structure of X°.

As we will see below, there is a map between the "toric" Kähler deformations of X and the "polynomial" complex structure deformations of X°, and this map allows for precise and direct tests of mirror symmetry. More generally, the "stringy Hodge numbers" $h^{1,1}(X)$ and $h^{1,d-2}(X^\circ)$, which equal to, respectively, the dimensions of the moduli spaces $\mathcal{M}_{cK}(X)$ and $\mathcal{M}_{c\text{-}x}(X)$, are determined by combinatorics of Δ and Δ° and satisfy $h^{1,1}(X) = h^{1,d-2}(X^\circ)$.[14] This is just the first indication that X and X° may be mirror symmetric, and there is by now very strong evidence that these geometries lead to pairs of isomorphic mirror-symmetric SCFTs. The evidence is reviewed in [40, 120, 237, 396].

Whew. We reviewed, in some detail, what is perhaps the most standard construction of a large class of Calabi-Yau manifolds. This is by no means the only one, nor is it the only one that uses toric varieties. The reader can find discussions of other constructions in [35, 53, 55, 71, 112, 120, 241, 272]. Recently, a new construction method has been proposed in [16]. It would be interesting to find whether these new models have a simple linear sigma model construction; some steps in this direction, as well as a search for mirrors were taken in [72].

The machinery that we developed will be invaluable in unraveling the physics of linear sigma models; on the other hand, the physics will lead to a profound perspective on the geometries. We turn to that structure next.

5.4 Phases of (2,2) Linear Sigma Models

The first application of our new-found knowledge of toric geometry will be to linear sigma models with (2,2) supersymmetry. To construct these we will specialize the general (0,2) Lagrangian so that it has (2,2) supersymmetry.

5.4.1 The (2,2) Lagrangian

The most economical way to present the theory is through a (2,2) superspace approach [388]. There are (2,2) chiral matter multiplets Φ_{22}^i, $i = 1, \ldots, n$ and (2,2) real vector multiplets $\mathcal{V}_{22}$, and they are acted upon by four superspace derivatives: $\mathcal{D}$, $\overline{\mathcal{D}}$, and their left-moving siblings $\mathcal{D}'$ and $\overline{\mathcal{D}}'$. As in our (0,2) discussion, we define the twisted superspace derivatives $\mathcal{D}_v$, $\overline{\mathcal{D}}_v$, $\mathcal{D}'_v$, and $\overline{\mathcal{D}}'_v$. Chiral multiplets then satisfy $\overline{\mathcal{D}}'_v \Phi_{22} = 0$ and $\overline{\mathcal{D}}_v \Phi_{22} = 0$. The minimal and superpotential couplings

---

[14]The adjective "stringy" is meaningful when X fails to be smooth but is instead a Gorenstein orbifold; it can be shown in that setting that the deformation theory is still governed by the dimensions of sheaf cohomology groups that generalize $H^1(X, T_X^*)$ and $H^1(X, T_X)$, and their dimensions yield the stringy Hodge numbers. In simple cases these results can be understood in terms of an orbifold SCFT, but the result is more general.

can be written as

$$S_{\min,22} = \frac{1}{2\pi} \int d^2z \mathcal{D}\overline{\mathcal{D}}\overline{\mathcal{D}}'\mathcal{D}' \sum_{i=1}^{n} \frac{1}{4}\overline{\Phi}_{22}^i \exp[2Q_i^\alpha V_{22\alpha}]\Phi_{22}^i ,$$

$$S_{W,22} = \frac{1}{2\pi} \int d^2z \frac{1}{2}\mathcal{D}\mathcal{D}'W(\Phi_{22}) + \text{h.c.} \tag{5.4.1}$$

The action is invariant under gauge transformations

$$\Phi_{22}^i \to e^{\Lambda_\alpha Q_i^\alpha}\Phi_{22}^i , \qquad V_\alpha \to V_\alpha - \tfrac{1}{2}(\Lambda_\alpha + \overline{\Lambda}_\alpha) , \tag{5.4.2}$$

where Λ_α is a (2,2) chiral multiplet, provided that $W(\Phi_{22})$ is a gauge-invariant function.

The most remarkable feature of the (2,2) theory compared to the general (0,2) Lagrangian is that the gauge field-strength multiplet is twisted chiral[15]:

$$\mathcal{S}_{22\alpha} = \mathcal{D}'\overline{\mathcal{D}}V_{22\alpha} \tag{5.4.3}$$

It has lowest component a boson σ_α and satisfies $\mathcal{D}'\mathcal{S}_{22\alpha} = \overline{\mathcal{D}}\mathcal{S}_{22\alpha} = 0$. The (2,2) Fayet-Iliopolous coupling resides in a twisted chiral superpotential:

$$S_{\text{Fayet-Iliopoulos},22} = \int d^2z\, \overline{\mathcal{D}}'\mathcal{D} \sum_{\alpha=1}^{n_G} \tau^\alpha \mathcal{S}_{22\alpha} + \text{h.c.} \tag{5.4.4}$$

More generally, the theory has two types of superpotential couplings: the chiral superpotential $W(\Phi)$, and the twisted chiral superpotential $\widetilde{W}(\mathcal{S})$. As long as quantum corrections preserve (2,2) supersymmetry, there is a fundamental non-renormalization theorem: $W(\Phi)$ and $\widetilde{W}(\mathcal{S})$ can only depend on, respectively, chiral and twisted chiral superfields. In particular, the parameters that reside in the chiral superpotential cannot renormalize the twisted chiral superpotential, and vice versa.

We have seen this before in the splitting of moduli of a (2,2) SCFT into $\mathcal{M}_{\text{ac}}$ and $\mathcal{M}_{\text{cc}}$ components; the former corresponds to twisted chiral parameters and the latter to the chiral parameters. Thus, mirror symmetry naturally involves an exchange of chiral and twisted chiral data. T-duality naturally exchanges the two types of multiplets [333], an observation that was pursued in [312] and applied in [236] to mirror symmetry.

---

[15]Twisted chiral multiplets were introduced in [189]. More general (2,2) multiplets are discussed in, for instance, [286].

Returning to the mundane details of the action, after fixing to Wess-Zumino gauge[16] we decompose the (2,2) multiplets into their (0,2) components:

$$\Phi_{22}^i \to \Phi^i , \ \Gamma^i \qquad\qquad \mathcal{S}_{22\alpha} \to \mathcal{S}_\alpha , \ \Upsilon_\alpha . \qquad (5.4.5)$$

The (0,2) chiral multiplet $\mathcal{S}_\alpha$ and its conjugate have the component expansions[17]

$$\mathcal{S}_\alpha = \sigma_\alpha + \sqrt{2}\theta\chi + \theta\bar\theta\bar\partial\sigma_\alpha \qquad \overline{\mathcal{S}}_\alpha = \bar\sigma_\alpha - \sqrt{2}\bar\theta\overline\chi - \theta\bar\theta\bar\partial\bar\sigma_\alpha . \qquad (5.4.6)$$

A (0,2) theory with this multiplet structure has (2,2) supersymmetry for a special choice of the E and J couplings:

$$E^i = \tfrac{1}{2}\Phi^i \sum_\alpha Q_i^\alpha \mathcal{S}_\alpha , \qquad\qquad J_i = \frac{\partial W}{\partial \Phi^i} . \qquad (5.4.7)$$

The $E \cdot J = 0$ supersymmetry constraint holds as a consequence of gauge invariance: under a gauge transformation $\delta_\lambda \Phi^i = \lambda_\alpha Q_i^\alpha \Phi^i$ we have

$$0 = \delta_\lambda W = \lambda_\alpha \sum_{i=1}^n Q_i^\alpha \Phi^i \frac{\partial W}{\partial \Phi^i} = 2 \sum_{i=1}^n E^i J_i \bigg|_{\mathcal{S}_\alpha = \lambda_\alpha} . \qquad (5.4.8)$$

Written in terms of the (0,2) fields, the action is a sum of several terms: the gauge kinetic terms, the minimal coupling terms, and the superpotential. For completeness we give these. We group the $\mathcal{S}_\alpha$ with the Υ; this is sensible, since the scalars σ_α naturally have dimensions of mass.

$$S_{\text{gauge}} = \frac{1}{2\pi e^2} \int d^2z \mathcal{D}\overline{\mathcal{D}} \sum_{\alpha=1}^{n_G} \left[\tfrac{1}{4}\overline{\mathcal{S}}_\alpha\partial\mathcal{S}_\alpha - \tfrac{1}{4}\mathcal{S}_\alpha\partial\overline{\mathcal{S}}_\alpha - \tfrac{1}{2}\overline{\Upsilon}_\alpha\Upsilon_\alpha \right]$$

$$= \frac{1}{2\pi e^2} \int d^2z \sum_{\alpha=1}^{n_G} \left[\bar\partial\sigma_\alpha\partial\bar\sigma_\alpha + \bar{v}_\alpha\bar\partial v_\alpha + \overline\chi_\alpha\bar\partial\chi_\alpha + \frac{1}{4}\left(F_{12\alpha}^2 - D_\alpha^2\right) \right] ,$$

$$(5.4.9)$$

---

[16]Note that the (2,2) gauge superfield Λ consists of a (0,2) chiral multiplet and a (0,2) fermi multiplet—the (2,2) supergauge invariance is therefore larger than that of a (0,2) theory. This motivates an alternative presentation of the E-couplings in terms of a "fermionic" gauge symmetry [143].

[17]In much of the literature on the subject, the $\mathcal{S}_{22\alpha}$ and $\mathcal{S}_\alpha$ are denoted by the slightly exasperating Σ_α. By departing from that common notation, we trade one exasperation for another.

$$S_{\min} = \frac{1}{2\pi} \int d^2z \mathcal{D}\overline{\mathcal{D}} \sum_{i=1}^{n} e^{2V_{0\alpha}Q_i^{\alpha}} \left\{ \tfrac{1}{4}\left[\overline{\Phi}^i \partial \Phi^i - \Phi^i \partial \overline{\Phi}^i + V_{1\alpha}\overline{\Phi}^i Q_i^{\alpha} \Phi^i\right]\right.$$

$$\left. - \tfrac{1}{2}\overline{\Gamma}^i \Gamma^i \right\}$$

$$= \frac{1}{2\pi} \int d^2z \sum_{i=1}^{n} \left\{ \tfrac{1}{2}(\overline{\nabla\phi}^i \nabla\phi^i + \nabla\overline{\phi}^i \overline{\nabla}\phi^i) + \overline{\psi}^i \nabla\psi^i + \overline{\gamma}^i \overline{\nabla}\gamma^i - \overline{G}^i G^i \right.$$

$$+ \overline{E}^i E^i - \tfrac{1}{2}\upsilon_\alpha Q_i^\alpha \overline{\phi}^i \psi^i - \tfrac{1}{2}\overline{\psi}^i \phi^i Q_i^\alpha \overline{\upsilon}_\alpha - \tfrac{1}{4}D_\alpha Q_i^\alpha \overline{\phi}^i \phi^i$$

$$\left. - \overline{E}^i_{,j}\overline{\psi}^j \gamma^i - \overline{\gamma}^i E^i_{,j}\psi^j - \overline{E}^i_{,\alpha}\overline{\chi}^\alpha \gamma^i - \overline{\gamma}^i E^i_{,\alpha}\chi^\alpha \right\} .$$

$$(5.4.10)$$

Finally, the superpotential terms are as before:

$$S_J = \frac{1}{2\pi\sqrt{2}} \int d^2z \sum_{i=1}^{n} \left\{ \mathcal{D}(\Gamma^i J_i) + \overline{\mathcal{D}}(\overline{\Gamma}^i \overline{J}_i) \right\}$$

$$= \frac{1}{2\pi} \int d^2z \sum_{i=1}^{n} \left\{ G^i J_i + \overline{G}^i \overline{J}_i - \gamma^i J_{i,j}\psi^j - \overline{\psi}^j \overline{J}_{i,j}\overline{\gamma}^i \right\} ,$$

$$S_{\text{Fayet-Iliopoulos}} = \frac{i}{2\sqrt{2}} \int d^2z \left\{ \tau^\alpha \mathcal{D}\Upsilon_\alpha - \overline{\tau}^\alpha \overline{\mathcal{D}}\overline{\Upsilon}_\alpha \right\}$$

$$= \frac{1}{2\pi} \int d^2z \left\{ \tfrac{1}{2}D_\alpha r^\alpha \right\} + i\frac{\vartheta^\alpha}{2\pi}\int F_\alpha . \qquad (5.4.11)$$

After integrating out the auxiliary fields and using (5.4.7), we obtain the scalar potential

$$U_{\text{bos}} = \tfrac{1}{4}\sum_{i=1}^{n} |\phi^i|^2 \left|\sum_{\alpha=1}^{r} Q_i^\alpha \sigma_\alpha\right|^2 + \sum_{i=1}^{n} \left|\frac{\partial W}{\partial \phi^i}\right|^2 + \frac{1}{4e^2}\sum_{\alpha=1}^{r} D_\alpha^2 , \qquad (5.4.12)$$

with

$$D_\alpha = -\tfrac{e^2}{2}\left(\sum_{i=1}^{n} Q_i^\alpha |\phi^i|^2 - 2r^\alpha\right) . \qquad (5.4.13)$$

5.4.2 Symmetries and Anomalies

Since for every charged chiral multiplet Φ^i there is a fermi multiplet Γ^i with identical gauge charges, the gauge anomaly is automatically zero in any (2,2) linear sigma model. The (2,2) multiplet structure tightly constrains the symmetry action: the $U(1)_L \times U(1)_R$ charges of $\mathcal{S}_\alpha$ and Υ_α are fixed by the charges of the (2,2) superspace coordinates θ' and θ, while those of Φ^i and Γ^i are tied together. The continuous symmetries of the Lagrangian that commute with (2,2) supersymmetry have a maximal abelian subgroup generated by a subgroup of $U(1)_A \times U(1)_B \times \prod_{i=1}^{n} U(1)_i$, with charges

	θ'	θ	Φ^i	Γ^i	W	E^i	$\mathcal{S}_\alpha$	Υ_α
$U(1)_A$	$\frac{1}{2}$	$\frac{1}{2}$	q_i	$q_i - \frac{1}{2}$	1	q_i	0	$\frac{1}{2}$
$U(1)_B$	$-\frac{1}{2}$	$\frac{1}{2}$	0	$\frac{1}{2}$	0	1	1	$\frac{1}{2}$
$U(1)_j$	0	0	δ_{ij}	δ_{ij}	0	δ_{ij}	0	0

$$(5.4.14)$$

This is an over-parametrization of the continuous abelian global symmetries for three reasons:

i. $U(1)_A$ requires W to be a quasi-homogeneous function: $W(t^{q_i} \Phi_i) = t W(\Phi)$;
ii. $U(1)_B$ is anomalous unless $\sum_i Q_i^\alpha = 0$ for all α;
iii. $U(1)_j$ include the global gauge symmetries—the quotient $\prod_j U(1)_j/G$ is exactly the quotient of the "big torus" by the gauge group, and it leads to a continuous symmetry group $U(1)^{n-n_G}$; these symmetries are also in general broken by the superpotential to a discrete (often trivial) subgroup.

When $U(1)_B$ is non-anomalous and W is quasi-homogeneous and generic enough to eliminate the "small torus" $U(1)^{n-n_G}$ symmetries, the UV theory has a $U(1)_L \times U(1)_R$ symmetry with currents

$$\mathcal{J}_L = \mathcal{J}_A - \mathcal{J}_B , \qquad \mathcal{J}_R = \mathcal{J}_A + \mathcal{J}_B \qquad (5.4.15)$$

and charge assignments

	θ'	θ	Φ^i	Γ^i	W	E^i	$\mathcal{S}_\alpha$	Υ_α
$U(1)_L$	1	0	q_i	$q_i - 1$	1	$q_i - 1$	-1	0
$U(1)_R$	0	1	q_i	q_i	1	$1 + q_i$	1	1

$$(5.4.16)$$

As in the general discussion in Sect. 5.2.2, these are ambiguous up to shifts by the gauge charges, have zero mixed anomaly, and lead to

$$c = \bar{c} = 3 \left\{ \sum_{i=1}^{n}(1 - 2q_i) - n_G \right\} . \qquad (5.4.17)$$

The condition for non-anomalous $U(1)_B$

$$\sum_{i=1}^{n} Q_i^\alpha = 0 \qquad \text{for all } \alpha \qquad (5.4.18)$$

means the expression for c is well-defined under shifts of the $q_i \rightarrow q_i + \lambda_\alpha Q_i^\alpha$.

Even in this (2,2) setting, the necessary and sufficient conditions for this expression to be reliable are not known in general. With the most optimistic assumptions, we might expect the theory to flow to a compact SCFT if the bosonic potential confines the fields to a compact subspace, and if the R-charges of gauge-invariant chiral operators obey the unitarity bounds $0 < \overline{q} \leq \overline{c}/3$. These conditions lead to combinatorial constraints that can be phrased in the language of toric geometry, but the general solution to these constraints remains elusive, even when restricted to theories where it is possible to choose $q_i \in \{0, 1\}$. When the R-charges q_i are integral, there is a conjectured form for the necessary and sufficient condition given in terms of combinatorial data [35].

5.4.3 Phases—Classical Analysis

But enough bemoaning our general ignorance! We now take a look at the case that we do understand and which corresponds to the toric constructions we described above.

The Quintic Mother of Invention

We start the discussion by building the linear sigma model for the quintic hypersurface in $\mathbb{P}^4$. It is easy to write a linear model for the toric variety $\mathbb{P}^4$ by generalizing the $\mathbb{P}^1$ example of Sect. 5.2.2: we take 5 (2,2) matter fields (Φ^i, Γ^i), with charge $Q = 1$ under U(1) and observe that if the Fayet-Iliopoulos parameter $r > 0$, then the classical $e \to \infty$ limit leads to a low energy description for a (2,2) non-linear sigma model with target space $\mathbb{P}^4$ equipped with the Fubini-Study metric. The massless fermions couple to the tangent bundle over $\mathbb{P}^4$, described by the Euler sequence (5.3.80).

There is a more physical way to think about the low energy limit. Instead of taking $e \to \infty$, we observe that the physics is the two-dimensional Higgs mechanism controlled by the value of the Fayet-Iliopoulos parameter r.[18] When $r \gg 0$, the (2,2) gauge multiplet is Higgsed by the expectation values of the matter fields at a mass scale of order $e\sqrt{r}$. Therefore we can integrate out the gauge fields at a scale much larger than that of the strong coupling dynamics and obtain a reliable description of the low energy physics in terms of the (2,2) non-linear sigma model with target space $\mathbb{P}^4$.

---

[18]The reader may want to review the classic discussion of the Abelian Higgs model [118].

Exercise 5.26 Work in a unitary gauge, where some field, say Φ^1, has expectation value $2r$. Show that as promised, the gauge degrees of freedom acquire the mass of order $e\sqrt{r}$. $\qquad\qquad\square$

The linear sigma model for $\mathbb{P}^4$ is an example of what we will call a "V-model"—this is a (2,2) linear sigma model that for some choice of the Fayet-Iliopoulos parameters has a classical low energy limit given by the non-linear sigma model with target space a compact toric variety V. The $\mathbb{P}^4$ theory has an anomalous $U(1)_B$ symmetry—$\sum_i Q_i = 5$—and we cannot introduce a gauge-invariant polynomial superpotential term. Both of these defects are eliminated if we introduce an additional (2,2) matter multiplet (Φ^0, Γ^0) with charge -5. In particular, we now can write a superpotential of the form

$$W = e\Phi_0 P(\Phi_1, \ldots, \Phi_5)\,, \qquad (5.4.19)$$

where P is a generic quintic polynomial

$$P = \Phi_1^5 + \Phi_2^5 + \Phi_3^5 + \Phi_4^5 + \Phi_5^5 + \ldots\,. \qquad (5.4.20)$$

In writing these polynomials it is useful to lower the index on the Φ^i to avoid stultifying expressions like $(\Phi^1)^2(\Phi^3)^3$. We will do so frequently; we will also often absorb the scale e into a definition of the coefficients in P.

The superpotential is quasi-homogeneous, and we can assign R-charges $q_0 = 1$ and $q_i = 0$ for $i = 1, \ldots, 5$, so that (5.4.17) yields $c = \bar{c} = 9$. The resulting linear sigma model is the example of an "X-model"—a (2,2) linear sigma model that for some choice of the Fayet-Iliopoulos parameters and generic coefficients in P has a classical low energy limit given by the non-linear sigma model with target space $X = \{P = 0\} \subset V$.

When $r \gg 0$, the low energy limit is well-described by a non-linear sigma model on the $U(1)$ quotient of the zero locus of the bosonic potential (no sum on i)

$$-5|\phi_0|^2 + |\phi_1|^2 + \ldots + |\phi_5|^2 = r\,, \quad Q_i|\phi_i|^2\sigma = 0\,, \quad \phi_0\frac{\partial P}{\partial \phi_i} = 0\,, \quad P = 0\,.$$
$$(5.4.21)$$

When $r \gg 0$, we recognize the first equation as the moment map for the symplectic quotient of the $U(1)$ action; the resulting space is smooth and can also be described as the holomorphic quotient

$$V_+ = \{\mathbb{C}^6 \setminus F\}/\mathbb{C}^*\,, \qquad (5.4.22)$$

where $F = \{\phi_1 = \phi_2 = \ldots = \phi_5 = 0\}$, and $\mathbb{C}^*$ acts with charges $(-5, 1, 1, 1, 1, 1)$. This is the manifold $V_+ = \mathcal{O}(-5) \to \mathbb{P}^4$ (compare to exercise 5.17). Since the ϕ_i are not simultaneously vanishing, the second set of equations implies $\sigma = 0$. Finally, the last conditions carve out a subvariety of V_+: the solution set has two components:

$$X = \{\phi_0 = 0 , P = 0\} , \quad X' = \{\partial P / \partial \phi_i = 0 , \ i = 1, \ldots, 5\} . \quad (5.4.23)$$

The hypersurface $\{P = 0\} \subset \mathbb{P}^4$ is singular if and only if X' is non-empty; therefore generically $X' = \emptyset$, and we obtain the X-model for the quintic hypersurface in $\mathbb{P}^4$. The Kähler class of X is proportional to r, which means that in the $r \gg 0$ limit the non-linear sigma model will be weakly coupled. In the IR we can reasonably expect the theory to flow to an SCFT, with $r \to \infty$ corresponding to the (infinite distance) large radius limit. From the discussion in Sect. 4.6 we expect the SCFT to have a moduli space $\mathcal{M}_{ac} \times \mathcal{M}_{cc}$, and in this case it is not hard to identify a parametrization of these: the complexified Fayet-Iliopoulos coupling is a coordinate on $\mathcal{M}_{ac}$, while the 126 complex coefficients in P overparameterize the 101-dimensional $\mathcal{M}_{cc}$.

We glossed over a subtlety: the bosonic potential for a non-singular hypersurface requires σ and ϕ^0 to have zero expectation value, but that by itself does not mean the fields are massless. Fortunately, this is the case: the bosonic potential is quadratic in σ and ϕ^0, which means these fields are massive. The fermi multiplets also follow suit, and the massless fermions couple to T_X, described explicitly by the cohomology of the complex (5.3.118).

The classical V-model for $\mathbb{P}^4$ breaks supersymmetry when $r < 0$.[19] That is not the case in the X-model: when $r \ll 0$ the moment map leads to the holomorphic quotient

$$V'_+ = \{\mathbb{C}^5 \setminus F\}/\mathbb{C}^* , \quad (5.4.24)$$

where the exceptional set is now $F = \{\phi_0 = 0\}$. The expectation value $\phi_0 \neq 0$ breaks the gauge symmetry $U(1) \to \mathbb{Z}_5$, where the $\mathbb{Z}_5$ acts by roots of unity ζ_5: $\phi_i \mapsto \zeta_5 \phi_i$ for $i = 1, \ldots, 5$. So, put another way, $V'_+ = \mathbb{C}^5/\mathbb{Z}_5$. Again, σ is massive and has zero expectation value, but now the superpotential term confines the expectation values ϕ_i to $X' \subset V'_+$. When $\{P = 0\} \subset \mathbb{P}^4$ is non-singular X' consists of a single point—the origin in $\mathbb{C}^5/\mathbb{Z}_5$. However, the massless excitations are no longer described by a non-linear sigma model, but rather by a Landau-Ginzburg theory with (0,2) superpotential

$$\mathcal{W} = \sum_{i=1}^{5} \Gamma^i \frac{\partial P}{\partial \Phi^i} . \quad (5.4.25)$$

Recall from Chap. 3 that $\mathcal{W}$ defines a compact Landau-Ginzburg theory if and only if the vanishing locus of the J_A—in this case $J_i = \frac{\partial P}{\partial \Phi^i}$—consists of a finite set of points. Thus $X' \subset V'_+$ leads to a compact Landau-Ginzburg theory if and only if

---

[19]This is a classical chimera, and we will discuss the fate of supersymmetry in the quantum theory below.

$X \subset V_+$ is a smooth hypersurface. Applying c-extremization we see that the $U(1)_L$ charges are

$$\overline{q}(\Phi^i) = \tfrac{1}{5}, \qquad\qquad \overline{q}(\Gamma^i) = \tfrac{1}{5}, \qquad\qquad (5.4.26)$$

and these lead to $c = \overline{c} = 3$. This charge assignment has a simple interpretation from the $r \gg 0$ point of view: recall that the R-symmetry assignment is ambiguous up to the gauge symmetry charges. We can use the ambiguity to set the R-charge of Φ^0 to be 0. This is very convenient for understanding the physics of a phase where Φ^0 is massive and has a non-zero expectation value.

We should not forget about the left-over discrete $\mathbb{Z}_5$ gauge symmetry that acts on the fields via $(\Phi^i, \Gamma^i) \to (\zeta_5 \Phi^i, \zeta_5 \Gamma^i)$. This means the low energy limit for $r \ll 0$ is a Landau-Ginzburg orbifold. The $\mathbb{Z}_5$ projection is just right so that gauge-invariant states will carry integral R-charges. The orbifold also has another important feature: the twisted sectors contain an (a,c)-ring marginal operator, and we identify the corresponding marginal deformation with the coordinate $\mathcal{M}_{ac}$; it is also easy to see that the superpotential parameters again overparameterize a 101-dimensional space of $\mathcal{M}_{cc}$ deformations.[20]

Finally, we can also consider $r = 0$. In this case the zero locus of the bosonic potential includes $\phi_0 = \phi_i = 0$. At that point the gauge theory is no longer Higgsed, and we also see that σ has a flat direction. This suggests that at least classically theories with $r = 0$ are singular, at least in the sense that the ground state is no longer normalizable.

We therefore have a picture of the secondary fan, where each phase corresponds to a low energy limit of the linear model for the quintic in $\mathbb{P}^4$:

LG/$\mathbb{Z}_5$	???	$X \subset V$
$r \ll 0$	$r = 0$	$r \gg 0$

$$(5.4.27)$$

This is a beautiful and compelling picture, and it is remarkable to what degree it generalizes to a wide class of theories. Let us make a few comments on some important lessons to take away from the example.

1. By an astute choice of combinatorics, we ensure that for generic values of the parameters, in this case r and the complex coefficients in P, we obtain a theory where the bosonic potential vanishes on a compact subset in the space of bosonic zero modes, and we find descriptions of the low energy physics in terms of some recognizable limit when the parameters are taken to be deep inside the phase.

---

[20]These statements can be made precise and greatly generalized: the A/2 and B/2 topological heterotic rings of a (0,2) Landau-Ginzburg orbifold theory can be constructed algorithmically. This technology was introduced for (2,2) theories in [259] and generalized to (0,2) theories [145, 266]. A review can be found in [143], and recent applications include [37, 41, 81, 84, 293].

More precisely, the "recognizable limit" has the correct topological data—a smooth compact Calabi-Yau manifold X equipped with a Kähler metric. It does not lead to the Ricci-flat Calabi-Yau metric on X. That would be a remarkable coincidence and, at any rate, is not even what we want to obtain an SCFT. The expectation is that the toric metric obtained from the construction differs from that of the metric defining an SCFT associated to X by irrelevant terms. The non-linear sigma model should flow to the "right metric," but we expect the RG flow to preserve the complex structure and Kähler class.[21]

2. The theory is singular for special values of the parameters—these special values constitute the singular locus of the theory. In the classical analysis one component of the singular locus is $\mathcal{Z}_B$—this is when $\{P = 0\} \subset V$ is a singular hypersurface, and it depends holomorphically on the coefficients of P. The analysis of the bosonic potential shows that the singularity is accompanied by a new flat direction: if there is a point in $\mathbb{P}^4$ where $\partial P / \partial \phi^i = 0$ for $i = 1, \ldots, 5$, then at that point ϕ^0 can take on arbitrarily large expectation values. Because of the IR divergences of two-dimensional scalars, this strongly suggests that the QFT no longer has a normalizable ground state.

3. There is another field that, at least classically, can acquire a non-compact direction—the σ boson in the (2,2) vector multiplet. This takes place on $\mathcal{Z}_{A,\mathrm{cl}}$, which is where the σ develops a flat direction. In our example $\mathcal{Z}_{A,\mathrm{cl}}$ is the locus $r = 0$.

4. Phases need not have an obvious geometric interpretation. The Landau-Ginzburg orbifold is perhaps the simplest example of such a non-geometric phase. The different phases, although separated by the classical singularity at $r = 0$, share many features. For instance, we argued that $\mathcal{Z}_B$ is exactly the same in both phases, and the $\mathcal{M}_{cc}$ is described by deforming the superpotential in each of the phases.

Exercise 5.27 The bi-cubic in $\mathbb{P}^2 \times \mathbb{P}^2$. Consider the rank $n_G = 2$ gauge theory with charges

$$Q = \begin{pmatrix} -3 & 1 & 1 & 1 & 0 & 0 & 0 \\ -3 & 0 & 0 & 0 & 1 & 1 & 1 \end{pmatrix} \tag{5.4.28}$$

and potential

$$W = \Phi_0 P(\Phi_1, \Phi_2, \ldots, \Phi_6), \tag{5.4.29}$$

---

[21]In theories with more supersymmetry, there are quotient constructions that lead to Ricci-flat metrics: the classic example is that of the ALE spaces. These define non-compact (4,4) supersymmetric non-linear sigma models, and the corresponding metrics can be obtained by a hyper-Kähler quotient [277].

where P is a generic bi-cubic polynomial. Perform the classical phase analysis to reproduce the following phase diagram:

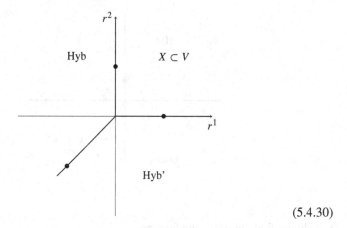

$$(5.4.30)$$

Each hybrid phase is a Landau-Ginzburg orbifold fibered over a base manifold $\mathbb{P}^2$, and the two are related by interchanging $\Phi_{1,2,3}$ with $\Phi_{4,5,6}$. Show that the phase with $r^2 - r^1 > 0$ and $r^1 < 0$ has the following description: the base manifold is $B = \mathbb{P}^2$ parameterized by the homogeneous coordinates $[\phi_4, \phi_5, \phi_6]$; the full geometry of the massless fields is then described as $\mathcal{O}(-1)^{\oplus 3} \rightarrow B$, and the fiber fields (Φ_1, Φ_2, Φ_3) have superpotential interactions of the form

$$W \supset \alpha(\Phi_4, \Phi_5, \Phi_6) M(\Phi_1, \Phi_2, \Phi_3),\qquad(5.4.31)$$

where the monomials M are cubic, and the coefficients α are holomorphic sections of a bundle over the base; in this case of $\mathcal{O}(3) \rightarrow \mathbb{P}^2$. Since $\mathcal{O}(-1)^{\otimes 3} = \mathcal{O}(-3)$, the full superpotential is well-defined. Show that the hybrid theory is compact precisely when $\{P = 0\} \subset \mathbb{P}^2 \times \mathbb{P}^2$ is compact in the large radius phase. In-depth discussion of hybrid phases is given in [82, 84, 388]. □

Exercise 5.28 The octic: a hypersurface in a resolved weighted projective space. Consider another $U(1)^2$ example, this time with charges

$$Q = \begin{pmatrix} -4 & 0 & 0 & 1 & 1 & 1 & 1 \\ 0 & 1 & 1 & 0 & 0 & 0 & -2 \end{pmatrix}\qquad(5.4.32)$$

Show that there are four phases arranged as follows:

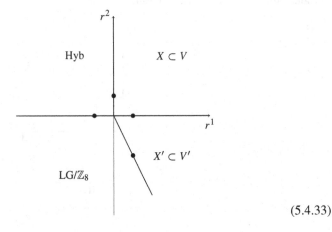

$$(5.4.33)$$

You should find the following features.

1. $X \subset V$ is a hypersurface in a smooth toric variety V with excluded set

$$F = \{\phi_1 = \phi_2 = 0\} \cup \{\phi_3 = \phi_4 = \phi_5 = \phi_6 = 0\} . \qquad (5.4.34)$$

   This famous example was one of the first SCFTs with a two-dimensional $\mathcal{M}_{ac}$ that was studied by mirror symmetry [111, 238].
   Show that a Fermat-type polynomial

$$P = (\Phi_1^8 + \Phi_2^8)\Phi_6^4 + \Phi_3^4 + \Phi_4^4 + \Phi_5^4 \qquad (5.4.35)$$

   defines a non-singular hypersurface in V.

2. $X' \subset V'$ is a hypersurface in a singular toric variety V' with excluded set

$$F = \{\phi_6 = 0\} \cup \{\phi_1 = \phi_2 = \phi_3 = \phi_4 = \phi_5 = 0\} . \qquad (5.4.36)$$

   Show that V' is a weighted projective space $\mathbb{P}^4_{11222}$ with homogeneous coordinates

$$[\phi_1, \phi_2, \phi_3, \phi_4, \phi_5] \sim [t\phi_1, t\phi_2, t^2\phi_3, t^2\phi_4, t^2\phi_5] . \qquad (5.4.37)$$

   A $\mathbb{Z}_2$ subgroup generated by $t = -1$ is unbroken on the surface $\{\phi^1 = \phi^2 = 0\} \subset V'$, and this surface intersects the hypersurface X' in a curve.

3. The Landau-Ginzburg theory described by a (2,2) quasi-homogeneous superpotential W in $\Phi_1, \ldots, \Phi_5$, with charges

$$q_{1,2} = \tfrac{1}{8} , \qquad q_{3,4,5} = \tfrac{1}{4} , \qquad W(t^q \Phi) = t W(\Phi) . \qquad (5.4.38)$$

Confirm that $c = \bar{c} = 9$, and that the $\mathbb{Z}_8$ action is $\Phi_{1,2} \mapsto \zeta_8 \Phi_{1,2}$, and $\Phi_{3,4,5} \mapsto \zeta_8^2 \Phi_{3,4,5}$. In other words, the orbifold projection is indeed onto integral R-charges.

4. The hybrid phase is a Landau-Ginzburg $\mathbb{Z}_4$ orbifold with potential

$$W = \alpha(\Phi_1, \Phi_2)\Phi_6^4 + \Phi_3^4 + \Phi_4^4 + \Phi_5^4 \qquad (5.4.39)$$

fibered over a base $\mathbb{P}^1$ parameterized by the Φ^1, Φ^2 fields; the Landau-Ginzburg fields carry R-charge 1/4. The total space for the hybrid model is $\mathcal{O}(-2) \oplus \mathcal{O}^{\oplus 3} \to \mathbb{P}^1$. □

Hypersurfaces in Toric Varieties

All of the examples we just gave belong to the class of hypersurfaces in Gorenstein Fano toric varieties we described above. Concretely, let Δ be a reflexive d-dimensional polytope with dual Δ° and Σ a maximal projective subdivision of the normal fan taken over the faces of Δ°. This defines a toric variety V with at worst Gorenstein orbifold singularities. The V-model is then the linear sigma model with n matter fields Φ_ρ, Γ_ρ, for $\rho \in \Sigma(1) \subset \Delta^\circ \cap N \setminus \{0\}$ and gauge group $G = U(1)^{n_G} \times H$, where $n_G = n - d$, and H is a discrete abelian group we defined when we described the holomorphic quotient and the Audin-Cox homogeneous coordinate ring of a simplicial toric variety. In many examples $H = 1$, but this is not always the case. From the results above, the generic hypersurface defined by the vanishing of

$$P = \sum_{m \in \Delta \cap M} \alpha_m \prod_\rho Z_\rho^{\langle m, \rho \rangle + 1} \qquad (5.4.40)$$

is a Gorenstein Calabi-Yau variety; it is smooth when $d = 4$, but may have orbifold singularities for $d > 4$.

We can then build the X model in two steps. First, consider the V_+ model, which is the total space of the canonical bundle $\mathcal{O}(K_V) \to V$. We described the construction of this non-compact toric variety above, so we know just what to do: we add a (2,2) multiplet with Φ^0, Γ^0 with G-transformations that are inverse to those of the monomial $\prod_\rho \Phi_\rho$.[22] Now

$$W = \Phi_0 P(\Phi_1, \ldots, \Phi_n) \qquad (5.4.41)$$

is a gauge-invariant (2,2) chiral superpotential interaction that can be added to the V_+ model Lagrangian: the result is the X-model. We are guaranteed that there is

---

[22]The monomial $\prod_\rho \Phi_\rho$ is special in our construction: it corresponds to the unique interior point of the polytope Δ.

some phase of this theory that is described as a non-linear sigma model on $X \hookrightarrow V$, and by going deep inside the phase, we obtain a large radius geometric limit when X is smooth. We therefore expect the corresponding linear sigma model to define a family of (2,2) SCFTs that includes this large radius limit. We assign $U(1)_L$ charges $q_0 = 1$ and $q_\rho = 0$ to the multiplets, so that (5.4.17) yields

$$c = \bar{c} = 3(n - 1 - (n - d)) = 3(d - 1) , \qquad (5.4.42)$$

the expected value.

If X is smooth for some generic choice of coefficients in P, then we obtain a family of linear sigma models: varying the Fayet-Iliopoulos parameters changes the Kähler class, and varying the complex coefficients in P changes the complex structure of X. The latter can lead to singularities. In the simplest case, when V is smooth, X will develop a B-type singularity when there is a point $p \in V$ such that $\frac{\partial P}{\partial \phi_\rho}(p) = 0$ for all ρ. Quasi-homogeneity of P implies that $P(p) = 0$, so this is just the usual way in which a hypersurface becomes singular. Just as in the quintic example, this is exactly where the zero locus of the bosonic potential fails to be compact: ϕ_0 is a flat direction. We let $\mathcal{Z}_B(X)$ be the locus of all α_m values where this is the case. $\mathcal{Z}_B(X)$ is a complicated reducible variety. It has one component that will be important for us: $\mathcal{Z}_B^{pri}$ is the irreducible component of the singular locus where the singular point $p \in X \in V$ is contained in the dense torus $(\mathbb{C}^*)^d \subset V$.

When X is not smooth for generic P and suitable Fayet-Iliopoulos parameters— a situation that is unavoidable for general examples with $d > 4$, we should be concerned that the non-linear sigma model description may be less sensible than usual. This should be contrasted with an orbifold SCFT construction, of a well-defined theory (say a non-linear sigma model on a torus) by a global symmetry. Unlike that case, the X model's orbifold singularities will typically be local in field space: X is not a global orbifold of some smooth manifold. Nevertheless, there is good reason to believe that the linear sigma model still defines a sensible family of SCFTs: there is no obvious pathology associated to non-compact field space, and, as we will see, there are computations of the A and B model observables that pass many checks. The sensibility of the stringy Hodge numbers and deformation theory at the level of algebraic geometry is another such check.

It turns out that there is no need to restrict to X defined by a single equation in V; more generally, it is possible to construct complete intersection varieties $X \subset V$ that share many features of the hypersurface examples such as the duality of the string Hodge numbers. This involves the language of NEF partitions. However, as the next example shows, this too does not exhaust the class of SCFTs with a linear sigma model formulation.

Exercise 5.29 A non-NEF example. The following linear sigma model is taken from [35]. Set $n_G = 1$ with eight fields $\Phi_{-1}, \Phi_0, \Phi_1, \ldots, \Phi_6$ and charges

$$Q = (-3, -4, 2, 1, 1, 1, 1, 1) \qquad (5.4.43)$$

and superpotential

$$W = \Phi_{-1}(\Phi_2^3 + \Phi_3^3 + \Phi_4^3 + \Phi_5^3 + \Phi_6^3) + \Phi_0(\Phi_1^2 + \Phi_2^4 + \Phi_3^4 + \Phi_4^4 + \Phi_5^4 + \Phi_6^4).$$

$$(5.4.44)$$

Show that the zero locus of the bosonic potential is compact when $r > 0$, and that it does not intersect the singular point of the corresponding V-model $(1, 0, \ldots, 0) \subset \mathbb{P}^5_{211111}$. We expect this theory to flow to a sensible SCFT, but, as discussed in [35], it does not correspond to a NEF partition, and $\mathbb{P}^5_{211111}$ is not a Gorenstein variety. $\qquad\square$

Bad Hybrids

The example of the previous exercise also has a hybrid phase when we take $r \ll 0$. At first blush, this hybrid model is no worse than some of the others we encountered: it appears to be a Landau-Ginzburg orbifold fibered over a weighted $\mathbb{P}^1_{3,4}$ that is parameterized by the Φ_{-1} and Φ_0. On closer inspection, it has some peculiar features. The base coordinates are charged with respect to the R-symmetry, and unlike in the previous cases, this cannot be improved by shifting the R-charges by the gauge charges. The fiber over a generic point in $\mathbb{C}^* \subset \mathbb{P}^1_{3,4}$ is a Landau-Ginzburg theory without any additional quotient by a gauge symmetry. On the other hand, the fiber over the point $\Phi_{-1} = 0$ has a $\mathbb{Z}_4$ gauge symmetry that acts on the fibers and the base, while that over the point $\Phi_0 = 0$ has a $\mathbb{Z}_3$ gauge symmetry that acts on the fiber and the base.

The techniques that have been developed for the study of hybrid theories are applicable to the "good hybrids," where the R-symmetry does not act on the base. Bad hybrids are different in two ways. First, on a technical level, good hybrid technology does not apply to them. There is a more profound distinction. When we take the limit deep in a good hybrid phase, the base geometry decompactifies, and we obtain a singularity at infinite distance, much as in a large radius limit. The expectation is that a good hybrid is as good as a non-linear sigma model, and one can imagine, for instance, computing nonperturbative corrections due to worldsheet instantons supported on the base. This is not the case for a bad hybrid: taking $r \to -\infty$ corresponds to a singular point at finite distance in the $\mathcal{M}_{ac}$ moduli space [35, 36]. There is some good news: the bad hybrid singularity is still detected by a non-compactness of the zero locus of the bosonic potential.

One may well guess that a generic phase of a generic linear sigma model will be a bad hybrid, so this is not some abstruse construction in a special example but rather an illustration of a general phenomenon.[23]

---

[23]The SCFT singularities just described should have an interpretation in the full string theory; some aspects of this for the type II string have been explored in [36], but the spacetime perspective should be studied further.

Summary and a Puzzle

Returning to the tamer world of a smooth Calabi-Yau hypersurface $X \subset V$, we now have the following picture from the linear sigma model perspective. There is a suitable large radius phase of the linear model, and we identify a relationship (which we will make more precise below) between the complex structure deformations—or equivalently motion on $\mathcal{M}_{cc}(X)$—and the parameters α_m in the chiral superpotential. Similarly, there Fayet-Iliopoulos parameters allow us to change the Kähler class of X. By reversing the roles of the polytopes Δ and $\Delta°$, we obtain a mirror linear sigma model that describes $X° \subset V°$. Just as in four-dimensional Seiberg duality, while the UV theories are distinct, their IR limits are conjectured to be identical, with an isomorphism $\mathcal{M}_{ac}(X) = \mathcal{M}_{cc}(X°)$.

On the face of it, this presents a puzzle. The complex structure moduli space is naturally complex, and while there is a singular locus of codimension at least one, where the linear sigma model develops non-compact directions in field space and the SCFT degenerates, there are no real codimension one walls that separate the moduli space. On the other hand, the phases picture we have been describing indicates that there are walls in the Kähler moduli space where the theory is ill-behaved. Even in the quintic example, we saw that $r = 0$ leads to a non-compact σ direction.

The resolution of this seeming contradiction can be understood both directly in the linear sigma model [311, 388], as well as by using mirror symmetry [38, 39]. First, we note that the boundaries of the Kähler cone are precisely where the gauge dynamics is no longer controlled by the Higgs mechanism: we expect strong coupling phenomena to matter. Second, we should remember that the r^α parameters are complexified by the ϑ^α, and the τ^α parameterize motion in the complexified Kähler moduli space $\mathcal{M}_{cK}(X)$. The upshot is that $\mathcal{M}_{cK}(X)$ has singularities, but they are complex codimension one and in fact mirror to those in $\mathcal{M}_{c\text{-}x}(X°)$. The phase boundaries are therefore just a semi-classical illusion.

To understand how that works, we will first study some properties of the simpler V model.

5.4.4 RG Flow for the V Model

Consider the V-model corresponding to any compact smooth toric variety, such as $\mathbb{P}^n$ or $\mathbb{F}_p$. From the symplectic quotient construction we know that the solution set to $D_\alpha = 0$ is empty outside the pointed cone $K_{cl}(V) \subset H^2(V, \mathbb{R})$ generated by the n vectors Q_ρ^α, where ρ labels the one-dimensional cones in the toric fan. We now have a physical interpretation for this statement: supersymmetry is classically broken if the vector of the Fayet-Iliopoulos parameters $r \notin K_{cl}(V)$.

In the quantum theory the Fayet-Iliopoulos parameters are renormalized. That is, if the theory is defined at some renormalization mass scale μ, we have

$$\mu \frac{\partial}{\partial \mu} r = R \qquad \Longrightarrow \quad r(\mu) = r(\mu_0) + R \log(\mu/\mu_0) \,, \qquad (5.4.45)$$

where the vector $\boldsymbol{R}$ has components

$$R^\alpha = \sum_\rho Q_\rho^\alpha . \tag{5.4.46}$$

$\boldsymbol{R} \subset H^2(V, \mathbb{R})$ is the anti-canonical class that corresponds to the anti-canonical divisor $-K_V = \sum_\rho D_\rho$.

Before we derive this statement, let us contemplate its consequences. For any compact V and its associated linear model, the classical phase diagram will fall into one of two qualitatively different cases (we indicate K_{cl} by the shadowed region):

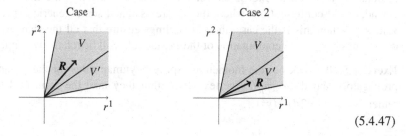

$$\tag{5.4.47}$$

The RG flow $\mu \to 0$ will drive r along the vector $-\boldsymbol{R}$. Thus, in either case, in the IR limit r will leave the cone K_{cl} of classical supersymmetric vacua. This sharpens the classical supersymmetry puzzle: not only is there a region in parameter space where supersymmetry is classically broken, but it is also unavoidable! All compact V models are driven outside of K_{cl} by the RG flow.

On the other hand, in the UV limit, as $\mu \to \infty$, we have a distinction between the cases: in the first case the UV limit is deep in the V phase. Therefore, in Case 1 the linear sigma model is a UV completion of a non-linear sigma model with target space V. We can expect RG-invariant quantities to be describable in terms of the geometry of the smooth toric manifold V and its associated non-linear sigma model. In Case 2, where $\boldsymbol{R} \notin K(V)$, the linear sigma model is a UV completion of a non-linear sigma model with target space V', and it will be the geometry V' that describes the UV phase and is appropriate for a weak-coupling analysis of RG-invariant quantities. When V' is singular, the geometric description is not straightforward. The linear sigma models associated to toric Fano varieties always fall into Case 1.[24]

The discussion sharpens the supersymmetry puzzle. We have the classic result that in a CP-even supersymmetric sigma model on a compact smooth manifold V the Witten index is given by the Euler characteristic: $\text{Tr}(-1)^F = \chi(V)$ [380].

---

[24]There are also some non-generic cases such as NEF Fano varieties where $\boldsymbol{R}$ is parallel to a face of $K(V)$; in this situation we expect either of the adjoining phases to offer a good description of the UV limit. This is the linear sigma model realization of a flop transition [38].

As we saw above, for compact toric varieties $\chi(V) > 0$, so we do not expect supersymmetry to be broken.[25]

A One-Loop Computation

To tackle the puzzle, we perform a simple one-loop calculation described in [388]. We would like to compute the one-loop correction to the classical expectation value $-\frac{2}{e^2}D_\alpha = \sum_\rho Q_\rho^\alpha |\phi_\rho|^2 - 2r^\alpha$, i.e. a tadpole $\delta\langle -\frac{2}{e^2}D_\alpha \rangle$. There is just one relevant divergent Feynman diagram due to the vertex $D_\alpha |\phi_\rho|^2$. To carry out the perturbative calculation we need to choose expectation values for the bosonic fields. Let us consider field configurations where the σ_α are assigned some generic expectation values.[26] When this is the case, the E-couplings ensure that all the charged fields ϕ_ρ are massive, so the computation of the D_α tadpole will be relatively simple.

Exercise 5.30 Work out the momentum space Feynman rules for the ϕ_ρ and D_α propagators and the $\phi_\rho\overline{\phi}_\rho D_\alpha$ vertex. Show that they take the form (it helps to remember $d^2z = 2dy^1dy^2$)

$$\frac{4\pi}{k^2 + |m_\rho|^2} \qquad\qquad -2\pi e^2$$

$$\frac{1}{4\pi}Q_\rho^\alpha \,, \tag{5.4.48}$$

where the masses m_ρ are determined by the σ expectation values:

$$m_\rho = \sum_\beta Q_\rho^\beta \sigma_\beta \,. \tag{5.4.49}$$

□

Using these rules, we compute the basic diagram

$$= \frac{Q_\rho^\alpha}{4\pi} \int_{k^2 \leq \Lambda^2} \frac{d^2k}{(2\pi)^2} \frac{4\pi}{k^2 + |m_\rho|^2} = \frac{Q_\rho^\alpha}{4\pi} \log\left(\frac{\Lambda^2}{|m_\rho|^2}\right) + O\left(\frac{|m_\rho|^2}{\Lambda^2}\right) \,. \tag{5.4.50}$$

---

[25]Every V-model has a non-anomalous $U(1)_A$ symmetry and therefore a $\mathcal{R}$ multiplet; we need not worry about the subtleties of deformed current algebra and the $\mathcal{S}$ multiplet that we discussed in Chap. 1.

[26]We will refer to such field configurations as the σ branch, and talk of σ-vacua. That is a good terminology for generalizing results from (2,2) to (0,2) theories, but in fact for (2,2) theories it is reasonable to speak of this as the Coulomb branch, just as in $\mathcal{N} = 2$ $d = 4$ field theories where the vector multiplet also contains scalars.

The UV cut-off Λ regulates the high energy divergence; the masses m_ρ regulate the IR divergence. Now we obtain the correction to the expectation value of D_α:

$$\langle -\frac{2}{e^2} D_\alpha \rangle = -2r^\alpha - \frac{2}{e^2} \times (-2\pi e^2) \times \sum_\rho \left\{ \frac{Q_\rho^\alpha}{4\pi} \log \left(\frac{\Lambda^2}{|m_\rho|^2} \right) + O\left(\frac{|m_\rho|^2}{\Lambda^2} \right) \right\}$$

$$= -2r^\alpha + \sum_\rho \left\{ Q_\rho^\alpha \log \left(\frac{\Lambda^2}{|m_\rho|^2} \right) + O\left(\frac{|m_\rho|^2}{\Lambda^2} \right) \right\}. \qquad (5.4.51)$$

The cut-off Λ can be absorbed into a counter-term for the Fayet-Iliopoulos parameter, but the logarithm inevitably means that to do so we must introduce a renormalization scale μ: this is a two-dimensional dimensional transmutation, and it implies the promised running of the r^α in (5.4.45). Once we write the expressions in terms of the renormalized $r^\alpha(\mu)$ couplings, we can take the $\Lambda \to \infty$ limit and obtain

$$\langle -\frac{2}{e^2} D_\alpha \rangle = -2r^\alpha(\mu) + \sum_\rho Q_\rho^\alpha \log \left(\frac{\mu^2}{|m_\rho|^2} \right). \qquad (5.4.52)$$

This simple result immediately suggests a possible resolution to our supersymmetry puzzle: since the σ expectation values determine the m_ρ, it may be possible to find supersymmetric vacua where the right-hand-side is zero, by balancing these expectation values against the Fayet-Iliopoulos terms. We will now argue that this is a self-consistent interpretation.

If supersymmetry is unbroken outside of $K_{cl}(V)$, then we should be able to interpret the right-hand-side of (5.4.52) as arising from an effective superpotential. This is the case. Consider

$$\mathcal{W}_{\text{eff}} = \sum_\alpha \Upsilon_\alpha \tilde{\mathcal{J}}_\alpha(\mathcal{S}), \qquad \tilde{\mathcal{J}}_\alpha(\mathcal{S}) = \frac{1}{4\pi\sqrt{2}} \log \left[q_\alpha^{-1} \prod_\rho \left(\frac{Q_\rho^\beta \mathcal{S}_\beta}{\mu} \right)^{Q_\rho^\alpha} \right],$$

$$(5.4.53)$$

where

$$q_\alpha = e^{2\pi i \tau^\alpha(\mu)} = e^{-r^\alpha(\mu) + i\vartheta^\alpha}. \qquad (5.4.54)$$

Exercise 5.31 Show that $\mathcal{D}\mathcal{W}_{\text{eff}} + \overline{\mathcal{D}\mathcal{W}}_{\text{eff}}$ contains the following couplings:

$$\mathcal{D}\mathcal{W}_{\text{eff}} + \overline{\mathcal{D}\mathcal{W}}_{\text{eff}} \supset \frac{1}{8\pi} \sum_\alpha D_\alpha \left(2r^\alpha + \sum_\rho Q_\rho^\alpha \log \frac{|m_\rho|^2}{\mu^2} \right)$$

$$+ \frac{i}{8\pi} \sum_\alpha F_{12\alpha} \left(2\vartheta^\alpha + \sum_\rho Q_\rho^\alpha \log \left(\frac{\overline{m}_\rho}{m_\rho} \right) \right). \qquad (5.4.55)$$

Also verify that $\tilde{J}_\alpha(\mathcal{S})$ can be derived from a (2,2) superpotential function $\tilde{J}_\alpha = \frac{\partial \tilde{W}_{\text{eff}}}{\partial \mathcal{S}^\alpha}$, where

$$\tilde{W}_{\text{eff}} = \frac{1}{4\pi\sqrt{2}} \sum_\alpha \mathcal{S}_\alpha \log\left[\frac{1}{q_\alpha} \prod_\rho \left(\frac{Q_\rho^\beta \mathcal{S}_\beta}{\exp(1)\mu}\right)^{Q_\rho^\alpha}\right]. \tag{5.4.56}$$

$\square$

The exercise shows that we can produce the desired tadpole (5.4.52) from a (2,2) superpotential. The superpotential also predicts a variety of other one-loop effects that can be calculated with just a slight elaboration of the preceding technology.

A Holomorphic Perspective

The symmetries of the V-model are

	θ'	θ	Φ_ρ	Γ^ρ	$\mathcal{S}_\alpha$	Υ_α	q_α
$U(1)_A$	$\frac{1}{2}$	$\frac{1}{2}$	0	$-\frac{1}{2}$	0	$\frac{1}{2}$	0
$U(1)_B$	$-\frac{1}{2}$	$\frac{1}{2}$	0	$\frac{1}{2}$	1	$\frac{1}{2}$	$\sum_\rho Q_\rho^\alpha$
$U(1)_{\rho'}$	0	0	$\delta_{\rho\rho'}$	$\delta_{\rho\rho'}$	0	0	0

$$\tag{5.4.57}$$

$U(1)_B$ is anomalous, but we can choose counter-terms so that the non-invariance of the effective action can be interpreted as a shift of the ϑ^α. Looking back to (2.5.21), we see that a $U(1)_B$ transformation with parameter ϵ shifts the theta angles by

$$\vartheta^\alpha \rightarrow \vartheta^\alpha + \epsilon \sum_\rho Q_\rho^\alpha. \tag{5.4.58}$$

Thus, we can keep track of the anomaly by assigning the transformation rules to the q_α as shown in the table. This is a powerful idea, especially when combined with constraints from holomorphy [24, 248], large field behavior, and classical limits. Using these constraints, it is not hard to see that the quantum corrections to $\mathcal{W}$ are tightly constrained: the only allowed terms are of the form $\Upsilon_\alpha \tilde{J}_\alpha^{\text{gen}}(q, \mathcal{S}, \mu)$, where the most general superpotential $\tilde{J}_\alpha^{\text{gen}}$ is of the form

$$\tilde{J}_\alpha^{\text{gen}} = \frac{1}{4\pi\sqrt{2}} \log\left(\frac{P_\alpha(\mathcal{S}/\mu)}{q_\alpha}\right), \tag{5.4.59}$$

and P_α is a homogeneous polynomial of degree $\sum_\rho Q_\rho^\alpha$. This already shows that the RG running of the Fayet-Iliopoulos parameter is one loop exact. The one loop analysis also determines the polynomial.

Reliability and a Conjecture on Strongly Coupled Dynamics

Supersymmetry and the one-loop computation tell us that as long as we are in a regime where the fields of the original Lagrangian continue to describe the physical degrees of freedom, and it is consistent to integrate out the charged matter fields, then (5.4.53) describes the most general quantum-corrected superpotential for the V-model. As usual, we have no such control over the D-terms in the action, but if we assume these do not develop a singularity, we can still study aspects of the V-model in terms of the Landau-Ginzburg theory with chiral multiplets S_α and chiral fermi multiplets Υ_α, interacting via a superpotential $\mathcal{W}_{\text{eff}}$. For example, we have an explicit expression for the σ-vacua of the theory: the σ_α must be solutions to

$$\prod_\rho \xi_\rho^{Q_\rho^\alpha} = q_\alpha , \qquad (5.4.60)$$

where $\xi_\rho = \mu^{-1} \sum_\beta Q_\rho^\beta \sigma_\beta$.

When is $\mathcal{W}_{\text{eff}}$ a reliable description of the physics? To think about this, we note that for any charges Q_ρ^α there is a $GL(n_G, \mathbb{Z})$ transformation to new charges $Q_{s\rho}^\alpha$ such that the vector $\boldsymbol{R}$ that governs the RG flow is transformed to $\boldsymbol{R}_s$ with a particularly nice form:

$$R_s^1 = \left| \gcd\left(R^1, R^2, \ldots, R^{n_G} \right) \right| , \qquad R_s^\alpha = 0 \qquad \text{for } \alpha > 1 . \qquad (5.4.61)$$

The $GL(n_G, \mathbb{Z})$ transformation will not in general be a symmetry of the theory, but it will just modify the kinetic terms of the gauge fields in a simple and presumably irrelevant fashion. This nice basis makes a few facts readily apparent.

First, from (5.4.58) there is a discrete subgroup $\mathbb{Z}_{2R_s^1} \subset U(1)_B$ that is a symmetry of the UV theory.[27] More precisely, the global symmetry group of the UV theory is

$$\frac{U(1)_A \times \mathbb{Z}_{2R_s^1}}{\mathbb{Z}_2^f} \times \frac{\prod_\rho U(1)_\rho}{G} , \qquad (5.4.62)$$

where the $\mathbb{Z}_2$ in the first quotient acts by $(-1)^F$ and is naturally a subgroup of $U(1)_A$ and $\mathbb{Z}_{2R_s^1}$.

Second, in the special basis $|q_1| = e^{-r^1}$ sets the scale for the σ vacuum expectation values. Now it is clear that in the UV, as $r^1 \to 0$, the $|\sigma_\alpha|$ are driven to zero. This means it is not legitimate to integrate out the charged matter fields, as indicated by the IR divergence of the one-loop integral above. This is not a surprise: when $r^1 \gg 0$, then the linear sigma should be well-described by the Higgs

---

[27] The extra factor of 2 arises from the charge $1/2$ assigned to the fermions.

mechanism and therefore a non-linear sigma model with a toric target space (which, as we saw above, will not be the variety V unless V is Fano or NEF Fano).

On the other hand, when $r^1 \ll 0$, i.e. the theory is deep in the IR, it is sensible to conjecture that $\mathcal{W}_{\text{eff}}$ is a good description of the vacuum structure. This is by no means a rigorous statement, but it is consistent with expectations based on asymptotic freedom and large-N calculations (for theories like $\mathbb{P}^N$) for compact non-linear sigma models and two-dimensional abelian gauge theories.[28] We will assume this conjecture in what follows and study its consequences.

An immediate consequence is a universal description of the IR physics for any compact V-model: (2,2) supersymmetry is preserved, and for generic q_α there are massive σ-vacua, where the bosonic fields take expectation values given by (5.4.60). The σ vacua spontaneously break the first factor in (5.4.62) to $U(1)_A$, and the symmetries of the second factor must decouple from the IR physics—this is two-dimensional confinement.

This universality is in marked contrast to the geometric interpretation of the UV phase: depending on whether V is Fano or not, the UV phase is or is not described by the toric geometry V. Another feature is the promised lack of singularity at phase boundaries. While classically the linear sigma model develops a flat σ-direction on any phase boundary, these flat directions are lifted by quantum corrections, at least for generic values of the q_α. There is a smooth RG flow from $r^1 \gg 0$ to $r^1 \ll 0$; in the former limit we describe the dynamics in terms of the Higgs mechanism and a non-linear sigma model; in the latter limit we describe the vacuum structure with $\mathcal{W}_{\text{eff}}$.

UV, IR, and Intermediate Phases of the V-model

Exercise 5.32 Work out the σ-vacua for the Hirzebruch surfaces $\mathbb{F}_p$ with charges and secondary fan

$$Q = \begin{pmatrix} p & 0 & 1 & 1 \\ 1 & 1 & 0 & 0 \end{pmatrix}$$

$$\tag{5.4.63}$$

A convenient way to do this is to set $\sigma_2 = z\sigma_1$ and then express (5.4.60) in terms of z and $\eta = \sigma_1/\mu$. It helps to treat odd p and even p separately, and that should

---

[28] There can be surprises: the famous bosonic $\mathbb{P}^1$ theory has been argued to be gapped at $\vartheta = 0$ but on the other hand to have massless degrees of freedom at $\vartheta = \pi$; see, for instance [6].

lead to

$$p = 2m \implies \eta_1^2 = \frac{q_2}{(p+z)z} , \qquad q_1 z^{m+1} - q_2^{m+1}(p+z)^{m-1} = 0 ,$$

$$p = 2m + 1 \implies \eta_1 = \frac{q_1 z^{m+1}}{q_2^{m+1}(p+z)^m} , \qquad q_1^2 z^{p+2} - q_2^{p+2}(p+z)^{p-2} = 0 .$$

$$(5.4.64)$$

Thus, for generic q_1, q_2, the number of σ-vacua is

$$\#(\sigma) = \begin{cases} 4 & p \le 1 , \\ p+2 & p > 1 \end{cases} . \qquad (5.4.65)$$

$\square$

If the Landau-Ginzburg description of the IR physics is sensible, the number of σ vacua should yield $\mathrm{Tr}(-1)^F$ [283]. On the other hand, if we can describe the theory as a non-linear sigma model with target space V, we should have $\mathrm{Tr}(-1)^F = \chi(V)$. In the case of the $\mathbb{F}_p$, we have $\chi(\mathbb{F}_p) = 4$ because the toric fan has 4 full-dimensional cones. Thus, for $p \le 2$, we find a pleasant agreement: $\#(\sigma) = \chi(\mathbb{F}_p)$. On the other hand, for $p > 2$ there are many more σ-vacua! The origin of the discrepancy is clear from (5.4.47): $\mathbb{F}_0$ and $\mathbb{F}_1$ are Fano, while $\mathbb{F}_2$ is NEF Fano. Therefore the associated linear models fall into Case 1, and we expect the geometric analysis based on the weakly coupled gauge theory to be reliable. On the other hand for $p > 2$, the UV phase has geometry $\mathbb{P}^2/\mathbb{Z}_p$, and we need to compute $\mathrm{Tr}(-1)^F$ for that singular geometry.[29]

These features are not special to $\mathbb{F}_p$. More generally, many features of the toric geometry of a Fano variety are reflected in the IR dynamics of the linear sigma model. For example, $\#(\sigma) = \chi(V)$ for any Fano or NEF Fano variety V [152]. We will see other correspondences shortly when we study computations of RG-invariant correlation functions.

The σ vacua also play a role in understanding the physics of intermediate phases. Quite generally, if one takes a limit that is deep in the cone of a particular phase, it should be possible to compute various RG-invariant quantities by adding contributions from the Higgs vacua and from the σ-vacua. We will not discuss this in great detail, but let us consider the example of $\mathbb{F}_3$ with a simple goal in mind: we would like to understand $\mathrm{Tr}(-1)^F = 5$ by working in the smooth geometric

---

[29]Note that the topological orbifold Euler characteristic reviewed in [227] yields $\chi(\mathbb{P}^2/\mathbb{Z}_p) = \frac{1}{p}(p+2)$, which matches the number of sigma vacua up to the $1/p$ prefactor. This is curious and perhaps can motivate a calculation of $\mathrm{Tr}(-1)^F$ for such a non-linear sigma model. Because we have a linear sigma model as the UV completion, it is perfectly possible to compute $\mathrm{Tr}(-1)^F$ directly from the Lagrangian gauge theory; indeed, the full elliptic genus can be computed by localization. The result for $\mathrm{Tr}(-1)^F$ matches the σ-vacua. We leave those computations to the references.

phase. In order to take that limit, we must scale the $q_{1,2}$ parameters so that the corresponding Fayet-Iliopoulos parameters are deep in the cone and far from the boundaries. We can take the limit as follows:

$$q_2 \to 0, \qquad x \to 0, \qquad q_1 = xq_2^3. \qquad (5.4.66)$$

In this limit we will clearly obtain the Higgs vacua described by the target space $\mathbb{F}_3$, with $\chi(\mathbb{F}_3) = 4$. However, we also have a σ surprise, since the equations reduce to

$$\eta_1 = \frac{xq_2z^2}{(3+z)}, \qquad x^2q_2z^5 - (3+z) = 0. \qquad (5.4.67)$$

The second equation has five roots, but four of them are not reliable, in the sense that in the limit, for those four roots the σ expectation values are driven to 0. On the other hand, there is one reliable σ vacuum, where

$$z \approx -3 \quad , \qquad \eta_1 \approx \frac{1}{x(-3)^5}. \qquad (5.4.68)$$

Using the familiar identity $\chi(\mathbb{F}_3) + 1 = 4 + 1 = 5$, we declare victory.

It would be interesting to develop this sort of analysis further, and it is clear that many computations can be carried out by combining contributions from Higgs vacua and σ-vacua in various phases. Such an analysis has not been carried out systematically, although there are developments inspired by D-branes and open topological string theory that have some overlap with these ideas—see, for instance, [225, 265, 319].

Singularities

The description of the vacuum structure of the V-model based on $\mathcal{W}_{\mathrm{eff}}(\mathcal{S})$ is simple: generically there are a number of isolated massive σ-vacua, and the expectation values σ are meromorphic functions of the parameters q_α. Each vacuum is (2,2) supersymmetric and has an unbroken $U(1)_A$ R-symmetry. For generic values of the parameters we therefore have a simple picture of the vacua.[30]

How might the description break down? Intuition from Landau-Ginzburg theory that we studied in Chap. 3 suggests there is essentially one way: singularities can develop when some of the σ vacua run off to infinity. This will be automatic if the theory develops a flat σ direction, but it is sufficient for some of the vacua to escape to infinity to lead to a singularity. If all vacua wander off to infinity, then we might

---

[30]The low energy physics beyond the vacuum, which includes the study of supersymmetric solitons, is extremely rich. The reader will find a wide-ranging treatment in [185].

guess that supersymmetry is broken. This does not take place in (2,2) theories, but it can happen in (0,2) deformations.

For example, in the $\mathbb{F}_2$ example the σ vacua satisfy

$$\eta_1^2 = \frac{q_2}{(2+z)z}, \qquad\qquad z^2 = q_1^{-1}q_2^2. \qquad\qquad (5.4.69)$$

Generically there are four vacua, but there is a singular locus $\mathcal{Z}_A = \{q_2^2 - 4q_1 = 0\}$ where two of the vacua wander off to infinity. Note that for parameters in K_{cl} this is corresponds to points that are "close" to the phase boundary between the phases, but it is fundamentally different from $\mathcal{Z}_{A,\mathrm{cl}} = \{2r^2 - r^1 = 0\}$ classical wall: the quantum-corrected singularity is in complex codimension one and does not separate the phases.

We also note that $\mathcal{Z}_A$ extends to the regime where q_1 and q_2 are small, in other words, back into the regime where a Higgs interpretation should be valid. Thus, we should be able to describe the singularities from the Higgs analysis as well. Since the singularity will be close to the phase boundary, the natural suggestion is that the effect is due to a mixed Higgs-σ branch,[31] where the σ expectation values give masses to some of the charged matter, while the remaining charged fields are in a Higgs phase.

A full construction of the singular locus with all of its components is complicated. Fortunately, there is a well-developed combinatorial machinery of the GKZ discriminant [191] that associates components of the singular locus to faces of the polytope Δ_+° of the V_+ model [311]. Its application to a much wider class of linear sigma models is presented in [35, 42].

5.4.5   Phases—Quantum Analysis for the X-model

Having studied the RG flow of the V-model, we now turn to the X model for a hypersurface in V. The key difference is that the RG flow vector is zero because $Q_0^\alpha = -\sum_\rho Q_\rho^\alpha = -R^\alpha$. This modifies the analysis as follows.

1. Because $R = 0$, the complexified Fayet-Iliopoulos parameters τ^α do not run.
2. There is a non-anomalous $U(1)_L \times U(1)_R$ symmetry.
3. The secondary fan is complete, and there is no analogue of the V-model's supersymmetry puzzle.
4. There are no isolated σ vacua: σ vacua either correspond to a mixed Higgs-σ branch, or they belong to a flat σ-direction.

We expand on the last point. We identify the pure σ-branch as before: we assume that the σ expectation values are generic and large, so that all of the charged matter

---

[31] If not for our (0,2) goals, we would call this a mixed Higgs-Coulomb branch.

fields can be integrated out. In that case the light degrees of freedom interact through
the effective superpotential as in (5.4.53), but now with

$$\tilde{J}_\alpha(\mathcal{S}) = \frac{1}{4\pi\sqrt{2}} \log\left[q_\alpha^{-1}(-R^\beta \mathcal{S}_\beta)^{-R^\alpha} \prod_\rho (Q_\rho^\beta \mathcal{S}_\beta)^{Q_\rho^\alpha} \right]. \tag{5.4.70}$$

The renormalization scale μ drops out of the general expression, and the equations
for the supersymmetric σ-vacua are

$$\prod_\rho (Q_\rho^\beta \sigma_\beta)^{Q_\rho^\alpha} = q_\alpha(-R^\beta \sigma_\beta)^{R^\alpha}. \tag{5.4.71}$$

Since the equations are quasi-homogeneous with respect to $U(1)_L$, if $(\sigma_1, \sigma_2, \ldots, \sigma_{n_G})$ is a non-zero reliable solution, then so is $(t\sigma_1, t\sigma_2, \ldots, t\sigma_{n_G})$ for all t. For
generic q_α we expect no solutions to the equations. On the other hand, whenever the
q_α are chosen so that there are reliable solutions, we expect the theory to be singular
due to the non-compact σ direction.

We will denote the locus for which there are reliable solutions by $\mathcal{Z}_A^{pri}$—it is the
"primary component" of the full singular locus $\mathcal{Z}_A(X)$. The "reliability" criterion is
the same as that for the V-model: the σ-expectation values should become very large
and should be generic enough so that all the charged matter fields can be consistently
integrated out. As in the V model, the full discriminant $\mathcal{Z}_A(X)$ will have additional
components due to mixed Higgs-σ branches developing non-compact directions.[32]
All of these components have at least complex codimension 1, so that the phases
identified in the classical analysis of the X model belong to families of smoothly
connected theories. So, as promised, the quantum corrections resolve the puzzle of
the Kähler walls.

Exercise 5.33 Find $\mathcal{Z}_A^{pri}$ for the X-model examples presented above. You should
obtain

$$\mathcal{Z}_A^{pri}(\text{quintic}) = \{1 + (5)^5 q = 0\},$$

$$\mathcal{Z}_A^{pri}(\text{bicubic}) = \{3^9(q_1 + q_2)^3 + 3^7(q_1 + q_2)^2 - 3^9 q_1 q_2 + 3^4(q_1 + q_2) + 1 = 0\},$$

$$\mathcal{Z}_A^{pri}(\text{octic}) = \{(1 - 2^8 q_1)^2 - 2^{18} q_1^2 q_2 = 0\}. \tag{5.4.72}$$

---

[32]There is one obvious component of $\mathcal{Z}_A(X)$ that never intersects $\mathcal{Z}_A^{pri}$: in the strict large radius
limit the target space of the non-linear sigma model is non-compact. This component of the
discriminant (and many others) lie at infinite distance in the SCFT moduli space; this is not the
case for $\mathcal{Z}_A^{pri}$, which is a finite distance away from a generic point. A bad hybrid limit point is
another example of a singular points that does not intersect $\mathcal{Z}_A^{pri}$ but is at finite distance in the
SCFT moduli space.

The first is ridiculously easy; the second can be obtained by a resultant for the two equations in terms of $z = \sigma_1/\sigma_2$. The last illustrates a subtlety of reliability. The resultant yields a product

$$\mathcal{Z}_A^{\text{res}}(\text{octic}) = \left\{ (1 - 4q_2)^3 \left[(1 - 2^8 q_1)^2 - 2^{18} q_1^2 q_2 \right] = 0 \right\} . \tag{5.4.73}$$

Show that the vacua associated with the first component, where $1 - 4q_2 = 0$, are either not reliable because $\sigma_1 = 0$, so that many of the charged fields remain massless, or they are contained in the intersection of $1 - 4q_2 = 0$ and $\mathcal{Z}_A^{\text{pri}}(\text{octic})$.

□

Amoebas

The singular locus is an example of a beautiful structure with deep connections to symplectic and tropical geometries that goes under the name of an amoeba—a review can be found in [306]. To illustrate this, we consider again the bicubic and rewrite the σ equations of motion as parametric equations for the singular locus in terms of an affine parameter $z = \sigma_2/\sigma_1$:

$$q_1(z) = \frac{1}{3^3(1+z)^3} , \qquad q_2(z) = \frac{z^3}{3^3(1+z)^3} , \tag{5.4.74}$$

from which we obtain

$$r_1 = \tfrac{3}{2} \log \left(3^2 (1+z)(1+\bar{z}) \right) , \qquad r_2 = r_1 - \tfrac{3}{2} \log(z\bar{z}) . \tag{5.4.75}$$

When we set $z \in \mathbb{R}$, the parametric curve looks like the outline of an amoeba:

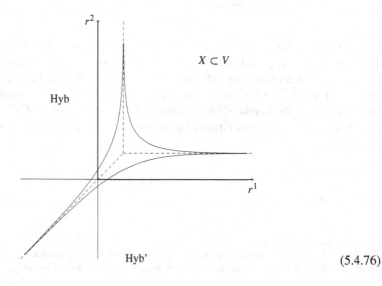

$$\tag{5.4.76}$$

The dashed lines indicate the amoeba's asymptotes:

i. $q_1 = 3^{-3}$ and $q_2 \to 0$;
ii. $q_2 = 3^{-3}$ and $q_1 \to 0$;
iii. $q_1 = q_2 \to \infty$.

These asymptotes are the quantum-corrected versions of the Kähler walls obtained in the classical analysis, but now a "wall" is just an artifact of the projection of a complex variety into a real subspace.[33]

The restriction to $z \in \mathbb{R}$ allows us to find the intersection of $\mathcal{Z}_A^{\text{pri}}$ with $q_1, q_2 \in \mathbb{R}$. More generally, by taking $z \in \mathbb{C}$, we can find the projections to the (r^1, r^2) plane of all the singular points. The reader can convince herself that these will fill in the interior of the amoeba and will be contained by the outline obtained with $z \in \mathbb{R}$.

Exercise 5.34 Construct the amoeba for the octic example, and show that there is a missing asymptote: there is no asymptote that resembles the boundary between the smooth and orbifold phases. This defect is corrected by another component of the singular locus where $1 - 4q_2 = 0$. Details can be found in [311].   □

$\mathcal{M}_{\text{ac}}$(quintic)

The linear sigma model analysis for the quintic leads to a simple picture of the moduli space $\mathcal{M}_{\text{ac}}$ parameterized by the single coordinate q:

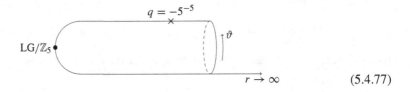

$$ (5.4.77) $$

Here the $\times$ indicates the singular locus $\mathcal{Z}_A^{\text{pri}}$—a point with $\vartheta = \pi$ and $r^1 = 5 \log 5$. The $r \to \infty$ is the large radius limit; it too is a singular point but at infinite distance. Finally, the Landau-Ginzburg orbifold at $q = \infty$ is a point at finite distance in the moduli space. This last fact is not obvious directly in the linear sigma model, but if we believe the Landau-Ginzburg theory to define a sensible SCFT, then it is plausible; it can be verified directly by determining the special Kähler metric on $\mathcal{M}_{\text{ac}}$.

---

[33]In this example we obtain an asymptote for each of the walls identified classically. When this is not the case, it is a sign that $\mathcal{Z}_A(X)$ has extra components due to the mixed branches.

5.4.6 Toric and Polynomial Deformations

In this section we consider a linear sigma model for a smooth Calabi-Yau hypersurface $X = \{P = 0\} \subset V$, with V a Gorenstein Fano toric variety of dimension d. Our discussion generalizes almost verbatim to the situation where X has Gorenstein orbifold singularities inherited from the ambient space, but we will stick to the smooth case. This is sufficient to describe the important case of Calabi-Yau threefolds, and it allows us to avoid the slight complications of working with stringy Hodge numbers.

By construction, there is a phase of the theory that is well-described by a weakly coupled non-linear sigma model on X. The linear sigma model has $n + 1$ charged chiral fields Φ_0 and Φ_ρ, with $\rho \in \Delta^\circ \cap M$, as well as the associated fermi multiplets; in addition there are the $n_G = n - d$ S_α chiral fields in 1:1 correspondence with the vector multiplets for the gauge group G. The UV theory depends on the complex parameters q_α and α_m encoded in the superpotential:

$$W = \sum_{\alpha=1}^{n_G} \Upsilon_\alpha \left(-\frac{1}{4\pi\sqrt{2}} \log q_\alpha \right) + \Gamma^0 P + \sum_{\rho \in \Sigma(1)} \Gamma^\rho \frac{\partial P}{\partial \Phi^\rho} ,$$

$$P = \sum_{\alpha \in \Delta \cap M} \alpha_m \prod_\rho \Phi_\rho^{\langle m, \rho \rangle + 1} . \tag{5.4.78}$$

(2,2) supersymmetry implies that under the RG flow the q_α parameters should be related to motion on $\mathcal{M}_{ac}(X)$, while the α_m should be related to motion on $\mathcal{M}_{cc}(X)$. This is, roughly speaking, correct, but in general the relationship is neither 1:1 or onto. Let us make this more precise.

Toric Divisors and the q_α

The toric variety V has an $(n - d)$-dimensional Kähler cone and Kähler class $\omega = \sum_\alpha r^\alpha \eta_\alpha$, where the η_α form a basis for $H^2(X, \mathbb{Z})$. The Kähler class on X is then obtained by pulling back ω to the hypersurface $\iota : X \hookrightarrow V$; explicitly, the Kähler class is $[\iota^*(\omega)] \in H^2(X, \mathbb{R})$. Not every variation of the r^α leads to a change in $[\iota^*(\omega)]$. This is most easily seen in terms of the corresponding divisor classes D_ρ. The hypersurface X is linearly equivalent to the anti-canonical divisor $-K_V$, and every divisor D_ρ leads to a divisor on X, which we can realize in V as the intersection $(-K_V) \cdot D_\rho$. This intersection will be trivial if D_ρ does not meet the generic hypersurface, and when this happens, $(-K_V) \cdot D_\rho$ will be trivial in $A_{d-2}(V)$. This in turn implies that a shift of $\omega \to \omega + \epsilon \xi_\rho$ preserves $[\iota^*(\omega)] \in H^2(X, \mathbb{R})$.

We can see this explicitly in an example that we already met many times:

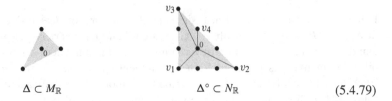

$$\Delta \subset M_\mathbb{R} \qquad\qquad \Delta^\circ \subset N_\mathbb{R} \qquad\qquad (5.4.79)$$

The toric variety V has the charge matrix

$$Q = \begin{pmatrix} 1 & 1 & 1 & 0 \\ 0 & 2 & 1 & -3 \end{pmatrix}, \qquad\qquad (5.4.80)$$

and therefore toric classes

$$\xi_1 = \eta_1, \quad \xi_2 = \eta_1 + 2\eta_2, \quad \xi_3 = \eta_1 + \eta_2, \quad \xi_4 = -3\eta_2. \qquad (5.4.81)$$

These satisfy the Stanley-Reisner relations $\xi_1\xi_4 = 0$ and $\xi_2\xi_3 = 0$. The anti-canonical class is $\boldsymbol{R} = \sum_\rho \xi_\rho = 3\eta_1$, and $\boldsymbol{R}\xi_4 = 0$; equivalently, $-K_V$ does not intersect the exceptional divisor D_4.

The example generalizes: the exceptional classes ξ_ρ for ρ strictly contained in the interior of a facet of Δ° will not contribute to the Kähler class on X.

To state the result precisely, we introduce a little bit of polytopal notation. For any closed subset $\Theta \subset M_\mathbb{R}$ we set $\ell(\Theta)$ to be the number of lattice points in Θ, and we set $\ell^*(\Theta)$ to be the number of points contained in the relative interior of Θ. For example, if Θ is the convex hull of v_1 and v_2 in the example, $\ell(\Theta) = 4$, while $\ell^*(\Theta) = 2$.

Now take $V = V_{\max}$, a maximal projective subdivision, so that $\Sigma(1) = (\Delta^\circ \setminus \{0\}) \cap N$. The exceptional classes then correspond to lattice points that are strictly contained in the interiors of facets—the codimension 1 faces—of Δ°, and we therefore set

$$h^{1,1}_{\text{toric}}(X) = \ell(\Delta^\circ) - 1 - d - \sum_{\text{facets } \varphi^\circ} \ell^*(\varphi^\circ). \qquad (5.4.82)$$

This is not the whole story because some of the toric divisors become reducible when restricted to X. That is, there are directions in the Kähler cone of X that cannot be described by varying the Kähler class on V. The full expression for the dimension of the Kähler cone is still combinatorial [52, 120]:

$$h^{1,1}(X) = h^{1,1}_{\text{toric}}(X) + \sum_{\text{codim 2 } \Theta^\circ} \ell^*(\Theta^\circ)\ell^*(\widehat{\Theta^\circ}), \qquad (5.4.83)$$

where the sum is over the codimension two faces Θ° of Δ°, and $\widehat{\Theta^\circ} \subset \Delta$ is the two-dimensional face dual to Θ°.

Thus, at any point the tangent space $T_{\mathcal{M}_{ac}}$ has a decomposition

$$T_{\mathcal{M}_{ac}} = T_{\mathcal{M}_{ac}}^{\text{toric}} \oplus T_{\mathcal{M}_{ac}}^{\text{non-toric}}, \tag{5.4.84}$$

and the q_α of the linear sigma model, or, equivalently, τ^α over-parameterize the toric subspace of deformations. That is, if ξ_ρ is an exceptional class, then we expect the IR physics to be unchanged by a shift $\delta\tau^\alpha = Q_\rho^\alpha \epsilon$. We will see below that the precise form of the decoupling is corrected by instanton effects.

The situation is much simpler if we take $V = V_{\min}$, a minimal projective subdivision. Now every τ^α counts because $\Sigma(1)$ excludes the exceptional classes. There are also intermediate cases, such as the addition of the generator v_4 in the example.

In our discussion of (0,2) Landau-Ginzburg theories we met a wide class of theories with a large class of redundant operators. We could understand the decoupling of these operators by studying the classical equations of motion of the theory, or, equivalently, field redefinitions. The redundancy in the τ^α is different: it cannot be understood in terms of the field redefinitions of the UV theory.

In a (2,2) SCFT the (a,c) deformations correspond to marginal chiral operators that carry $U(1)_L \times U(1)_R$ charges $q = -1$ and $\bar{q} = 1$. When the SCFT arises as the low energy limit of a linear sigma model, there should be a way to describe these marginal chiral operators in terms of chiral gauge-invariant operators constructed from the linear sigma model fields. A look at (5.4.16) shows that the σ_α are the natural candidates. In the (2,2) theory this is clear: σ_α is the lowest component of the twisted chiral superfield $\mathcal{S}_{22\alpha}$, and the τ^α coupling (5.4.4) is then exactly of the form $\tau^\alpha\{\overline{Q}', [Q, \sigma_\alpha]\}$+c.c. The redundancy in the τ^α now has a clear interpretation: some combinations of the σ_α flow to operators that are trivial in the (a,c) ring. By a small abuse of notation, we will refer to such operators as redundant.[34]

The computation of the Hodge numbers also indicates that in general the SCFT will have (a,c) operators that do not have a simple description in terms of the UV fields. Finding a description for these operators in terms of the UV data is an important open problem. It may not be possible to give a simple global form that makes sense in all of the phases—that perhaps is the truly magical aspect of the σ_α—but it should be possible to find some clues by working in a large radius phase and exploiting the combinatoric structure.

---

[34]In standard terminology in field theory, a redundant operator is zero in correlation functions away from other insertions; in Lagrangian field theories such operators are easily constructed from the equations of motion of the theory. In our case, a redundant σ_α is not zero by the equations of motion; instead, it is equivalent to a $\overline{Q}$-exact operator by the equations of motion. So, while it is fair to call it redundant if we project to $\overline{Q}$-cohomology, in the physical theory it would perhaps be better to call it redundantly irrelevant. We will let that pedantry remain in this footnote.

Polynomial Deformations

The infinitesimal complex structure deformation space of X is isomorphic to

$$H^1_{\bar\partial}(X, T_X) \simeq H^{d-2,1}(X, \mathbb{C}) \tag{5.4.85}$$

and are unobstructed. Some of these can obtained by varying the coefficients of the defining hypersurface, but there may also be deformations that cannot be represented while X is embedded as a hypersurface in the particular toric variety V. The Hodge number $h^{d-2,1}(X)$ is given by [52, 120]

$$h^{d-2,1}(X) = h^{d-2,1}_{\text{poly}}(X) + \sum_{\text{codim } 2\,\Theta} \ell^*(\Theta)\ell^*(\widehat{\Theta}) , \tag{5.4.86}$$

where Θ are co-dimension two faces of Δ, and $\widehat{\Theta} \subset \Delta^\circ$ is the face dual to Θ, and

$$h^{d-2,1}_{\text{poly}}(X) = \ell(\Delta) - 1 - d - \sum_{\text{facet } \varphi} \ell^*(\varphi) . \tag{5.4.87}$$

The terms in $h^{d-2,1}_{\text{poly}}(X)$ are easy to understand: there are $\ell(\Delta)$ monomials in P, but rescaling the defining equation of the hypersurface does not change the complex structure. The automorphisms of the toric variety have an action on the coefficients α_m, and whenever two sets of coefficients α_m and α'_m are related by such a redefinition, the corresponding $X = \{P = 0\}$ and $X' = \{P' = 0\}$ are isomorphic as complex manifolds.

We discussed the automorphism group of a toric variety above and showed that its connected component is generated by the action of the torus T_N, which has dimension d, and the "non-linear" redefinitions described in (5.3.61). It turns out [38] that these have a particularly simple form for our Fano V. From (5.3.61) we see that we are interested in characterizing automorphisms with infinitesimal form

$$\Phi_\rho \to \Phi_\rho + \epsilon \mathsf{M} , \qquad \mathsf{M} = \prod_{\rho'} \Phi_{\rho'}^{a_{\rho'}} , \tag{5.4.88}$$

where $\mathsf{M} \neq \Phi_\rho$ is any monomial in the remaining fields with the same gauge charge as Φ_ρ. From Exercise 5.13 it follows that there is a lattice point $m \in M$ such that

$$a_\rho = \langle m, \rho \rangle + 1 , \qquad a_{\rho'} = \langle m, \rho' \rangle \quad \text{for } \rho' \neq \rho . \tag{5.4.89}$$

But, we know that $a_\rho = 0$; otherwise M/Φ_ρ would descend to a non-constant holomorphic function on V. Thus,

$$\langle m, \rho \rangle = -1 , \qquad \langle m, \rho' \rangle > -1 \quad \text{for } \rho' \neq \rho . \tag{5.4.90}$$

If ρ is not a vertex of Δ°, then the second set of inequalities has a unique solution $m = 0$, which is incompatible with the first equation. On the other hand, if ρ is a vertex of Δ°, then the dual facet $\varphi = \widehat{\rho} \subset \Delta$ is defined by the inequalities

$$\langle m, \rho \rangle = -1, \qquad \langle m, \rho \rangle \geq -1 \qquad \text{for all other vertices } \rho'. \tag{5.4.91}$$

When the inequalities are strict, m must be in the relative interior of the facet φ. It follows that the "non-linear" automorphisms are in 1:1 correspondence with the lattice points $m \in \text{relint}\varphi$ of the facets $\varphi \subset \Delta$: we label these as

$$\mathsf{M}_{\rho,m} = \prod_{\rho' \neq \rho} \Phi_\rho^{\langle m, \rho' \rangle}. \tag{5.4.92}$$

This is the origin of the last term in (5.4.87).

Thus, we obtain a parallel, one may even say mirror, structure for the polynomial deformations as for the toric ones. We have

$$T_{\mathcal{M}_{cc}(X)} = T_{\mathcal{M}_{cc}(X)}^{\text{poly}} \oplus T_{\mathcal{M}_{cc}(X)}^{\text{non-poly}}, \tag{5.4.93}$$

and the first set of deformations is over-parameterized by the UV parameters α_m. The α_m are naturally paired with chiral gauge-invariant operators with $U(1)_L \times U(1)_R$ charges $q = \bar{q} = 1$:

$$\alpha_m \leftrightarrow \mathcal{O}_m = \phi_0 \prod_\rho \phi_\rho^{\langle m, \rho \rangle + 1}, \tag{5.4.94}$$

and the deformation of the Lagrangian can be written as $\sum_m \alpha_m \{\bar{Q}', [Q, \mathcal{O}_m]\} +$ c.c..

There is a redundancy in the $\mathcal{O}_m$, but this time it can be understood in terms of the classical Lagrangian. From the form of the (2,2) superpotential and the general action of the supercharges in the linear sigma model (5.2.32), we have (no sum on ρ)

$$\overline{Q} \cdot \overline{\gamma}_0 \phi_0 = \phi_0 P(\phi), \quad \overline{Q} \cdot \overline{\gamma}_\rho \phi_\rho = \phi_0 \phi_\rho \frac{\partial P}{\partial \phi_\rho}, \quad \overline{Q} \cdot \overline{\gamma}_\rho \mathsf{M}_{\rho,m} = \phi_0 \mathsf{M}_{\rho,m} \frac{\partial P}{\partial \phi_\rho}. \tag{5.4.95}$$

Each left-hand-side is a $\overline{Q}$ action on a gauge-invariant operator, and therefore the right-hand-sides are $\overline{Q}$-exact. Thus, there are relations on the $\mathcal{O}_m$, and they are in 1:1 correspondence with the structure of the automorphisms:

i. $\overline{Q} \cdot \overline{\gamma}_0 \phi_0 = \phi_0 P(\phi)$ corresponds to the rescaling of P, or, equivalently, of Φ_0;
ii. Gauge-invariance of P implies $\sum_\rho Q_\rho^\alpha \phi_\rho \frac{\partial P}{\partial \phi_\rho} = R^\alpha P$, so that the second set of relations leads to d-independent linear relations on the $\mathcal{O}_m$.
iii. Finally, for every vertex ρ and m a lattice point in the interior of a dual facet φ we find another independent relation of the last type.

It follows that for every monomial associated to a lattice point in the interior of a facet, there is a linear combination of the $\mathcal{O}_m$ that is irrelevant and flows to a trivial element of the (c,c) ring.

Exercise 5.35 Let us see how this works in an example by considering, once again, the torus in $\mathbb{P}^2$ with

$$\Delta \subset M_\mathbb{R} \qquad\qquad \Delta^\circ \subset N_\mathbb{R} \qquad\qquad (5.4.96)$$

Use the general definition to show that the polynomial takes the form

$$P = \alpha_1 \Phi_1^3 + \alpha_2 \Phi_2^3 + \alpha_3 \Phi_3^3 + \alpha_0 \Phi_1 \Phi_2 \Phi_3$$
$$+ \alpha_4 \Phi_1^2 \Phi_2 + \alpha_5 \Phi_1 \Phi_2^2 + \alpha_6 \Phi_2^2 \Phi_3 + \alpha_7 \Phi_2 \Phi_3^2 + \alpha_8 \Phi_1 \Phi_3^2 + \alpha_9 \Phi_1^2 \Phi_3 .$$
$$(5.4.97)$$

Suppose P takes the Fermat form, where $\alpha_{1,2,3} = 1$, and all other coefficients are zero. Show that the redundant operators $\phi_0 \mathcal{O}_m$ are all $\overline{Q}$-exact; moreover, the set of chiral operators is spanned by $\phi_0 \phi_1 \phi_2 \phi_3$.   □

Mirror Symmetry and the Monomial-Divisor Mirror Map

When we exchange the roles of Δ and Δ° in the construction, we obtain a beautiful exchange of geometric data:

M	$\longleftrightarrow$	N
Δ	$\longleftrightarrow$	Δ°
V	$\longleftrightarrow$	V°
X	$\longleftrightarrow$	X°

$$(5.4.98)$$

with $h^{1,1}(X) = h^{d-2,1}(X^\circ)$ and $h^{d-2,1}(X) = h^{1,1}(X^\circ)$. The exchange preserves the toric/non-toric and polynomial/non-polynomial split:

$$h_{\text{toric}}^{1,1}(X) = h_{\text{poly}}^{d-2,1}(X^\circ) , \qquad\qquad h_{\text{poly}}^{d-2,1}(X) = h_{\text{toric}}^{1,1}(X^\circ) . \qquad\qquad (5.4.99)$$

This isomorphism is direct at the level of the polytopes and shows that to each toric divisor in V there is a corresponding monomial in P°, the polynomial that defines $X^\circ \subset V^\circ$. This motivates the monomial-divisor mirror map [38]. We will be interested in the interpretation given in [311], which is particularly suited to the linear sigma model.

The construction is complicated by the structure of $\text{Aut}(V)$. As we discussed, this has an action on the parameters α_m in P, so, naively, we might want to describe the parameter space as follows: the α_m live in $Y_B = \mathbb{C}^{\ell(\Delta)}$, and there is a singular locus $\mathcal{Z}_B(X) \subset Y_B$, where the theory fails to be compact. Thus, we might want to study $\{Y_B \setminus \mathcal{Z}_B(X)\}/\{\text{Aut}(V) \times \mathbb{C}^*\}$, where the $\mathbb{C}^*$ factor accounts for rescaling P, or, equivalently, Φ_0. We met such quotients in a much simpler guise in Chap. 3, where they helped us to identify accidental symmetries in (0,2) Landau-Ginzburg RG flow. In general taking such a quotient is not a good way to get a moduli space: the trouble is the action is not reductive, so that even the machinery of GIT does not apply directly. These difficulties arise from the redundant divisors that correspond to lattice points strictly contained in facets of Δ°, or, equivalently, their "mirror" redundant monomials and corresponding non-linear redefinitions in $\text{Aut}(V^\circ)$.

A simple way to avoid these troubles is to restrict attention to polytopes that have no interior lattice points contained in facets. We say that a polytope Δ is reflexively plain if it reflexive and neither Δ or Δ° have a facet with any interior lattice points. Reflexively plain polytopes exist: among the $d = 4$ reflexive polytopes there are $6, 677, 743$ reflexively plain pairs and 5518 self-dual reflexive polytopes [276]. This is merely 3% of the database, but is enough to provide plenty of examples. In $d = 2$ there is a unique reflexively plain polytope, and it is self-dual:

$$(5.4.100)$$

The reader will recognize it as the polytope for $V = dP_3$.

In the reflexively plain case every minimal projective subdivision is also maximal, and we can write the following correspondence for the pair of linear sigma models:

$$
\begin{array}{cccc}
\text{polytopes} & \Delta \subset M & \longleftrightarrow & \Delta^\circ \subset N , \\
\text{chiral matter fields} & \Phi_{\dot{m}} , \dot{m} \in \Delta^\circ \cap N & \longleftrightarrow & \Phi_m^\circ , m \in \Delta \cap M , \\
\text{UV parameters} & q_\alpha , \alpha_m & \longleftrightarrow & q_{\dot{\alpha}}^\circ , \alpha_{\dot{m}}^\circ . \quad (5.4.101)
\end{array}
$$

The gauge charges are defined as follows:

$$
Q_0^\alpha = - \sum_{\dot{m} \neq 0} Q_{\dot{m}}^\alpha , \qquad Q_0^{\circ\dot{\alpha}} = - \sum_{m \neq 0} Q_m^{\circ\dot{\alpha}} . \qquad (5.4.102)
$$

Furthermore, taking a look back at the definition of the gauge group (5.3.40), we recall that the Q_m^α with $m \neq 0$ are an integral basis for the relations among the

$\dot{m} \in N$ and satisfy

$$\sum_{\dot{m} \neq 0} Q_{\dot{m}}^{\alpha} \dot{m} = 0 \in N . \tag{5.4.103}$$

Similarly, the $Q_m^{\circ \dot{\alpha}}$ satisfy

$$\sum_{m \neq 0} Q_m^{\circ \dot{\alpha}} m = 0 \in M . \tag{5.4.104}$$

The field redefinitions that induce an action on the α_m and $\alpha_{\dot{m}}^{\circ}$ parameters arise from the "big torus" action in $\widetilde{\text{Aut}}(V)$ and the rescaling of Φ_0. In fact, we can usefully combine the two as follows. The non-compact toric variety V_+ has a "big torus" $(\mathbb{C}^*)^{n+1}$, which includes the rescaling of Φ_0. Now $\widetilde{\text{Aut}}(V_+)$ is complicated because V_+ is non-compact and therefore supports many non-constant holomorphic functions. However, we can restrict attention to automorphisms that preserve the fiber, or, equivalently, to field redefinitions that preserve the $U(1)_L \times U(1)_R$ symmetry and supersymmetry. Let us call this group $\widetilde{\text{Aut}}'(V_+)$ and restrict attention to the component connected to the identity, denoted by the rococo $\widetilde{\text{Aut}}_0'(V_+)$. Fortunately, while its name is complicated, its action is simple:

$$\alpha_m \mapsto \alpha_m \prod_{\dot{m}} t_{\dot{m}}^{\langle m, \dot{m} \rangle + 1} , \qquad \alpha_{\dot{m}}^{\circ} \mapsto \alpha_{\dot{m}}^{\circ} \prod_{m} (t_m^{\circ})^{\langle m, \dot{m} \rangle + 1} . \tag{5.4.105}$$

The proposal for the mirror map is then contained in the following exercise.

Exercise 5.36 Verify that the induced action of $\widetilde{\text{Aut}}_0'(V_+)$ on the parameters is (5.4.105). Next, prove that the combinations

$$\kappa_{\dot{\alpha}} = \prod_{m \in \Delta \cap M} \alpha_m^{Q_m^{\circ \dot{\alpha}}} , \qquad \kappa_{\alpha}^{\circ} = \prod_{\dot{m} \in \Delta^{\circ} \cap N} (\alpha_{\dot{m}}^{\circ})^{Q_{\dot{m}}^{\alpha}} \tag{5.4.106}$$

are invariant under (5.4.105).   □

Now the UV parameters for the pair of linear models take a much more symmetric form

$$\text{UV parameters} \qquad q_{\alpha} , \kappa_{\dot{\alpha}} \qquad \longleftrightarrow \qquad q_{\dot{\alpha}}^{\circ} , \kappa_{\alpha}^{\circ} , \tag{5.4.107}$$

and there is a natural suggestion for the map between parameters:

$$q_{\alpha} = \kappa_{\alpha}^{\circ} , \qquad \kappa_{\dot{\alpha}} = q_{\dot{\alpha}}^{\circ} . \tag{5.4.108}$$

This is the global monomial-divisor mirror map.

The result is remarkably simple and undoubtedly correct. For instance, it correctly reproduces the behavior in large radius limit, which is essentially tied to the correct monodromies in the ϑ when these are interpreted as components of the B-field [38, 120]. More generally, the map leads to isomorphic three-point functions between the toric (a,c) marginal operators on one side and the polynomial (c,c) operators on the mirror [54, 311, 355]. This is known in some generality and goes beyond the hypersurface examples we discuss here. It is an open problem to establish it directly at the level of the linear sigma model. Efforts in this direction can be found in [236, 312].

We will discuss a proposal that generalizes this map to a class of (0,2) linear sigma models, and we will show that the proposal satisfies the following basic consistency check. When we set $q_\alpha = \kappa_\alpha^\circ$, we will obtain

$$Z_A^{pri}(X) = Z_B^{pri}(X^\circ) . \tag{5.4.109}$$

What is to be done about pairs that are not reflexively plain? There are essentially two approaches: we might either work with the full quotient that includes the field non-linear field redefinitions and, correspondingly, the redundant monomials in P, or we can restrict to a subfamily in the parameter space that sets the redundant q_α, q_α°, and the corresponding redundant monomials in P and P° to zero. It is then consistent to restrict the automorphisms to the toric ones, and we can apply the results above.

Exercise 5.37 Check the claims for the mirror quintic. The polytope for $\mathbb{P}^4$ is simple; in fact it is a simplex, with vertices

$$(\Delta^\circ)_{vertices} = \begin{pmatrix} 1 & 0 & 0 & 0 & -1 \\ 0 & 1 & 0 & 0 & -1 \\ 0 & 0 & 1 & 0 & -1 \\ 0 & 0 & 0 & 1 & -1 \end{pmatrix} \tag{5.4.110}$$

Show that the polytope Δ has the vertices

$$(\Delta)_{vertices} = \begin{pmatrix} 4 & -1 & -1 & -1 & -1 \\ -1 & 4 & -1 & -1 & -1 \\ -1 & -1 & -1 & 4 & -1 \\ -1 & -1 & -1 & 4 & -1 \end{pmatrix} . \tag{5.4.111}$$

Prove that every additional non-zero lattice point in Δ is contained in the interior of a facet and corresponds to a redundant monomial on the quintic side and a redundant divisor on the mirror quintic side. Thus, there is a unique minimal model for the mirror, and it is given by the vanishing of

$$P^\circ = \alpha_1 \Phi_1^5 + \alpha_2 \Phi_2^5 + \alpha_3 \Phi_3^5 + \alpha_4 \Phi_4^5 + \alpha_5 \Phi_5^5 + \alpha_0 \Phi_1 \Phi_2 \Phi_3 \Phi_4 \Phi_5 \tag{5.4.112}$$

in $\mathbb{P}^4/(\mathbb{Z}_5 \times \mathbb{Z}_5 \times \mathbb{Z}_5)$ with projective coordinates $[\Phi_1, \ldots, \Phi_5]$ and $\mathbb{Z}_5 \times \mathbb{Z}_5 \times \mathbb{Z}_5$ action

$$g(\zeta_5^{(1)}, \zeta_5^{(2)}, \zeta_5^{(3)}) : [\Phi_1, \ldots, \Phi_5] \mapsto [\zeta_5^{(1)}\Phi_1, \zeta_5^{(2)}\Phi_2, \zeta_5^{(3)}\Phi_3,$$
$$(\zeta_5^{(1)}\zeta_5^{(2)}\zeta_5^{(3)})^4\Phi_4, \Phi_5] \,. \qquad (5.4.113)$$

Show that the singular locus for this P° is

$$\mathcal{Z}_{\mathrm{B}}^{\mathrm{pri}}(X^\circ) = \{5^5\alpha_1 \cdots \alpha_5 + \alpha_0^5 = 0\} \,. \qquad (5.4.114)$$

Show that under the monomial-divisor mirror map this is isomorphic to $\mathcal{Z}_{\mathrm{A}}^{\mathrm{pri}}(X)$ in (5.4.72). $\qquad\qquad \Box$

There is a discrete subtlety: there are further generators of $\mathrm{Aut}(V)$ that correspond to generators of the discrete symmetry group of the fan. A quotient by these additional elements leads to further discrete identifications on the space of the toric and polynomial parameters. We will continue to gloss over this subtlety and therefore work over what might be called covering space of $\mathcal{M}_{\mathrm{ac}}^{\mathrm{toric}}$ and $\mathcal{M}_{\mathrm{cc}}^{\mathrm{poly}}$ parameterized by the q_α and the $\kappa_{\dot\alpha}$.

5.5   The A-model of the V Model and Related Matters

In this section we will show how topological twisting of the (2,2) linear sigma models introduced in the previous section, together with basic localization technology, leads to computations of the A/2 and B/2 topological heterotic rings. In the next section we will see that these results generalize to (0,2) linear sigma models.

5.5.1   Cohomological Topological Field Theory

We will give a quick sketch of the basic ideas to set the stage for our applications to linear sigma models and leave the general story to the excellent references, which include [134, 237, 387, 390]. The original references [385, 386] are remarkable.

Topological Formalities

Consider a quantum field theory defined on a curved compact (as always for us, this is a short-hand for compact without boundary) Riemann surface Σ equipped with a background metric g_Σ. We say the theory is a topological quantum field theory if the correlation functions are independent of small variations δg_Σ of the background metric.

In a cohomological topological quantum field theory this independence arises as follows. There is a quantum field theory with fermion number, an energy-momentum tensor $T_{\mu\nu}$ and a conserved fermionic worldsheet-scalar operator Q_{top} that satisfies $Q_{\text{top}}^2 = 0$ and annihilates the ground state, such that

$$T_{\mu\nu} = \{Q_{\text{top}}, G_{\mu\nu}\} \tag{5.5.1}$$

for some well-defined local operator $G_{\mu\nu}$. Let $\mathcal{H}_{\text{top}}$ denote the cohomology of Q_{top}. By analogy with BRST cohomology, we will often refer to representatives of classes in $\mathcal{H}_{\text{top}}$ as observables. As above, we will use the short-hand

$$Q_{\text{top}} \cdot \mathcal{O} = \begin{cases} \{Q_{\text{top}}, \mathcal{O}\}, & \mathcal{O} \text{ is fermionic} \\ [Q_{\text{top}}, \mathcal{O}], & \mathcal{O} \text{ is bosonic}. \end{cases} \tag{5.5.2}$$

Then for any collection of Q_{top}-closed local operators $\mathcal{O}_i \in \mathcal{H}_{\text{top}}$ we have several formal statements that the reader can easily formally prove:

i. $\langle \mathcal{O}_{i_1}(x_1) \cdots \mathcal{O}_{i_n}(x_n) \rangle$ only depends on the cohomology classes $[\mathcal{O}_i] \in \mathcal{H}_{\text{top}}$;
ii. $\frac{\delta}{\delta g_\Sigma^{\mu\nu}} \langle \mathcal{O}_{i_1}(x_1) \cdots \mathcal{O}_{i_n}(x_n) \rangle = 0$;
iii. the correlation functions of local operators are independent of the insertion positions;
iv. for any local operator $\mathcal{O}_i(x) \in \mathcal{H}_{\text{top}}$ we have $d\mathcal{O}_i = Q_{\text{top}} \cdot \mathcal{O}_i^{(1)}$, where $\mathcal{O}_i^{(1)}$ is a 1-form operator.

To show the last two statements we recall that translations are generated by the momentum operator in the sense $\partial_\mu \mathcal{O}(x) = -[P_\mu, \mathcal{O}(x)]$, and the momentum operator is derived from $T_{\mu\nu}$.

The last result is the first step in topological descent (perhaps ascent would be better terminology): given a local k-form operator $\mathcal{O}^{(k)}$, we have $d\mathcal{O}^{(k)} = Q_{\text{top}} \cdot \mathcal{O}^{(k+1)}$. This leads to two developments. First, in addition to the local topological observables $\mathcal{O} \in \mathcal{H}_{\text{top}}$, we naturally have non-local observables for any 1-cycle $\gamma \in \Sigma$

$$\mathcal{O}[\gamma] = \int_\gamma \mathcal{O}^{(1)}, \qquad \mathcal{O}[\Sigma] = \int_\Sigma \mathcal{O}^{(2)}. \tag{5.5.3}$$

Second, we can package these conveniently in terms of topological superfields. Introduce anti-commuting variables θ_{top}^μ, and for any local bosonic 0-form operator $\mathcal{O}$ define

$$\Phi_\mathcal{O}(x, \theta_{\text{top}}) = \mathcal{O}(x) + \theta_{\text{top}}^\mu \mathcal{O}_\mu^{(1)}(x) + \theta_{\text{top}}^\mu \theta_{\text{top}}^\nu \mathcal{O}_{\mu\nu}^{(2)}(x). \tag{5.5.4}$$

Introducing a superspace derivative

$$\mathcal{D}_{\text{top}} = \mathcal{Q}_{\text{top}} - \theta^\alpha \frac{\partial}{\partial x^\alpha}\,, \tag{5.5.5}$$

we summarize descent by the statement $\mathcal{D}_{\text{top}}\Phi_\mathcal{O} = 0.$[35] In all of our topological discussions we will stick to bosonic local operators; this will be sufficient for most of our linear sigma model goals and will ease up the notation.

We can use descent to define a deformation of the theory in a fashion that closely resembles conformal perturbation theory. Indeed, this is the case in which one can define the latter structure with some degree of rigor and generality. We set

$$\langle\ \cdots\ \rangle(\lambda) = \langle\ \cdots\ \exp\left[\lambda^a \mathcal{O}_a^{(2)}[\Sigma]\right]\rangle\,. \tag{5.5.6}$$

We expect a reasonable theory to have well-defined two-point functions of local bosonic operators $\eta_{ab} = \langle \mathcal{O}_a \mathcal{O}_b \rangle_{\Sigma=\mathbb{P}^1}$: η_{ab} should be symmetric and invertible, and we expect it to be complete, in the sense that gluing a handle into the Riemann surface can be replaced by an insertion of $\mathcal{O}_a\rangle\eta^{ab}\langle\mathcal{O}_b$ in the correlation functions; for instance, we might pinch a torus and arrive at

$$\langle \mathcal{O}_{a_1}\mathcal{O}_{a_2}\mathcal{O}_{a_3}\mathcal{O}_{a_4}\rangle_{T^2} = \sum_{b_1 b_2}\langle \mathcal{O}_{a_1}\mathcal{O}_{a_2}\mathcal{O}_{b_1}\rangle_{\mathbb{P}^1}\,\eta^{b_1 b_2}\langle \mathcal{O}_{b_1}\mathcal{O}_{a_2}\mathcal{O}_{a_4}\rangle_{\mathbb{P}^1}\,. \tag{5.5.7}$$

When this holds it is not hard to argue that any correlation function on any Σ reduces to sums and products of the three-point functions on the sphere:

$$C_{abc} = \langle \mathcal{O}_a \mathcal{O}_b \mathcal{O}_c \rangle_{\mathbb{P}^1}\,. \tag{5.5.8}$$

Since the local operators include the identity operator $\mathcal{O}_0 = 1$, these correlation functions include $C_{ab0} = \eta_{ab}$. The C_{abc} are completely symmetric in the indices and define a commutative and associative product:

$$\mathcal{O}_a \cdot \mathcal{O}_b = \sum_c C_{ab}{}^c \mathcal{O}_c\,, \tag{5.5.9}$$

where $C_{ab}{}^c = \sum_d C_{abd}\eta^{dc}$. All of this generalizes to the family of theories defined by (5.5.6) and is reminiscent of the structures we discussed in Chap. 2.

---

[35]Up to a sign same structure works for fermionic $\mathcal{O}$ as well.

Topological Twists of an SCFT

Consider a compact unitary (2,2) SCFT with $c = \bar{c}$ and weights and $U(1)_L \times U(1)_R$ charges satisfying the requirements familiar from our discussion of topological heterotic rings in Sect. 3.4:

i. the spins of the fields are integral or half-integral: $h - \bar{h} \in \frac{1}{2}\mathbb{Z}$;
ii. $q - \bar{q} \in \mathbb{Z}$, and the fermion number satisfies $(-1)^F = (-1)^{q-\bar{q}}$;
iii. $h - \bar{h}$ is integral for bosonic operators and half-integer for fermionic operators.

We would like to study this theory on a curved worldsheet, and, just as in Sect. 3.6, we can do so by modifying the coupling to worldsheet gravity by twisting the Lorentz current, or, equivalently, by turning on a background gauge field for either $U(1)_A$ or $U(1)_B$ symmetry. Taking, the first perspective, we construct modified Lorentz currents

$$\mathcal{J}^{B\,\text{twist}} = \mathcal{J}_\Lambda + \mathcal{J}_B \, , \qquad \mathcal{J}^{A\,\text{twist}} = \mathcal{J}_\Lambda + \mathcal{J}_A \qquad (5.5.10)$$

where $\mathcal{J}_B = \frac{1}{2}(\mathcal{J}_R - \mathcal{J}_L)$ and $\mathcal{J}_A = \frac{1}{2}(\mathcal{J}_R + \mathcal{J}_L)$.[36] We then have the following assignments for the supercharges and operators in the c,c and a,c rings:

	Q	$\bar{Q}$	Q'	$\bar{Q}'$	$\mathcal{O}_{cc}^{q,\bar{q}}$	$\mathcal{O}_{ac}^{q,\bar{q}}$
spin	$-\frac{1}{2}$	$-\frac{1}{2}$	$\frac{1}{2}$	$\frac{1}{2}$	$\frac{q-\bar{q}}{2}$	$-\frac{q+\bar{q}}{2}$
$U(1)_A$	$-\frac{1}{2}$	$\frac{1}{2}$	$-\frac{1}{2}$	$\frac{1}{2}$	$\frac{q+\bar{q}}{2}$	$\frac{q+\bar{q}}{2}$
$U(1)_B$	$-\frac{1}{2}$	$\frac{1}{2}$	$\frac{1}{2}$	$-\frac{1}{2}$	$\frac{\bar{q}-q}{2}$	$\frac{\bar{q}-q}{2}$
spin^A	-1	1	0	0	q	0
spin^B	-1	0	$+1$	0	0	$-q$

$$(5.5.11)$$

We have two candidate operators for the topological twist: the twisted right-moving supercharge $\bar{Q}$, and the twisted left-moving supercharge $\bar{Q}'$. In our chiral algebra discussion in Chap. 3 we already observed that the worldsheet $\bar{\partial}$ is trivialized on the $\bar{Q}$ cohomology. In exactly the same way, ∂ is trivialized on the $\bar{Q}'$ cohomology. Thus, if we set

$$\mathcal{Q}_B = \bar{Q} + \bar{Q}' \, , \qquad (5.5.12)$$

then $d = \partial + \bar{\partial}$ will act trivially on the cohomology $\mathcal{H}_B$. Moreover, we see that the space of local operators in $\mathcal{H}_B$ is isomorphic to the vector space of (c,c) operators

---

[36]This definition is consistent with (5.4.14) and differs by a factor of 2 from the one used in Sect. 3.6, where the twist has a compensating $\frac{1}{2}$, so that twisted Lorentz currents $\mathcal{J}^{B\,\text{twist}}$ agree.

of the untwisted SCFT. The resulting topological theory is the B-model associated to the SCFT.

Exercise 5.38 Show that the SCFT energy-momentum tensor is Q_B-exact, and that the modification of the Lorentz generator amounts to redefining the energy-momentum tensor to have components

$$T_B = T + \tfrac{1}{2}\partial J_L , \qquad\qquad \overline{T}_B = T + \tfrac{1}{2}\bar{\partial} J_R . \qquad (5.5.13)$$

Verify that:

i. T_B and $\overline{T}_B$ have no conformal anomaly;
ii. T_B and $\overline{T}_B$ have non-singular OPEs with $\mathcal{O}_{cc}^{q\bar{q}}$.

Finally, compute the flat space correlators

$$\langle T_B(z_1) J_L(z_2)\rangle = -\frac{c}{6}\frac{1}{z_{12}^3} , \qquad \langle \overline{T}_B(\bar{z}_1) J_R(\bar{z}_2)\rangle = -\frac{\bar{c}}{6}\frac{1}{z_{12}^3} . \qquad (5.5.14)$$

□

The $U(1)_A$ symmetry with current $\mathcal{J}_A = \tfrac{1}{2}(\mathcal{J}_R + \mathcal{J}_L)$ assigns charge $+1/2$ to Q_B and therefore grades the $\mathcal{H}_B$. In analogy with the ghost symmetry of standard BRST treatments, we define the ghost symmetry by the current $J_{gh} = 2J_A$, which assigns charge $+1$ to Q_B. The last computation from the exercise, which the reader should compare to the discussion at the end of Sect. 2.5, shows that J_{gh} is anomalous in a background curved geometry, which leads to the ghost number selection rule on a genus g Riemann surface [134]:

$$\langle \mathcal{O}_{a_1}\cdots\mathcal{O}_{a_k}\rangle_{\Sigma_g} = 0 \qquad (5.5.15)$$

unless

$$\sum_i q_{gh,i} = \tfrac{c}{3}(1 - g) . \qquad (5.5.16)$$

Our interest is primarily in genus 0.

We can deform the topological theory by adding two-form topological descendants as indicated above. The deformations that preserve the ghost symmetry are precisely the (c,c) marginal operators, while deformations of descendants of operators with $q_{gh} \neq 2$ define what is known as the "thickened moduli space." Correlation functions in such theories will have modified ghost-number selection rules. This has some remarkable properties; for example, there are preferred coordinates in which the two-point function η_{ab} is flat. This is the seed that leads to tt^* geometry and the special geometry of (2,2) SCFT moduli spaces [77, 113, 236].

Suppose that the theory has integral R-charges. In that case $q + \bar{q} \in Z$ as well, and the theory admits an A-twist, with $\mathcal{J}_{\Lambda'}^{\text{A twist}} = \mathcal{J}_\Lambda + J_A$, topological charge $Q_A = Q' + \overline{Q}$ and ghost number current $J_{\text{gh}} = 2J_B$. As a vector space the local observables are now isomorphic to the (a,c) ring of the untwisted theory. This is the A-model associated to the SCFT.

Twists and RG Flow

In the SCFT the A and B models are arbitrary names: flipping the sign of the $U(1)_L$ charge exchanges the two notions. On the other hand, suppose that we have an accident-free flow from a UV quantum field theory to the SCFT of the sort just described. In that case we have all of the ingredients necessary to make the topological twists and projections already in the definition of the UV theory. If the formal topological field theory properties hold for a twist of the UV theory, then it must be a scale-invariant theory, and, in particular, it must agree with the topologically twisted SCFT.

All of this should sound familiar. We made much the same statements in Chap. 3 when discussing the chiral algebra of a (0,2) theory. In (2,2) theories twisting isolates finite-dimensional sub-algebras of the chiral algebra. The A/2 and B/2 heterotic rings that played a prominent role in the preceding chapters have associated twists that generalize the topological twists of (2,2) theories.

The bottom line is that the topological A and B models can be used to check any mirror proposal. If the RG fixed points of two theories T and T°, such as a pair of linear sigma models, are mirror isomorphic (i.e. related by the sign of the $U(1)_L$ charge), then it must be that

$$A(T) = B(T^\circ) , \qquad\qquad B(T) = A(T^\circ) . \qquad\qquad (5.5.17)$$

This, and its topological string theory generalization, provide the most extensive evidence for mirror symmetry.

We will focus on topological theories with genus 0 worldsheet, where all of the computations boil down to three-point functions of the local observables. When the topological theory arises from a twisted SCFT, the topological three-point functions are related to the OPE coefficients of the (a,c) or (c,c) ring operators. The corresponding correlation functions are zero in the physical theory due to the $U(1)_L \times U(1)_R$ symmetry. The twist introduces a background charge that allows for non-zero topological correlation functions that obey the selection rule (5.5.16).

In theories with spectral flow the twists can be interpreted directly as insertions of spectral flow operators. For $E_8 \times E_8$ heterotic compactification using a $c = \bar{c} = 9$ theory with integral $U(1)_L \times U(1)_R$ charges, this leads to the famous correspondences between spacetime superpotential couplings and topological correlation functions:

$$27_i \cdot 27_j \cdot 27_k \supset (\mathbf{16}_{+1/2})_i \cdot (\mathbf{10}_{-1})_j \cdot (\mathbf{16}_{+1/2})_k \propto \langle \mathcal{O}^{1,1}_{cc,i} \mathcal{O}^{1,1}_{cc,j} \mathcal{O}^{1,1}_{cc,k} \rangle^{\text{B-model}}_{\mathbb{P}^1},$$

$$\overline{27}_{i'} \cdot \overline{27}_{j'} \cdot \overline{27}_{k'} \supset (\overline{\mathbf{16}}_{-1/2})_{i'} \cdot (\mathbf{10}_{+1})_{j'} \cdot (\overline{\mathbf{16}}_{-1/2})_{k'}$$

$$\propto \langle \mathcal{O}^{-1,1}_{ac,i'} \mathcal{O}^{-1,1}_{ac,j'} \mathcal{O}^{-1,1}_{ac,k'} \rangle^{\text{A-model}}_{\mathbb{P}^1}. \tag{5.5.18}$$

The proportionality constants depend on the normalization of the spacetime fields. They can be fixed on the $(2,2)$ locus, but the dependence on more general $(0,2)$ marginal deformations has not been computed.

5.5.2 The Topological Sector of the V-model

The V-models we described above do not, as far as we know, flow to non-trivial SCFTs. The interpretation we gave above, based on the σ-vacua in the IR implies a gapped theory. Gapped, however, does not imply trivial: the V-models have a rich topological sector at zero energy, and by the previous discussion, we can probe that sector via a topological twist. This is not particular to linear sigma models: cohomological topological field theories were introduced in part to give a field-theoretic framework for a number of deep problems in symplectic geometry [386] that now go under the name of Gromov-Witten theory. However, the linear models are particularly tractable examples of this general structure.

Our motivation to study these is two-fold: first, we will be able to describe the topological sector of a wide class of quantum field theories; second, we will see in the sequel that V-model computations can be used effectively to obtain results for linear sigma models that do flow to non-trivial SCFTs.

Local Observables in the A-twisted V-model

Let us now consider the action and symmetries of the A-twisted V-model. The topological charge is

$$Q_A = Q' + \overline{Q}, \tag{5.5.19}$$

and, just as in Exercise 5.6, it is not hard to see that most terms in the linear sigma model action are Q_A-exact, so that the A-twisted theory can only depend on the holomorphic Fayet-Iliopoulos couplings τ^α, or, equivalently, on $q_\alpha = e^{2\pi i \tau^\alpha}$.

Clearly, the next step is to construct the cohomology $\mathcal{H}_A$ of Q_A. This is not as straightforward as in the $(0,2)$ Landau-Ginzburg theories we discussed in Chap. 3 due to the gauge symmetry of the theory. Strangely, to the author's knowledge this has not been carried out. It would be useful to work out these details for general linear sigma models. For example, such results would undoubtedly cast light on the somewhat mysterious nature of non-toric and non-polynomial deformations.

Fortunately, it is easy to identify a non-trivial class of operators in $\mathcal{H}_A$: the σ_α are gauge-invariant Q_A-closed operators that represent non-trivial cohomology classes in $\mathcal{H}_A$. In the topological theory these operators have a non-singular OPE, so that we can consider polynomials $P(\sigma) \in \mathbb{C}[\sigma_1, \ldots, \sigma_{n_G}]$ and focus on the topological correlators

$$\langle \sigma_{\alpha_1} \sigma_{\alpha_2} \cdots \sigma_{\alpha_k} \rangle_A(q) . \tag{5.5.20}$$

In a V-model for a NEF Fano toric variety the σ_α generate the Q_A-cohomology of local operators.[37] This can be understood as follows. First, since the UV phase is described by a smooth geometry, we can apply the A-twist to the non-linear sigma model with target space V. For a compact V the set of local A-model observables are in 1:1 correspondence with the de Rham cohomology $H^\bullet(V, \mathbb{R})$—see, for instance, [390]. Second, we know that the cohomology of the Fano toric variety is generated by the classes $\eta_\alpha \in H^2(V, \mathbb{R})$ modulo the Stanley-Reisner relations discussed in (5.3.72), so for each of these we obtain an operators $\mathcal{O}[\eta_\alpha]$ in the A-twisted non-linear sigma model. Finally, topological descent relates the σ_α to the complexified Fayet-Iliopoulos parameters τ^α, while in the non-linear model, topological descent relates the operators $\mathcal{O}[\eta_\alpha]$ to t^α—the complexified Kähler parameters.

It follows that the σ_α generate the A-twisted linear model's cohomology of local operators. Up to a change of coordinates relating the t^α of the non-linear model to the τ^α of the linear theory, these operators have a geometric interpretation: the σ_α correspond to classes in $H^{1,1}(V, \mathbb{C})$. In the classical $q_\alpha \to 0$ limit we expect to recover classical geometry, i.e. the correlation functions should just be determined by classical intersection theory. For instance, for $V = \mathbb{P}^n$ we expect

$$\langle \sigma^k \rangle = \delta_n^k + O(q) . \tag{5.5.21}$$

Since $H^\bullet(V, \mathbb{R})$ is finite-dimensional, we also expect that $\mathcal{H}_A$ should be finite-dimensional. As we will now describe, this turns out to be correct, and the resulting structure is known as the quantum cohomology of V [51]. It takes the form

$$\mathcal{H}_A = \frac{\mathbb{C}[\sigma_1, \ldots, \sigma_{n_G}]}{\mathcal{I}(V, q)} \xrightarrow[q \to 0]{} H^\bullet(V, \mathbb{R}) , \tag{5.5.22}$$

---

[37] For toric varieties that are not NEF Fano, the situation is less clear [115, 296, 297].

where $\mathcal{I}(V, q)$ is the ideal of quantum cohomology relations. In the example of $V = \mathbb{P}^n$ we will show

$$\mathcal{H}_A = \frac{\mathbb{C}[\sigma]}{\sigma^{n+1} - q} . \tag{5.5.23}$$

The idea of quantum cohomology is not special to toric varieties: it makes sense for any A-twisted non-linear sigma model for a compact manifold and forms a cornerstone of Gromov-Witten theory. A discussion of various quantum cohomology theories defined in Gromov-Witten theory can be found in chapter 8 of [120].What makes the Fano toric case so nice is that it is so readily computable. Remarkably, if we work in the linear sigma model basis, and in particular, in the natural coordinates q_α, the same turns out to be true for the toric part of the quantum cohomology of Calabi-Yau hypersurfaces we will discuss below.

To compute the quantum cohomology of the V model we will need to delve a little bit into the details of the twisting and then use localization. Fortunately, the reader will find that much of the work has already been done.

Twisting the Fields

We start with the field content of the V model and its symmetries. With a look ahead, we will describe not just the A-twist—our main interest, but also the B-twist and the half-twist. In the latter we shift the spins by the R-symmetry: $\mathcal{J}^{1/2} = \mathcal{J}_A + \frac{1}{2}\mathcal{J}_R$. We organize the results by (2,2) multiplets.

Matter multiplet with $U(1)_L \times U(1)_R$-charge $q = \bar{q} = 0$:

	ϕ	$\bar{\phi}$	ψ	$\bar{\psi}$	γ	$\bar{\gamma}$
spin	0	0	$-\frac{1}{2}$	$-\frac{1}{2}$	$\frac{1}{2}$	$\frac{1}{2}$
$U(1)_A$	0	0	$-\frac{1}{2}$	$\frac{1}{2}$	$-\frac{1}{2}$	$\frac{1}{2}$
$U(1)_B$	0	0	$-\frac{1}{2}$	$\frac{1}{2}$	$\frac{1}{2}$	$-\frac{1}{2}$
spin^A	0	0	-1	0	0	1
spin^B	0	0	-1	0	1	0
spin$^{1/2}$	0	0	-1	0	1/2	1/2

$$\tag{5.5.24}$$

Matter multiplet with $U(1)_L \times U(1)_R$ charge $q = \bar{q} \neq 0$:

	ϕ	$\bar{\phi}$	ψ	$\bar{\psi}$	γ	$\bar{\gamma}$
spin	0	0	$-\frac{1}{2}$	$-\frac{1}{2}$	$\frac{1}{2}$	$\frac{1}{2}$
$U(1)_A$	q	$-q$	$\frac{2q-1}{2}$	$\frac{1-2q}{2}$	$\frac{2q-1}{2}$	$\frac{1-2q}{2}$
$U(1)_B$	0	0	$-\frac{1}{2}$	$\frac{1}{2}$	$\frac{1}{2}$	$-\frac{1}{2}$
spin^A	q	$-q$	$q-1$	$-q$	q	$1-q$
spin^B	0	0	-1	0	1	0
$\text{spin}^{1/2}$	$q/2$	$-q/2$	$\frac{q-2}{2}$	$-\frac{q}{2}$	$\frac{q+1}{2}$	$\frac{1-q}{2}$

$$(5.5.25)$$

Gauge multiplet:

	σ	$\bar{\sigma}$	χ	$\bar{\chi}$	υ	$\bar{\upsilon}$	$F_{12} \mp D$
spin	0	0	$-\frac{1}{2}$	$-\frac{1}{2}$	$\frac{1}{2}$	$\frac{1}{2}$	0
$U(1)_A$	0	0	$-\frac{1}{2}$	$\frac{1}{2}$	$\frac{1}{2}$	$-\frac{1}{2}$	0
$U(1)_B$	1	-1	$\frac{1}{2}$	$-\frac{1}{2}$	$\frac{1}{2}$	$-\frac{1}{2}$	0
spin^A	0	0	-1	0	1	0	0
spin^B	0	0	0	-1	1	0	0
$\text{spin}^{1/2}$	$\frac{1}{2}$	$-\frac{1}{2}$	$-\frac{1}{2}$	$-\frac{1}{2}$	1	0	0

$$(5.5.26)$$

A few things are worth noting about these spin assignments. First, F_{12} and D remain worldsheet scalars (not so the auxiliary fields G and $\bar{G}$ that we did not explicitly list in the table). Second, the fields with non-trivial spin after the twist naturally assemble into differential forms. For instance, in the A-twisted gauge multiplet the $\bar{\chi}$ and $\bar{\upsilon}$ are worldsheet scalars, and we have the fermionic 1-form

$$\upsilon dz + \chi d\bar{z} . \qquad (5.5.27)$$

Let K_Σ be the canonical bundle of the worldsheet, so that in the twisted theory

$$\bar{\upsilon} \in \mathscr{A}^0(\Sigma) , \quad \bar{\chi} \in \mathscr{A}^0(\Sigma) , \quad \upsilon \in \mathscr{A}^0(\Sigma, K_\Sigma) , \quad \chi \in \mathscr{A}^0(\Sigma, \overline{K}_\Sigma) . \qquad (5.5.28)$$

For an A-twisted matter multiplet with zero R-charge, we assemble the spin ± 1 fermions as

$$\bar{\gamma} dz + \psi d\bar{z} , \qquad (5.5.29)$$

but this is a little misleading, since the expression is obviously not gauge-covariant. We will be interested in global properties of our fields on the worldsheet, and for the matter fields this requires that we specify a background gauge field in a topological sector specified by the instanton numbers

$$N_\alpha = -\frac{1}{2\pi} \int F_\alpha \,. \tag{5.5.30}$$

The matter field ϕ_ρ is then a section of a holomorphic line bundle $\mathcal{L}_\rho \to \mathbb{P}^1$, with

$$\mathcal{L}_\rho = \mathcal{O}(d_\rho) \,, \qquad\qquad c_1(\mathcal{L}_\rho) = d_\rho \,, \tag{5.5.31}$$

where the integer d_ρ is determined by the gauge charges and instanton number:

$$d_\rho = \sum_\alpha Q_\rho^\alpha N_\alpha \,. \tag{5.5.32}$$

We will often denote the instanton numbers and degrees collectively by

$$\vec{d} \in \mathbb{Z}^n \,, \qquad\qquad \vec{N} \in \mathbb{Z}^{n_G} \,. \tag{5.5.33}$$

As we will soon see, $\vec{N} \in H_2(V, \mathbb{Z})$, and $\vec{N}$ will have an interpretation as the homology class of a worldsheet instanton.

We can now describe the twisted fields globally:

$$\phi_\rho \in \mathscr{A}^0(\Sigma, \mathcal{L}_\rho) \,, \qquad\qquad \overline{\phi}_\rho \in \mathscr{A}^0(\Sigma, \overline{\mathcal{L}}_\rho) \,,$$
$$\psi_\rho \in \mathscr{A}^0(\Sigma, \overline{K}_\Sigma \otimes \overline{\mathcal{L}}_\rho^*) \,, \qquad\qquad \overline{\psi}_\rho \in \mathscr{A}^0(\Sigma, \overline{\mathcal{L}}_\rho) \,,$$
$$\gamma_\rho \in \mathscr{A}^0(\Sigma, \mathcal{L}_\rho) \,, \qquad\qquad \overline{\gamma}_\rho \in \mathscr{A}^0(\Sigma, K_\Sigma \otimes \mathcal{L}_\rho^*) \,. \tag{5.5.34}$$

In making this assignment we use a trick we met before in discussing (0,2) non-linear sigma models, where we used the metric on a Hermitian bundle to "lower the index" on the $\overline{\gamma}^i$ left-moving fermions. We do the same here for the $\overline{\gamma}_\rho$ and ψ_ρ fields. This will be convenient for discussing the zero modes of the kinetic operators for these fields, which will correspond to global holomorphic/anti-holomorphic sections of the displayed line bundles.

Exercise 5.39 Show that the A-twist of a matter multiplet with $U(1)_L \times U(1)_R$ charge $q = \overline{q} = 1$ leads to the following assignment of sections. (Again, with an astute use of the Hermitian metric on the bundle $\mathcal{L}_0 \to \Sigma$.)

$$\phi_0 \in \mathscr{A}^0(\Sigma, K_\Sigma \otimes \mathcal{L}_0) \,, \qquad\qquad \overline{\phi}_0 \in \mathscr{A}^0(\Sigma, \overline{K}_\Sigma \otimes \overline{\mathcal{L}}_0) \,,$$
$$\psi_0 \in \mathscr{A}^0(\Sigma, \overline{\mathcal{L}}_0^*) \,, \qquad\qquad \overline{\psi}_0 \in \mathscr{A}^0(\Sigma, \overline{K}_\Sigma \otimes \overline{\mathcal{L}}_0) \,,$$
$$\gamma_0 \in \mathscr{A}^0(\Sigma, K_\Sigma \otimes \mathcal{L}_0) \,, \qquad\qquad \overline{\gamma}_0 \in \mathscr{A}^0(\Sigma, \mathcal{L}_0^*) \,. \tag{5.5.35}$$

This will be important in our discussion of the A-twist for the X-model.

Finally, we note that the V model has additional symmetries that include the non-chiral torus symmetries of the toric variety. These can in principle be used to define a wider class of twisted theories. Our interest is in situations where these additional symmetries will be broken, either by deformations that break the left-moving supersymmetry, or by turning on chiral superpotential couplings. Thus, we will stick to the A/B twists just described.

Localization and σ Correlators—UV Phase

Our goal is to compute the σ correlation functions. In this section we will discuss localization in the UV phase, following [311]. In the next section we will verify the conjecture by a localization computation in the IR phase.

Just as in our localization analysis for Landau-Ginzburg theories in Sect. 3.6, the first order of business is to identify the zero modes for the various fields. Working in a fixed instanton background, this is easily read off from (5.5.34), (5.5.28) and the kinetic terms. In particular, the zero modes of the fermions are easily counted, and give a quick way to arrive at the ghost number selection rule (5.5.16). The formal argument is simple:

$$\langle \sigma_{\alpha_1} \cdots \sigma_{\alpha_k} \rangle_A(q) = \int D[\text{fields}] e^{-S} \sigma_{\alpha_1} \cdots \sigma_{\alpha_k} \tag{5.5.36}$$

The action is invariant under a $U(1)_B$ rotation with $\sigma \mapsto e^{i\beta}\sigma$, and all other fields transforming accordingly. As usual, while the non-zero-mode part of the measure can be regulated in a $U(1)_B$-invariant fashion, the same is not true for the zero modes, which leads to the following transformation for the measure:

$$D[\text{fields}] \mapsto e^{iA\beta} D[\text{fields}] , \tag{5.5.37}$$

where

$$A = n_G(\tfrac{1}{2} + \tfrac{1}{2}) + \sum_{\rho}\left[-\tfrac{1}{2}(\#_0(\overline{\psi}_\rho) - \#_0(\psi_\rho)) - \tfrac{1}{2}(\#_0(\gamma_\rho) - \#_0(\overline{\gamma}_\rho)) \right] . \tag{5.5.38}$$

The first term is due to the $\overline{v}_\alpha$ and $\overline{\chi}_\alpha$ zero modes.

Exercise 5.40 Use Kodaira-Serre duality and the Riemann-Roch theorem to compute

$$A = -d + \sum_{\alpha} R^\alpha N_\alpha , \tag{5.5.39}$$

where

$$d = n - n_G , \qquad\qquad R^\alpha = \sum_\rho Q^\alpha_\rho . \qquad (5.5.40)$$

On a genus g Riemann surface, these two theorems are

$$H^0_{\bar\partial}(\Sigma_g, \mathcal{L}) \simeq \overline{H^1_{\bar\partial}(\Sigma_g, \mathcal{L}^* \otimes K_{\Sigma_g})} ,$$

$$\chi(\Sigma_g, \mathcal{L}) = h^0(\Sigma_g, \mathcal{L}) - h^1(\Sigma_g, \mathcal{L}) = c_1(\mathcal{L}) + 1 - g . \qquad (5.5.41)$$

$\square$

The $\vec{N}$-dependent term in A is canceled by assigning $U(1)_B$ charge to the q_α, and we obtain a selection rule along the lines of (5.5.16):

$$\langle \sigma_{\alpha_1} \cdots \sigma_{\alpha_k} \rangle_A(q) = e^{i(k-d)\beta} \langle \sigma_{\alpha_1} \cdots \sigma_{\alpha_k} \rangle_A(e^{iR^\alpha\beta} q_\alpha) . \qquad (5.5.42)$$

The path integral will localize on configurations annihilated by $Q_A = Q' + \overline{Q}$. Because of the $U(1)_A$ symmetry that assigns opposite charges to Q' and $\overline{Q}$, this means we are interested in configurations that are annihilated by both Q' and $\overline{Q}$. Taking a look at (5.2.32), (5.2.33), we see that $\overline{Q}$ requires

$$\overline{\nabla}\phi_\rho = 0 , \quad \bar\partial\sigma_\alpha = 0 , \quad E^\rho = \sum_\alpha \phi_\rho \sigma_\alpha Q^\alpha_\rho = 0 , \quad (F_{12} + D)_\alpha = 0 . $$

$$(5.5.43)$$

In order to be annihilated by Q', the field configuration should also satisfy $\nabla\overline{\phi}_\rho = 0$ and $\partial\sigma_\alpha = 0$. These do not lead to new constraints: $\nabla\overline{\phi}_\rho = 0$ is the conjugate of $\overline{\nabla}\phi_\rho = 0$, and since σ_α is just a complex scalar field, on a compact Riemann surface $d\sigma_\alpha = 0$ is equivalent to $\bar\partial\sigma_\alpha = 0$. This will be important for us: the A-twisted and half-twisted path integrals of the linear sigma model localize onto the same field configurations.

The Bogomolnyi bound from exercise (5.7) then shows that in each topological sector the dependence on the q_α is solely through the prefactor, so that

$$\langle \sigma_{\alpha_1} \cdots \sigma_{\alpha_k} \rangle_A = \sum_{\vec{N}} \prod_\alpha q_\alpha^{N_\alpha} \mathcal{N}^{\vec{N}}_{\alpha_1\cdots\alpha_k} , \qquad (5.5.44)$$

where the coefficients $\mathcal{N}$ are to be computed by localization in each of the instanton sectors. We expect the sum to converge in the UV limit, and the ghost number selection rule implies that the instantons that contribute to any fixed correlation function lie in a hyperplane

$$\sum_\alpha R^\alpha N_\alpha = k - d . \qquad (5.5.45)$$

Let us now discuss the field configurations of (5.5.43) further. Each $\phi_\rho \in H^0(\Sigma, \mathcal{L}_\rho)$ is a holomorphic section of the corresponding bundle, and these are characterized as follows: when $d_\rho < 0$ there are no holomorphic sections and $\phi_\rho = 0$. On the other hand, when $d_\rho \geq 0$, there is a space of sections of dimension $d_\rho + 1$. If we take z to be an affine coordinate on $\Sigma = \mathbb{P}^1$, then we can expand the field as

$$\phi_\rho = \phi_{\rho 0} + \phi_{\rho 1} z + \phi_{\rho 2} z^2 + \ldots + \phi_{\rho d_\rho} z^{d_\rho} . \tag{5.5.46}$$

The remaining constraint from the vanishing of $(F_{12} + D)_\alpha$ is

$$\frac{e^2}{4\pi} \left(\sum_\rho Q_\rho^\alpha |\phi_\rho|^2 - 2r^\alpha \right) - \frac{1}{2\pi} F_{12\alpha} = 0 . \tag{5.5.47}$$

We want to study the moduli space of solutions to these equations modulo the $U(1)^{n_G}$ gauge invariance with z-dependent gauge transformation parameters. We will call the resulting instanton moduli space $\mathcal{M}_{\vec{N}}$. The equation and its gauge invariance is strikingly reminiscent of the symplectic quotient construction discussed above—it is "just" an infinite-dimensional generalization thereof. This can be made precise [97], and the result is that (5.5.47) can be treated as a gauge-fixing condition for the $G_\mathbb{C}$-invariance of the remaining equations. Following [311, 388], we now summarize the results.

1. Consider the holomorphic quotient construction for V in the UV phase:

$$V = \frac{Y \setminus F}{G_\mathbb{C}} . \tag{5.5.48}$$

The collection of the $\{\phi_\rho\}$ determine a holomorphic map $f : \Sigma = \mathbb{P}^1 \to Y$. The set of all such maps is the space

$$Y_{\vec{N}} = \oplus_\rho H_{\bar{\partial}}^0(\Sigma, \mathcal{L}_\rho) \tag{5.5.49}$$

of dimension

$$\dim Y_{\vec{N}} = \sum_\rho h^0(\Sigma, \mathcal{L}_\rho) . \tag{5.5.50}$$

2. If the image of the map is contained in F, then its $G_\mathbb{C}$ orbit will not intersect the solution set to (5.5.47). Thus, to construct $\mathcal{M}_{\vec{N}}$ as a holomorphic quotient, we must remove all maps with image contained in F. This is the exceptional set $\mathcal{F}_{\vec{N}}$, and it is constructed as follows. For each primitive collection I for the fan Σ_V, F contains the exceptional set $\cap_{\rho \in I} \{\phi_\rho = 0\}$, and

$$\mathcal{F}_{\vec{N}} \supset \cap_{\rho \in I} \{\phi_{\rho 0} = 0\} \cap \{\phi_{\rho 1} = 0\} \cap \cdots \cap \{\phi_{\rho d_\rho} = 0\} . \tag{5.5.51}$$

3. For each map $f \in Y_{\vec{N}} \setminus \mathscr{F}_{\vec{N}}$ (5.5.47) determines a unique connection, so that the $G_{\mathbb{C}}$ invariance is fixed up to constant gauge transformations. These act on Y and preserve $\mathscr{F}_{\vec{N}}$, and, finally,

$$\mathscr{M}_{\vec{N}} = \frac{Y_{\vec{N}} \setminus \mathscr{F}_{\vec{N}}}{G_{\mathbb{C}}}. \tag{5.5.52}$$

Exercise 5.41 Prove the following useful results for $\mathscr{M}_{\vec{N}}$:

1. $\mathscr{M}_{\vec{0}} = V$.
2. $\dim \mathscr{M}_{\vec{N}} = d + \sum_{\rho} d_{\rho} - \sum_{\rho \mid d_{\rho} \leq -2}(d_{\rho} + 1)$.
3. $\mathscr{M}_{\vec{N}}$ is empty unless $\vec{N} \in K^*(V) \subset H_2(V, \mathbb{R})$—the closure of the cone dual to the the Kähler cone $K(V) \subset H^2(V, \mathbb{R})$.[38]
4. For a fixed ρ the toric divisors $D_{\rho 0}, D_{\rho 1}, \ldots D_{\rho d_{\rho}}$ are linearly equivalent and correspond to the same class in $H^2(\mathscr{M}_{\vec{N}}, \mathbb{Z})$. We will denote it by $\xi_{\rho, \vec{N}}$. □

Localization of the path integral therefore reduces the computation of $\mathcal{N}_{a_1 \cdots a_k}^{\vec{N}}$ to an integral over the $\mathscr{M}_{\vec{N}}$. This integral should be over a top form on $\mathscr{M}_{\vec{N}}$. A natural guess, based on the relation between the σ_{α} and cohomology classes of V, is that each σ_{α} insertion corresponds to a $(1,1)$ form on $\mathscr{M}_{\vec{N}}$. From the explicit description of the cohomology of a toric variety, the most natural choice is the class $\eta_{\alpha, \vec{N}} \in H^2(\mathscr{M}_{\vec{N}}, \mathbb{Z})$.

To see how this works more explicitly, we focus on the zero mode sector. The fermion zero modes are then contained in the Yukawa couplings in the action (5.4.10)

$$S \supset \frac{1}{2\pi} \int d^2z \left\{ -\tfrac{1}{2} \overline{\psi}^{\rho} \phi^{\rho} Q_{\rho}^{\alpha} \overline{v}_{\alpha} - \overline{E}_{,\alpha}^{\rho} \overline{\chi}^{\alpha} \gamma^{\rho} - \overline{E}_{,\rho'}^{\rho} \overline{\psi}^{\rho'} \gamma^{\rho} - \overline{\gamma}^{\rho} E_{,\rho'}^{\rho} \psi^{\rho'} \right\}. \tag{5.5.53}$$

The integral for $\mathcal{N}_{\alpha_1, \ldots, \alpha_k}^{\vec{N}}$ will have the form

$$\mathcal{N}_{\alpha_1, \ldots, \alpha_k}^{\vec{N}} \propto \int D[\text{fields}] \sigma_{\alpha_1} \cdots \sigma_{\alpha_k} e^{-S}, \tag{5.5.54}$$

and the fermion zero modes must be brought down from the action to obtain a non-zero answer. The first two Yukawa couplings are σ-independent, and each of them must be brought down n_G times to absorb the $\overline{\chi}^{\alpha}$ and $\overline{v}_{\alpha}$ zero modes. The zero modes that contribute to the last Yukawa coupling are counted by holomorphic sections of $K_{\Sigma} \otimes \mathcal{L}_{\rho}^* = \mathcal{O}(-2 - d_{\rho})$. On $\mathbb{P}^1$ there are no such zero modes if $d_{\rho} \geq -1$, but if $d_{\rho} \leq -2$, there are $-d_{\rho} - 1$ zero modes. Thus, the last Yukawa coupling

---

[38]The dual of the Kähler cone is known as the Mori cone: it is generated by the effective curves in V.

introduces a factor of

$$\prod_{\rho|d_\rho\leq-2} \left(\overline{\gamma}^\rho E_{,\rho'}^\rho \psi^{\rho'}\right)^{-d_\rho-1} = \prod_{\rho|d_\rho\leq-2} \left(\overline{\gamma}^\rho \psi_\rho Q_\rho^\alpha \sigma_\alpha\right)^{-d_\rho-1}. \tag{5.5.55}$$

When (and only when) (5.5.45) holds, it is possible to saturate the fermi zero modes by bringing down powers of the remaining Yukawa coupling in such a way as to obtain a $U(1)_B$-invariant function of the σ_α under the integral. The fermi zero modes lead to an additional σ-dependent factor proportional to

$$\text{Obs}_{\vec{N}} = \prod_{\rho|d_\rho\leq-2} \xi_\rho^{-d_\rho-1}, \tag{5.5.56}$$

where ξ_ρ is the frequently met linear combination $\xi_\rho = \sum_\alpha Q_\rho^\alpha \sigma_\alpha$.

This should then make plausible the following formula for the $\mathcal{N}$:

$$\mathcal{N}_{\alpha_1\cdots\alpha_k}^{\vec{N}} = \int_{\mathcal{M}_{\vec{N}}} \eta_{\alpha_1,\vec{N}} \wedge \cdots \wedge \eta_{\alpha_k,\vec{N}} \wedge \text{Obs}_{\vec{N}}, \quad \text{Obs}_{\vec{N}} = \bigwedge_{\rho|d_\rho\leq-2} \wedge^{-1-d_\rho} \xi_{\rho,\vec{N}}, \tag{5.5.57}$$

with the understanding that the right-hand-side is zero whenever the form is not of top degree. The integral obviously has an interpretation in terms of the intersection numbers of toric divisors on the toric variety $\mathcal{M}_{\vec{N}}$ and can be evaluated by toric methods (5.3.74).[39] By summing these as in (5.3.74), we obtain a complete expression for the A-twisted correlation functions of local observables in the V-model. We did not present a rigorous localization computation of the result, but we did sketch out the essential features that should be incorporated in such a derivation. A modern discussion can be found in [115].

Exercise 5.42 Show that $\mathcal{N}_{\alpha_1\cdots\alpha_k}^{\vec{N}} = 0$ unless (5.5.45) holds because the form is not top degree . $\qquad\square$

Applications and Examples of the Instanton Sum

Let us start with a simple example of $V = \mathbb{P}^n$.

Exercise 5.43 Apply the machinery of the previous section to $V = \mathbb{P}^n$ to evaluate $\langle \sigma^k \rangle_A$. Show that

1. $K^*(V)$ is the one-dimensional cone with $N \geq 0$;
2. For any $N \in K^*(V)$ $\mathcal{M}_{\vec{N}} = \mathbb{P}^{n+N(n+1)}$;

---

[39]One should use that interpretation when $\mathcal{M}_{\vec{N}}$ is singular.

3. $\mathcal{N}_k^N = 1$ if $k = n + N(n + 1)$ and zero otherwise;
4. thus, the non-zero correlation functions are

$$\langle \sigma^{n+N(n+1)} \rangle_{\mathrm{A}}(q) = q^N , \qquad (5.5.58)$$

and they satisfy the quantum cohomology relation

$$\sigma^{n+1} - q = 0 . \qquad (5.5.59)$$

$\square$

To get an idea for the relationship between the gauge instantons and the worldsheet instantons, consider the exercise for the special case $n = N = 1$. In this case $\mathcal{M}_1 = \mathbb{P}^3$, and it is enlightening to consider the maps $f \subset \mathcal{M}_1$ in detail. Let $[z_1, z_2]$ be projective coordinates on $\Sigma = \mathbb{P}^1$. In that case, the map f takes the form

$$\phi_1 = \phi_{11}z_1 + \phi_{12}z_2 , \qquad \phi_2 = \phi_{21}z_1 + \phi_{22}z_2 , \qquad (5.5.60)$$

and the exceptional set is $\mathscr{F} = \{\phi_{11} = \phi_{12} = \phi_{21} = \phi_{22} = 0\}$. There is a subvariety $\mathscr{P} \subset \mathcal{M}_1$ defined by the vanishing of

$$\phi_{11}\phi_{22} - \phi_{12}\phi_{21} = 0 , \qquad (5.5.61)$$

where the image of the map $f \in \mathscr{P}$ contains a point in F, the exceptional set in $V = \{\mathbb{C}^2 \setminus F\}/\mathbb{C}^*$. On the other hand, when $f \notin \mathscr{P}$, it is simply a holomorphic map $f : \Sigma \to \mathbb{P}^1$, i.e. a worldsheet instanton of the corresponding non-linear sigma model with target space V.

This illustrates two general points. First, the moduli space of worldsheet instantons in a non-linear sigma model is non-compact: in our example it is $\mathcal{M}_1 \setminus \mathscr{P}$. This is the origin of the difficulty of Gromov-Witten theory, which, at the end of the day, is reduced to intersection theory of the moduli space of holomorphic curves. Second, the moduli space of the linear sigma model gauge instantons contains the worldsheet instantons and in fact yields a compactification of the worldsheet instanton moduli space—the compact space $\mathcal{M}_{\vec{N}}$.

More generally, for any V with exceptional set F and gauge instanton moduli space $\mathcal{M}_{\vec{N}}$ we define $\mathscr{P} \subset \mathcal{M}_{\vec{N}}$ to be the maps f where $f(\Sigma) \cap F \neq \emptyset$. Following [388] we call such maps point-like instantons. The non-linear sigma model instantons are then the maps $\mathcal{M}_{\vec{N}} \setminus \mathscr{P}$. When $\mathscr{P}$ is in positive co-dimension, we can expect that the linear sigma model correlation functions agree with those computed by the A-twisted non-linear sigma model with the canonical identification of the complexified Kähler coordinates t^α and the linear sigma model parameters τ^α. This is the case for Fano toric varieties.

Exercise 5.44 Prove that for any Fano toric variety a correlation function $\langle \sigma_{\alpha_1} \cdots \sigma_{\alpha_k} \rangle_{\mathrm{A}}(q)$ is a Laurent polynomial in the q_α. That is, only a finite number of instantons contribute. To do so, it helps to rewrite (5.5.45) as

$$(\boldsymbol{R}, \vec{N}) = k - d , \tag{5.5.62}$$

where $(\cdot, \cdot)$ is the natural pairing between $H^2(V, \mathbb{R})$, which contains the Kähler cone, and $H_2(V, \mathbb{R})$, which contains $K^*(V)$.

In particular, the correlation functions cannot diverge unless some q_α goes to zero or to infinity. □

Now let us see an example with infinite instanton sums.

Exercise 5.45 Sum the instantons for $\mathbb{F}_2$. Working in the basis of Exercise 5.32, compute the correlation function $\langle \sigma_1 \sigma_1 \rangle_A (q_1, q_2)$. From exercise 5.32 we know that the correlation function vanishes classically in the $q_1 \to 0$ limit, so only non-trivial instanton numbers need to be considered.

Show that the contributing instanton numbers are $\vec{N} = (m, -2m)$, with $m > 0$,

$$Y_{m,-2m} = \mathbb{C}^{3+2m} ,$$
$$\mathscr{F}_{m,-2m} = \{\phi_{10} = 0\} \cup \{\phi_{30} = \phi_{31} \cdots \phi_{3m} = \phi_{40} \cdots = \phi_{4m} = 0\} . \tag{5.5.63}$$

Taking the quotient by $G_{\mathbb{C}}$ and using $\phi_2 = 0$, show that

$$\mathcal{M}_{m,-2m} = \text{pt} \times \mathbb{P}^{1+2m} , \tag{5.5.64}$$

where the first factor is associated to the non-zero ϕ_{10}, and the second corresponds to the remaining coordinates. Thus, the cohomology on $\mathcal{M}_{m,-2m}$ with $m > 0$ is determined by

$$\xi_1 = 2\eta_1 + \eta_2 = 0 , \quad \xi_3^{m+1} \xi_4^{m+1} = \eta_2^{2m+2} = 0 , \quad \int_{\mathcal{M}_{m,-2m}} \eta_2^{2m+1} = 1 . \tag{5.5.65}$$

Now sum the instantons to obtain

$$\langle \sigma_1 \sigma_1 \rangle_A = \frac{2q_1}{4q_1 - q_2^2} . \tag{5.5.66}$$

Note that this correlation function diverges on the $\mathcal{Z}_A$ computed for $\mathbb{F}_2$ below Eq. (5.4.69) from the σ-vacua. □

The exercise shows that for NEF Fano varieties, where $\boldsymbol{R}$ lies in a boundary of the Kähler cone, an infinite number of instantons can contribute to a correlation function. However, the instanton sum converges for suitable q_α to a rational function of the parameters, and there is no issue in analytically continuing the result to other regions of the parameter space. We will argue below that the rational dependence on the q_α is a general feature of the A-model correlation functions for projective toric varieties. However, if V is not Fano, the instanton sum can lead to some surprises, as the next example shows.

Exercise 5.46 Study the instanton sums for $V = \mathbb{F}_3$. First work in the UV phase, where $V = \mathbb{P}^2/\mathbb{Z}_3$, and show that

$$\langle \sigma_1 \sigma_1 \rangle_A = \frac{1}{3} , \qquad \langle \sigma_2^2 \rangle = 0 , \qquad \langle \sigma_1 \sigma_2 \rangle = 0 . \tag{5.5.67}$$

You should find that the instantons that obey the selection rule (5.5.62) have $\vec{N} = (-2m, 5m)$ with $m \geq 0$, but such $\vec{N} \notin K^*(\mathbb{P}^2/\mathbb{Z}_3)$ unless $m = 0$. The $m = 0$ result is easily evaluated using (5.3.74).

This result is very bizarre from the point of view of the smooth phase, since there we showed in Exercise 5.16 that the $\eta_1 \wedge \eta_1 = 0$ on $\mathbb{F}_p$. Instanton "corrections" cannot change 0 into 1/3; evidently, the corrections are not small. This is not surprising in view of our discussion of σ vacua and intermediate phases, but it is worthwhile to study it further.

Towards that noble end, compute the instanton contributions to $\langle \sigma_1 \sigma_1 \rangle_A$ in the $\mathbb{F}_3$ phase. Show that the instanton numbers that satisfy the selection rule and $\vec{N} \in K^*(\mathbb{F}_3)$ are $\vec{N} = (2m, -5m)$ with $m \geq 0$. Since we already know that the $m = 0$ contribution to the correlation function is zero, it suffices to study the $m > 0$ case, where

$$\mathscr{M}_m = \mathbb{P}^m \times \mathbb{P}^{4m+1} , \qquad \mathrm{Obs}_m = (\eta_{2,m})^{5m-1} . \tag{5.5.68}$$

The hyperplane classes on the two projective spaces are $\xi_{1,m} = (3\eta_{1,m} + \eta_{2,m})$ and $\xi_{3,m} = \eta_{2,m}$. Use these to show

$$\mathcal{N}_{11}^m = (-3)^{4m-1} \binom{5m-1}{m} . \tag{5.5.69}$$

The binomial coefficient grows quickly with m, and it follows that the instanton sum has zero radius of convergence in $q_1^2 q_2^{-5}$. $\square$

5.5.3 Localization and σ Correlators in the IR

The structure of the V-model classical Higgs vacua depends sensitively on the combinatorics of V, and computations of twisted correlation functions involve a fairly elaborate intersection theory on the instanton moduli spaces. By contrast, the σ-vacua are much simpler to describe in a universal fashion as we did in (5.4.60). This suggests that there should be a simpler expression for the correlation functions based on a localization to the σ vacua. A formal localization argument [297], which has recently been extended and made more rigorous [115], shows this to be the case. These developments were anticipated in algebraic geometry, where they were applied to extend the techniques of [311] to mirror symmetry in the setting of complete intersections in toric varieties [54, 262, 355]. Those algebraic geometry techniques have informed many efforts in contemporary localization computations.

The formal argument is simple. Study the path integral around a σ vacuum, say $\sigma_\alpha(z, \bar{z}) = \sigma_{\alpha,*}$. The non-zero modes generate the effective potential as in (5.4.53), repeated here for convenience:

$$W_{\text{eff}} = \sum_\alpha \Upsilon_\alpha \tilde{J}_\alpha(\mathcal{S}) , \qquad \tilde{J}_\alpha(\mathcal{S}) = \frac{1}{4\pi\sqrt{2}} \log\left[q_\alpha^{-1} \prod_\rho \left(\frac{Q_\rho^\beta \mathcal{S}_\beta}{\mu} \right)^{Q_\rho^\alpha} \right] .$$

$$(5.5.70)$$

The σ_*^α must be a zero of $\tilde{J}_\alpha(\mathcal{S})$. We now expand in fluctuations around the $\sigma_\alpha = \sigma_{\alpha,*}$ and $\phi_\rho = 0$ vacuum in A-twisted theory, and since all of the excitations are massive, the path integral reduces to a Gaussian for the fluctuations. Just as in the (0,2) Landau-Ginzburg localization from Chap. 3, the fluctuations of Υ_α and $\mathcal{S}_\alpha$ contribute a factor of

$$W_{\Upsilon, \mathcal{S}} = \frac{1}{\det_{\alpha,\beta} \tilde{J}_{\alpha,\beta}} .$$

$$(5.5.71)$$

The gauge bundle is trivial, so that the matter fields' fluctuations make a multiplicative contribution from the terms

$$S \supset \frac{1}{2\pi} \int d^2z \left\{ \overline{E}^\rho E^\rho - \overline{E}^\rho_{,\rho'} \overline{\psi}^{\rho'} \gamma^\rho \right\} , \qquad E^\rho = \xi_\rho \phi_\rho , \qquad \xi_\rho = \sum_\alpha Q_\rho^\alpha \sigma_\alpha .$$

$$(5.5.72)$$

$$W_{\Phi, \Gamma} = \prod_\rho \frac{1}{\xi_\rho} .$$

$$(5.5.73)$$

In writing these factors we chose a convenient normalization for the overall multiplicative constant that is not determined by these considerations.

Finally, summing over all of the σ vacua, we obtain the σ-branch formula for the correlation functions:

$$\langle \sigma_{\alpha_1} \cdots \sigma_{\alpha_k} \rangle_A (q) = \sum_{\sigma_*} \left\{ \sigma_{\alpha_1} \cdots \sigma_{\alpha_k} W_{\Upsilon, \mathcal{S}}(\sigma) W_{\Phi, \Gamma}(\sigma) \right\}\Big|_{\sigma=\sigma_*} .$$

$$(5.5.74)$$

The equations for the σ-vacua involve the renormalization scale μ, but since this is the only scale that arises in the topological theory, we can just absorb it into a redefinition of σ. Thus, the σ_* are the solutions to the relations

$$\prod_\rho \xi_\rho^{Q_\rho^\alpha} = q_\alpha , \qquad \xi_\rho = \sum_\alpha Q_\rho^\alpha \sigma_\alpha .$$

$$(5.5.75)$$

Let us make a few comments on this expression.

1. The result (5.5.74) leads to an efficient way to compute the V-model correlation
   functions, especially in cases with $n_G < 3$, where the result can be easily recast
   as a simple residue integral. This is facilitated by the observation

$$\tilde{J}_{\alpha,\beta} = \sum_{\rho} \frac{Q_\rho^\alpha Q_\rho^\beta}{\xi_\rho} \qquad (5.5.76)$$

   and the use of the selection rule.
2. More generally, the correlation function is a sum of rational expressions in the
   roots of a system of polynomials. The resulting expression will therefore be a
   rational function of the parameters of the polynomial system, namely the q_α.
3. In this form it is obvious that the correlation functions obey a set of operator
   relations, which we can write as

$$\prod_{\rho|Q_\rho^\alpha>0} \xi_\rho^{Q_\rho^\alpha} - q_\alpha \prod_{\rho|Q_\rho^\alpha<0} \xi_\rho^{-Q_\rho^\alpha} = 0 . \qquad (5.5.77)$$

   For smooth projective V, these are the toric quantum cohomology relations
   introduced in [51], and this means we can think of them as deformations of the
   ring structure on $H^\bullet(V, \mathbb{R})$ given by the Stanley-Reisner ideal.
4. It is not hard to generalize the A-model result to a Riemann surface with genus
   g—the formulas merely change by replacing the weight factors by $W_{\Upsilon,\mathcal{S}}^{1-g}$ and
   $W_{\Phi,\Gamma}^{1-g}$.

Exercise 5.47 Apply (5.5.74) to reproduce the correlation functions of the $\mathbb{F}_2$ and
$\mathbb{F}_3$ exercises from the previous sections.   □

5.6   The X-model and Quantum Restriction

We return to the X-model for a hypersurface in a d-dimensional Fano toric variety
V. The main differences between the V-model and the X-model are as follows.
First, the X-model has a non-anomalous U(1)$_B$, or, equivalently, there is a U(1)$_L$ ×
U(1)$_R$ chiral symmetry. Second, there are new superpotential couplings encoded in
the coefficients of the defining polynomial P. The first point implies that there are
now two possible twists for the X-model. We will now discuss the A-twist and later
touch on the B-twisted theory.

   The A-twisted X-model has a simple class of local observables—the σ_α. Modulo
subtleties of redundancy, of which more anon, these parameterize a subspace
$\mathcal{H}_A^{\text{toric}}$ of the Q_A cohomology. We will focus on the correlation functions of
these observables on the sphere. The A-twisted path integral for the X-model has

additional fields in the Φ^0 and Γ^0 multiplets. From Exercise 5.5.35 we have the global description of these fields in an instanton background as

$$
\begin{aligned}
\phi_0 &\in \mathscr{A}^0(\Sigma, K_\Sigma \otimes \mathcal{L}_0)\,, & \overline{\phi}_0 &\in \mathscr{A}^0(\Sigma, \overline{K}_\Sigma \otimes \overline{\mathcal{L}}_0)\,, \\
\psi_0 &\in \mathscr{A}^0(\Sigma, \overline{\mathcal{L}}_0^*)\,, & \overline{\psi}_0 &\in \mathscr{A}^0(\Sigma, \overline{K}_\Sigma \otimes \overline{\mathcal{L}}_0)\,, \\
\gamma_0 &\in \mathscr{A}^0(\Sigma, K \otimes \mathcal{L}_0)\,, & \overline{\gamma}_0 &\in \mathscr{A}^0(\Sigma, \mathcal{L}_0^*)\,.
\end{aligned}
\tag{5.6.1}
$$

The bundle is given by $\mathcal{L}_0 = \otimes_\rho \mathcal{L}_\rho^* = \mathcal{O}(-\sum_\rho d_\rho)$.

Since the X-model is supposed to flow to an SCFT, by identifying the twist of the massive theory with that of the SCFT we expect the ghost number selection rule (5.5.16) to hold with $c = (d-1)$. This is the case: following the same logic that led to (5.5.38), and using the appropriate $U(1)_B$ charges for the extra Φ^0 and Γ^0 multiplets, we have—note that the q_α no longer transform

$$
\langle \sigma_{\alpha_1} \cdots \sigma_{\alpha_k} \rangle_A(q) = e^{i(k-A)\alpha} \langle \sigma_{\alpha_1} \cdots \sigma_{\alpha_k} \rangle_A(q)\,,
\tag{5.6.2}
$$

where

$$
\begin{aligned}
A &= n_G(\tfrac{1}{2} + \tfrac{1}{2}) + \sum_\rho \left[-\tfrac{1}{2}(\#_0(\overline{\psi}_\rho) - \#_0(\psi_\rho)) - \tfrac{1}{2}(\#_0(\gamma_\rho) - \#_0(\overline{\gamma}_\rho)) \right] \\
&\quad - \tfrac{1}{2}(\#_0(\overline{\psi}_0) - \#_0(\psi_0)) - \tfrac{1}{2}(\#_0(\gamma_0) - \#_0(\overline{\gamma}_0)) \\
&= n_G - \sum_\rho (1 + c_1(\mathcal{L}_\rho)) + (1 + c_1(\mathcal{L}_0)) = n_G - n + 1 = -(d-1)\,.
\end{aligned}
\tag{5.6.3}
$$

Thus, the correlation function vanishes unless $k = d - 1$, the dimension of X. We will be interested in a relationship between the X-model correlation functions and those of the V-model. To distinguish the two, we will denote the X-model correlation functions by

$$
\langle\!\langle \sigma_{\alpha_1} \cdots \sigma_{\alpha_{d-1}} \rangle\!\rangle_A(q)\,.
\tag{5.6.4}
$$

These have a formal instanton expansion just the correlation functions of the V-model:

$$
\langle\!\langle \sigma_{\alpha_1} \cdots \sigma_{\alpha_{d-1}} \rangle\!\rangle_A(q) = \sum_{\vec{N}} \prod_\alpha q_\alpha^{N_\alpha} \mathcal{N}_{\alpha_1 \cdots \alpha_{d-1}}^{\vec{N}, X}\,.
\tag{5.6.5}
$$

5.6.1   Classical Restriction

Consider the X-model where we choose a basis for the charges Q^α_ρ such that $q_\alpha \to 0$ is in the Calabi-Yau phase, where at low energy the linear sigma model reduces to a large radius non-linear sigma model on $X \subset V$. As a first step, we would like to evaluate the A-model correlation functions in the $q_\alpha \to 0$ limit. In this limit the calculation reduces to classical geometry. Each σ_α corresponds to a class in $\eta_\alpha \in H^{1,1}(X, \mathbb{Z})$, and in the non-linear sigma model the correlation function reduces to

$$\langle\!\langle \sigma_{\alpha_1} \cdots \sigma_{\alpha_{d-1}} \rangle\!\rangle_A(q) = \#(\eta_{\alpha_1} \eta_{\alpha_2} \cdots \eta_{\alpha_{d-1}})_X + O(q) \, . \qquad (5.6.6)$$

The classical term is just the intersection number (which can also be thought of as an integral of a top form on X), and the $O(q)$ terms are the worldsheet instanton corrections.

Let $\iota : X \hookrightarrow V$ be a smooth compact subvariety of V, a smooth d-dimensional compact manifold, and let D_X be the corresponding divisor in $A_{d-1}(V, \mathbb{Z})$. A divisor $D \in A_{d-1}(V, \mathbb{Z})$ can be pulled back to a divisor $\iota^*(D) \in A_{d-2}(X, \mathbb{Z})$.[40] The pulled-back divisors need not generate $A_{d-2}(X, \mathbb{Z})$ because an irreducible divisor on V may pull back to a reducible one on X; this is the origin of the non-toric Kähler deformations we discussed above. On the other hand, if we restrict to the class of divisors on X that do pull back from V, then we have a simple formula for the intersection numbers[41]:

$$\#(\iota^*(D_1) \cdot \iota^*(D_2) \cdots \iota^*(D_{d-1}))_X = \#(D_1 \cdot D_2 \cdots D_{d-1} \cdot D_X)_V \, . \qquad (5.6.7)$$

Following [311], we call this the classical restriction formula.

The classical restriction formula implies a relationship between the X and V model's correlation functions:

$$\langle\!\langle \sigma_{\alpha_1} \cdots \sigma_{\alpha_{d-1}} \rangle\!\rangle_A(q) = \langle \sigma_{\alpha_1} \cdots \sigma_{\alpha_{d-1}} R \rangle_A(q) + O(q) \, , \qquad (5.6.8)$$

where the anti-canonical class is

$$R = \sum_\rho \xi_\rho \, . \qquad (5.6.9)$$

---

[40]Recall that in this smooth setting the divisors are Cartier, and so correspond to line bundles. So, in completely analogous fashion we can pull back the corresponding line bundles from V to X.

[41]There is a suitable generalization to the intersection of $\mathbb{Q}$-Cartier divisors that naturally arise in the more general setting where X and V have Gorenstein orbifold singularities. In our toric setting this just amounts to keeping track of the fractions that arise in (5.3.74).

The quantum restriction formula will be a version of this relationship that takes into account the q-dependence. We obtain it by a localization argument following [311].

5.6.2 Localization in the X-model

Consider the X-model in a Calabi-Yau phase, and focus on the A-twisted path integral with instanton number $\vec{N}$. The path integral will localize onto field configurations annihilated by $\overline{Q}$:

$$\overline{\nabla}\phi_\rho = 0, \quad \bar{\partial}\sigma_\alpha = 0, \quad E^\rho = 0, \quad (F_{12} + D^+)_\alpha = 0, \quad P = 0, \quad \phi_0 \frac{\partial P}{\partial \phi^\rho} = 0,$$
$$(5.6.10)$$

where we include ϕ_0 by an abuse of notation we already saw above: $\rho = 0$ corresponds to the unique interior point in Δ°, and the D-term D^+ includes the contribution of ϕ_0.[42] The first three equations imply that the ϕ_ρ are holomorphic sections of appropriate holomorphic bundles, and $\sigma_\alpha = 0$. The remaining equations then show that the $\overline{Q}$-fixed locus, $\mathcal{M}_{\vec{N},X}$ is a subvariety

$$\mathcal{M}_{\vec{N},X} = \{P = 0, \phi_0 \frac{\partial P}{\partial \phi^\rho} = 0\} \subset \mathcal{M}_{\vec{N}}^+ ,$$
$$(5.6.11)$$

where $\mathcal{M}_{\vec{N}}^+ = (Y_{\vec{N}}^+ \setminus F_{\vec{N}}^+)/G_{\mathbb{C}}$ is a toric variety constructed in the same fashion as the instanton moduli space $\mathcal{M}_{\vec{N}} = (Y_{\vec{N}} \setminus F_{\vec{N}})/G_{\mathbb{C}}$ of the V-model.

The affine spaces are related in a simple way:

$$Y_{\vec{N}}^+ = H_{\bar{\partial}}^0(\Sigma, K_\sigma \otimes \mathcal{L}_0) \oplus Y_{\vec{N}} ,$$
$$(5.6.12)$$

and in a Calabi-Yau phase the exceptional sets are the same: $F_{\vec{N}}^+ = F_{\vec{N}}$—a look back to Exercise 5.25 will immediately show why this is the case. This in turn implies that the map described by the collection $\{\phi^\rho, \phi^0\} : \Sigma \to V_+$ only meets the exceptional set F of the V_+ variety at a finite number of points on Σ. Therefore, when, as we will assume, P is a smooth hypersurface in V, ϕ_0 is zero at a generic point on Σ. Having shown that $\phi_0 = 0$, we see that the instanton moduli space of the X-model is a subvariety in the instanton moduli space of the V-model:

$$\mathcal{M}_{\vec{N},X} = \{P = 0\} \subset \mathcal{M}_{\vec{N}} = \{\phi_0 = 0\} \subset \mathcal{M}_{\vec{N}}^+ .$$
$$(5.6.13)$$

---

[42]Classically we already saw this structure in our discussion of phases: the D-term constraints modulo G determine the toric variety $V_+ = \mathrm{tot}\{\mathcal{O}(-K_V) \to V\}$, and the superpotential constraints force $\phi_0 = 0$ and $\{P = 0\} \subset V$.

Just as the in the V-model $\mathcal{M}_{\vec{N}}$ will be empty unless $\vec{N} \in K^*(V)$. Because V is Fano,

$$d_0 = -\sum_{\rho \neq 0} d_\rho \leq 0 . \tag{5.6.14}$$

P "Does Not Matter"

The path integral over $\mathcal{M}_{\vec{N},X}$ is formally independent of the couplings encoded in the chiral superpotential. More precisely, all of the P-dependent couplings in the action are of the form

$$S_P = \{Q', V_P\} + \{\overline{Q}, V_{\overline{P}}\} , \tag{5.6.15}$$

where V_P and $V_{\overline{P}}$ carry $U(1)_R$ charge 0 and -1 respectively; $(2,2)$ supersymmetry implies $\{\overline{Q}, V_P\} = 0$ and $\{Q', V_{\overline{P}}\} = 0$.[43] It follows that we can study the A-twisted theory by setting $P = 0$. We now have a path integral over the V-model moduli space.[44]

The path integral is not yet that of the V-model. In particular, since $d_0 \leq 0$ (5.6.1) shows that $\overline{\gamma}^0$ and ψ_0 each have $1 - d_0$ zero modes, but we recognize this! These zero modes, and the only possible couplings in the action that can absorb them, are of the same form as those contributing to (5.5.55). Thus, performing the integral over the fermi zero modes leads to an insertion of

$$\mathrm{Obs}_{\vec{N},X} = \xi_{0,\vec{N}}^{1-d_0} \mathrm{Obs}_{\vec{N}} . \tag{5.6.16}$$

Assuming, as we did for the V-model, that the integral over the non-zero modes is $\vec{N}$-independent,

$$\mathcal{N}_{\alpha_1 \cdots \alpha_{d-1}}^{\vec{N},X} = C \#(\eta_{\alpha_1,\vec{N}} \cdots \eta_{\alpha_{d-1},\vec{N}} \xi_{0,\vec{N}}^{1-d_0} \mathrm{Obs}_{\vec{N}})_{\mathcal{M}_{\vec{N}}} , \tag{5.6.17}$$

where C is a constant. Since $\xi_0 = -\sum_\rho \xi_\rho = -R$, we see that at $\vec{N} = 0$ we will reproduce the classical restriction result if the constant $C = -1$. Taking this value, we also observe

$$\#(\eta_{\alpha_1,\vec{N}} \cdots \eta_{\alpha_{d-1},\vec{N}} \xi_{0,\vec{N}}^s \mathrm{Obs}_{\vec{N}})_{\mathcal{M}_{\vec{N}}} = 0 \qquad \text{unless} \qquad s = 1 - d_0 . \tag{5.6.18}$$

---

[43] This easily seen in $(2,2)$ superspace, where $V_P = \int d^2z \, \mathcal{D}W(\Phi_{22})$, and $[\overline{Q}, W] = $ total derivative.

[44] In the untwisted theory $P = 0$ is a pathological limit, since the theory is non-compact: the zero mode of ϕ_0 is not constrained by any term in the potential. This is not the case once we restrict to the instanton sector of interest: $d_0 \leq 0$, and therefore $\phi_0 \in \mathscr{A}(\Sigma, \mathcal{O}(-2+d_0))$ has no zero modes.

This means that we can replace the insertion $\xi_{0,\vec{N}}^{1-d_0} \rightarrow \sum_{s=1}^{\infty}(\xi_{0,\vec{N}})^s$ and formally sum up the s series. The virtue of this is that now the insertion that reduces the X-model computation to the V-model computation is independent of instanton number:

$$-\#(\eta_{\alpha_1,\vec{N}} \cdots \eta_{\alpha_{d-1},\vec{N}} \xi_{0,\vec{N}}^{1-d_0} \mathrm{Obs}_{\vec{N}}).\mathscr{M}_{\vec{N}} = -\#(\eta_{\alpha_1,\vec{N}} \cdots \eta_{\alpha_{d-1},\vec{N}} \frac{R}{1+R} \mathrm{Obs}_{\vec{N}}).\mathscr{M}_{\vec{N}} .$$

$$(5.6.19)$$

Finally, performing the instanton sum, we obtain the (2,2) quantum restriction formula

$$\langle\!\langle \sigma_{\alpha_1} \cdots \sigma_{\alpha_{d-1}} \rangle\!\rangle_{\mathrm{A}}(q) = \langle \sigma_{\alpha_1} \cdots \sigma_{\alpha_{d-1}} \frac{R}{1+R} \rangle_{\mathrm{A}}(q) .$$

$$(5.6.20)$$

The R is now interpreted as the operator $R = \sum_{\rho \neq 0,\alpha} Q_\rho^\alpha \sigma_\alpha$ in the quantum cohomology of the V-model. Our formal manipulation of the series in $\xi_{0,\vec{N}}$ will then make sense as long as the operator $1 + R$ is invertible in the V-model.

5.6.3 Applications and Aspects of Quantum Restriction

The quantum restriction formula is remarkable: a set of correlation functions in a complicated interacting SCFT can be determined provided we know all of the correlation functions of a vastly simpler topological field theory associated to a gapped quantum field theory. Together with the σ-branch localization, it gives a powerful way to do computations and to draw some general conclusions.

Explicit Computations

Combining quantum restriction and the σ-branch localization, we have

$$\langle\!\langle \sigma_{\alpha_1} \cdots \sigma_{\alpha_{d-1}} \rangle\!\rangle_{\mathrm{A}}(q) = \sum_{\sigma_*} \sigma_{\alpha_1} \cdots \sigma_{\alpha_k} W_{\Upsilon,\mathcal{S}}(\sigma) W_{\Phi,\Gamma}(\sigma) \frac{R}{1+R}\bigg|_{\sigma=\sigma_*} ,$$

$$(5.6.21)$$

where σ_* are the V-model σ vacua.

Exercise 5.48 Apply the technology to the quintic and show that

$$\langle\!\langle \sigma\sigma\sigma \rangle\!\rangle_{\mathrm{A}} = \frac{5}{1+5^5 q} .$$

$$(5.6.22)$$

Now apply the tricks to the bicubic. First show that every correlator can be written as

$$\langle\!\langle \sigma_{\alpha_1} \sigma_{\alpha_2} \sigma_{\alpha_3} \rangle\!\rangle = \frac{1}{3} \sum_{\sigma_*} \frac{\sigma_{\alpha_1} \sigma_{\alpha_2} \sigma_{\alpha_3} (\sigma_1 + \sigma_2)}{\sigma_1^2 \sigma_2^2 (1 + 3\sigma_1 + 3\sigma_2)}\Bigg|_{\sigma = \sigma_*} , \tag{5.6.23}$$

where the σ_* are solutions to $\sigma_1^3 = q_1$ and $\sigma_2^3 = q_2$.

Evaluate these sums. There are a number of ways to do so. One of the simplest is to use recall that to sum an expression $f(z)$ over the roots z_* of a polynomial $P(z)$ we can write

$$\sum_{z_*} f(z_*) = \oint_C \frac{dz}{2\pi i} \frac{f(z) P'(z)}{P(z)} , \tag{5.6.24}$$

where C is chosen to enclose the roots of $P(z)$ and no other poles. Applying this here and then closing the contour on the remaining poles, you should find

$$\langle\!\langle \sigma_1 \sigma_1 \sigma_1 \rangle\!\rangle_A = \frac{3^5 q_1 (-2 + 3^3 q_1 + 3^3 q_2)}{\mathcal{T}_1} ,$$

$$\langle\!\langle \sigma_1 \sigma_1 \sigma_2 \rangle\!\rangle_A = \frac{3}{\mathcal{T}_1} \left(1 + 3^3 (q_1 + q_2) \right) \left(1 + 3^3 q_2 - 2 \cdot 3^3 q_1 \right) , \tag{5.6.25}$$

where

$$\mathcal{T}_1 = 3^9 (q_1 + q_2)^3 + 3^7 (q_1 + q_2)^2 - 3^9 q_1 q_2 + 3^4 (q_1 + q_2) + 1 . \tag{5.6.26}$$

Next, carry out the same manipulations for the octic example. Here you should find[45] [297]

$$\langle\!\langle \sigma_1 \sigma_1 \sigma_1 \rangle\!\rangle_A = \frac{8}{\mathcal{T}_2} ,$$

$$\langle\!\langle \sigma_1 \sigma_1 \sigma_2 \rangle\!\rangle_A = \frac{4(1 - 2^8 q_1)}{\mathcal{T}_2} ,$$

$$\langle\!\langle \sigma_1 \sigma_2 \sigma_2 \rangle\!\rangle_A = \frac{8 q_2 (2^9 q_1 - 1)}{(1 - 4q_2) \mathcal{T}_2} ,$$

$$\langle\!\langle \sigma_2 \sigma_2 \sigma_2 \rangle\!\rangle_A = \frac{4 q_2 (1 + 4q_2 - 2^8 q_1 - 3 \cdot 2^{10} q_1 q_2)}{(1 - 4q_2)^2 \mathcal{T}_2} , \tag{5.6.27}$$

---

[45]If using residue tricks, watch out for the residue at ∞.

where

$$\mathcal{T}_2 = (1 - 2^8 q_1)^2 - 2^{18} q_1^2 q_2 . \tag{5.6.28}$$

Check that in all of the examples the $q_\alpha \to 0$ limit is consistent with the classical restriction formula.   □

Relation to the Singular Locus

Besides the nice workings of the machinery, the examples illustrate a general feature: comparing the correlation functions to the singular loci $\mathcal{Z}_A^{\mathrm{pri}}(X)$ obtained in (5.4.72), we observe that in each case the correlators diverge on the singular locus. We can understand this by working in the special basis introduced in (5.4.61), where $R^\alpha = 0$ for $\alpha > 1$ and $R^1 = \Delta$, a positive integer. The reader can easily show that in this basis the correlation functions take the form

$$\langle\!\langle \sigma_{\alpha_1} \cdots \sigma_{\alpha_{d-1}} \rangle\!\rangle_A (q) = \sum_{\sigma_*} \sigma_{\alpha_1} \cdots \sigma_{\alpha_k} W_{\Upsilon,S}(\sigma) W_{\Phi,\Gamma}(\sigma) \left. \frac{\Delta\sigma_1}{1 - (-\Delta\sigma_1)^\Delta} \right|_{\sigma=\sigma_*} , \tag{5.6.29}$$

The equations for the σ-vacua of the V-model can be rewritten as

$$1 = q_1 \prod_\rho \xi_\rho^{-Q_\rho^1} , \qquad \prod_\rho \xi_\rho^{Q_\rho^\alpha} = q_\alpha \qquad \text{for } \alpha > 1 , \tag{5.6.30}$$

while the σ-vacua for the X-model (which only exist for $q \in \mathcal{Z}_A^{\mathrm{pri}}(X)$) satisfy

$$\xi_0^{Q_0^\alpha} \prod_\rho \xi_\rho^{Q_\rho^\alpha} = q_\alpha , \tag{5.6.31}$$

and in the special basis we split them up as

$$(-\Delta\sigma_1)^{-\Delta} \prod_\rho \xi_\rho^{Q_\rho^1} = q_1 , \qquad \prod_\rho \xi_\rho^{Q_\rho^\alpha} = q_\alpha \qquad \text{for } \alpha > 1 . \tag{5.6.32}$$

Comparing the two sets, we see that precisely when $q \in \mathcal{Z}_A^{\mathrm{pri}}$ the correlation function will diverge due to the pole $1/(1 - (-\Delta\sigma_1)^\Delta)$. Thus, precisely when $1 + R$ fails to be invertible, we have $q \in \mathcal{Z}_A^{\mathrm{pri}}(X)$.

As the last example illustrates, the correlation functions can have other poles associated to other components of $\mathcal{Z}_A(X)$, which can be studied by the methods described in [35].

Quantum Cohomology and Redundant Parameters

Just as for the V-model, there is a notion of quantum cohomology for the correlation functions of the X-model, although the relations are not related in as simple a fashion to the σ-vacua. We leave the general discussion for the references [120, 311], but we will explore one point related to this structure.

In the discussion of mirror symmetry and the monomial-divisor mirror map, we touched on the notion of redundant q_α, which correspond to blowing up singularities in V that do not meet the hypersurface X. On one hand, it seems sensible that these should decouple from the IR physics and therefore from the topological sector. On the other hand, blessed with the quantum restriction formula (5.6.20), we can ask: how does this decoupling take place?

Recall that in a mirror description, the structure is reasonably clear: there is a field redefinition of the Lagrangian that shifts the coupling into the D-terms, and we expect those to be irrelevant. We would like to understand this directly in terms of the q_α, so we will consider an instructive example.

We will study the Calabi-Yau hypersurface in a resolution of the projective space $\mathbb{P}^4_{11226}$. The polytope Δ° contains eight lattice points, counting the origin. The non-zero points are the columns in

$$\Delta^\circ = \begin{pmatrix} -1 & 1 & 0 & 0 & 0 & 0 & 0 \\ -2 & 0 & 1 & 0 & 0 & -1 & 0 \\ -2 & 0 & 0 & 1 & 0 & -1 & 0 \\ -6 & 0 & 0 & 0 & 1 & -3 & -1 \end{pmatrix}. \tag{5.6.33}$$

We can understand the resolution as follows. The first five points are the vertices of a simplex in $N_\mathbb{R}$, and its five facets, which we label by their vertices are

$$\varphi^\circ_{1234}, \qquad \varphi^\circ_{1235}, \qquad \varphi^\circ_{1245}, \qquad \varphi^\circ_{1345}, \qquad \varphi^\circ_{2345}. \tag{5.6.34}$$

The simplex defines a simplicial but singular toric variety—the weighted projective space.

We observe that $v_3 = (v_1 + v_2)/2$ is a lattice point that lies on an edge of the simplex. We introduce the corresponding ray and resolve the singularity. Now the full-dimensional cones in the fan are taken over

$$\varphi^\circ_{1346}, \quad \varphi^\circ_{2346}, \quad \varphi^\circ_{1356}, \quad \varphi^\circ_{2356}, \quad \varphi^\circ_{1456}, \quad \varphi^\circ_{2456}, \quad \varphi^\circ_{1345}, \quad \varphi^\circ_{2345}. \tag{5.6.35}$$

The resulting toric variety is still not smooth, since φ°_{1234} contains an interior point

$$v_7 = \frac{1}{3}(v_3 + v_4 + v_6) = \frac{1}{3}\left(v_3 + v_4 + \frac{1}{2}(v_1 + v_2)\right). \tag{5.6.36}$$

This is the source of our redundant coupling. Adding v_7, we find the full-dimensional cones over

$$\varphi^\circ_{1347}, \quad \varphi^\circ_{1367}, \quad \varphi^\circ_{1467}, \quad \varphi^\circ_{2347}, \quad \varphi^\circ_{2367}, \quad \varphi^\circ_{2467},$$
$$\varphi^\circ_{1356}, \quad \varphi^\circ_{2356}, \quad \varphi^\circ_{1456}, \quad \varphi^\circ_{2456}, \quad \varphi^\circ_{1345}, \quad \varphi^\circ_{2345}. \tag{5.6.37}$$

The reader can practice toric thinking by showing that the exceptional set is therefore

$$F = \{346\} \cup \{12\} \cup \{57\}, \tag{5.6.38}$$

and the gauge group $G_{\mathbb{C}} = (\mathbb{C}^*)^3$ has a basis of charges

$$Q = \begin{pmatrix} 0 & 0 & 1 & 1 & 0 & 1 & -3 \\ 1 & 1 & 0 & 0 & 0 & -2 & 0 \\ 0 & 0 & 0 & 0 & 1 & 0 & 1 \end{pmatrix}, \tag{5.6.39}$$

and the smooth phase is the positive orthant $r_\alpha > 0$.

It follows that the quantum cohomology relations of the V-model include

$$\sigma_3(\sigma_3 - 3\sigma_1) = q_3, \tag{5.6.40}$$

and $R = 2\sigma_3$. But now we use quantum restriction:

$$\langle\!\langle \sigma_\alpha \sigma_\beta (\sigma_3 - 3\sigma_1) \rangle\!\rangle = \langle \sigma_\alpha \sigma_\beta (\sigma_3 - 3\sigma_1) \frac{R}{1+R} \rangle = \langle \sigma_\alpha \sigma_\beta \frac{2q_3}{1+R} \rangle. \tag{5.6.41}$$

On the other hand, upon expanding $1/(1 + R)$, the only terms that contribute to the correlator are of the form R^{2+2k}, which means we have

$$\langle \sigma_\alpha \sigma_\beta \frac{2q_3}{1+R} \rangle = \langle \sigma_\alpha \sigma_\beta \frac{2q_3(R)^2}{1+R} \rangle = \langle \sigma_\alpha \sigma_\beta \frac{2q_3(2\sigma_3)(R)}{1+R} \rangle, \tag{5.6.42}$$

and therefore

$$\langle\!\langle \sigma_\alpha \sigma_\beta (\sigma_3 - 3\sigma_1) \rangle\!\rangle = \langle\!\langle \sigma_\alpha \sigma_\beta (4q_3\sigma_3) \rangle\!\rangle. \tag{5.6.43}$$

Since this holds for all α, β, we just found a linear quantum cohomology relation for the X-model:

$$(1 - 4q_3)\sigma_3 - 3\sigma_1 = 0. \tag{5.6.44}$$

When $q_3 = 0$ this reduces to the classical statement that the hypersurface does not intersect the exceptional divisor $\xi_7 = \eta_3 - 3\eta_1$. More generally, this shows that the

A-model correlation functions will be invariant under an infinitesimal shift

$$\delta q_1 = -3\epsilon q_1 \,, \qquad \delta q_2 = 0 \,, \qquad \delta q_3 = \epsilon(q_3 - 4q_3^2) \,. \qquad (5.6.45)$$

This example as well as the mirror perspective show that this will be the general structure: there will be linear quantum cohomology relations that will allow us to eliminate the redundant couplings.

Point-Like Instantons in the X-model

We solved two A-twisted theories, the V-model and the X-model, by what appear to be very similar techniques. In the case of the V-model, there was a clear relationship between the linear model and the non-linear theory: the two compute a very similar integral of a top form over closely related instanton moduli spaces: the gauge instanton moduli space is a compactification of the non-linear model's instanton moduli space, and the two differ in positive co-dimension. It is therefore not surprising that the two should match in a simple fashion: the parameters of the non-linear model t^{α} are identified with the τ^{α} of the linear theory.

This is not the case for the X-model and the corresponding non-linear sigma model: the generic gauge instanton in $\mathcal{M}_{\tilde{N},X}$ is point-like (i.e. is not a worldsheet instanton) [388]. In this case, the parameters are related in a more complicated way. Indeed, the relationship $t(\tau)$ is essentially the mirror map. Recent advances have shown that $t(\tau)$ can be computed without invoking mirror symmetry [200, 254] by evaluating the sphere partition function of the linear sigma model using the localization techniques of [68, 156]. On the other hand, it would be instructive to understand how $t(\tau)$ arises through renormalization. The idea is that one should be able to integrate out the point-like instantons to obtain a low energy theory that localizes on the worldsheet instanton moduli space. Steps in this direction may be found in [287] .

5.6.4   The B-model for the X-model

We do not discuss the B-twist for the V-model because in that case $U(1)_B$ is anomalous. We will not discuss the B-twist for the X-model in depth for two reasons: it is classical, and it is not so simple.

The first part is easily seen. When we localize the path integral after a B-twist, the field configurations must obey (5.6.10), which follows from $\overline{Q}$-closure, but they must also be annihilated by $\overline{Q}'$. This constrains the matter fields to constant values and requires $F_{12\alpha} = 0$. Furthermore the q-dependence is Q_B-exact, so that all computations can be done in the smooth phase with the techniques of toric residues—see[54, 355] and references therein for more details and a rigorous treatment of the "toric mirror conjecture," which in our language is the statement $A(X) = B(X^\circ)$: the correlation functions of the toric/polynomial operators agree via the monomial-divisor mirror map. The B-model observables that correspond to the polynomial deformations are the gauge-invariant polynomials $\phi_0 M(\phi)$.

We mention one important simplifying case where the computations are straightforward: whenever the X-model has a Landau-Ginzburg phase, we can use the q-independence to evaluate the B-model correlators using the Landau-Ginzburg techniques described in Chap. 3.

Much more can be said of the wonders of (2,2), but we move on: (0,2) awaits! We return to it next, but with a much improved toolkit.

5.7 Deformations of (2,2) Theories

Our study of (2,2) theories was carried out in a slightly perverse way: the (0,2) structure of the theory was made explicit, and the left-moving supersymmetries Q' and $\overline{Q}'$ and their consequences often seemed to be an afterthought. We will now reap the benefits of this strange approach, because many of the results obtained so far will be seen to follow just from the (0,2) structure and will generalize in a simple way to (0,2) theories deformed away from the (2,2) locus.

5.7.1 Deformations of the X-model

We begin with the X-model of the previous section and consider Lagrangian deformations that preserve the right-moving supersymmetry and the $U(1)_L \times U(1)_R$ symmetry. We will keep the D-terms in their canonical form since we have no control over them, but we start with the most general deformation of the holomorphic couplings encoded in the E that accounts for the non-chirality of the fermi multiplets, and the most general deformation of the (0,2) superpotential.

Taking a look back at the charge assignments for the X-model (5.4.16), which with explicit split between the (Φ^0, Γ^0) multiplets and the rest, takes the form

	θ	Φ^0	Γ^0	Φ^ρ	Γ^ρ	$\mathcal{S}_\alpha$	Υ_α	
$U(1)_L$	0	1	0	0	-1	-1	0	
$U(1)_R$	1	1	1	0	1	1	1 ,	(5.7.1)

we see that the symmetry-preserving deformations that are polynomial in the fields are limited. We find:

1. the Υ_α couplings remain unchanged;
2. $E^0 = -\frac{1}{2}\Phi^0 \mathfrak{R}^\alpha \mathcal{S}_\alpha$ for a constant vector $\mathfrak{R}^\alpha$;

3. $E^\rho = \frac{1}{2} \mathrm{m}^\rho_{\rho'} \Phi^{\rho'} + \mathrm{n}^\rho$, with

   i. $\mathrm{m}^\rho_{\rho'} = S_\alpha \mathrm{m}^\rho_{\alpha\rho'}$, for $\mathrm{m}^\rho_{\alpha\rho'}$ constant parameters,

   ii. $\mathrm{n}^\rho = S_\alpha \mathrm{n}^\alpha_\rho$, and n^α_ρ a polynomial in the Φ without any linear terms;

4. $J_0 = P(\Phi)$, where $P(\Phi)$ has gauge charges $-Q^0_\alpha$;
5. $J_\rho = \Phi^0 J'_\rho(\Phi)$, where $J'_\rho(\Phi)$ has gauge charges $-Q^\rho_\alpha - Q^0_\alpha$.

The E and J deformations are further constrained by (5.2.37):

$$E^0 J_0 + E^\rho J_\rho = 0 \qquad \Longleftrightarrow \qquad -\tfrac{1}{2}\mathfrak{R}^\alpha S_\alpha P + E^\rho J'_\rho = 0 . \tag{5.7.2}$$

The $(2,2)$ locus is recovered by setting

$$\mathfrak{R} = R , \qquad \mathrm{m}^\rho_{\rho'} = \delta^\rho_{\rho'} Q^\alpha_\rho S_\alpha , \qquad \mathrm{n}^\rho = 0 , \qquad J'_\rho = \frac{\partial P}{\partial \Phi^\rho} . \tag{5.7.3}$$

In what follows, we will find it convenient to drop the $'$ on J'_ρ. That is, we write the $(0,2)$-deformed J superpotential term as

$$
\begin{aligned}
S_J &= \frac{1}{2\pi\sqrt{2}} \int d^2z\, \mathcal{D}\left\{ \Gamma^0 P(\Phi) + \Gamma^\rho \Phi^0 J_\rho(\Phi) \right\} + \text{h.c.} \\
&= \frac{1}{2\pi\sqrt{2}} \int d^2z \sum_{i=1}^{n} \left\{ \mathcal{D}(\Gamma^i J_i) + \overline{\mathcal{D}}(\overline{\Gamma}^i \overline{J}_i) \right\} \\
&= \frac{1}{2\pi} \int d^2z \left\{ G^0 P + \overline{G}^0 \overline{P} + G^\rho \phi_0 J_\rho + \overline{G}^\rho \overline{\phi}_0 \overline{J}_\rho \right. \\
&\qquad\qquad \left. - \gamma^0 P_{,\rho} \psi^\rho - \overline{\psi}^\rho \overline{P}_{,\rho} \overline{\gamma}^0 - \gamma^\rho \phi_0 J_{\rho,\rho'} \psi^{\rho'} - \overline{\psi}^{\rho'} \overline{\phi}_0 \overline{J}_{\rho,\rho'} \overline{\gamma}^\rho \right\} ,
\end{aligned}
\tag{5.7.4}
$$

and the supersymmetry constraint is

$$-\tfrac{1}{2}\mathfrak{R}^\alpha S_\alpha P + \tfrac{1}{2} \left(\mathrm{m}^\rho_{\rho'} \Phi^{\rho'} + \mathrm{n}^\rho \right) J_\rho = 0 . \tag{5.7.5}$$

A Geometric Interpretation

When we work in the smooth phase of the linear model and keep the E and J parameters generic, this structure has a straightforward geometric interpretation: the zero locus of the bosonic potential remains the same as in the $(2,2)$ theory: $\phi_0 = 0$, $\sigma_\alpha = 0$, and the ϕ_ρ parametrize $X = \{P = 0\} \subset V$. Supersymmetry implies that the massless right-moving fermions couple to (the pullback of) T_X described by (5.3.118). On the other hand, the massless left-moving fermions coupled to sections of $\mathcal{E}$, the deformed tangent bundle $\mathcal{E} \to X$ described by the cohomology of the complex

$$0 \longrightarrow H_2(V, \mathbb{Z}) \otimes \mathcal{O}_X \xrightarrow{\;E\;} \oplus_\rho \mathcal{O}_V(D_\rho)|_X \xrightarrow{\;J\;} \mathcal{O}_V([X])|_X \longrightarrow 0 \; . \qquad (5.7.6)$$

Since this is a smooth deformation of a (2,2) theory, this (0,2) non-linear sigma model will satisfy $c_1(\mathcal{E}) = c_1(T_X) = 0$ and $c_2(\mathcal{E}) = c_2(T_X)$. As long as we stick to the monadic deformations that satisfy the supersymmetry constraint, the deformations are obviously classically unobstructed.[46]

Just as for the (2,2) theories, we can describe components of the singular locus of the (0,2) theory by searching for non-compact directions of the zero locus of the bosonic potential. Working in a smooth phase, the classical story is as follows.

The σ_α are massive unless there is a point $p \in X$ where $E^\rho(p) = 0$ for all ρ. Unlike in the (2,2) case, this need not happen at a wall of the Kähler cone. The locus of all parameters for which such a p exists is $\mathcal{Z}_{\mathrm{cl},A/2}(E, r)$. As in the (2,2) case this will receive quantum corrections; we will compute these below.

It is also possible for the ϕ_0 to develop a flat direction. This occurs whenever there is a point $p \in X$ where $J_\rho(p) = 0$ for all ρ; we will refer to the corresponding locus in the parameter space as $\mathcal{Z}_{\mathrm{cl},B/2}(J)$. Unlike in the (2,2) case, this non-compactness is unrelated to the smoothness of X! This is remarkable from the point of view of classical geometry, but it is not so surprising when we recall that the separation of deformations into "complex structure" and "bundle" is rather artificial in (0,2) theories.[47] We will argue below that at least in some (perhaps all?) cases, this singular locus does not receive quantum corrections.

Away from the singular locus we expect the theory to flow to a $c = \bar{c} = 9$ (0,2) SCFT with $U(1)_L \times U(1)_R$ symmetry and integral charges. As in the case of (2,2) theories, we expect that the q_α, E, J parameters both under- and over-parameterize the (0,2) SCFT moduli space. There is now an additional complication, since a marginal (0,2) deformation need not be exactly marginal. We will return to the parameter space when we discuss (0,2) mirror symmetry.

The A/2 and B/2 Rings

We examined at some length the A-model and at considerably shorter length the B-model on the (2,2) locus. The deformed theory does not have these topological sectors, but it does have the $\overline{Q}$ cohomology and the corresponding half-twisted theory. As we discussed in Chap. 3, on general grounds we expect the half-twisted

---

[46]For small deformations the bundle will also remain stable.

[47]We do not suggest that the non-linear sigma model description is useful in this situation. The reader should recall that we met such features before: in the general construction of (2,2) mirror pairs we came across singular geometries that could not be recast as global orbifolds but that seemingly defined well-behaved SCFTs. This may be considered a more extreme example of the same phenomenon.

theory to be conformally invariant. The $U(1)_L \times U(1)_R$ symmetry implies the existence of the A/2 and B/2 subrings that reduce to the rings of the A-model and B-model on the (2,2) locus. In the linear sigma model we continue to represent the corresponding observables by, respectively, the σ_α, and the $\phi_0 M(\phi)$ operators.

As we will now discuss, the computation of the ring structure is no more difficult than on the (2,2) locus. There are, however, crucial differences between the twisted and half-twisted theories. In the (0,2) case there is no direct correspondence between, say, the local A/2 observables σ_α and the deformation parameters τ^α. The observables will transform as a section of a bundle over the moduli space, and there is not a canonical trivialization in any particular patch.[48] We can work in the natural variables of the linear model and compute correlation functions, but we should bear in mind that the results we obtain are only determined up to an overall normalization.

The σ Vacua

After the great applications of the σ-vacua in the (2,2) theory, it would be foolish to leave these unexplored in the deformed theory. Recall that the σ-branch is obtained by assuming that the σ_α have large expectation values that are generic enough to give masses to all of the matter multiplets. The most general E couplings of the X-model include more complicated bosonic interactions than the simple quartic ones of the (2,2) theory: these are encoded in the "non-linear" n^ρ terms above. These couplings remain a bit mysterious. On one hand, there is evidence and arguments that these non-linear terms do not affect the IR physics of a V-model when the linear terms in E^ρ are generic [116, 152, 153, 276]; on the other hand, they do modify the large field behavior of the theory, and may play a role when the linear terms are tuned to special values. We will make the simplifying assumption that

$$n^\rho = 0 . \qquad (5.7.7)$$

In some cases this is forced on us by the combinatorics: $n^\rho = 0$ when Δ has no interior lattice points in any facet.[49]

With that simplification, the computation of $\mathcal{W}_{\text{eff}}$ of Sect. 5.4.4 goes through almost verbatim. The slight complication is that the mass term $m^\rho_{\rho'}$ need not be diagonal. To deal with this, we group the multiplets according to charge:

$$\{\Phi^\rho, \rho \in \Sigma_V(1)\} \rightarrow \cup_a \{\Phi^{\rho_\alpha}_{(a)} \mid \rho_\alpha \in I_a\} , \qquad (5.7.8)$$

---

[48]This is in contrast to the tangent bundle to the moduli space, for which, once we pick a set of coordinates in a patch, we have the canonical basis furnished by the derivatives with respect to the coordinates.

[49]The astute reader will note that there is a close relationship between the E^ρ couplings and the automorphisms of the toric variety. We will discuss this in some more detail below.

where I_a is a subset of the fields that have identical gauge charge, which we will denote by $Q^\alpha_{(a)}$. For example, for the bi-cubic we have

$$Q = \begin{pmatrix} -3 & 1 & 1 & 1 & 0 & 0 & 0 \\ -3 & 0 & 0 & 0 & 1 & 1 & 1 \end{pmatrix}, \tag{5.7.9}$$

and therefore three collections $I_1 = \{0\}$, $I_2 = \{1, 2, 3\}$, and $I_3 = \{4, 5, 6\}$. Let $n_a = |I_a|$, and organize the Γ^ρ in the same fashion. The E couplings then take the form

$$E^0 = -\tfrac{1}{2}\Phi^0 \mathfrak{R}^\alpha S_\alpha, \qquad E^{(a)} = \tfrac{1}{2}\mathfrak{m}_{(a)}\Phi_{(a)}, \tag{5.7.10}$$

where the $n_a \times n_a$ matrix $\mathfrak{m}_{(a)}$ is linear in the S_α. Now the bosonic mass matrix for the fields belonging to I_a is proportional to $\mathfrak{m}^\dagger_{(a)}\mathfrak{m}_{(a)}$. Diagonalizing the mass term and integrating out the massive fields at one loop leads to the effective superpotential for the remaining fields

$$W_{\text{eff}} = \sum_\alpha \Upsilon_\alpha \tilde{J}_\alpha(S), \qquad \tilde{J}_\alpha(S) = \frac{1}{4\pi\sqrt{2}} \log\left[q_\alpha^{-1} \prod_a \left(\frac{\det \mathfrak{m}_{(a)}}{\mu} \right)^{Q^\alpha_{(a)}} \right]. \tag{5.7.11}$$

The σ_α equations of motion are therefore, quite generally,

$$\prod_a \left(\frac{\det \mathfrak{m}_{(a)}}{\mu} \right)^{Q^\alpha_{(a)}} = q_\alpha, \tag{5.7.12}$$

and for the X-model they simplify to

$$(-\sigma_\alpha \mathfrak{R}^\alpha)^{Q^\alpha_0} \prod_{a>1} \left(\det \mathfrak{m}_{(a)} \right)^{Q^\alpha_{(a)}} = q_\alpha. \tag{5.7.13}$$

These equations are more complicated than the (2,2) relations we studied before, but the complication comes solely through the dependence on the E-parameters: the structure of the equations is really much the same. For instance, for the X-model they are still quasi-homogeneous of degree 0 in the σ_α, and therefore, for generic values of the parameters we expect no solution. On the other hand, the locus where the equations do have a solution is the quantum-corrected version of $\mathcal{Z}_{\text{cl},A/2}(X)$. As in the (2,2) case, there will be additional components to the singular locus where only some of the matter fields have been integrated out, but we will focus on this principal component that generalizes $\mathcal{Z}^{\text{pri}}_A(X)$, and we will call it $\mathcal{Z}^{\text{pri}}_{A/2}(X)$.

$\mathcal{Z}^{\text{pri}}_{A/2}(X)$ has codimension 1 in the parameter space, and it also shows the feature we remarked on before with regards to $\mathcal{Z}_{B/2}(X)$: a classically singular bundle need not lead to problems for the (0,2) SCFT. Moreover, the singularity structure mixes the "Kähler" parameters with the "bundle" E-parameters. We again see that there is no reasonable split of the (0,2) parameter space into categories motivated either by (2,2) loci or classical geometry considerations.

Exercise 5.49 Describe $\mathcal{Z}^{\text{pri}}_{A/2}(X)$ for the bicubic theory in the following parametrization. Set

$$m_{(1)} = \mathbb{1}_{3\times3}\sigma_1 + \tfrac{i}{2\pi}A\sigma_2 , \quad m_{(2)} = \mathbb{1}_{3\times3}\sigma_2 + \tfrac{i}{2\pi}B\sigma_1 , \quad \mathfrak{R} = -3(\sigma_1 + \sigma_2) . \tag{5.7.14}$$

Using the familiar

$$\det(\mathbb{1} + \tfrac{i}{2\pi}A) = 1 + \sum_i c_i(A) \tag{5.7.15}$$

identity, where c_i are the trace combinations that define the Chern classes, you should find that $\mathcal{Z}^{\text{pri}}_{A/2}(X)$ is the resultant of the following two equations in $z = \sigma_2/\sigma_1$:

$$(1 + c_1(A)z + c_2(A)z^2 + c_3(A)z^3) + 3^3 q_1(1 + z)^3 = 0 ,$$
$$(z^3 + c_1(B)z^2 + c_2(B)z + c_3(B)) + 3^3 q_2(1 + z)^3 = 0 . \tag{5.7.16}$$

Eliminating z leads to a very un-enlightening expression, but we can check that it reduces to the (2,2) bicubic singular locus when we set $A = B = 0$.   □

Our next goal is to study the A/2 heterotic ring in the half-twisted theory. Toward that end, we will, as in the (2,2) discussion, begin with the simpler V-model.

5.7.2   Deformations of the V-model

The (2,2) V-model has (0,2) deformations described by deforming the tangent bundle T_V to a more general bundle $\mathcal{E}$; the monadic deformations, i.e. those that are represented by deforming the map in the Euler sequence are precisely the deformations that can be realized in the (0,2) linear sigma model Lagrangian E-couplings. We will again restrict attention to deformations with $n^\rho = 0$. In the case of the X-model there was not much sense in placing any further restrictions on the E-couplings, but this is not so in the V-model, since the latter has, in addition to the U(1)$_B$ symmetry, the toric symmetries that simultaneously rotate a pair Φ^ρ, Γ^ρ. Thus, it is natural to consider the "toric" bundle deformations that preserve those symmetries: they are precisely the deformations where the $m^\rho_{\rho'}$ matrix is taken to be diagonal. The resulting bundle $\mathcal{E} \to V$ is then a torus-equivariant (or toric) bundle.[50]

When V is a projective toric variety, the deformed theory has, just as the original (2,2) V-model, a supersymmetry puzzle: there are no classical vacua for $r^\alpha \notin K_{\mathrm{cl}}(V)$. The natural resolution is also the same: there are σ-vacua in the IR, and we can describe the vacua in terms of the effective potential (5.7.11). Thus for generic parameters (0,2) supersymmetry remains unbroken, and there are a number of massive vacua. On the other hand, the (0,2) theory does have a richer structure: the singularities now depend on the parameters in m as well as on q_α.

For a simple example of the richer structure we can consider the toric case, where

$$\mathrm{m}^\rho_{\rho'} = S_\alpha \mathfrak{M}^\alpha_\rho \delta^\rho_{\rho'} , \tag{5.7.17}$$

and the constant $n_G \times n$ matrix $\mathfrak{M}$ encodes the E-deformations. For generic and fixed q_α it is possible to obtain a singular theory, where some of the σ-vacua wander off to infinity by deforming $\mathfrak{M}$. Even more dramatically, whenever $\mathfrak{M}$ is not full rank, the theory is singular because there is a decoupled linear combination of the σ_α. On the other hand, supersymmetry is now spontaneously broken since there are no solutions to the σ-equations of motion that follow from (5.7.11) [152, 184, 366].

Exercise 5.50 The singular locus for $V = \mathbb{P}^1 \times \mathbb{P}^1$. Fix the charge matrix

$$Q = \begin{pmatrix} 1 & 1 & 0 & 0 \\ 0 & 0 & 1 & 1 \end{pmatrix} , \tag{5.7.18}$$

and consider toric deformations. Show that for generic q_α the singular locus is defined by the vanishing of

$$\mathcal{T} = \zeta_{13}\zeta_{14}\zeta_{23}\zeta_{24} , \tag{5.7.19}$$

where ζ_{ij} is the determinant of the minor of $\mathfrak{M}^\alpha_\rho$ obtained by taking the i-th and j-th column. Thus, in this case the degeneration happens just where the bundle degenerates (the sheaf defined by the Euler sequence fails to be locally free). □

Still, as long as we stay away from the singular locus, there is a sensible (0,2) supersymmetric theory with a non-anomalous $U(1)_A$ symmetry, and we should be able to define and compute its A/2 ring in the half-twisted theory. These observations were made in the seminal work [2] that posed the problem and performed the first computations of this structure. These ideas were developed further and led to the definition of the heterotic A/2 and B/2 rings [3], and a more general setting for the computations in terms of the notion of quantum sheaf cohomology introduced and developed in [153, 154, 216, 264].

---

[50]Such bundles have a nice combinatorial description—see [268, 270, 323] for the formalism and [152] for some applications to the deformed V-model.

Classical and Quantum Sheaf Cohomology

To set the V-model computations we are about to describe in a more general setting, consider $\mathcal{E} \to V$—a rank k holomorphic bundle over a smooth compact complex manifold of dimension d satisfying

$$c_1(\mathcal{E}) = c_1(T_V) \,, \qquad\qquad c_2(\mathcal{E}) = c_2(T_V) \,. \tag{5.7.20}$$

The second of these conditions we recognize: it means we can define a $(0,2)$ non-linear sigma model with this target space. The first condition guarantees the existence of $U(1)_A$ symmetry and is also crucial for defining a class of correlation functions.

The Dolbeault cohomology groups $H_{\bar{\partial}}^q(V, \wedge^p \mathcal{E}^*)$ have a natural product structure:

$$H_{\bar{\partial}}^q(V, \wedge^p \mathcal{E}^*) \times H_{\bar{\partial}}^{q'}(V, \wedge^{p'} \mathcal{E}^*) \to H_{\bar{\partial}}^{q+q'}(V, \wedge^{p+p'} \mathcal{E}^*) \,, \tag{5.7.21}$$

and since $c_1(\mathcal{E}) = c_1(T_V)$, we also see that $\wedge^k \mathcal{E}^* = \wedge^d T_V^*$ and therefore also an isomorphism $H_{\bar{\partial}}^d(V, \wedge^k \mathcal{E}^*) = H^{d,d}(V, \mathbb{C})$. This means that these cohomology groups have the structure of a Frobenius algebra: there is an identity element, the multiplication is associative, and there is a non-degenerate pairing (or a trace) defined on the product by integration. That is in fact exactly the algebraic structure of the quantum cohomology rings, and the structure specializes to the classical cohomology when $\mathcal{E} = T_V$.

A precise mathematical formulation of quantum sheaf cohomology in this general setting still does not exist. However, the idea is fairly clear [154, 264]: one should construct a sheaf cohomology theory on the worldsheet instanton moduli spaces, lift representatives from $H_{\bar{\partial}}^q(V, \wedge^p \mathcal{E}^*)$ to classes on the moduli space, perform the computation, and sum up the instantons! We will not embark on this enterprise and instead will return to the tamer toric setting of deformations of $(2,2)$ theories, where all of these goals can be realized concretely.

We note one important point that that does emerge from this discussion: the trace that defines the classical pairing and, correspondingly, the classical correlation functions, does not have a canonical normalization—we met this point when we discussed the A/2 and B/2 rings above, and here this is an issue already at the level of classical geometry.[51]

An important observation is that we can generalize this structure from holomorphic bundles to more general sheaves. This is useful and necessary, because we already saw that the bundle can degenerate without leading to singularities in the physical theory.

---

[51] A detailed discussion of this point in the non-linear sigma model context can be found in [186].

Quantum Sheaf Cohomology for Toric Varieties

Now we return to our tamer world: V is a projective toric variety, and $\mathcal{E}$ is a monadic deformation of T_V with $\mathfrak{n}^\rho = 0$. In this case it can be shown that the relevant sheaf cohomology groups are concentrated in $H_{\bar\partial}^p(V, \wedge^p \mathcal{E}^*)$ [154], and the classes in $H_{\bar\partial}^1(V, \mathcal{E}^*)$ are represented by the σ_α.

We now turn to the A/2 half-twisted theory, and it is not very different from the A-model of the V-model discussed above. The fields continue to transform as sections of the bundles in (5.5.34), and on $\Sigma = \mathbb{P}^1$ the supercharge $\overline{Q}$ localizes the field configurations to the same locus as Q_A. The selection rules for the correlation functions, which follow from anomaly considerations, are unaffected by the deformation.

We can evaluate the correlation functions by working in the IR phase. The arguments that lead to (5.5.74) go over with minor modifications to incorporate the E-parameters and diagonalize the mass matrices $\mathfrak{m}_{(a)}^\dagger \mathfrak{m}_{(a)}$.[52] The result is

$$\langle \sigma_{\alpha_1} \cdots \sigma_{\alpha_k} \rangle_{A/2}(q, \mathfrak{m}) = \sum_{\sigma_*} \left\{ \sigma_{\alpha_1} \cdots \sigma_{\alpha_k} W_{\Upsilon,\mathcal{S}}(\sigma) W_{\Phi,\Gamma}(\sigma) \right\} \big|_{\sigma=\sigma^*} , \quad (5.7.22)$$

where

$$W_{\Upsilon,\mathcal{S}} = (\det \det_{\alpha,\beta} \tilde{J}_{\alpha,\beta})^{-1} , \qquad W_{\Phi,\Gamma} = \prod_a \frac{1}{\det \mathfrak{m}_{(a)}} , \quad (5.7.23)$$

and σ_* are the zeroes of

$$\prod_a \left(\det \mathfrak{m}_{(a)} \right)^{Q_{(a)}^\alpha} = q_\alpha . \quad (5.7.24)$$

The latter equations are then also interpreted as the quantum sheaf cohomology relations, i.e. operator relations in the half-twisted correlators

$$\prod_{a \mid Q_a^\alpha > 0} \left(\det \mathfrak{m}_{(a)} \right)^{Q_{(a)}^\alpha} = q_\alpha \prod_{a \mid Q_a^\alpha < 0} \left(\det \mathfrak{m}_{(a)} \right)^{-Q_{(a)}^\alpha} . \quad (5.7.25)$$

These results can also be obtained by working in the UV phase and summing instantons, but at considerably greater effort. The early work [216, 264] was fairly heroic and not easy to generalize, but a complete solution is now available [153, 154]. It has the advantage of showing that the non-linear $\mathfrak{n}^\rho$ deformations do not affect these half-twisted correlation functions, something that is not immediately obvious in the σ-branch computation.

---

[52]The required unitary transformation acts identically on the Γ^ρ as well as the Φ^ρ. It is non-chiral and does not affect the path integral measure.

Exercise 5.51 A simple and classic example is offered by $V = \mathbb{P}^1 \times \mathbb{P}^1$. It is not hard to solve the general case, but the algebra is a little bit tedious. So, we consider the deformations with

$$\mathfrak{m}_{(1)} = \mathbb{1}_{2 \times 2} \sigma_1 + A \sigma_2 \,, \qquad \mathfrak{m}_{(2)} = \mathbb{1}_{2 \times 2} \sigma_2 + B \sigma_1 \,, \qquad (5.7.26)$$

and furthermore restrict to traceless A and B matrices. Let $\det A = a$ and $\det B = b$ and show that the non-zero correlation functions are

$$\langle \sigma_1^{2s+1} \sigma_2^{2t+1} \rangle_{A/2} = \frac{(q_1 - a q_2)^s (q_2 - b q_1)^t}{(1 - a_1 a_2)^{1+s+t}} \,. \qquad (5.7.27)$$

The correlation function reduces to the right values on the (2,2) locus, has a pole when $\mathcal{E}$ degenerates, and only depends on A and B through the determinants.

Now repeat the exercise for $\mathbb{F}_2$ with

$$Q = \begin{pmatrix} 2 & 0 & 1 & 1 \\ 1 & 1 & 0 & 0 \end{pmatrix} \,, \quad \mathfrak{m}_{(1)} = 2\sigma_1 + \sigma_2 \,, \quad \mathfrak{m}_{(2)} = \sigma_2 \,, \quad \mathfrak{m}_{(3)} = \mathbb{1}\sigma_1 + A\sigma_2 \,, \qquad (5.7.28)$$

where A is a traceless matrix with $\det A = a$. Show that

$$\langle \sigma_1^2 \rangle_{A/2} = \frac{2(q_1 - a q_2^2)}{4 q_1 + (1 - 4a) q_2^2} \,. \qquad (5.7.29)$$

Compare the singular locus to the classical degeneration of the bundle: prove that the bundle degenerates if and only if $a = 1/4$: the rank of the sheaf jumps up at a point $[1, 0, \phi_3, \phi_4]$. The lesson here is that for generic q_α the classical degeneration of the bundle does not lead to any trouble in the correlation functions. $\qquad \square$

5.7.3 The Deformed X-model

We return to the X-model with the goal of determining the correlation functions $\langle\!\langle \sigma_{\alpha_1} \cdots \sigma_{\alpha_{d-1}} \rangle\!\rangle_{A/2}$ and $\langle\!\langle \phi_0 \mathsf{M}_1 \phi_0 \mathsf{M}_2 \cdots \phi_0 \mathsf{M}_{d-1} \rangle\!\rangle_{B/2}$.

The A/2 Correlation Functions via Quantum Restriction

We work in a Calabi-Yau phase and instanton sector $\vec{N}$. The structure of the A/2 twisted fields is exactly the same as on the (2,2) locus: they continue to be sections of the same bundles as indicated in (5.5.35), (5.5.34). Moreover, the path integral continues to localize onto $\mathcal{M}_{\vec{N},X}$. However, we cannot immediately argue that the correlation functions will be independent of either P or J_ρ: while $\overline{P}$ and $\overline{J}_\rho$ couplings are $\overline{Q}$-exact, the same is not true of the holomorphic couplings.

Fortunately, we still have enough control to reduce the problem to V-model correlation functions. We argue following [292]. First, we do a partial localization to $\mathcal{M}_{\tilde{N}}$ and keep all of the terms in the action.[53] Now we can use the $\overline{Q}$-exactness of $\overline{P}$ and $\overline{J}_\rho$ terms in the action. Since ϕ_0 has no zero modes, the $\overline{P}, \overline{J}_\rho \to 0$ limit does not lead to a pathology in the path integral, and the correlation functions are unaffected by taking the limit. In order to obtain a non-zero answer it is still necessary to soak up the $\overline{\gamma}^0$ and ψ^0 zero modes by bringing down terms from the action. There are two types of such terms:

$$S \supset \frac{1}{2\pi} \int d^2z \left[+\tfrac{1}{2} \overline{\gamma}^0 \mathfrak{R} \psi^0 - \gamma^0 \frac{\partial P}{\partial \phi^\rho} \psi^\rho \right] . \tag{5.7.30}$$

A moment's thought shows that the only way to soak up the zero modes is to bring down the first term $1 - d_0$ times. Now we are done, because the correlation function is reduced to an integral over the V-model moduli space with a particular insertion. It follows that we can perform exactly the same summing tricks as in the (2,2) case and arrive at

$$\langle\!\langle \sigma_{\alpha_1} \cdots \sigma_{\alpha_{d-1}} \rangle\!\rangle_{A/2}(q, \mathfrak{m}) = \langle \sigma_{\alpha_1} \cdots \sigma_{\alpha_{d-1}} \frac{\mathfrak{R}}{1 + \mathfrak{R}} \rangle_{A/2}(q, \mathfrak{m}) . \tag{5.7.31}$$

We therefore conclude that the correlation functions vary holomorphically with $q, \mathfrak{m}$ parameters, diverge on the $\mathcal{Z}_{A/2}^{\text{pri}}(X)$ locus (when $1 + \mathfrak{R}$ fails to be invertible), and are independent of P and J_ρ.

The last point is less impressive than it might seem: as we will soon discuss, field redefinitions act on all of the parameters at once, so this result should not be interpreted as some sort of canonical split in the moduli space. A closely related point is that the normalization of the correlation functions is also not canonical. A particular split of the parameters is a choice of gauge, and the correlation functions will depend on that choice.

These results were also obtained and generalized to non-abelian linear sigma models via post-modern localization in [116].

The B/2 Twist

On the (2,2) locus it is a triviality to argue that the B-twisted theory is independent of the q_α. Since we just established that the A/2 correlation functions are independent of P and J_ρ, one might be tempted to conjecture that the B/2 correlation

---

[53] We saw such partial localization tricks in our treatment of the B/2 Landau-Ginzburg correlation functions.

functions might be independent of the q_α and m. It is easy to show that, up to an overall normalization, this holds in many examples. The q_α independence can often be established by degree considerations, and once that is established, the m independence will hold in any theory with a Landau-Ginzburg orbifold phase. More generally, the situation is less clear [292, 347]. The expectation is that even when degree considerations allow for a non-zero top form on the instanton moduli space, the form turns out to be exact, but to the author's knowledge this has not been shown in general.

We will illustrate how this independence works in some special circumstances. In the B/2 twisted theory the fields in the Φ^0, Γ^0 multiplets get twisted in the same way as the other multiplets, and $\phi_0 \in \mathscr{A}(\Sigma, \mathcal{L}_0)$. Working in a geometric phase, and $\vec{N} \neq 0$, we then know that the path integral localizes onto field configurations with $\phi_0 = 0$. On the other hand, the B/2 observables are the gauge-invariant operators $\phi_0 M(\phi_\rho)$, and we expect the computation of the B/2 correlator to reduce to an integral over the zero modes. It follows that the correlators will vanish unless ϕ_0 has zero modes, but this requires $d_0 \geq 0$. Since in a geometric phase $d_0 \leq 0$, we obtain a vanishing theorem: the B/2 correlation functions only receive contributions from instantons for which $d_0 = 0$. Since $d_0 = -(\boldsymbol{R}, \vec{N})$, and $\vec{N} \in K^*(V)$, this is only possible if $\boldsymbol{R}$ lies in the boundary of the Kähler cone. For any Fano V $\boldsymbol{R}$ is contained in the interior of the Kähler cone, and therefore $d_0 = 0$ implies $\vec{N} = 0$, and therefore the correlation functions are independent of q_α.[54]

Once we know that the B/2 theory is q_α-independent, then we can evaluate the correlation functions in any phase without changing the result. If there is a Landau-Ginzburg orbifold phase, then the evaluation is particularly simple. We know that deep in such a phase the gauge and $\mathcal{S}$ multiplets are lifted: their masses are proportional to the m couplings and the expectation values of the n_G matter fields that acquire large expectation values in that phase. Integrating out the massive multiplets, i.e. the vector multiplets, the $\mathcal{S}$, and the n_G matter and fermi multiplets, we see that any m dependence can only appear as an overall multiplicative factor. This was shown in examples in [292].

In [292] it was assumed that the m terms that lift the $\mathcal{S}$ and fermi multiplets in the Landau-Ginzburg phase are non-degenerate. When this is the case, a field redefinition can often be used to scale them to their (2,2) values, and in that gauge it is then easy to see in the examples that the correlation functions are independent of the remaining E-parameters. On the other hand, [292] neglected the possibility that those mass terms can degenerate, in which case the Landau-Ginzburg phase will be a singular point due to the additional massless fields and a non-compact σ direction.

To get an idea of what this possibility entails, we will examine the simplest example: the quintic in its Landau-Ginzburg orbifold phase. Along the way we will correct some algebra mistakes made in section 5.4.1 of [292]. We work with the quintic theory in $\vec{N} = 0$ sector, and we expand around field configurations with

---

[54]For some NEF Fano varieties more refined arguments can also be used to show $\vec{N} = 0$, but a general argument is not known to the author.

$\sigma = 0$ and $\phi_0 = -r/5$. The Lagrangian for the zero modes then takes the form ($i = 1, \ldots, 5$)

$$\mathcal{L}_0 = \tfrac{5}{2}\overline{\psi}_0\phi_0\overline{\upsilon} - \tfrac{1}{2}(\overline{\psi}_i\phi_i)\overline{\upsilon} - \tfrac{1}{2}\overline{\gamma}_0 m_0^0\phi_0\chi - \tfrac{1}{2}\overline{\gamma}_i m_j^i\phi_j\chi + P\overline{P} - \overline{\psi}_i\overline{P}_{,i}\overline{\gamma}_0$$
$$-\overline{\psi}_0 J_i\overline{\gamma}_i + \phi_0\overline{\phi}_0 J_i\overline{J}_i - \phi_0\overline{\psi}_i\overline{J}_{j,i}\overline{\gamma}_j \ . \tag{5.7.32}$$

To work in the Landau-Ginzburg phase, we need to send $r \to -\infty$. Since in this limit some of the terms in $\mathcal{L}_0$ will diverge, we perform a field redefinition:

$$\phi_i = (\phi_0)^{-1/4}\phi_i', \qquad \overline{\phi}_i = (\phi_0)^{-1/4}\overline{\phi}_i', \qquad \overline{\gamma}_i = (\phi_0)^{-1/4}\overline{\gamma}_i,$$
$$\overline{\upsilon} = \phi_0^{-1}\overline{\upsilon}', \qquad\qquad \chi = \phi_0^{-1}\chi'. \tag{5.7.33}$$

This change affects both the action and the measure. The latter transforms as

$$D[\text{fields}] = \phi_0^{3/4} D[\text{fields}'] , \tag{5.7.34}$$

while the former becomes

$$\mathcal{L}_0 = \tfrac{5}{2}\overline{\psi}_0\overline{\upsilon}' - \tfrac{1}{2}\overline{\gamma}_0 m_0^0\chi'$$
$$- \tfrac{1}{2}\phi_0^{-5/4}(\overline{\psi}_i\phi_i')\overline{\upsilon}' - \tfrac{1}{2}\phi_0^{-3/2}\overline{\gamma}_i' m_j^i\phi_j'\chi'$$
$$+ \phi_0^{-5/2}P(\phi')\overline{P}(\overline{\phi}') - \phi_0^{-1}\overline{\psi}_i\overline{P}_{,i}(\overline{\phi}')\overline{\gamma}_0 - \phi_0^{-5/4}\overline{\psi}_0\overline{J}_i(\overline{\phi}')\overline{\gamma}_i'$$
$$+ J_i(\phi')\overline{J}_i(\overline{\phi}') - \overline{\psi}_i\overline{J}_{j,i}(\overline{\phi}')\overline{\gamma}_j' \ . \tag{5.7.35}$$

We recognize the last line: it is the same as the zero model action for a B/2-twisted Landau-Ginzburg theory with potential $J_i(\phi')$. The middle two lines go to zero in the $r \to -\infty$ limit, and the first line is used to soak up the $\overline{\gamma}_0$, $\overline{\psi}_0$, $\overline{\upsilon}'$ and χ' zero modes. Finally, the insertion is of the form

$$\phi_0 M_1(\phi)\phi_0 M_2(\phi)\phi_0 M_3(\phi) = \phi_0^{-3/4} M_1(\phi')M_2(\phi')M_3(\phi') , \tag{5.7.36}$$

and therefore

$$\langle\!\langle\phi_0 M_1(\phi)\phi_0 M_2(\phi)\phi_0 M_3(\phi)\rangle\!\rangle_{B/2} \propto m_0^0 \langle M_1(\phi')M_2(\phi')M_3(\phi')\rangle_{B/2,LG} . \tag{5.7.37}$$

As claimed, the correlation functions have only an over-all dependence on m_0^0. The reader can verify that a similar result holds for other examples with a Landau-Ginzburg phase, such as the (0,2) deformed octic.

What Are We Computing?

We described a class of A/2 and B/2 twisted correlation functions that can be computed in a (0,2) linear sigma model with a (2,2) locus. The E and J couplings encode deformations that take the theory off the (2,2) locus, and it appears from the point of view of the half-twisted theories that it does so smoothly: the dimensions of the half-twisted rings do not jump, and the correlation functions vary holomorphically with the (0,2) deformation parameters.

All of this suggests the following sensible tale for a heterotic compactification based on such a family of (0,2) theories with $c = \bar{c} = 9$: the A/2 correlation functions encode a subset of the $\overline{27} \cdot \overline{27} \cdot \overline{27}$ couplings, while the B/2 correlation functions encode a subset of the $27 \cdot 27 \cdot 27$ couplings. More precisely, these couplings are determined up to overall normalizations, and they vary holomorphically on the (0,2) moduli space.

It is peculiar that this structure should vary so smoothly: the so-called singlets—the (0,2) marginal deformations away from the (2,2) locus—need not be exactly marginal; even if they are exactly marginal, they can lead to mass terms for the charged matter of the form $1 \cdot 27 \cdot \overline{27}$, but we see no evidence of the jumping. It turns out that these peculiar properties are due to the special structure of the linear sigma model: roughly speaking, the (0,2) deformations that can be realized as parameters in the UV linear sigma model, and the A/2 and B/2 subrings that can be realized in terms of the UV fields as we have shown, are protected.

5.7.4 (0,2) Mirror Symmetry

Mirror symmetry originated in the study of Gepner models, a class of exactly solvable (2,2) SCFTs. There are a number of analogous constructions of exactly solvable (0,2) SCFTs [74, 87, 181], and some of these can be used to build mirror pairs that follows the (2, 2) logic: there are exactly solvable theories with an orbifold action that leads to an isomorphic theory up to the flip of the sign of $U(1)_L$ charge [88]. For many examples it is possible to construct an embedding of the theories in a (0,2) linear sigma model and therefore obtain examples of (0,2) mirror pairs. It would be interesting to find a general framework, perhaps even with some combinatorial restrictions, that captures and generalizes these examples, and to write down an explicit map that relates the A/2 and B/2 topological heterotic rings on the two sides of the mirror. We will discuss some of the challenges of this problem in the concluding section below.

To get an idea of what a general (0,2) mirror map may look like, we can consider the case where we are guaranteed success: the theories with a (2,2) locus. In such a theory conformal perturbation theory implies that the existence of a (0,2) mirror map is a tautology: if two (2,2) SCFTs are isomorphic, then by definition their spectra and correlation functions agree on the (2,2) locus. In particular, the space of exactly marginal (0,2) deformations must be the same, and the (0,2) SCFT moduli spaces, each equipped with the Zamolodchikov metric, must be isometric. It is also true that the spectra of the irrelevant operators and the obstruction theory for marginally irrelevant operators must also agree on the two sides of the mirror, but we will focus on the space of (0,2) exactly marginal deformations.

To make the tautology useful, we must be able to identify the abstract SCFT deformation spaces with some concrete realizations. We saw that in the case of the (2,2) linear sigma models for hypersurfaces in Fano toric varieties we had a concrete realization for the toric and polynomial subspaces of deformations, and under the mirror isomorphism the two are related by the monomial-divisor mirror map. On the other hand, we also discussed the (0,2) deformations of the linear sigma models and found the following features:

1. the deformations preserved the dimension of the "toric" A/2 ring spanned by the σ_α, while the ring structure was found to depend on the q_α and m;
2. similarly, the deformations preserved the dimension of the "polynomial" B/2 ring spanned by the R-charge 1 gauge-invariant monomials $\phi_0 M$, while the ring structure was found to depend on the complex parameters in P and J_ρ;
3. any (2,2)-inspired split of the parameters into subspaces is suspect due to normalization factors and the supersymmetry constraint (5.7.5) tying the m, J_ρ, and P.

Thus, in the best of all possible worlds we might expect a correspondence

$$H_{A/2}^{\text{toric}}(X, q, \text{m}, P, J) \qquad \longleftrightarrow \qquad H_{B/2}^{\text{poly}}(X^\circ, q^\circ, \text{m}^\circ, P^\circ, J^\circ) \qquad (5.7.38)$$

via an isomorphism, akin to the mirror map, of the toric+polynomial parameter spaces. Despite all other evidence to the contrary, the study of (0,2) linear sigma model shows that we do not live in the best of all possible worlds: the dimensions of the parameter spaces, when counted modulo (0,2) field redefinitions, do not match for (0,2) deformations of generic Batyrev mirror pairs [276]. Furthermore, counting parameters in such a fashion seems to depend sensitively on whether we take a minimal, maximal, or some intermediate projective subdivision in constructing the linear model.

We emphasize that this does not constitute a spontaneous violation of a fact[55]: we are not claiming that the SCFT moduli spaces are non-isomorphic; merely that certain UV representations of the SCFT deformations do not match up. This suggests that, unlike in the (2,2) case, with its miraculous and mirror symmetric split of "toric+non-toric" and "polynomial+non-polynomial" deformations, in the (0,2) case no such split exists in general.

Fortunately, the study of the combinatorics reveals the culprit: the difficulties arise due to interior lattice points in polytope facets. On the (2,2) locus these lead to a headache, but for (0,2) deformations they seem to cause more significant problems. We can therefore proceed in two ways: first, we can restrict to the reflexively plain

---

[55]Strangely enough, though perhaps not these days, that is a scientific term [114].

case, where these nasty features are absent; second, in a more general theory we can always choose a sub-family where the unpleasant couplings are set to zero by hand and the offending field redefinitions are eliminated by a gauge fixing. We will describe the results from the first perspective, but the reader should bear in mind that the statements will extend to subfamilies in more general examples.

Parameters and Field Redefinitions: The Reflexively Plain Case

Let X, X° be a Batyrev mirror pair as above, with Δ, Δ° the pair of reflexive polytopes that are reflexively plain: neither polytope has an interior lattice point in any facet. We will now follow [276, 298] and the same logic as in the (2,2) case in considering the space of (0,2) UV parameters modulo the (0,2) field redefinitions.

Consider the $X = \{P = 0\} \subset V$ model, where Δ is the Newton polytope for P:

$$P = \sum_{m \in \Delta \cap M} \alpha_m \mathsf{M}_m \,, \qquad \mathsf{M}_m = \prod_{\dot{m} \in \Delta^\circ \cap N \backslash 0} \Phi_{\dot{m}}^{\langle m, \dot{m} \rangle + 1} \,. \tag{5.7.39}$$

Though it is not strictly necessary, we will assume that the α_m are all non-zero. This will considerably simplify the following discussion. As in our discussion of the (2,2) monomial-divisor mirror map, we set $n = |\Delta^\circ \cap N \backslash 0|$ to be the number of homogeneous coordinates for V, and we denote the lattice points in Δ° by $\dot{m} \in N$. To simplify notation, a product or sum with index $\dot{m}$ will be over the n non-zero lattice points in Δ°.

Similarly, we parameterize the J_ρ couplings:

$$\Phi_{\dot{m}} J_{\dot{m}} = \sum_{m \in \Delta \cap M} j_{m\dot{m}} \mathsf{M}_m \,, \tag{5.7.40}$$

where the $j_{m\dot{m}}$ obey

$$j_{m\dot{m}} = 0 \qquad \text{when} \qquad \langle m, \dot{m} \rangle = -1 \,. \tag{5.7.41}$$

As with the α_m, we make a simplifying non-vanishing assumption: $j_{0\dot{m}} \neq 0$ for all $\dot{m}$.

The most general E-couplings are

$$E^0 = -\tfrac{1}{2} S \cdot \mathfrak{R} \Phi_0 = -\tfrac{1}{2} S_\alpha \mathfrak{R}^\alpha \Phi_0 \,, \qquad E^{\dot{m}} = \tfrac{1}{2} S \cdot \mathfrak{M}_{\dot{m}} \Phi_{\dot{m}} = \tfrac{1}{2} S_\alpha \mathfrak{M}_{\dot{m}}^\alpha \Phi_{\dot{m}} \,, \tag{5.7.42}$$

where $\mathfrak{M}$ is the $n_G \times n$ constant matrix of parameters we met in discussing toric deformations of the bundle T_V, and $\mathfrak{R}$ is the (0,2) generalization of the anti-canonical class that we encountered before. In this notation the supersymmetry constraint (5.7.5) takes the form

$$-\alpha_m \mathfrak{R} + \sum_{\dot{m}} j_{m\dot{m}} \mathfrak{M}_{\dot{m}} = 0 \qquad \text{for all } m \in \Delta \cap M .\qquad (5.7.43)$$

The field redefinitions compatible with the (0,2) superfield structure of a reflexively plain theory are a subgroup of $(\mathbb{C}^*)^{n+1} \times (\mathbb{C}^*)^{n+1} \times \mathrm{GL}(n_G, \mathbb{C})$ that act on the fields via

$$\Phi_0 \mapsto u_0 \Phi_0 , \quad \Phi_{\dot{m}} \mapsto u_{\dot{m}} \Phi_{\dot{m}} , \quad \Gamma^0 \mapsto v_0 \Gamma^0 , \quad \Gamma^{\dot{m}} \mapsto v_{\dot{m}} \Gamma^{\dot{m}} , \qquad (5.7.44)$$

and

$$S \mapsto u_0^{-1} v_0 S H , \qquad (5.7.45)$$

where $H \in \mathrm{GL}(r, \mathbb{C})$. We chose the $u_0^{-1} v_0$ prefactor for convenience. These redefinitions induce an action on the holomorphic couplings.

Exercise 5.52 Show that the induced action on the parameters is

$$\mathfrak{R} \mapsto H\mathfrak{R} , \qquad\qquad \mathfrak{M}_{\dot{m}} \mapsto \frac{u_{\dot{m}} v_{\dot{m}}^{-1}}{u_0 v_0^{-1}} H \mathfrak{M}_{\dot{m}} ,$$

$$\alpha_m \mapsto \alpha_m \times v_0 \prod_{\dot{m}} u_{\dot{m}}^{\langle m, \dot{m}\rangle + 1} ,$$

$$j_{m\dot{m}} \mapsto j_{m\dot{m}} \times u_0 v_{\dot{m}} u_{\dot{m}}^{-1} \prod_{\dot{m}'} u_{\dot{m}'}^{\langle m, \dot{m}'\rangle + 1} . \qquad (5.7.46)$$

$\square$

In addition, the redefinitions will introduce an overall normalization factor in the path integral. There is, however, one additional surprise. Because the redefinitions are chiral, i.e. they do not act symmetrically on the gauge-charged fermions, there will be an anomalous shift in the measure that can be interpreted as an action on the q_α. Supersymmetry implies we can determine the form of this shift by restricting to transformations valued in $\mathrm{U}(1)^n \times \mathrm{U}(1)^n$, and using (2.5.21) we conclude that the induced action is

$$q_\alpha \mapsto q_\alpha \times \prod_{\dot{m}} \left[\frac{u_{\dot{m}} v_0}{u_0 v_{\dot{m}}} \right]^{Q_{\dot{m}}^\alpha} . \qquad (5.7.47)$$

Exercise 5.53 As a check of the sensibility of this transformation, evaluate the quantum sheaf cohomology relations for the corresponding V-model, and show that they take the form

$$\prod_{\dot{m}} (\sigma \cdot \mathfrak{M}_{\dot{m}})^{Q_{\dot{m}}^{\alpha}} = q_{\alpha} \, . \tag{5.7.48}$$

Verify that these relations are preserved by the field redefinitions.   □

Looking back to the monomial-divisor mirror map in Sect. 5.4.6, the next step is to construct some combinations of these parameters that are invariant under the redefinitions. To that end, we recall that the mirror gauge charges, $Q_{\dot{m}}^{\circ\dot{\alpha}}$ are an integral basis for the relations among the $\dot{m} \in N$. Equivalently—see ((5.4.103), (5.4.104))—these charges are an integral basis for the cokernel of the $n^{\circ} \times n$ pairing matrix

$$\pi_{m\dot{m}} = \langle m, \dot{m} \rangle \, . \tag{5.7.49}$$

We then observe that

$$\kappa_{\dot{\alpha}}^{\circ} = \prod_{m} \alpha_{m}^{Q_{m}^{\circ\dot{\alpha}}} \tag{5.7.50}$$

are invariant under the redefinitions, as are

$$\kappa_{\alpha} = q_{\alpha} \prod_{\dot{m} \neq 0} \left(\frac{j_{0\dot{m}}}{\alpha_{0}} \right)^{Q_{\dot{m}}^{\alpha}} \tag{5.7.51}$$

and

$$b_{m\dot{m}} = \frac{\alpha_{0} j_{m\dot{m}}}{\alpha_{m} j_{0\dot{m}}} - 1 \, . \tag{5.7.52}$$

Since the right-hand-side of the last expression is trivially zero for $m = 0$, we define $b_{m\dot{m}}$ for $m \neq 0$, so that b is a $n^{\circ} \times n$ matrix. On the (2,2) locus b reduces to the intersection matrix π, and, as a result of (5.7.41), $b_{m\dot{m}} = -1$ whenever $\pi_{m\dot{m}} = -1$.

Finally, we end our fraktur masochism with

$$\mathfrak{G}_{\dot{m}} = \frac{j_{0\dot{m}}}{\alpha_{0}} \mathfrak{M}_{\dot{m}} \, . \tag{5.7.53}$$

The remaining field redefinitions act solely on $\mathfrak{R}$ and $\mathfrak{G}$ via

$$\mathfrak{R} \mapsto H\mathfrak{R} \, , \qquad\qquad \mathfrak{G}_{\dot{m}} \mapsto H\mathfrak{G}_{\dot{m}} \, , \tag{5.7.54}$$

and the supersymmetry constraint reduces to

$$-\alpha_0 \mathfrak{R} + \sum_{\dot{m}} \alpha_0 \mathfrak{G}_{\dot{m}} = 0 ,$$

$$-\alpha_m \mathfrak{R} + \sum_{\dot{m}} \alpha_m (b_{m\rho} + 1) \mathfrak{G}_{\dot{m}} = 0 \qquad \text{for} \quad m \neq 0 . \tag{5.7.55}$$

Our simplifying assumption that the α_m are non-vanishing now leads to

$$\mathfrak{R} = \sum_{\dot{m}} \mathfrak{G}_{\dot{m}} , \qquad \sum_{\dot{m}} b_{m\dot{m}} \mathfrak{G}_{\dot{m}} = 0 . \tag{5.7.56}$$

At the (2,2) locus the matrix b has rank $d = n - n_G$, and for a small deformation the rank will not change. Thus, $\mathfrak{G}$ is in the n_G-dimensional kernel of b with a basis $\{B_{\dot{m}}^{\alpha}\}_{\alpha=1,\ldots,n_G}$:

$$\mathfrak{G}_{\dot{m}}^{\alpha} = \mathcal{G}_{\beta}^{\alpha} B_{\dot{m}}^{\beta} . \tag{5.7.57}$$

Exercise 5.54 Trace back all of the redefinitions to show that if the $n_G \times n_G$ matrix $\mathcal{G}_{\beta}^{\alpha}$ is not full rank, then the theory is singular because of a decoupled $\mathcal{S}_{\alpha}$ multiplet. □

Using the exercise, we therefore see that away from the singular locus $\mathfrak{G}$ is completely determined; we can use the $\text{GL}(n_G, \mathbb{C})$ redefinition to set $\mathfrak{G}_{\dot{m}}^{\alpha} = Q_{\dot{m}}^{\alpha}$, and this has a trivial stabilizer in $\text{GL}(n_G, \mathbb{C})$.

We can generalize the statement away from small deformations: if $b_{m\dot{m}}$ has rank larger than d, then rank $\mathfrak{G} < n_G$, and there will again be a singularity due to an unconstrained $\mathcal{S}_{\alpha}$ multiplet. We will argue below that the theory will also be singular when rank $b < d$, so that for non-singular theories the $\widehat{\kappa}_{\dot{\alpha}}$, κ_{α}, and the matrix $b_{m\dot{m}}$ give a full set of invariant coordinates for the quotient of the UV parameters by the field redefinitions, and we can reasonably expect these to correspond to a subset of marginal deformations of the (0,2) SCFT.

What is the dimension of this subset of marginal deformations? It is easy enough to count

$$\#(\kappa) = n_G = n - d , \qquad \#(\kappa^{\circ}) = n_G^{\circ} = n^{\circ} - d . \tag{5.7.58}$$

What about the parameters in $b_{m\dot{m}}$? We use a result from basic algebraic geometry [221]: the set of $p \times q$ matrices of rank at most d is an irreducible subvariety of codimension $(p - d)(q - d)$ in the space of $p \times q$ matrices. In addition to the rank constraint, b must also satisfy a number of linear constraints. Let $K(\Delta, \Delta^{\circ})$ be the set of pairs of non-zero lattice points $m \subset \Delta$, $\dot{m} \subset \Delta^{\circ}$ with $\pi_{m\dot{m}} = \langle m, \dot{m} \rangle = -1$; for each such pair $b_{m\dot{m}} = -1$. Thus, the number of parameters in b is then

$$\#(b) = d(n + n^{\circ} - d) - |K(\Delta, \Delta^{\circ})| . \tag{5.7.59}$$

This parametrization suggests a simple conjecture for the $(0,2)$ linear sigma model mirror map: the mirror linear sigma model is obtained by exchanging the polytopes Δ, Δ°, transposing the matrix b, and exchanging the roles of κ_α and $\kappa^\circ_{\dot\alpha}$.

On the $(2,2)$ locus $b_{m\dot m} = \pi_{m\dot m}$, and $\kappa_\alpha = q_\alpha$; therefore the proposed $(0,2)$ mirror map reduces to the monomial-divisor mirror map on the $(2,2)$ locus.

A Check of the Mirror Map

As a check of the proposal, we will prove that the principal components of the singular loci match across the two sides of the mirror. In the process, we will recover, as a special case, the promised verification (5.4.109) of the $(2,2)$ monomial-divisor mirror map. To achieve this, we will first discuss $\mathcal{Z}^{\text{pri}}_{\text{A}/2}(X)$ in the reflexively plain case, and we will then describe $\mathcal{Z}^{\text{pri}}_{\text{B}/2}(X)$.

The σ-branch singular locus, $\mathcal{Z}^{\text{pri}}_{\text{A}/2}(X)$, where the σ expectation values are generic enough to lift all of the matter fields, is the zero locus of (5.7.13), and for the reflexively plain case this simplifies to

$$\prod_{\dot m} \left[\frac{\sigma \cdot \mathfrak{M}_{\dot m}}{-\sigma \cdot \mathfrak{R}} \right]^{Q^\alpha_{\dot m}} = q_\alpha \, , \tag{5.7.60}$$

but we can also recast it in terms of the invariant coordinates as

$$\prod_{\dot m \neq 0} \left[\frac{\sigma \cdot \mathfrak{G}_{\dot m}}{-\sigma \cdot \mathfrak{R}} \right]^{Q^\alpha_{\dot m}} = \kappa_\alpha \, , \tag{5.7.61}$$

The singular locus $\mathcal{Z}^{\text{pri}}_{\text{A}/2}(X)$ is then the set of all $\mathfrak{G}$ and $\mathfrak{R}$ modulo the $\text{GL}(n_G, \mathbb{C})$ action, and κ_α, for which there is a solution to these equations with $\sigma \neq 0$. These singularities can be compared to the classical geometric singularities associated to the degeneration of the E-parameters in the $\kappa_\alpha \to 0$ limit, where $\Phi_0 = 0$ for generic J_ρ and P. Specializing (5.7.6) to the reflexively plain case, we have the bundle $\mathcal{E} \to X$ as the cohomology of the complex

$$0 \longrightarrow H_2(V, \mathbb{Z}) \otimes \mathcal{O}_X \xrightarrow{\mathfrak{M}_{\dot m}\Phi_{\dot m}} \oplus_{\dot m}\mathcal{O}_V(D_\rho)|_X \xrightarrow{J} \mathcal{O}_V([X])|_X \longrightarrow 0 \, . \tag{5.7.62}$$

In the classical limit a solution to (5.7.61) implies there exists a point $\Phi_* \in V$ and $\sigma \neq 0$ such that

$$\sigma \cdot \mathfrak{G}_{\dot m} \Phi_{*\dot m} = 0 \qquad \text{(no sum on } \dot m\text{)} \, , \tag{5.7.63}$$

and the reader can then verify that the supersymmetry constraint implies $\Phi_* \in X$. This in turn implies that the rank of $\mathcal{E}$ jumps up at Φ_*. In this way we see that (5.7.61) interpolates between two familiar notions: the (2,2) $\mathcal{Z}_A^{pri}(X)$ and classical bundle singularities due to the degeneration of the first map in (5.7.62).

In analogy with the (2,2) case and the so-called principal discriminant of [191, 260], we define $\mathcal{Z}_{B/2}^{pri}(X)$ as the locus in the $\widehat{\kappa}_{\dot{\alpha}}$ and b parameter space, where for some

$$\Phi_* \in T_N = (\mathbb{C}^*)^d \subset V \tag{5.7.64}$$

we have

$$P(\Phi_*) = 0 , \qquad\qquad J_{\dot{m}}(\Phi_*) = 0 . \tag{5.7.65}$$

When this holds, then in a Calabi-Yau phase the field Φ_0 develops a flat direction. As we saw above, our understanding of the B/2 model parameter space is still incomplete: there is no proof that it is q_α independent. We will assume q_α-independence, in which case $\mathcal{Z}_{B/2}^{pri}(X)$ must be a component of the singular locus of the theory.

Our next goal is to prove that $\mathcal{Z}_{A/2}^{pri}(X) = \mathcal{Z}_{B/2}^{pri}(X^\circ)$ under the proposed mirror map. To do so, we will characterize $\mathcal{Z}_{B/2}^{pri}(X)$ in a more transparent fashion following [260, 298]. First, we note that for Φ_* to be contained in the dense torus, it must be that $\Phi_{*\dot{m}} \neq 0$ for all $\dot{m}$. Next, the supersymmetry constraint simplifies our problem: we just need to find $\Phi_* \in T_N$ such that for all $\dot{m}$

$$\Phi_{*\dot{m}} J_{\dot{m}}(\Phi_*) = 0 . \tag{5.7.66}$$

If such a point exists, then $P(\Phi_*) = 0$ will hold due to the supersymmetry constraint. Using (5.7.40), we then arrive at

$$\alpha_0 M_0(\Phi_*) + \sum_{m \neq 0} \alpha_m M_m(\Phi_*)(b_{m\dot{m}} + 1) = 0 \qquad \text{for all } \dot{m} \neq 0 . \tag{5.7.67}$$

The problem is easily solved if we consider the monomials M_m to be independent variables: choose a basis $\mathfrak{G}_m^{\circ\dot{\alpha}}$ for the cokernel of $b_{m\dot{m}}$ and set

$$\alpha_m M_m = \sigma^\circ \cdot \mathfrak{G}_m^\circ , \quad m \neq 0 , \qquad \alpha_0 M_0 = -\sigma^\circ \cdot \mathfrak{R}^\circ , \tag{5.7.68}$$

where

$$\mathfrak{R}^\circ = \sum_{m \neq 0} \mathfrak{G}_m^\circ . \tag{5.7.69}$$

The monomials are not independent but instead for any Φ with $\Phi_{\dot{m}} \neq 0$ for all $\dot{m}$ satisfy the constraints

$$\prod_{m \neq 0} \left[\frac{\alpha_m \mathsf{M}_m}{\alpha_0 \mathsf{M}_0} \right]^{Q^{\circ\dot{\alpha}}_m} = \kappa^{\circ}_{\dot{\alpha}} , \tag{5.7.70}$$

where the $Q^{\circ\dot{\alpha}}_m$ is an integral basis for the cokernel of $\pi_{m\dot{m}}$. So, plugging in the result of the linear problem, we see that the desired singular point Φ_* exists only if

$$\prod_{m \neq 0} \left[\frac{\sigma^{\circ} \cdot \mathfrak{G}^{\circ}_m}{-\sigma^{\circ} \cdot \mathfrak{R}^{\circ}} \right]^{Q^{\circ\dot{\alpha}}_m} = \kappa^{\circ}_{\dot{\alpha}} . \tag{5.7.71}$$

Exercise 5.55 Show that if (5.7.71) holds then the requisite Φ_{*m} exists. Hint: take the logarithm of both sides and write linear equations for $\log \Phi_{*\dot{m}}$.   □

Comparing (5.7.61) and (5.7.71), we see that we reached our goal: the (0,2) mirror map implies

$$\mathcal{Z}^{\mathrm{pri}}_{\mathrm{A}/2}(X) = \mathcal{Z}^{\mathrm{pri}}_{\mathrm{B}/2}(X^{\circ}) . \tag{5.7.72}$$

The reader who computed some of these loci explicitly in the preceding exercises will agree that this is a non-trivial check of the proposal. We also observe that on the (2,2) locus this reduces to the promised $\mathcal{Z}^{\mathrm{pri}}_{\mathrm{A}}(X) = \mathcal{Z}^{\mathrm{pri}}_{\mathrm{B}}(X^{\circ})$.

Much more has been established in the (2,2) theory. Not only do the other components of the singular locus agree [311], but the $A(X)$ and $B(X^{\circ})$ genus zero correlation functions are exchanged as well [54, 262, 355]. It would be useful to generalize the toric residue techniques that lead to the (2,2) results to the (0,2) setting. This has not been carried out, but recently, a comparison of $A/2$ twisted and mirror $B/2$ twisted correlation functions was carried out in cases where the $B/2$ theory could be solved in a Landau-Ginzburg phase [79], and it was found that the correlation functions agree up to an overall normalization.

5.8   Singlets and Their Spacetime Superpotential

We have seen some remarkable structures possessed by (0,2) theories with a (2,2) locus. Do these describe a (0,2) superconformal theory? If they do, is the same true of more generic (0,2) linear sigma models? In this section we will review the arguments [57, 80, 350] that, with some caveats, yield affirmative answers to these questions.

Consider an abelian (0,2) linear sigma model with a $U(1)_L \times U(1)_R$ symmetry as in Sect. 5.2.2 that has a geometric phase in the $q_\alpha \to 0$ limit described by a smooth stable rank r bundle $\mathcal{E} \to X$, where X is a 3-dimensional Calabi-Yau manifold. Anomaly cancelation implies $c_1(\mathcal{E}) = c_1(T_X) = 0$ and $c_2(\mathcal{E}) = c_2(T_X)$, and $c = r + 6$, and $\bar{c} = 9$. In this limit we can evaluate the massless spectrum and focus on the spacetime gauge singlets; in the limit we can separate the "Kähler moduli" from the remaining massless singlets—we will schematically label the corresponding $N = 1$ four-dimensional chiral superfields by T and the remaining singlets by Z. The spacetime superpotential then takes the schematic form[56]

$$W(T, Z) = g(Z) + f(Z) \exp[2\pi i T] \dots , \qquad g(Z) = g_{ijk} Z_i Z_j Z_k + \dots ,$$

$$(5.8.1)$$

where $T = (ir + \vartheta)/2\pi$, and $\operatorname{im} T \to \infty$ is the large radius limit, and the $\exp[2\pi i T]$ is meant to invoke non-perturbative worldsheet instanton corrections, but in a schematic fashion: we expect the terms to be at least as complicated as the instanton sums we met above, and we will assume that the instanton expansion converges. Recall that the ϑ are components of the heterotic B-field reduced on $H^{1,1}(X, \mathbb{C})$, and in the perturbative α' expansion of the non-linear model the shifts $\vartheta \to \vartheta + \Delta\vartheta$ are symmetries. These symmetries forbid any terms in W that are polynomial in the T multiplets. On the other hand, the non-perturbative terms are weighted by the action evaluated on the instanton configuration and are invariant under $\vartheta \to \vartheta + 2\pi$.

If we set $T = 0$, then the condition for a four-dimensional supersymmetric Minkowski vacuum are $g(Z) = 0$ and $\partial g/\partial Z = 0$. We discussed in Chap. 4 that the spacetime massless spectrum can jump, and in the large radius limit the jumping is determined by algebraic geometry, and we also mentioned that even in (2,2) theories there are terms of the form $Z^2 \exp[2\pi i T]$, which means the jumping in the massless spectrum can depend on worldsheet instantons [37, 41]. These ingredients suggest, probably correctly, that in general W is an exceedingly complicated object.[57]

It has been known for some time that individual instantons in generic (0,2) theories make contributions to W, and these can Z-independent [139, 140] or make contributions to Z masses and masses for gauge-charged fields [73]. It is possible to fortuitously choose the bundle and geometry so that such instanton contributions vanish [142, 144], but that is a highly non-generic situation.

The special properties of the linear sigma model realization lead to the following structure. Split the marginal deformations according to whether they are realized through UV parameters of the linear model or not, i.e. $T \to T^{\text{lin}}, T^{\text{nlin}}$, and similarly for the Z. With some combinatoric constraints, it is possible to prove the

---

[56]We work at string tree-level and neglect the dilaton; we will also not be concerned with global properties of the superpotential because we are here just interested in establishing its vanishing in some open neighborhood in the classical parameters.

[57]Recent discussion of the superpotential in the supergravity approximation and its connection to the geometric constraints from Chap. 4 can be found in [28, 291].

following vanishing theorem:

$$\frac{\partial^3 W}{\partial T_\alpha^{\text{lin}} \partial T_\beta^{\text{lin}} \partial T_\gamma^{\text{lin}}} = 0 \qquad \text{for all} \qquad T^{\text{lin}}, Z^{\text{lin}}. \tag{5.8.2}$$

In this case, although individual instantons still make non-zero contributions to the potential, the instanton sums conspire to cancel in the indicated terms. This is linear sigma model magic.[58]

Even if the magic result holds, it does not imply that the superpotential is zero: there may be additional Kähler parameters T^{nlin} that have a potential that drives $\text{im} \, T^{\text{nlin}} \to \infty$. To avoid such a pathology, we can further restrict to theories where there are no non-linear Kähler deformations. In that case, sometimes known as "favorable," we can be reasonably confident that the linear theory flows to a (0,2) superconformal theory with expected symmetries, central charges, topological heterotic rings, and the full chiral algebra, and with a subset of exactly marginal deformations—T^{lin} and Z^{lin}—that have a direct representation in the UV Lagrangian theory.

5.8.1 The Deformed Hypersurface

To see how the computation works, we will first consider the case of (0,2) deformations of the (2,2) Calabi-Yau hypersurface theories we studied above. In the next section we will discuss the general case. The linear parameters τ^α are identified at large radius with the complexified Kähler moduli, and we wish to compute $\frac{\partial^3 W}{\partial \tau^3}$. We showed in Chap. 3 that marginal deformations of a (0,2) theory should be of the form

$$\Delta S_\tau = \int d^2 z \, \tau \{ \boldsymbol{Q}, \mathcal{O}_\tau \} + \text{h.c.}, \tag{5.8.3}$$

where $\mathcal{O}_\tau$ is some $\overline{\boldsymbol{Q}}$-closed operator in the $\overline{\text{NS}}$ sector. We need not search far for the operator. Taking a look back to the linear model action and supersymmetry transformations, we have

$$\Delta S_\tau = \frac{i}{4} \int d^2 z \, \tau (F_{12} - D) + \text{h.c.} = -\frac{i}{2} \int d^2 z \tau \{ \boldsymbol{Q}, \upsilon \}, \tag{5.8.4}$$

and $\{ \overline{\boldsymbol{Q}}, \upsilon \} = 0$. We conclude that $\mathcal{O}_\tau = -\frac{i}{2} \upsilon$.

---

[58]There are counter-examples outside of the linear setting [98, 100, 104]. The precise fate of such theories is not clear the author: it seems to be veiled in the details of the full superpotential: are there supersymmetric vacua with a finite and smooth geometry?

The simplest way to evaluate $\partial^3 W / \partial \tau^3$ is to directly compute the fermion–boson–fermion three-point function in tree-level perturbation theory at zero spacetime momentum. Leaving aside the spacetime terms of the vertex operators, the coupling is proportional to the internal SCFT correlation function $\langle \mathcal{O}_\tau^F \mathcal{O}_\tau \mathcal{O}_\tau^F \rangle$, where $\mathcal{O}_\tau^F$ is the $\overline{\mathrm{R}}$ operator obtained by half unit spectral flow.

In the language of string perturbation theory $\mathcal{O}_\tau$ is is the internal contribution to the vertex operator taken in the 0 picture, while $\mathcal{O}_\tau^F$ is the internal contribution to the vertex operator in the $-1/2$ picture. The operators have $\bar{h} = \bar{q}/2$, so that $\lim_{z \to 0} \mathcal{O}_\tau^F(z)|0\rangle$ creates a Ramond ground state. Therefore we can think of the three-point function we would like to compute in terms of a cylinder worldsheet: the ends correspond to the Ramond ground states, and there is a remaining $\mathcal{O}_\tau$ insertion in the bulk [390]. This observation shows that there is a direct relationship between the desired physical correlation function and the half-twisted correlation function $\langle \mathcal{O}_\tau(z_1) \mathcal{O}_\tau(z_2) \mathcal{O}_\tau(z_3) \rangle_{1/2}$: on the cylinder worldsheet the twisting is trivial (because the canonical bundle is trivial) and simply shifts the spins of the fields by the R-charge: $s \to s + \bar{q}/2$. But that is precisely the same as the half-unit spectral flow of the physical theory.

Thus, if we can show the vanishing of

$$C_{\alpha_1 \alpha_2 \alpha_3} = \langle \upsilon_{\alpha_1}(z_1) \upsilon_{\alpha_2}(z_2) \upsilon_{\alpha_3}(z_3) \rangle_{1/2} \tag{5.8.5}$$

in the half-twisted linear sigma model on $\Sigma = \mathbb{P}^1$, we will establish that $\partial^3 W / \partial \tau^3 = 0$. To make the argument, we will work in a smooth phase—the limit $q \to 0$—and the instanton expansion and use four properties of the half-twisted theory:

i. the zero mode structure is determined by the half-twisted spin assignments and instanton number;
ii. the path integral localizes onto a compact instanton moduli space;
iii. the dependence on anti-holomorphic parameters is $\overline{Q}$-exact;
iv. the zero mode integral converges when we set anti-holomorphic parameters to zero;
v. in that limit the theory has a selection rule that implies $C_{\alpha_1 \alpha_2 \alpha_3} = 0$.

We start with the details of the twist. Taking a look back to the twist tables (5.5.24), (5.5.25), (5.5.26), we have the following description of the fields in the half-twisted theory on Σ:

$$\phi_\rho \in \mathscr{A}^0(\Sigma, \mathcal{L}_\rho), \qquad\qquad \overline{\phi}_\rho \in \mathscr{A}^0(\Sigma, \overline{\mathcal{L}}_\rho),$$

$$\psi_\rho \in \mathscr{A}^0(\Sigma, \overline{K}_\Sigma \otimes \overline{\mathcal{L}}_\rho^*), \qquad \overline{\psi}_\rho \in \mathscr{A}^0(\Sigma, \overline{\mathcal{L}}_\rho),$$

$$\gamma_\rho \in \mathscr{A}^0(\Sigma, K_\Sigma^{1/2} \otimes \mathcal{L}_\rho), \qquad \overline{\gamma}_\rho \in \mathscr{A}^0(\Sigma, K_\Sigma^{1/2} \otimes \mathcal{L}_\rho^*),$$

$$\phi_0 \in \mathscr{A}^0(\Sigma, K_\Sigma^{1/2} \otimes \mathcal{L}_0), \qquad \overline{\phi}_0 \in \mathscr{A}^0(\Sigma, \overline{K}_\Sigma^{1/2} \otimes \overline{\mathcal{L}}_0),$$

$$\psi_0 \in \mathscr{A}^0(\Sigma, \overline{K}_{\Sigma}^{1/2} \otimes \overline{\mathcal{L}}_0^*) \,, \qquad \overline{\psi}_0 \in \mathscr{A}^0(\Sigma, \overline{K}_{\Sigma}^{1/2} \otimes \overline{\mathcal{L}}_0) \,,$$

$$\gamma_0 \in \mathscr{A}^0(\Sigma, K_{\Sigma} \otimes \mathcal{L}_0) \,, \qquad \overline{\gamma}_0 \in \mathscr{A}^0(\Sigma, \overline{\mathcal{L}}_0) \,,$$

$$\sigma \in \mathscr{A}^0(\Sigma, K_{\Sigma}^{1/2}) \,, \qquad \overline{\sigma} \in \mathscr{A}^0(\Sigma, \overline{K}_{\Sigma}^{1/2}) \,,$$

$$\chi \in \mathscr{A}^0(\Sigma, \overline{K}_{\Sigma}^{1/2}) \,, \qquad \overline{\chi} \in \mathscr{A}^0(\Sigma, \overline{K}_{\Sigma}^{1/2}) \,,$$

$$\upsilon \in \mathscr{A}^0(\Sigma, K_{\Sigma}) \,, \qquad \overline{\upsilon} \in \mathscr{A}^0(\Sigma, \mathcal{O}_{\Sigma}) \,. \qquad (5.8.6)$$

As in the other twists we discussed, we use the Hermitian metric on $\mathcal{L}_\rho$ and $\mathcal{L}_0$ to organize the fields so that the zero modes of the kinetic operators correspond to the holomorphic or anti-holomorphic sections of the respective bundles.

Exercise 5.56 Show that the measure of the half-twisted path integral carries $U(1)_B$ charge $-3/2$ on $\Sigma = \mathbb{P}^1$, which is just right to allow for a non-trivial υ 3-point function. It might be helpful to take a look back at the similar Exercise 5.40 for the V-model. □

Now we consider the instanton expansion in the Calabi-Yau phase. When we discussed the A-model or the A/2-model path integral localization in the Calabi-Yau phase, we remarked that the restriction to the fixed points of $\overline{Q}$ led to the instanton moduli space $\mathscr{M}_{\vec{N},X}$—the very same instanton moduli space as in the A-model. Just as in that discussion because $d_0 \leq 0$, ϕ_0 and $\overline{\phi}_0$ have no zero modes. A new feature of the half-twisted theory is that the fields $\sigma_\alpha, \overline{\sigma}_\alpha$ have no zero modes. Thus, not only is the dependence of the theory on the anti-holomorphic parameters $\overline{Q}$-exact, but also the limit $\overline{E} = 0$ and $\overline{J} = 0$ is smooth.

The final observation is that if we also set $J = 0$ and $E = 0$, the resulting theory, dubbed the O-model in [80], has a large symmetry group. This group of symmetries includes $U(1)_C$ a symmetry that assigns charges $+1$ to Φ_ρ, Γ^ρ for all ρ and charge 0 to all remaining fields.

Exercise 5.57 Show that $U(1)_C$ is non-anomalous in the untwisted O-model. This is important for our purposes, since it implies that in the half-twisted X-model the path integral measure will transform the same way in all topological sectors.

Next, show that the measure carries $U(1)_C$ charge $+n$.

Finally, show that the holomorphic E and J couplings of the half-twisted X-model carry non-negative charge with respect to $U(1)_C$. □

With the exercise results in hand, we conclude that $C_{\alpha_1\alpha_2\alpha_3} = 0$ as desired in three easy steps. First, the insertion is $U(1)_C$ neutral. Second, the measure carries a positive charge that must be canceled by bringing down negatively charged terms from the action. Finally, the only terms in the action that may possibly carry a negative charge are $\overline{Q}$-exact and decouple from the $\overline{Q}$-closed correlation function.

5.8.2 A Generalization

The vanishing theorem can be generalized to a much wider class of abelian linear sigma models [80]. It holds in a subclass of theories that were introduced following [388] in [145] and extensively studied by comparing exact computations in Landau-Ginzburg phases to large radius expectations [143, 266].

The class of theories is distinguished by having a large radius phase described by the geometry $\mathcal{E} \to X$, where $X \subset V$ is a complete intersection Calabi-Yau manifold in a NEF Fano toric variety V, and the bundle is built as the cohomology of a two-step complex, much like our construction of T_X and its deformation in the hypersurface case. We will now describe the combinatorics in detail.

The whole construction is still anchored on the familiar V-model toric construction: there are the chiral bosonic multiplets Φ_ρ, with ρ labeling the n_Φ one-dimensional cones in the fan for V, and these are coupled to the abelian gauge group G with (0,2) vector multiplets $\mathcal{V}_{0\alpha}, \mathcal{V}_{1\alpha}$ and field strengths Υ_α. The latter couple to the complexified Fayet-Iliopoulos parameters τ^α, and when the r^α are in the interior of the Kähler cone $K(V) \subset K_{cl}(V)$, the bosonic potential is minimized with ϕ_ρ modulo G taking values in V, a d-dimensional toric variety.

X will be constructed as the intersection of n_{hyp} hypersurfaces in V. Each hypersurface is the vanishing of a quasi-homogeneous polynomial $H_A(\Phi) = 0$, and H_A carries gauge charges $d_A^\alpha = -Q_A^\alpha$ and corresponds to an ample divisor $D_A \in A_{d-1}(V)$. To implement the constraint, we introduce fermi multiplets Λ^A with gauge charges Q_A^α and a superpotential interaction $W \supset \sum_{A=1}^{n_{hyp}} \Lambda^A H_A$.

To describe the bundle, we introduce n_Γ fermi multiplets Γ^I with gauge charges Q_I^α and n_S gauge-neutral chiral multiplets $\mathcal{S}_\mu$, which generalize the familiar $\mathcal{S}_\alpha$ of the (2,2) construction. In addition we also have $n_\mathcal{P}$ gauge-charged chiral multiplets $\mathcal{P}_i$ with gauge charges Q_i^α—these generalize the Φ_0 multiplet of our hypersurface examples. We now set the following E and J couplings:

$$\overline{\mathcal{D}}\Gamma^I = E^I(\mathcal{S}, \Phi) = \mathcal{S}_\mu E^{\mu I}(\Phi), \quad \overline{\mathcal{D}}\Lambda^A = E^A(\mathcal{S}, \mathcal{P}, \Phi) = \mathcal{S}_\mu \mathcal{P}_i E^{\mu i A}(\Phi),$$
(5.8.7)

and

$$\mathcal{W}_J = \sum_A \Lambda^A H_A + \sum_I \Gamma^I(\mathcal{P}_i J_I^i)$$
(5.8.8)

subject to the (0,2) supersymmetry requirement

$$\sum_A E^{\mu i A} H_A + \sum_I E^{\mu I} J_I^i = 0.$$
(5.8.9)

The final ingredient is a "spectator multiplet"—a chiral multiplet $\mathcal{X}$ and an fermi multiplet Ξ with $\overline{\mathcal{D}}\Xi = 0$. The two carry gauge charges, respectively, Q_{spct}^α and

$-Q^\alpha_\text{spct}$, with

$$Q^\alpha_\text{spct} = -\sum_\Phi Q^\alpha(\Phi) - \sum_\mathcal{P} Q^\alpha(\mathcal{P}) = -\sum_\rho Q^\alpha_\rho - \sum_i Q^\alpha_i , \qquad (5.8.10)$$

and a superpotential interaction $\mathcal{W}_\text{spct} = J_\text{spct} \mathcal{X} \Xi$.

The gauge symmetry should be non-anomalous, and we require the theory to possess $\text{U}(1)_\text{L} \times \text{U}(1)_\text{R}$ invariance with charges (we also list the gauge charges)

	$\mathcal{P}_i$	Φ_ρ	$\mathcal{S}_\mu$	Λ^A	Γ^I	$\mathcal{X}$	Ξ	Υ_α
$\text{U}(1)_\text{L}$	1	0	-1	0	-1	1	-1	0
$\text{U}(1)_\text{R}$	1	0	1	1	0	1	0	1
$\text{U}(1)_\alpha$	Q^α_i	Q^α_ρ	0	Q^α_A	Q^α_I	Q^α_spct	$-Q^\alpha_\text{spct}$	0

$$(5.8.11)$$

The E and J couplings can be taken to be the most general couplings consistent with these symmetries, and we will assume that they break any remaining continuous symmetries to a discrete group. Anomaly cancelation leads to the requirements as in Sect. 5.2.2:

$$\text{gauge:} \qquad \sum_i Q^\alpha_i Q^\beta_i + \sum_\rho Q^\alpha_\rho Q^\beta_\rho = \sum_A Q^\alpha_A Q^\beta_A + \sum_I Q^\alpha_I Q^\beta_I ,$$

$$\text{U}(1)_\text{L}: \qquad \sum_i Q^\alpha_i = -\sum_I Q^\alpha_I ,$$

$$\text{U}(1)_\text{R}: \qquad -\sum_\rho Q^\alpha_\rho = \sum_A Q^\alpha_A . \qquad (5.8.12)$$

The massive spectator multiplet $\mathcal{X}$, Ξ drops out of these considerations, as it should.

Once this is the case, then by the familiar machinations of c-extremization we find that, barring accidents, this theory should flow to an SCFT with

$$\bar{c} = 9 , \qquad\qquad\qquad c = 6 + r , \qquad (5.8.13)$$

where $r = n_\Gamma - n_\mathcal{S} - n_\mathcal{P}$ is the level of $\text{U}(1)_\text{L}$.

We now make a number of smoothness assumptions that should hold for generic coefficients in the defining polynomials in the K(V) phase.

1. The complete intersection $X = \cap_A\{H_A = 0\} \subset V$ is a smooth manifold of dimension $d - n_\text{hyp} = 3$. When this holds the Λ^A can be integrated out. When this fails, the non-linear sigma model interpretation of the low energy physics is complicated, and it is not clear that large radius supergravity results hold. Note, however, that this failure does not lead to non-compact bosonic directions.

2. $E^{\mu I}(\Phi)$ is full rank for all $p \in X$, and $n_\Gamma \geq n_S$. When this holds the σ_μ fields are massive and can be integrated out: the massive multiplets are the S_μ and the n_S linear combinations of Γ^I in the image of $E^{\mu I}$.

3. $J_I^i(\Phi)$ is full rank for all $p \in X$, and $n_\Gamma \geq n_P$. When this holds the $\mathcal{P}_i$ fields are massive and can be integrated out: the massive multiplets are the $\mathcal{P}_i$ and the n_P linear combinations Γ^I in the image of J_I^i.

4. The previous two assumptions imply that the massless left-moving fermions couple to the holomorphic rank r bundle $\mathcal{E} \to X$ defined by the cohomology of the complex

$$0 \longrightarrow \mathcal{O}_X^{\oplus n_S} \overset{E}{\longrightarrow} \oplus_I \mathcal{O}_V(D_I)|_X \overset{J}{\longrightarrow} \oplus_i \mathcal{O}_V(D_i)|_X \longrightarrow 0 \;, \qquad (5.8.14)$$

and the divisors D_I and D_i correspond to the classes $Q_I^\alpha \eta_\alpha$ and $Q_i^\alpha \eta_\alpha$ in $H^2(V, \mathbb{Z})$ in the usual way.

We assume this bundle is stable when the Kähler class belongs to a subcone in $K(V)$.

Exercise 5.58 Show that this structure implies that the non-linear sigma model geometry satisfies

$$c_1(T_X) = 0\,, \qquad c_1(\mathcal{E}) = 0\,, \qquad c_2(T_X) = c_2(\mathcal{E})\,. \qquad (5.8.15)$$

Equation (5.3.120) and discussion around it may be useful. □

After integrating out the massive fields, we will be left with a (0,2) non-linear sigma model on $\mathcal{E} \to X$. Clearly X is topologically a compact Calabi-Yau manifold, and classically has a Kähler metric inherited from the Kähler form on V. One-loop corrections to the effective action will produce the expected non-zero torsion [4], and we can expect that with our assumptions the theory will be a (0,2) SCFT to all orders in α' perturbation theory. Thus, by construction this set-up is designed so that the logic based on the vanishing of the spacetime superpotential applies. Therefore, in this much more general setting, if in the large radius limit the Kähler moduli are all inherited from the toric variety, and we can prove $\mathcal{C}_{\alpha_1 \alpha_2 \alpha_3} = 0$ as in the hypersurface case, we will have shown that this linear sigma model flows to the expected conformal fixed point.

There is no known vanishing theorem that applies to all linear sigma models in this broad class. However [80] identified additional sufficient conditions that guarantee the desired vanishing by precisely the same logic as we discussed for the hypersurface, and we now review the argument and results.

While the theorem leads to sufficient conditions, they are not necessary. There are indications that there may also be additional protection for at least some of the deformations that cannot be linearized through the UV Lagrangian [29, 33]. Therefore it would be useful to sharpen the vanishing result. One of the challenges in achieving a sharpening comes from the redundancy in the choice of a UV linear model. Even in the (2,2) context, there are examples of pairs of realizations $X_1 \subset V_1$

and $X_2 \subset V_2$ that flow to the same IR fixed point but have different realizations of the parameters, e.g. $h^{1,1}_{\text{toric}}(X_1) = h^{1,1}(X_2) > h^{1,1}_{\text{toric}}(X_2)$.

A Vanishing Theorem

The first observation is that in the half-twisted O-model, obtained by setting the E, H, J couplings to zero, the bosonic fields that may have non-compact zero modes are the lowest components $p_i \in \mathcal{P}_i$ and the lowest component of the spectator $x \in \mathcal{X}$. In the half-twisted theory in instanton sector $\vec{N}$

$$p_i \in \mathscr{A}^0(\Sigma, K_\Sigma^{1/2} \otimes \mathcal{L}_i), \qquad x \in \mathscr{A}^0(\Sigma, K_\Sigma^{1/2} \otimes \mathcal{L}_x), \qquad (5.8.16)$$

where the bundles on $\Sigma = \mathbb{P}^1$ are

$$\mathcal{L}_i = \mathcal{O}(Q_i^\alpha N_\alpha), \qquad \mathcal{L}_x = \mathcal{O}(Q_{\text{spct}}^\alpha N_\alpha). \qquad (5.8.17)$$

To discuss the absence of dangerous zero modes, we recall a few toric facts. The Kähler cone $K(V)$ and the "classical cone" $K_{\text{cl}}(V)$ that contains it are both pointed because V is a compact toric variety. Thus, p_i will have no zero modes for all $\vec{N} \in K^*(V)$ if and only if the class $\zeta_i = -\sum_\alpha Q_i^\alpha \eta_\alpha \in K(V)$. Once this holds, we proceed to the zero mode of the spectator x.

It is useful to consider the remaining bosons, the ϕ_ρ and x together. Working in the untwisted theory with $p_i = 0$, the holomorphic quotient produces a $d + 1$-dimensional toric variety $\widehat{V}$ with homogeneous coordinates $[\phi_1, \ldots, \phi_n, x]$. There are now three mutually exclusive possibilities.

i. If $\zeta_x = -Q_{\text{spec}}^\alpha \eta_\alpha \in K(V)$, then $\widehat{V}$ is non-compact, but x has no zero modes in the half-twisted theory.
ii. If $\zeta_x \notin K_{\text{cl}}(V)$, then $K_{\text{cl}}(\widehat{V})$ is pointed, which implies $\widehat{V}$ is a complete toric variety. In this case x will in general have zero modes, but the gauge instanton moduli space will nevertheless be compact.[59]
iii. Finally, if $\zeta_x \in K_{\text{cl}}(V)$, but $\zeta_x \notin K(V)$, then $\widehat{V}$ is not complete, and the zero modes of x are not controlled.

Exercise 5.59 To better understand the last two points, prove the following result. Let $K \in \mathbb{R}^k$ be a pointed cone with apex at the origin, and for a fixed non-zero $\zeta \in \mathbb{R}^k$ set

$$\widehat{K} = \{w + t(-\zeta) \,|\, w \in K, t \in \mathbb{R}_{\geq 0}\}. \qquad (5.8.18)$$

Prove that $\widehat{K}$ is pointed if and only if $\zeta \notin K$.[60]   □

---

[59]Compare this to the second half of Exercise 5.17.
[60]Recall that a cone is pointed if there is just one way to obtain the apex: the coefficients of all generators are identically zero.

So, in addition to the hefty combinatorial data, much of it implicit in the smoothness demands, that defines the (0,2) linear sigma model, we impose additional "ampleness" conditions

$$\zeta_i \in K(V), \quad \text{for all } i, \quad \text{and either} \quad \zeta_x \in K(V) \quad \text{or} \quad \zeta_x \notin K_{cl}(V).$$
(5.8.19)

These guarantee that the half-twisted O-model will have a compact instanton moduli space for any $\vec{N} \in K^*(V)$. These have a nice geometric interpretation: the ζ_i are a set of ample classes on V, and ζ_x can be expressed in terms of the canonical class $-R$ of V and the ζ_i:

$$\zeta_x = -R - \sum_i \zeta_i \,.$$
(5.8.20)

Help from Spectators

The next step is a search for a $U(1)_C$ symmetry that will be used for the selection rule. It should satisfy two criteria: in every topological sector the measure and the insertion should carry positive $U(1)_C$ charge, and all holomorphic couplings of the X-model should carry non-negative $U(1)_C$ charge. It is at this point that the spectator fields play a positive, albeit technical role. Without their inclusion it is difficult to find an appropriate $U(1)_C$ because the transformation of the measure in the half-twisted theory depends on the topological sector. On the other hand, including the spectators with the charges as above, there is a simple choice that does the trick:

	$\mathcal{P}_i$	Φ_ρ	$\mathcal{S}_\mu$	Λ^A	Γ^I	$\mathcal{X}$	Ξ	Υ_α	
$U(1)_C$	0	1	0	0	1	1	0	0	(5.8.21)

The measure and the insertion carry charge n_Φ, and the holomorphic couplings of the X-model carry non-negative charge. Thus, the vanishing theorem holds.

Exercise 5.60 Verify the assertions. First prove that $U(1)_C$ is a non-anomalous symmetry in the untwisted O-model.

Next, with obvious substitutions, the half-twist for the hypersurface (5.8.6) describes all of the fields in the half-twisted theory. Use that and the Riemann-Roch theorem to show that the measure carries $U(1)_C$ charge

$$-\sum_\rho \chi(K_\Sigma \otimes \mathcal{L}_\rho^*) - \sum_I \chi(K_\Sigma^{1/2} \otimes \mathcal{L}_I) - \chi(K_\Sigma^{1/2} \otimes \mathcal{L}_{\text{spct}}^*) = n_\Phi \,.$$
(5.8.22)

Finally, argue that the holomorphic couplings of the X-model all carry a non-negative $U(1)_C$ charge. □

The spectator fields used in the construction deserve another word. They were introduced in [145] to solve another technical puzzle in this general class of theories. Suppose we do not include the spectators. By construction, the models have a $U(1)_L \times U(1)_R$ symmetry and a large radius Calabi-Yau phase, with a Kähler class of X linear in the r^α of the linear sigma model. On the other hand, we know from Sect. 5.4.4 that there the r^α have a one-loop exact RG running:

$$r^\alpha(\mu) = r^\alpha(\mu_0) + \sum_{\text{chiral}} Q^\alpha(\text{chiral}) \log(\mu/\mu_0) . \qquad (5.8.23)$$

In our class of models we have

$$\sum_{\text{chiral}} Q^\alpha(\text{chiral}) = \sum_\Phi Q^\alpha(\Phi) + \sum_\mathcal{P} Q^\alpha(\mathcal{P}) = \sum_\rho Q^\alpha_\rho + \sum_i Q^\alpha_i . \qquad (5.8.24)$$

Evidently, the RG running is independent of the $U(1)_L \times U(1)_R$ symmetry: the anomaly-free conditions (5.8.12) do not imply the vanishing of the right-hand-side. A (2,2) theory (or its deformation) is an exception, and there are three equivalent conditions: no running of the r^α, a non-anomalous $U(1)_L$, a non-anomalous $U(1)_R$. We now understand the choice of Q^α_{spct}—the one-loop correction to D_α due to $\mathcal{X}$ just cancels the right-hand-side. Now the r^α, and therefore the τ^α do not run, and it makes sense to identify motion on the UV parameter space with $\tau^\alpha \to \tau^\alpha + \delta\tau^\alpha$ with deformations of the SCFT.

This is important for our vanishing purposes: if the τ^α run, then the identification of $\mathcal{O}_\tau$ with υ must be reconsidered. As remarked in [80], "this is a technical matter": if there are no instanton corrections in the presence of spectators, then the same is true once they are integrated out, but the argument can be much more complicated: one needs to use the correct half-twisted operators, and one has to use a different $U(1)_C$ symmetry.

It is perhaps worthwhile to contrast the (0,2) spectators with a (2,2) "improvement." Consider the (2,2) V-model with $V = \mathbb{P}^{n-1}$. The Fayet-Iliopoulos parameter runs and $U(1)_L \times U(1)_R$ is anomalous, but we can eliminate both defects by adding a Φ_0 (2,2) chiral multiplet with charge $-n$. The result is a very different theory: it is non-compact and flows to an SCFT with $c = \bar{c} = n$. At large radius the target space is the non-compact Calabi-Yau manifold $\mathcal{O}(-n) \to \mathbb{P}^{n-1}$, while taking $r \to -\infty$ we obtain the orbifold $\mathbb{C}^n/\mathbb{Z}_n$. We could also try to improve the situation by adding a (0,2) spectator multiplet as above, with $Q_{\text{spct}} = -n$. This eliminates the one-loop running of r but does not modify the $U(1)_L \times U(1)_R$ anomaly: there is no IR SCFT in sight. An optimist who wishes to think of τ as some marginal coupling of some sort of (0,2) SCFT must still be disappointed: the anomaly implies that the real part of τ, $\vartheta/2\pi$, is not a meaningful parameter. Spectators can help, but they do not perform miracles.

5.9 Concluding Remarks

Much more can, should, and has been said on the beautiful subject of (0,2) theories. Perhaps you, dear reader, will help to unravel some of the many questions that remain. I hope that this brief survey of the basic features and tools that are currently available will prove useful in such explorations. Throughout the text we noted many puzzles that arise even in the relatively tame parts of the (0,2) world that we explored: many of those, I hope, will interest the reader and lead them to explore the (0,2) world. We will now end our survey by mentioning just a few of the directions we did not pursue, but which will undoubtedly bring new connections between (0,2) theories and geometry and lead to new results for quantum fields and strings.

1. Conformal perturbation theory, and (0,2) conformal perturbation theory in particular should be developed further. To what extent can we lose the crutch of target space geometry and supergravity? When is the obstruction theory of a (0,2) SCFT described by a function that resembles the four-dimensional spacetime superpotential?
2. The class of (0,2) linear models surveyed in the previous section is remarkably large but remains poorly understood. Indeed, the vanishing result is one of the few general statements known to the author. In the case of (2,2) theories a lot of insight was gained into the structure of the theories through the Batyrev construction, which gave a set of sufficient conditions for the existence of a smooth (or mildly singular) large radius phase. It would be a great breakthrough to formulate analogous sufficient conditions for the (0,2) theories. Such conditions may inform our understanding of dualities and superconformal invariance of the models. Along the way, one would have to address the following question: what is the role of bundle stability? What is the relationship between unstable bundles and accidental symmetries in the IR?
3. Many of the (0,2) SCFTs constructed by linear sigma model technology will be connected by motion in the moduli space that may not be obvious in the (0,2) UV gauge theory. This might happen in a variety of ways. For instance, pairs of theories may be (0,2) mirror-symmetric or perhaps they may be identical points in a much large (0,2) duality web introduced in [146] and recently explored in [14, 86], or they may live on different branches, such as the various Higgs branches we discussed in the simple example of K3 compactification. Studying these constructions and getting a better understanding of the dualities can shed light on the space of (0,2) heterotic string vacua and leads to a host of additional string-theoretic questions concerning resolutions of singularities and connectedness.
4. Non-abelian (0,2) linear sigma models deserve more attention than we have been able to give them. Recent work on these theories points to a rich duality structure [67, 184, 215, 253, 280].
5. Throughout this work, "non-compact" was essentially synonymous with "singular." That is reasonable when the non-compactness of a target space is obtained in a limit in the parameter space. On the other hand, there are interesting theories

that are, so to speak, non-compact from the very start. These have no $SL(2, \mathbb{C})$ vacuum, so that many of the tools we developed do not apply. Nevertheless, such a theory can also have its own simplifications precisely in the non-compact region of the target space. String theory constructions and reduction of higher dimensional theories can be used to engineer examples of such non-compact theories with (0,2) supersymmetry [22, 23, 338].

6. Modern localization technology is bound to play an important role in future developments. Computations of elliptic genera and twisted and half-twisted correlation functions will clarify known results and uncover new ones; indeed, this has already been the case, and the computations have served as a quantitative check of duality proposals, dynamical supersymmetry breaking, and other applications. An exposition can be found in [324].

Ci falt la geste que Ilarion declinet.

Appendix A
Conformalities

A.1 Virasoro Algebra

In our review of basic CFT properties we stated a number of well-known facts. For completeness we now show how these may be deduced from our axioms.

1. *The vacuum state $|0\rangle$ is primary and has $(h, \overline{h}) = (0, 0)$.* It corresponds to the identity operator. There is nothing to prove here: we assume the existence of an $\mathfrak{sl}_2\mathbb{C}$-invariant vacuum.

2. *The central charge c is positive.* Consider the state $|T\rangle = L_{-2}|0\rangle$ that corresponds $T(z)$. Using the Virasoro algebra we find the norm

$$\||T\rangle\|^2 = \langle 0|[L_2, L_{-2}]|0\rangle = \frac{c}{2}\||0\rangle\|^2 .$$

It follows that $c > 0$ in a unitary theory with $T(z) \neq 0$.

3. *L_0 has a non-negative spectrum.* This is obvious for any quasi-primary state $|\Phi\rangle$ with $L_0|\Phi\rangle = h|\Phi\rangle$, since

$$\|L_{-1}|\Phi\rangle\|^2 = \langle\Phi|[L_1, L_{-1}]|\Phi\rangle = 2h\||\Phi\rangle\|^2 .$$

By assumption any state is a sum of quasi-primary states and their descendants, which are realized by repeated action of L_{-1} (or derivatives on the operators). Since the weight of a descendant is strictly larger than that of its quasi-primary ancestor, the result follows.

4. *Every state is a sum of primary states and their Virasoro descendants.* The argument follows the familiar one for finite-dimensional representations of a simple Lie algebra. For any primary state $|\Phi\rangle$ and an ordered partition of an

© Springer Nature Switzerland AG 2019
I. V. Melnikov, *An Introduction to Two-Dimensional Quantum Field Theory with (0,2) Supersymmetry*, Lecture Notes in Physics 951,
https://doi.org/10.1007/978-3-030-05085-6

integer K $[K] = (k_1, k_2, \cdots, k_p)$, with $k_i \geq k_{i+1}$, we define the descendant

$$L_{[K]}|\Phi\rangle = L_{-k_1} L_{-k_2} \cdots L_{-k_p}|\Phi\rangle .$$

Organize the states in the theory according to level, with level K states having weight $h = K + \epsilon$ for $0 \leq \epsilon < 1$. Unitarity implies that level $K = 0$ states are primary. Suppose every level K state is a sum of primary states and descendants, and let $|\Phi\rangle$ be a level $K + 1$ state that is orthogonal to all descendant states. For any $m \in \mathbb{Z}_{>0}$ define

$$|\Psi\rangle = L_m|\Phi\rangle .$$

This is a level $K + 1 - m$ state, and is therefore a linear combination of primaries and their descendants, so that in particular $L_{-m}|\Psi\rangle$ is a descendant state, and

$$\langle\Psi|\Psi\rangle = \langle\Phi|L_{-m}|\Psi\rangle = 0 .$$

Unitarity implies $|\Psi\rangle = 0$, and therefore $|\Phi\rangle$ must be primary. The result now follows by induction on K.

5. *An operator is anti-holomorphic if and only if it has weight $h = 0$. This follows from unitarity and*

$$2h\||\Phi\rangle\|^2 + \|L_1|\Phi\rangle\|^2 = \|L_{-1}|\Phi\rangle\|^2 = \|\partial\Phi\rangle\|^2 .$$

If $h = \bar{h} = 0$ then the operator must be position-independent, and therefore (in any local quantum field theory) a constant multiple of the identity.

A.2   Superconformal Current Algebras

In this section we review properties of N=1 superconformal current algebras (SCCAs) following [148, 149, 327]. We suppose that our CFT has a left-moving supercurrent G and a SCA

$$G(z_1)G(z_2) \sim \frac{2c/3}{z_{12}^3} + \frac{2T(z_2)}{z_{12}} ,$$

$$T(z_1)G(z_2) \sim \frac{3/2G(z_2)}{z_{12}^2} + \frac{\partial G(z_2)}{z_{12}} ,$$

$$T(z_1)T(z_2) \sim \frac{c/2}{z_{12}^2} + \frac{2T(z_2)}{z_{12}^2} + \frac{\partial T(z_2)}{z_{12}} , \qquad\qquad \text{(A.2.1)}$$

$$\{G_r, G_s\} = 2L_{r+s} + \tfrac{c}{12}(4r^2 - 1)\delta_{r+s,0}, \qquad [L_m, G_r] = \tfrac{m-2r}{2}G_{m+r},$$

$$[L_m, L_n] = (m - n)L_{m+n} + \tfrac{c}{12}(m^3 - m)\delta_{m+n,0}. \tag{A.2.2}$$

The vacuum is annihilated by $G_{\pm 1/2}$ and $L_{0,\pm 1}$.

A.2.1 Superconformal Multiplets

A pair of Virasoro primary operators Ψ, Φ, with weights h and $h + 1/2$ constitute a primary representation of $N = 1$ iff

$$G(z)\Psi(0) \sim \frac{\Phi(0)}{z}, \qquad \text{and} \qquad G(z)\Phi(0) \sim \frac{2h\Psi(0)}{z^2} + \frac{\partial\Psi(0)}{z}. \tag{A.2.3}$$

We will assume that Ψ is fermionic, while Φ is bosonic. If we mode expand Ψ and Φ as

$$\Psi = \sum_{r \in \mathbb{Z}-h} \Psi_r z^{-r-h}, \qquad \Phi = \sum_{m \in \mathbb{Z}-h-1/2} \Phi_m z^{-m-h-1/2}, \tag{A.2.4}$$

we find the following commutators from the OPEs:

$$[L_n, \Psi_r] = (n(h - 1) - r)\Psi_{r+n}, \qquad [L_n, \Phi_m] = (n(h - 1/2) - m)\Phi_{n+m},$$

$$\{G_r, \Psi_s\} = \Phi_{r+s}, \qquad [G_r, \Phi_m] = ((2h - 1)r - m)\Psi_{r+m}. \tag{A.2.5}$$

Now specialize to the case of interest, with $h = 1/2$, and $\Psi = \psi^a$, and $\Phi = j^a$.

Holomorphy and the assumption of a single-valued OPE imply

$$\psi^a(z)\psi^b(0) \sim \frac{g^{ab}}{z}, \quad j^a(z)j^b(0) \sim \frac{G^{ab}}{z^2} + \frac{J^{ab}(0)}{z}, \quad j^a(z)\psi^b(0) \sim \frac{\Psi^{ab}(0)}{z}. \tag{A.2.6}$$

g, G are constants, and Ψ^{ab}, J^{ab} are operators of dimensions $1/2$, 1. We will show that Ψ^{ab}, J^{ab} fit into an $N = 1$ primary multiplet and relate the constants. The OPEs are equivalent to

$$\{\psi_r^a, \psi_s^b\} = g^{ab}\delta_{r+s,0}, \quad [j_m^a, \psi_r^b] = \Psi_{m+r}^{ab},$$

$$[j_m^a, j_n^b] = mG^{ab}\delta_{m+n,0} + J_{n+m}^{ab}. \tag{A.2.7}$$

The Jacobi identity implies $\Psi^{ab} = -\Psi^{ba}$:

$$[j_m^a, \psi_r^b] + [j_m^b, \psi_r^a] = [\{G_{m-r}, \psi_r^a\}, \psi_r^b] + [\{G_{m-r}, \psi_r^b\}, \psi_r^a]$$

$$+ [\{\psi_r^a, \psi_r^b\}, G_{m-r}] = 0. \tag{A.2.8}$$

We will first show $g^{ab} = G^{ab}$ by using unitarity. Since the vacuum is annihilated by $G_{\pm 1/2}$, as well as all the raising modes and j_0, we have

$$G^{ab} = \langle 0|j_1^a j_{-1}^b|0\rangle = \langle 0|\psi_{1/2}^a G_{1/2} G_{-1/2} \psi_{-1/2}^b|0\rangle = \langle 0|\psi_{1/2}^a \psi_{-1/2}^b \, 0\rangle = g^{ab}. \tag{A.2.9}$$

The Ψ^{ab}, J^{ab} are clearly Virasoro primary. Moreover, the Jacobi identity for fermionic (F) and bosonic (B) operators

$$\{F_1, [F_2, B]\} + [B, \{F_1, F_2\}] + \{F_2, [F_1, B]\} = 0, \implies \{F, [F, B]\} = [F^2, B]$$

$$[F_1, \{F_2, F_3\}] + [F_3, \{F_1, F_2\}] + [F_2, \{F_3, F_1\}] = 0 \implies [F, \{F, F_3\}] = [F^2, F_3] \tag{A.2.10}$$

shows

$$\{G_s, \Psi_{m+r}^{ab}\} = \{G_s, [j_m^a, \psi_r^b]\} = J_{s+m+r}^{ab}. \tag{A.2.11}$$

Here we used $g = G$ to cancel the central terms. Similarly, we can use the Jacobi identity and anti-symmetry of Ψ^{ab} to show that

$$[G_s, J_m^{ab}] = -m\Psi_{m+s}^{ab}. \tag{A.2.12}$$

Thus, (Ψ^{ab}, J^{ab}) do form an $h = 1/2$ $N = 1$ primary multiplet, and as expected the algebra of the superconformal currents closes. Unitarity of the Kac-Moody algebra identifies $g^{ab} = k\delta^{ab}$, $\Psi^{ab} = if^{abc}\psi^c$ and implies

$$\{\psi_r^a, \psi_s^b\} = k\delta^{ab}\delta_{r+s,0}, \quad [j_m^a, \psi_r^b] = if^{abc}\psi_{m+r}^c,$$

$$[j_m^a, j_n^b] = mk\delta^{ab}\delta_{m+n,0} + if^{abc}j_{n+m}^c. \tag{A.2.13}$$

Let

$$\tilde{J}_\psi^a \equiv -\frac{i}{2k} f^{abc}\psi^b\psi^c . \tag{A.2.14}$$

These form a current algebra with level $h(g)$—the dual Coxeter number. Moreover, the current $\widetilde{J}^a = j^a - \widetilde{J}^a_\psi$ commutes with $\widetilde{J}^a_\psi$ and has level $k_b = k - h(g)$. A Sugawara construction for these two sets of currents leads to the total central charge

$$c_{\text{SKM}} = \left(\frac{k_b}{k} + \frac{1}{2} \right) \dim \mathfrak{g} \ . \tag{A.2.15}$$

There is a super-Sugawara construction as well.

A.2.2 Currents and (1,0) SUSY

In the previous section we saw how a KM current can naturally show up as the top component of an N=1 multiplet. Must this be so? To examine the point in detail, we suppose that we have the N=1 SCA and a level k KM current J. Our first concern is the action, determined by the OPE, of $G_{\mp 1/2}$ on $J(z)$. For any operator X

$$G_r \cdot X(w) \equiv [G_r, X(w)]_\pm = \frac{1}{2\pi i} \oint_{C(w)} dz \, z^{r+1/2} G(z) X(w) \ . \tag{A.2.16}$$

Now using $X = J$, we see that the most general holomorphic and single-valued G–J OPE

$$G(z)J(w) \sim \frac{\Psi(w)}{(z-w)^2} + \frac{\partial\Psi(w) - \mathcal{X}(w)}{z - w} \tag{A.2.17}$$

leads to

$$G_r \cdot J(w) = \partial \left[w^{r+1/2} \Psi(w) \right] - w^{r+1/2} \mathcal{X}(w)$$

$$\Longleftrightarrow [G_r, J_n] = -n\Psi_{r+n} - \mathcal{X}_{r+n} \ , \tag{A.2.18}$$

where we used mode expansions $\Psi(w) = \sum_r \Psi_r w^{-r-1/2}$ and $\mathcal{X}(w) = \sum_r \mathcal{X}_r w^{-r-3/2}$. Thus, we see that

$$\mathcal{X}_r = [J_0, G_r], \qquad \Psi_s = [G_{s+1}, J_{-1}] + [J_0, G_s] \ . \tag{A.2.19}$$

At least one of these is non-zero, since otherwise $G_{\pm 1/2}J_{-1}|0\rangle = 0$, an impossibility in a unitary theory:

$$0 = \langle 0|J_1 G_{1/2} G_{-1/2} J_{-1}|0\rangle = 2\langle 0|J_1 J_{-1}|0\rangle = 2k \, \||0\rangle\|^2 \ . \tag{A.2.20}$$

It is easy to show

$$[L_m, \mathcal{X}_r] = \frac{m - 2r}{2} \mathcal{X}_{m+r} , \qquad [L_m, \Psi_s] = \frac{-m - 2s}{2} \Psi_{m+s} . \qquad \text{(A.2.21)}$$

Hence $\Psi(z)$ and $\mathcal{X}(z)$ are Virasoro primary fields with weights $1/2$ and $3/2$ respectively. In fact, Ψ is also N=1 primary. To see that, it is simplest to work with the corresponding state $|\Psi\rangle \equiv \lim_{z \to 0} \Psi(z)|0\rangle = \Psi_{-1/2}|0\rangle$. This is N=1 primary iff $G_{1/2}|\Psi\rangle = 0$. We compute

$$G_{1/2}|\Psi\rangle = G_{1/2} \left[[G_{1/2}, J_{-1}] + [J_0, G_{-1/2}] \right] |0\rangle = G_{1/2}^2 J_{-1}|0\rangle$$

$$= L_1 J_{-1}|0\rangle = J_0|0\rangle = 0. \qquad \text{(A.2.22)}$$

The second equality follows because $G_{\pm 1/2}|0\rangle = 0$ and $J_0|0\rangle = 0$. Hence, we have

$$G(z)\Psi(w) = \frac{K(w)}{z - w} \qquad \text{(A.2.23)}$$

for a weight 1 operator $K(z) \equiv G_{-1/2} \cdot \Psi(z)$ with moding

$$K(z) = \sum_n K_n z^{-n-1}, \qquad K_n \equiv \{G_{-1/2}, \Psi_{n+1/2}\} = \{G_r, \Psi_{n-r}\}. \qquad \text{(A.2.24)}$$

It is a Virasoro primary field since the corresponding state

$$|K\rangle = \lim_{z \to 0} K(z)|0\rangle = \lim_{z \to 0} \left(z^{-1}\{G_{-1/2}, \Psi_{1/2}\} + \{G_{-1/2}, \Psi_{-1/2}\} \right) |0\rangle$$

$$= G_{-1/2}\Psi_{-1/2}|0\rangle \qquad \text{(A.2.25)}$$

satisfies

$$L_1|K\rangle = L_1 G_{-1/2}\Psi_{-1/2}|0\rangle = G_{1/2}|\Psi\rangle = 0 . \qquad \text{(A.2.26)}$$

This is a non-trivial object as long as $|\Psi\rangle \neq 0$:

$$\||K\rangle\|^2 = \langle 0|\Psi_{1/2}G_{1/2}G_{-1/2}\Psi_{-1/2}|0\rangle = \||\Psi\rangle\|^2 . \qquad \text{(A.2.27)}$$

From $G_{1/2}K_{-1}|0\rangle = \Psi_{-1/2}|0\rangle$ and $G_{-1/2}K_{-1}|0\rangle = L_{-1}\Psi_{-1/2}|0\rangle$, we also see

$$G(z)K(w) = \frac{\Psi(w)}{(z - w)^2} + \frac{\partial\Psi(w)}{z - w} \quad \Longleftrightarrow \quad [G_r, K_m] = -m\Psi_{r+m} . \qquad \text{(A.2.28)}$$

This is sufficient to show that (Ψ, K) form an N=1 superconformal multiplet, and in fact a component of a superconformal current algebra explored in the previous section.

The remaining structure is reasonably clear. Given the full set of currents $\{J^A\}$, we construct the SCCA with multiplets (Ψ^α, K^α) for each α such that $|\Psi^\alpha\rangle \neq 0$. We can then choose a basis such that the remaining currents, labeled by R^a, will commute with Ψ^α, K^α and will act on the supercharges:

$$R^a(z)G(w) = \frac{\mathcal{X}^a(w)}{z - w} \quad \Longleftrightarrow \quad [R^a_n, G_r] = \mathcal{X}^a_{n+r} . \tag{A.2.29}$$

We can now apply the classification of superconformal algebras referenced in the text to see that the R^a must generate the corresponding R-symmetry algebra.

A.3   Conformal Perturbation Theory Details

A.3.1   Anomalous Dimensions of Non-deforming Operators

In this section we give a short discussion of the wavefunction renormalization for the non-deforming operators. To start, we must choose how we will define the renormalized operators. We chose conventions so that the deforming operators have mass-dimension 2—this was a nice way to ensure that we keep dimensionless couplings, and that $\frac{\partial}{\partial g^i}$ brings down insertions of $\widehat{\mathcal{O}}_i$. It does not seem very natural to extend this to the remaining operators, and instead we can normalize them to have the naive mass-dimension. Following our example from the beginning of Sect. 2.6.1, we will do something a little bit more general and allow the non-deforming operators Φ_α to have anomalous dimensions γ_α. We will recover the familiar result: the renormalized two-point function takes the form

$$\cdot \langle\!\langle \widehat{\Phi}_\alpha(x)\widehat{\Phi}_\beta(0)\rangle\!\rangle^c = \frac{\delta_{\alpha\beta}}{|x|^{2(\Delta_\alpha - \gamma_\alpha)}} + O(g^2) . \tag{A.3.1}$$

The calculation will have one important technical lesson: the descendant operators mix with the quasi-primaries, and we will see this mixing explicitly.

We then write the general wave-function renormalization as

$$\widehat{\Phi}_\alpha = \ell^{\gamma_\alpha(g)-\Delta_\alpha}\left[Z^\mu_\alpha(g, \ell/a)\ell^{\Delta_\mu}\Phi_\mu + Z^i_\alpha(g, \ell/a)\ell^{\Delta_i}\mathcal{O}_i + \cdots\right] . \tag{A.3.2}$$

We are allowing the deforming operators $\mathcal{O}_i$ to mix into the undeforming ones, and the $\gamma_\alpha(g)$ factors are the "anomalous dimensions." We will assume that as $g \to 0$

$$\gamma_\alpha(g) \to 0 , \qquad Z^\beta_\alpha \to \delta^\beta_\alpha , \qquad Z^i_\alpha \to 0 . \tag{A.3.3}$$

The $\cdots$ denote contributions from descendant operators that will include shortly. To keep the discussion simple we will focus on quasi-primary bosonic Φ_α with integral spin, and we will just consider marginal deformations.

A.3.2 Preliminaries

To start, we normalize the operators in the usual way:

$$G_{\alpha\beta}(z_{12}, \bar{z}_{12}) = \langle \Phi_{\alpha 1} \Phi_{\beta 2} \rangle = \frac{\delta_{\alpha\beta}}{z_{12}^{2h_\alpha} \bar{z}_{12}^{2\bar{h}_\alpha}} . \tag{A.3.4}$$

With a little bit of foresight, we will also choose the basis such that whenever $h_\alpha = h_\beta$ and $\bar{h}_\alpha = \bar{h}_\beta$, then the symmetric matrix $C_{\alpha\beta i} g^i$ is diagonal:

$$C_{\alpha\beta i} g^i = \delta_{\alpha\beta} C_{\alpha i} g^i + O(g^2) \qquad \text{when } h_\alpha = h_\beta \text{ and } \bar{h}_\alpha = \bar{h}_\beta . \tag{A.3.5}$$

This decision is not so puzzling if we recall degenerate perturbation theory in quantum mechanics—to have a well-behaved perturbative expansion we need to diagonalize the first order perturbation in each degenerate block of the unperturbed theory. We will encounter exactly this issue below.

As we will see, renormalization will naturally involve the descendants $\Phi_\alpha^{m\bar{m}} = \partial^m \bar{\partial}^{\bar{m}} \Phi_\alpha$ as well. Their two-point functions are given by

$$\langle \Phi_{\alpha 1}^{m\bar{m}} \Phi_{\beta 2}^{n\bar{n}} \rangle = (-1)^{m+\bar{m}} \mathcal{P}_{h_\alpha, \bar{h}_\alpha}^{m+n, \bar{m}+\bar{n}} \frac{\delta_{\alpha\beta}}{z_{12}^{2h_\alpha + m + n} \bar{z}_{12}^{2\bar{h}_\alpha + \bar{m} + \bar{n}}} , \tag{A.3.6}$$

where we introduced a notation for the product of the rising Pochhammer symbols:

$$\mathcal{P}_{x, \bar{x}}^{m\bar{m}} \equiv \frac{\Gamma(2x + m) \Gamma(2\bar{x} + \bar{m})}{\Gamma(2x) \Gamma(2\bar{x})} . \tag{A.3.7}$$

We now include the descendants in our wavefunction renormalization:

$$\widehat{\Phi}_\alpha = \ell^{\gamma_\alpha(g) - \Delta_\alpha} \left[Z_\alpha^{\mu m \bar{m}}(g, \ell/a) \ell^{\Delta_\mu + m + \bar{m}} \Phi_\mu^{m\bar{m}} + Z_\alpha^i(g, \ell/a) \ell^2 \mathcal{O}_i + \cdots \right]$$

$$= \Phi_\alpha + g^i \delta_i^1 \Phi_\alpha + O(g^2) , \tag{A.3.8}$$

where

$$\delta_i^1 \Phi_\alpha = \gamma_{1\alpha i} \Phi_\alpha \log \ell + Z_{1\alpha i}^{\mu m \bar{m}} \ell^{\Delta_\mu + m + \bar{m} - \Delta_\alpha} \Phi_\mu^{m\bar{m}} + Z_{1\alpha i}^j \ell^{2 - \Delta_\alpha} \mathcal{O}_j . \tag{A.3.9}$$

We expanded $\gamma_\alpha(g)$ and $Z(g, \ell/a)$ in g and just kept the first order terms. There is an implied sum on m, $\overline{m}$, and Lorentz invariance requires $Z_{1\alpha i}^{\beta m \overline{m}} = 0$ unless the spins of the operators satisfy $s_\alpha = s_\beta + m - \overline{m}$, or, equivalently,

$$h_\beta - h_\alpha + m = \overline{h}_\beta - \overline{h}_\alpha + \overline{m}, \tag{A.3.10}$$

The two-point function can now be expanded as

$$x^{h_\alpha + h_\beta} \overline{x}^{\overline{h}_\alpha + \overline{h}_\beta} \langle\!\langle \widehat{\Phi}_\alpha(x) \widehat{\Phi}_\beta(0) \rangle\!\rangle^c = \delta_{\alpha\beta} + g^i \delta_i^1 G_{\alpha\beta} + O(g^2), \tag{A.3.11}$$

where

$$\delta_i^1 G_{\alpha\beta} = -x^{h_\alpha + h_\beta} \overline{x}^{\overline{h}_\alpha + \overline{h}_\beta} \int_{\mathbb{D}_{2,1}} d^2z \, \langle \Phi_\alpha(x) \Phi_\beta(0) \mathcal{O}_i(z) \rangle^c + 2\delta_{\alpha\beta}\gamma_{\alpha i} \log \ell$$

$$+ x^{h_\alpha + h_\beta} \overline{x}^{\overline{h}_\alpha + \overline{h}_\beta} Z_{1\alpha i}^{\mu m \overline{m}} \ell^{\Delta_\mu + m + \overline{m} - \Delta_\alpha} \langle \Phi_\mu^{m\overline{m}}(x) \Phi_\beta(0) \rangle$$

$$+ x^{h_\alpha + h_\beta} \overline{x}^{\overline{h}_\alpha + \overline{h}_\beta} Z_{1\beta i}^{\mu m \overline{m}} \ell^{\Delta_\mu + m + \overline{m} - \Delta_\beta} \langle \Phi_\alpha(x) \Phi_\mu^{m\overline{m}}(0) \rangle. \tag{A.3.12}$$

Set

$$\mathcal{J}_{1\alpha\beta i} = x^{h_\alpha + h_\beta} \overline{x}^{\overline{h}_\alpha + \overline{h}_\beta} \int_{\mathbb{D}_{2,1}} d^2z \, \langle \Phi_\alpha(x) \Phi_\beta(0) \mathcal{O}_i(z) \rangle^c, \tag{A.3.13}$$

so that

$$\delta_i^1 G_{\alpha\beta} = -\mathcal{J}_{1\alpha\beta i} + 2\gamma_{\alpha i}\delta_{\alpha\beta} \log \ell$$

$$+ (-1)^{m+\overline{m}} \mathcal{P}_{h_\beta \overline{h}_\beta}^{m\overline{m}} Z_{1\alpha i}^{\beta m \overline{m}} \left(\frac{\ell}{x}\right)^{h_\beta - h_\alpha + m} \left(\frac{\ell}{\overline{x}}\right)^{\overline{h}_\beta - \overline{h}_\alpha + \overline{m}}$$

$$+ \mathcal{P}_{h_\alpha \overline{h}_\alpha}^{m\overline{m}} Z_{1\beta i}^{\alpha m \overline{m}} \left(\frac{\ell}{x}\right)^{h_\alpha - h_\beta + m} \left(\frac{\ell}{\overline{x}}\right)^{\overline{h}_\alpha - \overline{h}_\beta + \overline{m}}. \tag{A.3.14}$$

Using (A.3.10), the right-hand-side is explicitly spin-less:

$$\delta_i^1 G_{\alpha\beta} = -\mathcal{J}_{1\alpha\beta i} + 2\gamma_{\alpha i} \log \ell \delta_{\alpha\beta}$$

$$+ (-1)^{m+\overline{m}} \mathcal{P}_{h_\beta \overline{h}_\beta}^{m\overline{m}} Z_{1\alpha i}^{\beta m \overline{m}} \left(\frac{\ell}{|x|}\right)^{\Delta_\beta - \Delta_\alpha + m + \overline{m}}$$

$$+ \mathcal{P}_{h_\alpha \overline{h}_\alpha}^{m\overline{m}} Z_{1\beta i}^{\alpha m \overline{m}} \left(\frac{\ell}{|x|}\right)^{\Delta_\alpha - \Delta_\beta + m + \overline{m}}. \tag{A.3.15}$$

We absorb some distracting coefficients by defining $\widehat{Z}$ via

$$\delta_{s_\alpha,s_\beta+m-\overline{m}}\widehat{Z}^{\beta m\overline{m}}_{1\alpha i} = Z^{\beta m\overline{m}}_{1\alpha i}\mathcal{P}^{m\overline{m}}_{h_\beta \overline{h}_\beta}\left(\frac{\ell}{a}\right)^{\Delta_\beta-\Delta_\alpha-m-\overline{m}}. \tag{A.3.16}$$

With that, setting $\epsilon = a/|x|$, the first order correction takes on a manageable form

$$\delta^1_i G_{\alpha\beta} = -\mathcal{J}_{1\alpha\beta i} + 2\gamma_{\alpha i}\log\ell\delta_{\alpha\beta}$$

$$+ (-1)^{s_\beta-s_\alpha}\widehat{Z}^{\beta m\overline{m}}_{1\alpha i}\epsilon^{\Delta_\beta-\Delta_\alpha+m+\overline{m}}\delta_{s_\alpha,s_\beta+m-\overline{m}}$$

$$+ \widehat{Z}^{\alpha m\overline{m}}_{1\beta i}\epsilon^{\Delta_\alpha-\Delta_\beta+m+\overline{m}}\delta_{s_\beta,s_\alpha+m-\overline{m}}. \tag{A.3.17}$$

The integral is simplified via the substitution $z = xu$:

$$\mathcal{J}_{1\alpha\beta i}(\epsilon) = x^{h_\alpha+h_\beta}\overline{x}^{\overline{h}_\alpha+\overline{h}_\beta}\int_{\mathbb{D}_{2,1}}d^2z\,\langle\Phi_\alpha(x)\Phi_\beta(0)\mathcal{O}_i(z)\rangle^c$$

$$= \int_{\mathbb{D}_{2,1}}d^2z\,\frac{C_{\alpha\beta i}x\overline{x}d^2z}{(x-z)^{h_{\alpha\beta}+1}(\overline{x}-\overline{z})^{\overline{h}_{\alpha\beta}+1}(-z)^{h_{\beta\alpha}+1}(-\overline{z})^{\overline{h}_{\beta\alpha}+1}}$$

$$= \int_{\mathbb{D}_{2,1}}d^2u\,\frac{C_{\alpha\beta i}d^2u}{(1-u)^{h_{\alpha\beta}+1}(1-\overline{u})^{\overline{h}_{\alpha\beta}+1}(-u)^{h_{\beta\alpha}+1}(-\overline{u})^{\overline{h}_{\beta\alpha}+1}}, \tag{A.3.18}$$

where $h_{\beta\alpha} = h_\beta - h_\alpha$, etc. With that

$$\mathcal{J}_{\alpha\beta i} = C_{\alpha\beta i}\underbrace{\int_{\mathbb{D}_{2,1}}d^2u(1-u)^{h-1}(1-\overline{u})^{\overline{h}-1}(-u)^{-h-1}(-\overline{u})^{-\overline{h}-1}}_{=\mathcal{I}(h_{\beta\alpha},\overline{h}_{\beta\alpha},\epsilon)}. \tag{A.3.19}$$

We study the integral in detail below and summarize the results here. First, we have a symmetry: $\mathcal{I}(h,\overline{h},\epsilon) = \mathcal{I}(-h,-\overline{h},\epsilon)$, which means $\mathcal{J}_{\beta\alpha i}$ has the same symmetry in $\alpha \leftrightarrow \beta$ as $C_{\alpha\beta i}$.[1] Second, there are three qualitatively different behaviors of $\mathcal{I}$, depending on h and $\overline{h}$. Let $\Delta = h + \overline{h}$ and $s = h - \overline{h}$; note that $s \in \mathbb{Z}$.

1. $h \neq 0$ and $\overline{h} \neq 0$. This is the generic case, where $|\Delta| \neq |s|$.

$$\mathcal{I} = \sum_{0\leq k<(\Delta-|s|)/2}d_{2k}\epsilon^{2k+|s|-\Delta} + O(\epsilon^2); \tag{A.3.20}$$

---

[1] To see this we make a substitution $v = u - 1$.

the d_{2k} are (in general somewhat complicated) numeric coefficients. So, the integral has no finite contribution and only power law divergences.

2. $h = 0$ and $\bar{h} \neq 0$ or $\bar{h} = 0$ and $h \neq 0$. The integral is finite and given by

$$\mathcal{I} = \frac{2\pi}{|\Delta|} + O(\epsilon^2) . \tag{A.3.21}$$

3. $h = 0$ and $\bar{h} = 0$. There is a logarithmic divergence:

$$\mathcal{I} = -8\pi \log \epsilon + O(\epsilon^2) . \tag{A.3.22}$$

A.3.3   Renormalization

We will use these results to show that unless $\Delta_{\beta\alpha} = s_{\beta\alpha} = 0$, the first order correction can be set to zero by a choice of the renormalization coefficients; when $\Delta_{\beta\alpha} = s_{\beta\alpha} = 0$, the renormalization is encoded in non-zero values of the anomalous dimensions γ_α. Without loss of generality we suppose that $\Delta_{\beta\alpha} = \Delta_\beta - \Delta_\alpha \geq 0$, and we analyze (A.3.17) case by case.

Generic Case: $\Delta_{\beta\alpha} > |s_{\beta\alpha}|$

When this is the case, then up to $O(\epsilon^{\Delta_{\beta\alpha}})$ terms

$$\delta_i^1 G_{\alpha\beta} = -C_{\alpha\beta i} \sum_{0 \leq k < (\Delta_{\beta\alpha} - |s_{\beta\alpha}|)/2} d_{2k} \epsilon^{2k + |s_{\beta\alpha}| - \Delta_{\beta\alpha}} + \widehat{Z}_{1\beta i}^{\alpha m \bar{m}} \epsilon^{-\Delta_{\beta\alpha} + m + \bar{m}} \delta_{s_{\beta\alpha}, m - \bar{m}} . \tag{A.3.23}$$

If $s_{\beta\alpha} \geq 0$ we write $m = s_{\beta\alpha} + \bar{m}$ in the second term. We can then set $\delta_i^1 G_{\alpha\beta} = 0$ by taking

$$\widehat{Z}_{1\beta i}^{\alpha, s_{\beta\alpha} + k, k} = C_{\alpha\beta i} d_{2k} . \tag{A.3.24}$$

We can play a similar trick by solving for $\bar{m}$ when $s_{\beta\alpha} \leq 0$, so that $\delta_i^1 G_{\alpha\beta} = 0$ in this case.

Less Generic Case: $\Delta_{\beta\alpha} = |s_{\beta\alpha}| > 0$

Now up to positive powers of ϵ we have

$$\delta_i^1 G_{\alpha\beta} = -\frac{2\pi C_{\alpha\beta i}}{|\Delta_{\beta\alpha}|} + \widehat{Z}_{1\beta i}^{\alpha m \bar{m}} \epsilon^{-\Delta_{\beta\alpha} + m + \bar{m}} \delta_{s_\beta, s_\alpha + m - \bar{m}} . \tag{A.3.25}$$

We also know that $\Delta_{\beta\alpha} = |s_{\beta\alpha}| = M$ is a positive integer, so we set

$$\widehat{Z}_{1\beta i}^{\alpha M0} = \frac{2\pi C_{\alpha\beta i}}{M} \qquad \text{when } s_{\beta\alpha} > 0 \; ;$$

$$\widehat{Z}_{1\beta i}^{\alpha 0M} = \frac{2\pi C_{\alpha\beta i}}{M} \qquad \text{when } s_{\beta\alpha} < 0 \; . \qquad (A.3.26)$$

Least Generic Case: $\Delta_{\beta\alpha} = s_{\beta\alpha} = 0$

Finally, we consider this case, where we have (recall our choice of basis to diagonalize $C_{\alpha\beta i}$)

$$\delta_i^1 G_{\alpha\beta} = 8\pi C_{\alpha i} \delta_{\alpha\beta} \log \epsilon + 2\gamma_{\alpha i} \log \ell \delta_{\alpha\beta} + Z_{1\beta i}^{\alpha mm} \epsilon^{2m} + Z_{1\alpha i}^{\beta mm} \epsilon^{2m} \; . \qquad (A.3.27)$$

We set $Z_{1\beta i}^{\alpha 00} = -4\pi C_{\alpha i} \delta_{\alpha\beta} \log a/\ell$ and $\gamma_{\alpha i} = -4\pi C_{\alpha i}$, so that

$$\delta_i^1 G_{\alpha\beta} = 2\gamma_{\alpha i} \delta_{\alpha\beta} \log |x| \; . \qquad (A.3.28)$$

All together, we obtain the expected result: the renormalized two-point function takes the form

$$\langle\!\langle \widehat{\Phi}_\alpha(x) \widehat{\Phi}_\beta(0) \rangle\!\rangle^c = \frac{\delta_{\alpha\beta}}{|x|^{2(\Delta_\alpha - \gamma_\alpha)}} + O(g^2) \; . \qquad (A.3.29)$$

The technical lesson is that we must allow for descendant operators to mix with the quasi-primaries.

A.3.4   Integration

We collect some results for integrals that show up in first orders of conformal perturbation theory.

Preliminaries

There are two basic tools that simplify many of the integrals we encounter. First, we have the Stokes theorem. If v_z, $v_{\bar{z}}$ are the components of a complex 1 form, and $R \subset \mathbb{C}$ is some closed simply connected region with boundary $\partial R = C$, a counter-clockwise-oriented closed contour, then

$$\int_R d^2 z (\bar{\partial} v_z - \partial v_{\bar{z}}) = -i \oint_C (dz v_z + d\bar{z} v_{\bar{z}}) \; . \qquad (A.3.30)$$

For us a more relevant formula is the complement:

$$\int_{\mathbb{C}\backslash R} d^2z(\bar{\partial}v_z - \partial v_{\bar{z}}) = +i \oint_C (dz v_z + d\bar{z} v_{\bar{z}}) \ . \tag{A.3.31}$$

If R consists of several disjoint components with boundaries C_1, C_2, etc., then we get a sum over the C_i on the right-hand-side.

Another nice tool is the Heaviside step function $H(x)$:

$$H(x) = \begin{cases} 0 & x < 0 \\ 1 & x > 1 \ . \end{cases} \tag{A.3.32}$$

It satisfies $\partial_x H(x) = \delta(x)$, and we can use it to enforce the constraints on the integration domain $\mathbb{C} \to \mathbb{D}_{m,n}$.

Finally, it is useful to recall the relationship between holomorphic and polar coordinates. If $z = re^{i\theta}$, then

$$\frac{\partial}{\partial z} = \frac{1}{2z}\left(\frac{\partial}{\partial r} - i\frac{\partial}{\partial \theta}\right) \ , \qquad \frac{\partial}{\partial \bar{z}} = \frac{1}{2\bar{z}}\left(\frac{\partial}{\partial r} + i\frac{\partial}{\partial \theta}\right) \ . \tag{A.3.33}$$

We also recall that $d^2z = 2r dr d\theta$.

Divergence Structure of the Spin-Less Three-Point Function

In this section we study the basic integral of CPT:

$$\mathcal{I}(\epsilon^2) = \int_{\mathbb{D}_{2,1}} \frac{d^2u}{|1 - u|^{2\alpha+2}|u|^{2\beta+2}} \ . \tag{A.3.34}$$

Note that this integral is dimensionless, so that the UV (ϵ) and IR (L) cut-offs are also dimensionless: for instance, $\epsilon = a/|x|$, where a is the UV cut-off lengthscale, and $|x|$ is a finite length-scale in the correlation function. We will assume that $\alpha + \beta > -1$, so that the integral converges in the IR, and we can remove the IR cutoff $L \to \infty$ and set $\mathbb{D}_{2,1} = \mathbb{D}_{2,1}^{uv}$. The integral can be written as

$$\mathcal{I} = \mathcal{I}^{\mathrm{div}} + \mathcal{I}^{\mathrm{fin}} + O(\epsilon^2) \ . \tag{A.3.35}$$

Our interest is in the divergence structure of the integral, and we can extract that by computing $-\epsilon\frac{d}{d\epsilon}\mathcal{I}^{\mathrm{div}} = -\epsilon\frac{d}{d\epsilon}\mathcal{I} + O(\epsilon^2)$. Since

$$\mathcal{I}(\epsilon^2) = \int \frac{d^2u H(|u| - \epsilon)H(|u - 1| - \epsilon)}{|1 - u|^{2\alpha+2}|u|^{2\beta+2}} \ , \tag{A.3.36}$$

we have

$$-\epsilon \frac{d}{d\epsilon}\mathcal{I} = \int \frac{d^2u\epsilon\delta(|u|-\epsilon)\boldsymbol{H}(|u-1|-\epsilon)}{|1-u|^{2\alpha+2}|u|^{2\beta+2}} + \int \frac{d^2u\boldsymbol{H}(|u|-\epsilon)\epsilon\delta(|u-1|-\epsilon)}{|1-u|^{2\alpha+2}|u|^{2\beta+2}}.$$

(A.3.37)

In each of the integrals the Heaviside is now vacuous, so

$$-\epsilon \frac{d}{d\epsilon}\mathcal{I} = \underbrace{2\epsilon^{-2\alpha}\int_0^{2\pi} d\theta |1-\epsilon e^{i\theta}|^{-2\beta-2}}_{=\mathcal{J}(\beta+1)} +2\epsilon^{-2\beta}\int_0^{2\pi} d\theta |1-\epsilon e^{i\theta}|^{-2\alpha-2}.$$

(A.3.38)

The integral

$$\mathcal{J}(\gamma) = \int_0^{2\pi} d\theta |1-\epsilon e^{i\theta}|^{-2\gamma}$$

(A.3.39)

is easy to evaluate using the binomial expansion

$$(1-x)^{-r} = \sum_{k=0}^{\infty} \frac{\Gamma(r+k)}{k!\Gamma(r)} x^k$$

(A.3.40)

for the $(1-\epsilon e^{i\theta})^{-\gamma}$ and $(1-\epsilon e^{-i\theta})^{-\gamma}$ factors separately. We obtain

$$\mathcal{J} = \sum_{k,l=0}^{\infty} \frac{\Gamma(\gamma+k)}{k!\Gamma(\gamma)}\frac{\Gamma(\gamma+l)}{l!\Gamma(\gamma)}\epsilon^{k+l}\int_0^{2\pi} e^{i(k-l)\theta}$$

$$= 2\pi \sum_{k=0}^{\infty} \frac{\Gamma(\gamma+k)^2(\epsilon^2)^k}{(k!)^2\Gamma(\gamma)^2} = 2\pi\, {}_2F_1(\gamma,\gamma;1;\epsilon^2),$$

(A.3.41)

where ${}_2F_1(a_1, a_2; b; z)$ denotes the hypergeometric function with upper Pochhammer symbols for a_1 and a_2 and a lower Pochhammer symbol for b.

Thus, we have

$$-\epsilon \frac{d}{d\epsilon}\mathcal{I} = 4\pi\epsilon^{-2\alpha}\, {}_2F_1(\beta+1,\beta+1;1;\epsilon^2) + 4\pi\epsilon^{-2\beta}\, {}_2F_1(\alpha+1,\alpha+1;1;\epsilon^2)$$

$$= 4\pi\left(\epsilon^{-2\alpha}+\epsilon^{-2\beta}\right) + O(\epsilon^{2-2\alpha}) + O(\epsilon^{2-2\beta}).$$

(A.3.42)

Since we are not interested in terms that involve positive powers of ϵ, we can always truncate the expansion to a finite number of terms; the leading order truncation is the simple one we just illustrated.

When $\alpha = \beta = 0$

$$-\epsilon \frac{d}{d\epsilon} \mathcal{I} = 8\pi \ , \tag{A.3.43}$$

so that

$$\mathcal{I}^{\text{div}} = -8\pi \log \epsilon \ . \tag{A.3.44}$$

We will return to this case below and show that also $\mathcal{I}^{\text{fin}} = 0$.

Finally, for completeness, we compute the integral when it converges. This is standard and nicely presented in, for instance, [90]. The integral

$$\mathcal{I} = \int d^2 u |u|^{-2\alpha-2} |1 - u|^{-2\beta-2} \tag{A.3.45}$$

converges if $\alpha < 0$, $\beta < 0$ and $\alpha + \beta > -1$. In that case we represent

$$(|u|^2)^{-1-\alpha} = \frac{1}{\Gamma(1+\alpha)} \int_0^\infty dt\, t^\alpha e^{-|u|^2 t} \ , \tag{A.3.46}$$

and similarly for $|1 - u|^{-2\beta-2}$. Let $u = x + iy$, so that $d^2 u = 2dxdy$, and, with a little algebra

$$|u|^2 t + |1 - u|^2 s = (t + s) \left[(x - s/(s+t))^2 + y^2 \right] + \frac{st}{s+t} \ . \tag{A.3.47}$$

We then have, with $x' = x - s/(s+t)$,

$$\mathcal{I} = \frac{2}{\Gamma(1+\alpha)\Gamma(1+\beta)} \int_0^\infty dt\, t^\alpha \int_0^\infty ds\, s^\beta e^{-st/(s+t)} \int dx' dy\, e^{-(s+t)(x'^2+y^2)} \tag{A.3.48}$$

Carrying out the Gaussian integration on x' and y leads to

$$\mathcal{I} = \frac{2\pi}{\Gamma(1+\alpha)\Gamma(1+\beta)} \int_0^\infty dt\, t^\alpha \int_0^\infty ds\, s^\beta \frac{e^{-st/(s+t)}}{s+t} \ . \tag{A.3.49}$$

Now substitute $s = (t\xi)/(1 - \xi)$:

$$\int_0^\infty dt\, t^\alpha \int_0^\infty ds\, s^\beta \frac{e^{-st/(s+t)}}{s+t} = \int_0^\infty dt\, t^{\alpha+\beta} \int_0^1 d\xi\, \xi^\beta (1 - \xi)^{-\beta-1} e^{-t\xi} \ . \tag{A.3.50}$$

After the t integration we get a standard beta function integral:

$$\int_0^\infty dt\, t^{\alpha+\beta} \int_0^1 d\xi\, \xi^\beta (1-\xi)^{-\beta-1} e^{-t\xi} = \Gamma(1+\alpha+\beta) \int_0^1 d\xi\, \xi^{-\alpha-1}(1-\xi)^{-\beta-1}$$

$$= \frac{\Gamma(1+\alpha+\beta)\Gamma(-\alpha)\Gamma(-\beta)}{\Gamma(-\alpha-\beta)} . \tag{A.3.51}$$

Therefore, as quoted in the text, we obtain

$$\mathcal{I} = \frac{2\pi\,\Gamma(1+\alpha+\beta)\Gamma(-\alpha)\Gamma(-\beta)}{\Gamma(1+\alpha)\Gamma(1+\beta)\Gamma(-\alpha-\beta)} . \tag{A.3.52}$$

Computation of a Three-Point Function with Spin

In this section we study the integral

$$\mathcal{I} = \int_{\mathbb{D}_{2,1}} (1-u)^{h-1}(1-\overline{u})^{\overline{h}-1}(-u)^{-h-1}(-\overline{u})^{-\overline{h}-1} , \tag{A.3.53}$$

We assume, as above, that the difference in the spin is integral: $s = h - \overline{h} \in \mathbb{Z}$; on the other hand $\Delta = h + \overline{h}$ is unconstrained.[2]

Observe that if $h \neq 0$ then

$$\frac{1}{h}\frac{\partial}{\partial u}\left(\frac{1-u}{-u}\right)^h = (1-u)^{h-1}(-u)^{-h-1} . \tag{A.3.54}$$

This motivates us to consider the integral for three different possibilities:

1. $h \neq 0, \overline{h} \neq 0$;
2. $h = 0, \overline{h} \neq 0$ or $h \neq 0, \overline{h} = 0$;
3. $h = 0$ and $\overline{h} = 0$.

Case 1 $h \neq 0$ and $\overline{h} \neq 0$. We will show that in this case $\mathcal{I}^{\mathrm{fin}} = 0$, and $\mathcal{I}^{\mathrm{div}}$ is zero unless $\Delta = h + \overline{h} > |s| = |h - \overline{h}|$.

Since by (A.3.54) the integrand is a total derivative, we can use Stokes to write

$$\mathcal{I} = \underbrace{\frac{i}{h\overline{h}} \oint_{C(0,\epsilon)} du\, \frac{\partial}{\partial u}\left[\left(\frac{1-u}{-u}\right)^h \left(\frac{1-\overline{u}}{-\overline{u}}\right)^{\overline{h}}\right]}_{\mathcal{J}(h,\overline{h})} + \mathcal{J}(-h,-\overline{h}) . \tag{A.3.55}$$

---

[2] Since h and $\overline{h}$ are differences of operator weights, there are not any positive conditions on Δ.

Here $C(0, \epsilon)$ is a counter-clockwise contour along $|u| = \epsilon$, and $\mathcal{J}(-h, -\bar{h})$ is the contribution from a similar contour with $|u - 1| = \epsilon$.

Let $u = re^{i\theta}$ and

$$F(u, \bar{u}) = \left(\frac{1-u}{-u}\right)^h \left(\frac{1-\bar{u}}{-\bar{u}}\right)^{\bar{h}} . \tag{A.3.56}$$

Since $s = h - \bar{h} \in \mathbb{Z}$, $F(u, \bar{u})$ is a single-valued function of θ along the contour. We now have

$$\mathcal{J}(h, \bar{h}) = \frac{i}{h\bar{h}} \int_0^{2\pi} ird\theta e^{i\theta} \frac{1}{2e^{i\theta}} \left(\frac{\partial}{\partial r} - \frac{i}{r}\frac{\partial}{\partial \theta}\right) F(re^{i\theta}, re^{-i\theta})\Big|_{r=\epsilon} . \tag{A.3.57}$$

The contribution from the $\partial/\partial\theta$ derivative term is zero, and the $\partial/\partial r$ derivative can be pulled out of the θ integral. Therefore

$$\mathcal{J} = \frac{1}{2h\bar{h}} \left(-\epsilon\frac{d}{d\epsilon}\right) \mathcal{K}(h, \bar{h}) , \tag{A.3.58}$$

where, with a substitution $\theta \to \theta + \pi$, we get

$$\mathcal{K} = \int_0^{2\pi} d\theta \epsilon^{-h-\bar{h}} e^{-i\theta(h-\bar{h})} (1 + \epsilon e^{i\theta})^h (1 + \epsilon e^{-i\theta})^{\bar{h}}$$

$$= \epsilon^{-\Delta} \int_0^{2\pi} d\theta e^{-i\theta s} (1 + \epsilon e^{i\theta})^h (1 + \epsilon e^{-i\theta})^{\bar{h}} . \tag{A.3.59}$$

The most divergent term in $\mathcal{K}$ is $O(\epsilon^{|s|-\Delta})$, so in order for these to contribute to $\mathcal{J}$ (and therefore to $\mathcal{I}^{\mathrm{div}}$) we need $\Delta > |s|$; otherwise $\mathcal{J} \to 0$ as $\epsilon \to 0$.

Assume $\Delta > s \geq 0$; this implies $h \geq \bar{h} > 0$. Then

$$\mathcal{K} = \epsilon^{-\Delta} \sum_{k,l=0}^{\infty} \binom{h}{k}\binom{\bar{h}}{l} \epsilon^{k+l} \int_0^{2\pi} d\theta e^{-i\theta(s+l-k)} , \tag{A.3.60}$$

and with a few manipulations

$$\mathcal{K} = 2\pi\epsilon^{-2\bar{h}} \frac{\Gamma(h+1)}{s!\Gamma(\bar{h}+1)} \, {}_2F_1(-\bar{h}, -\bar{h}; s+1; \epsilon^2) . \tag{A.3.61}$$

If, on the other hand, $\Delta > -s \geq 0$, we can change the integration variable $\theta \to -\theta$ and find

$$\mathcal{K} = 2\pi\epsilon^{-2h} \frac{\Gamma(\bar{h}+1)}{(-s)!\Gamma(h+1)} \, {}_2F_1(-h, -h; -s+1; \epsilon^2) \tag{A.3.62}$$

We combine these cases into

$$\mathcal{K} = 2\pi \epsilon^{|s|-\Delta} \frac{\Gamma(\frac{|s|+\Delta}{2}+1)}{(|s|)!\Gamma(\frac{\Delta-|s|}{2}+1)} \, {}_2F_1(\frac{|s|-\Delta}{2}, \frac{|s|-\Delta}{2}; |s|+1; \epsilon^2)$$

(A.3.63)

and write the complete Case 1 answer

$$\mathcal{I}^{\text{div}} = -\frac{1}{2h\bar{h}} \epsilon \frac{d}{d\epsilon} \mathcal{K} \,.$$

(A.3.64)

Now We Look at the Case $h = 0$ and $\bar{h} \neq 0$

Now $\bar{h} = -s \in \mathbb{Z}$, and

$$\mathcal{I} = \frac{1}{h} \int_{\mathbb{D}_{2,1}} d^2 u \frac{\partial}{\partial \bar{u}} \left[\frac{1}{(1-u)(-u)} \left(\frac{1-\bar{u}}{-\bar{u}} \right)^{\bar{h}} \right]$$

$$= \frac{i}{h} \underbrace{\oint_{C(0,\epsilon)} du \left[\frac{1}{(1-u)(-u)} \left(\frac{1-\bar{u}}{-\bar{u}} \right)^{\bar{h}} \right]}_{\mathcal{J}(\bar{h})} + \mathcal{J}(-\bar{h}) \,.$$

(A.3.65)

Since on the contour $u = \epsilon e^{i\theta}$ and $\bar{u} = \epsilon e^{-i\theta} = \epsilon^2/u$, we have

$$\mathcal{J} = \frac{2\pi}{h} \oint_{C(0,\epsilon)} \frac{du}{2\pi i u} \frac{1}{1-u} \left(\frac{\epsilon^2 - u}{\epsilon^2} \right)^{\bar{h}} \,.$$

(A.3.66)

When $\bar{h} > 0$ the only pole enclosed by the contour is at $u = 0$, and we obtain

$$\mathcal{J} = \frac{2\pi}{h} \,.$$

(A.3.67)

On the other hand, if $\bar{h} < 0$, we write

$$\mathcal{J} = \frac{2\pi}{h} \oint_{C(0,\epsilon)} \frac{du}{2\pi i u} \frac{1}{1-u} \left(\frac{\epsilon^2}{\epsilon^2 - u} \right)^{-\bar{h}} \,.$$

(A.3.68)

Now there are two poles enclosed. It is not hard to directly show that their contributions cancel; we will instead convert the integral by setting $u = \epsilon/v$, so that

$$J = \frac{2\pi}{\bar{h}} \oint_{C(0,1)} \frac{dv}{2\pi i(v - \epsilon)} \left(\frac{\epsilon v}{\epsilon v - 1} \right)^{-\bar{h}} = \frac{2\pi}{\bar{h}} \left(\frac{\epsilon^2}{\epsilon^2 - 1} \right)^{-\bar{h}} . \tag{A.3.69}$$

This is zero in $\epsilon \to 0$ limit.

So, plugging this back into $\mathcal{I}$, we find

$$\mathcal{I} = \frac{2\pi}{|\bar{h}|} = \frac{2\pi}{|\Delta|} . \tag{A.3.70}$$

$h = 0$ and $\bar{h} = 0$

We showed in (A.3.44) that in this case $\mathcal{I}^{\mathrm{div}} = -8\pi \log \epsilon$. We will now show that there is also no finite contribution. Since

$$\frac{1}{u(1 - u)} = \frac{1}{u} - \frac{1}{1 - u} = \frac{\partial}{\partial u} \log \left[\frac{u\bar{u}}{(1 - u)(1 - \bar{u})} \right] , \tag{A.3.71}$$

we have

$$\mathcal{I} = \int_{\mathbb{D}_{2,1}} d^2 u \frac{\partial}{\partial \bar{u}} \left[\frac{1}{2} \frac{\partial}{\partial u} \left[\log \frac{u\bar{u}}{(1 - u)(1 - \bar{u})} \right]^2 \right] . \tag{A.3.72}$$

But now compare to the discussion below (A.3.57). By the same arguments as there, we now see that $\mathcal{I} = -\epsilon \frac{d}{d\epsilon} \int_0^{2\pi} d\theta F$, where F is a function of ϵ and θ:

$$\mathcal{I} = 2i \oint_{C(0,\epsilon)} du \frac{\partial}{\partial u} \frac{1}{2} \left(\log \frac{\epsilon^2}{|1 - u|^2} \right)^2 = -\epsilon \frac{d}{d\epsilon} \int_0^{2\pi} d\theta \left[\log^2(\epsilon^2) \right] + O(\epsilon^2) . \tag{A.3.73}$$

Carrying out the computation, we obtain

$$\mathcal{I} = -8\pi \log \epsilon + O(\epsilon^2) . \tag{A.3.74}$$

Improper Integrals

In this section we discuss some improper integrals that arose in the computations in the main text. We start with a warm-up example. Let

$$\mathcal{I} = \int_{\mathbb{D}_{2,1}} \frac{d^2 z}{(z-x)\bar{z}^2}, \qquad \mathcal{I}' = \int d^2 z \frac{(\bar{z}-\bar{x})}{(|z-x|^2+a^2)} \frac{z^2}{(z\bar{z}+a^2)^2},$$

$$\mathcal{I}'' = \int \frac{d^2 z}{(z-x)\bar{z}^2}. \tag{A.3.75}$$

The first two of these integrals are well-defined, while the third one is improper: for instance, if we write it in terms of the polar coordinates with $z = re^{i\theta}$, then the integrand diverges as $1/r$ near the origin; the order of the r and θ integration matters. $\mathcal{I}$ and $\mathcal{I}'$ can be thought of as giving meaning to $\mathcal{I}''$, i.e. we define $\mathcal{I}'' = \lim_{a\to 0} \mathcal{I}$ or $\mathcal{I}'' = \lim_{a\to 0} \mathcal{I}'$. If the integral is merely improper, then these different definitions must agree.

Suppose that $\lim_{a\to 0} \mathcal{I}$ exists. In that case $\lim_{a\to 0} \mathcal{I}'$ exists as well, and the two must be equal. $\mathcal{I}'$ is particularly suited as a nice definition of $\mathcal{I}''$ because

$$\partial\bar{\partial}\log(z\bar{z}+a^2) = \frac{a^2}{(z\bar{z}+a^2)^2} \tag{A.3.76}$$

offers a representation of the delta function $2\pi\delta^2(z,\bar{z})$ in the limit $a \to 0$. This gives meaning to the following formal manipulations of $\mathcal{I}$:

$$\mathcal{I} = -\int d^2 z \frac{1}{z-x} \frac{\partial}{\partial\bar{z}} \frac{1}{\bar{z}} = \int d^2 z \frac{1}{\bar{z}} \frac{\partial}{\partial\bar{z}} \frac{1}{z-x} = \int d^2 z \frac{1}{\bar{z}} 2\pi\delta^2(z-x) = \frac{2\pi}{\bar{x}}. \tag{A.3.77}$$

That is, we have

$$\mathcal{I}' = \int d^2 z \frac{\partial}{\partial\bar{z}} \left(\frac{(\bar{z}-\bar{x})}{(|z-x|^2+a^2)} \right) \frac{z}{z\bar{z}+a^2} = \int d^2 z \frac{a^2}{(|x-z|^2+a^2)^2} \frac{z}{z\bar{z}+a^2}. \tag{A.3.78}$$

In the $a \to 0$ the first term in the integrand becomes the desired δ-function, and we obtain the desired result.

The same reasoning applies to the $\mathcal{K}_1$ and $\mathcal{K}_2$ integrals from (2.6.128), (2.6.129). For convenience, we repeat their definition:

$$\mathcal{K}_1 = \int_{\mathbb{D}_{2,2}} \frac{d^2 z_3 d^2 z_4}{z_{34}^2 \bar{z}_{13}^2 \bar{z}_{24}^2}, \qquad \mathcal{K}_2 = \int_{\mathbb{D}_{2,2}} \frac{d^2 z_3 d^2 z_4}{z_{13}^2 z_{24}^2 \bar{z}_{14}^2 \bar{z}_{23}^2}. \tag{A.3.79}$$

In each case we map

$$\frac{1}{z_{ij}^2} \rightarrow \frac{\bar{z}_{ij}^2}{(z_{ij}\bar{z}_{ij} + a^2)^2} \tag{A.3.80}$$

and construct the $\mathcal{K}_{1,2}'$ integrals (the analogues of $\mathcal{I}'$); each of these is a definition of the improper integrals $\mathcal{K}_{1,2}''$, which are just $\mathcal{K}_{1,2}$ but taken over all of $\mathbb{C}$. The presentation in terms of $\mathcal{K}_{1,2}'$ justifies the following formal manipulations in $\mathcal{K}_{1,2}''$.

$$\mathcal{K}_1'' = \int \frac{d^2 z_3 d^2 z_4}{z_{34}^2 \bar{z}_{13} \bar{z}_{24}^2} = \int d^2 z_3 d^2 z_4 \frac{1}{\bar{z}_{24}^2} \partial_3 \frac{1}{z_{34}} \bar{\partial}_3 \frac{1}{\bar{z}_{31}}$$

$$= \int d^2 z_3 d^2 z_4 \frac{1}{\bar{z}_{24}^2} \bar{\partial}_3 \frac{1}{z_{34}} \partial_3 \frac{1}{\bar{z}_{31}} = \frac{(2\pi)^2}{\bar{z}_{12}^2} . \tag{A.3.81}$$

Similarly, $\mathcal{K}_2$ now factorizes as

$$\mathcal{K}_2 = \int \frac{d^2 z_3}{z_{13}^2 \bar{z}_{23}^2} \int \frac{d^2 z_4}{z_{14}^2 \bar{z}_{24}^2} , \tag{A.3.82}$$

and

$$\int \frac{d^2 z_3}{z_{13}^2 \bar{z}_{23}^2} = \int d^2 z_3 \bar{\partial}_3 \frac{1}{z_{13}} \partial_3 \frac{1}{\bar{z}_{23}} = (2\pi)^2 \delta^2(z_{12}) = 0 . \tag{A.3.83}$$

The last equality follows because we are interested in $z_{12} \neq 0$.

These kinds of manipulations may be used to make sense of many finite but improper integrals. Indeed, even for UV-divergent integrals we can imagine the basic map we just introduced

$$\frac{1}{z_{ij}} \rightarrow \frac{\bar{z}_{ij}}{z_{ij}\bar{z}_{ij} + a^2} \tag{A.3.84}$$

as an alternative regularization procedure. It seems that this would be effective in theories where the unperturbed correlation functions are rational in the z_{ij}, but we do not know how to implement it for general conformal field theories, where already the four-point function is not simply a rational function of the z_{ij}.

Appendix B
Elements of Geometry

In this appendix we collect some geometric notions that prove useful for (0,2) exploration. We have three aims here. First, we will establish our notation. Second, we will remind the reader of some probably familiar notions. Finally, we will attempt to introduce the more advanced material sufficiently so that the reader can follow its applications to our (0,2) purposes.

We assume that the reader is reasonably familiar with standard differential geometry notions of differential manifolds and Riemannian geometry. There are many excellent references geared towards physicists of much of this material, for instance the early review [161], as well as [207]. Aspects of complex/Hermitian/Kähler/toric geometry are perhaps a little bit less familiar, but fortunately they are also presented in a number of excellent texts. A list of geometric references that the author found particularly useful is the following.

1. The introductory chapters of Griffiths and Harris [210] contain fundamental notions of Hermitian and Kähler geometries and their relation to algebraic geometry.
2. Bott and Tu [94] provides a very readable introduction to vector bundles, differential topology and characteristic classes, and these notions are explored further in the context of symplectic geometry in [290]. Rudiments of symplectic geometry are to be found in the timeless Arnold [27], and a very elegant introduction to symplectic notions classic and modern is in [102].
3. Kodaira [271] is a pedagogical presentation of deformation theory grounded in differential geometry.
4. Voisin gives a comprehensive introduction to complex geometry and Hodge theory in [373]; an excellent shorter set of notes is to be found in [336]. A very readable discussion of the $\partial\bar{\partial}$-lemma and related geometric concepts is in the thesis of Angella [20].
5. Special and exceptional holonomy manifolds are described in the book by Joyce [255].

© Springer Nature Switzerland AG 2019
I. V. Melnikov, *An Introduction to Two-Dimensional Quantum Field Theory with (0,2) Supersymmetry*, Lecture Notes in Physics 951,
https://doi.org/10.1007/978-3-030-05085-6

6. The geometry of gauge theory is presented in many texts, including the classic
   works of Freed and Uhlenbeck [171]. The classic paper of Atiyah et al. [44] is
   quite readable and useful, as is the physicist-friendly review of [161].
7. Aspects of Kähler geometry is discussed in many reviews, e.g. [45, 310].
8. A review of Calabi-Yau geometry and its relation to (2,2) SCFT is given in [237];
   a very readable introduction especially geared towards physicists is given by
   Aspinwall in [32].
9. The geometry complex surfaces is discussed in the classic text of Barth et al. [49].
   K3 surfaces in particular are discussed in many references including Aspinwall's
   lecture notes [31] and Huybrecht's lecture notes [247].

B.1 Baby Homological Algebra

We use, almost without thinking, some basic ideas from homological algebra, such
as the concept of an exact sequence. We introduce the notation, and perhaps this
will be enough to get the reader through most of our uses for it; the uninitiated
would definitely benefit from at least skimming some of the basics from a text such
as [223]. We are often interested in a sequence of maps between some spaces, say
finitely generated abelian groups:

$$0 \longrightarrow G_1 \xrightarrow{f_1} G_2 \xrightarrow{f_2} G_3 \xrightarrow{f_3} \ldots . \tag{B.1.1}$$

The maps in this case are group homomorphisms, and, more generally, they respect
some nice structure on the objects. A sequence is exact at some particular spot,
say at G_2, if $\ker f_2 = \operatorname{im} f_1$; the sequence is exact if this is true at every spot. A
particularly nice example is offered by a short exact sequence, which consists of just
three non-trivial entries:

$$0 \longrightarrow G_1 \xrightarrow{f_1} G_2 \xrightarrow{f_2} G_3 \longrightarrow 0 . \tag{B.1.2}$$

For a short exact sequence f_1 is an injective map, and its image yields G_1 as a
subgroup of G_2; on the other hand, $\operatorname{im}(f_2)$ is isomorphic to G_3, and since $\ker f_2 =
\operatorname{im} f_1$, the short exact sequence describes a quotient $G_3 = G_2/G_1$. The virtue of
this description is that it explicitly keeps track of the maps.

B.2 Real Geometry

Let M be a smooth manifold of dimension n. When M is real n is the real dimension;
when M is complex, we will most often talk about its complex dimension, and the
two will be denoted by $\dim_{\mathbb{R}} M$ and $\dim_{\mathbb{C}} M$ respectively when there is a possibility

of confusion. Shockingly, they are related by $\dim_{\mathbb{R}} M = 2 \dim_{\mathbb{C}} M$. In this section we focus on the real case. To simplify various tensorial operations, in what follows we will use the summation convention on tensorial indices unless otherwise noted.

First Steps

A differentiable structure on M (which we will assume is Hausdorff) is a collection of open sets $\{U_\alpha\}_{\alpha \in A}$, with A a countable set and each $U_\alpha \simeq \mathbb{R}^n$, and smooth transition functions relating the coordinates across non-trivial overlaps with $\alpha \neq \beta$ and $U_{\alpha\beta} = U_\alpha \cap U_\beta \neq \emptyset$. We will usually implicitly assume that the cover is refined enough so that the overlaps $U_{\alpha\beta}$, as well as higher ones with $\alpha_1, \alpha_2, \ldots, \alpha_k$ distinct and $U_{\alpha_1 \cdots \alpha_k} = U_{\alpha_1} \cap U_{\alpha_2} \cap \cdots \cap U_{\alpha_k} \neq \emptyset$, are all diffeomorphic to $\mathbb{R}^n$. Such a cover is called a "good cover" [94].

Many of our manifolds will be compact without boundary; we will often just abbreviate this to "compact." M is compact if and only if it has a finite good cover. When we discuss covers below, we will mostly restrict to good covers and to good finite covers when M is compact.

Let $\mathcal{A}^0(M)$ denote the space of smooth functions on M. Similarly, for any open set $U \subset M$, $\mathcal{A}^0(U)$ is the space of smooth functions on U; this space has an obvious additive structure, and, since the zero function is always present, also the structure of an additive group.

In addition to functions, M has a plethora of tensor fields. For instance, we have the vector fields on M, which in a local patch $U \simeq \mathbb{R}^n$ with coordinates x^μ take the form

$$V = V^\mu(x) \frac{\partial}{\partial x^\mu} . \tag{B.2.1}$$

The space of all smooth (aka C^∞) vector fields is denoted by $\mathcal{X}(M)$, and these have a natural linear action on smooth functions: $V(f) = V^\mu \frac{\partial f}{\partial x^\mu} \in \mathcal{A}^0(M)$.

There is a natural dual space to $\mathcal{X}(M)$, denoted by $\mathcal{A}^1(M)$; this is the space of linear functionals $\omega : \mathcal{X} \to \mathcal{A}^0(M)$, which in local coordinates are the 1-forms

$$\omega = \omega_\mu(x) dx^\mu . \tag{B.2.2}$$

In the coordinate basis duality is expressed through the relation

$$dx^\mu \left(\frac{\partial}{\partial x^\nu} \right) = \delta^\mu_\nu , \tag{B.2.3}$$

where δ^μ_ν is the Kronecker delta.

More general tensors are constructed in a hopefully familiar way by taking tensor products of the $\frac{\partial}{\partial x^\mu}$ and dx^μ bases. Every tensor is a section of a vector bundle on M, a notion we explore next.

B.2.1   Vector Bundles and Their Generalizations

A rank k vector bundle $V \to M$ is a differentiable manifold of dimension $k+n$ that can be thought of as a family of vector spaces $V_x \simeq \mathbb{R}^k$ that are labeled by points $x \in M$ and fit together in a way that is compatible with the differentiable structure on M and the linear structure on the vector space.

More precisely, a smooth rank k vector bundle $V \to M$ is a surjective map $\pi : V \to M$ such that $\pi^{-1}(x) = V_x \simeq \mathbb{R}^k$ for every $x \in m$, and for some good cover $\{U_\alpha\}_{\alpha \in A}$ of M there are diffeomorphisms (local trivializations) $\phi_\alpha : \pi^{-1}(U_\alpha) \to U_\alpha \times \mathbb{R}^k$ that act linearly on the vector space fibers. Equivalently, a vector bundle $V \to M$ is specified by the transition functions $\lambda_{\alpha\beta} : U_{\alpha\beta} \to GL(k, \mathbb{R})$ that satisfy

$$\lambda_{\alpha\beta}(x)\lambda_{\beta\alpha}(x) = \mathbb{1} \quad \text{on } U_{\alpha\beta} , \quad \lambda_{\alpha\beta}(x)\lambda_{\beta\gamma}(x)\lambda_{\gamma\alpha}(x) = \mathbb{1} \quad \text{on } U_{\alpha\beta\gamma} . \tag{B.2.4}$$

The transition functions are determined by the local trivializations as $\lambda_{\alpha\beta} = \varphi_\alpha \varphi_\beta^{-1}$. The subgroup G_V of $GL(k, \mathbb{R})$ generated by the transition functions is the structure group of the bundle.

Two vector bundles $V \to M$ and $V' \to M$ defined over a cover $\{U_\alpha\}$ are isomorphic if and only if there exist $\xi_\alpha : U_\alpha \to GL(k, \mathbb{R})$ such that on every overlap

$$\lambda_{\alpha\beta} = \xi_\alpha \lambda'_{\alpha\beta} \xi_\beta^{-1} . \tag{B.2.5}$$

A vector bundle is said to be trivial if it is isomorphic to $M \times \mathbb{R}^k$, where the latter has all transitions functions given by identity: $\lambda_{\alpha\beta} = \mathbb{1}$ for all $U_{\alpha\beta}$.

Exercise B.1 Given a smooth map of manifolds $\zeta : N \to M$ and a vector bundle $V \to M$, there is a natural pull-back bundle $\zeta^*(V) \to N$. Let $\pi : V \to M$; describe the corresponding projection map for $\zeta^*(V) \to N$. If $\lambda_{\alpha\beta}$ denote the transition functions for $V \to M$, describe the transition functions for $\zeta^*(V) \to N$.   □

A section s of the bundle $V \to M$ over an open set $X \subset M$ is a map $s : X \to V$ such that $s(x) \in V_x$ for all $x \in X$. In terms of a trivialization with respect to an open cover $\{U'_\alpha\}_{\alpha \in A}$ of X, with $U'_\alpha = U_\alpha \cap X$, the section s is a collection of $s_\alpha : U'_\alpha \to \mathbb{R}^k$, such that the components patch by the transition functions on all non-empty overlaps $U'_{\alpha\beta}$:

$$(s_\alpha)^i = (\lambda_{\alpha\beta})^i_j (s_\beta)^j . \tag{B.2.6}$$

We denote the space of all smooth sections over X by $\mathscr{A}^0(X, V)$. This is a vector space over $\mathbb{R}$ because every vector bundle has a trivial global section $s_\alpha = 0$, and a linear combination of two sections is also a section.

A rank k vector bundle $V \to M$ is isomorphic to the trivial bundle $M \times \mathbb{R}^k$ if and only if it has k sections that are linearly independent at every point on M. This is easily seen by arranging the sections into a $k \times k$ matrix

$$S_\alpha = \begin{pmatrix} s_\alpha^1 & s_\alpha^2 & \cdots & s_\alpha^k \end{pmatrix} . \tag{B.2.7}$$

Clearly S_α is a map $S_\alpha : U_\alpha \to \mathrm{GL}(k, \mathbb{R})$, and for every overlap $U_{\alpha\beta}$

$$S_\alpha = \lambda_{\alpha\beta} S_\beta . \tag{B.2.8}$$

But, it then follows that every transition function can be expressed as $\lambda_{\alpha\beta} = S_\beta^{-1} S_\alpha$, and the bundle V is therefore isomorphic to the trivial bundle. Every vector bundle over $\mathbb{R}^n$ (or any topologically trivial manifold) is isomorphic to the trivial bundle.

Vector bundles inherit operations familiar from linear algebra. For instance, if $V_1 \to M$ and $V_2 \to M$ are vector bundles of rank, respectively, k_1 and k_2, then we can construct vector bundles $V_1 \oplus V_2 \to M$ and $V_1 \otimes V_2 \to M$ of rank, respectively, $k_1 + k_2$ and $k_1 k_2$ by taking a direct sum or product of the transition functions.

There is also the dual bundle V^*: its transition functions are inverses of those of V, and a section $\sigma \in \mathscr{A}^0(X, V^*)$ is a linear functional $\mathscr{A}^0(X, V) \to \mathscr{A}^0(X)$. With explicit components, we can write s_α^A for a section of V over U_α, and $\sigma_{\alpha A}$ for a section of V^* over U_α, and in that case $\sigma(s) = \sigma_{\alpha A} s_\alpha^A \in \mathscr{A}^0(U_\alpha)$ is a function.

Let $\pi : V \to M$ be a vector bundle with structure group G. There is an associated *fiber* bundle $\widehat{\pi} : \mathcal{P} \to M$, where $\widehat{\pi}^{-1}(U_\alpha) = U_\alpha \times G$, i.e. the fiber is the Lie group G itself. More details may be found in any standard reference, for instance [161], but we will give a quick presentation. By definition the transition functions $\lambda_{\alpha\beta}$ for V are valued in a representation r for the group G, and the representation is by construction faithful (otherwise the structure group would be smaller). We then define the space of sections of $\mathcal{P}$ to be

$$\sigma_\alpha : U_\alpha \to G , \tag{B.2.9}$$

subject to the relation on overlaps:

$$\sigma_\alpha = \lambda_{\alpha\beta} \sigma_\beta . \tag{B.2.10}$$

Since the representation is faithful, the definition makes sense at the level of the abstract group. Because the transition functions act from the left, we observe that there is a sensible right action of G on the space of sections: $\sigma_\alpha \to \sigma_\alpha g$ for any $g \in G$.

In addition to multiplication from the right and from the left, a group G also admits the action by conjugation: $\sigma \to g\sigma g^{-1}$. This allows us to construct the infinite-dimensional group of gauge transformations that act on $\mathcal{P}$. That is, we take $\mathrm{Aut}(\mathcal{P}) \to M$ to be the fiber bundle whose sections are also valued in G but with transition functions

$$\tau_\alpha = \lambda_{\alpha\beta} \tau_\beta \lambda_{\alpha\beta}^{-1} . \tag{B.2.11}$$

The main reason we introduce the associated bundle is that it provides a unified way to treat all vector bundles with structure group G. Given the principal G-bundle $\widehat{\pi} : \mathcal{P} \to M$ and any representation r of G, we automatically construct a vector bundle with fiber the vector space for the representation r by taking the transition functions to be valued in r.

We also mention the bundle $\mathrm{Aut}(V)$, the bundle where the fibers are valued in $\mathrm{GL}(k, \mathbb{R})$, and the sections $\xi_\alpha : U_\alpha \to \mathrm{GL}(k, \mathbb{R})$ have transition functions

$$\xi_\alpha = \lambda_{\alpha\beta}\xi_\beta\lambda_{\alpha\beta}^{-1} \ . \tag{B.2.12}$$

This bundle describes the action of G-gauge transformations on the representation V of G. Any section of $\mathrm{Aut}(\mathcal{P})$ defines a section of $\mathrm{Aut}(V)$ by taking its value in the appropriate representation.

Unlike a general fiber bundle, $\mathrm{Aut}(\mathcal{P})$ and $\mathrm{Aut}(V)$ have a global section $\xi_\alpha = 1$. Since we can always linearize a vector space automorphism, we also have the endomorphism bundle $\mathrm{End}(V) = V \otimes V^*$; its sections are $k \times k$ matrices that need not be invertible. Just like $\mathrm{Aut}(V)$, $\mathrm{End}(V)$ has a canonical non-zero section—the identity, so that we often talk of the "traceless" endomorphism bundle, which has rank $k^2 - 1$.

In a similar vein, the linearization of $\mathrm{Aut}(\mathcal{P})$ leads to adjoint vector bundle with fibers isomorphic to the Lie algebra $\mathfrak{g}$ of G. We will occasionally refer to this vector bundle as $\mathfrak{g}_\mathcal{P}$ or $\mathfrak{g}_V$.

This construction has a myriad of generalizations and special cases. For us, an important example of a specialization is the class of complex vector bundles over a complex manifold. An important example of a generalization is given by more general fiber bundles, where the fiber, instead of being a vector space, is taken to be a manifold with more complicated structure and perhaps topology; the principal bundle is an example.

The most "natural" bundles associated to a smooth manifold are the tangent bundle T_M and its dual T_M^*. The smooth sections of the former are the vector fields $\mathcal{X}(M)$, while the smooth sections of the latter are the differential 1-forms. Both of these bundles have rank n, and their tensor powers define the tensor bundles on M. The differential k-forms are sections of $\wedge^k T_M^*$, and we turn to these next.

B.2.2   Conventions for Differential Forms

Differential forms are essential tools of differential geometry and topology. In this section we review our conventions for these objects. A degree k form on M is a section of $\wedge^k T_M^*$. In local coordinates we normalize the coefficients as

$$\omega = \frac{1}{k!}\omega_{\mu_1\mu_2\cdots\mu_k}(x)dx^{\mu_1} \wedge dx^{\mu_2} \wedge \cdots \wedge dx^{\mu_k} \ . \tag{B.2.13}$$

We will frequently suppress the wedge symbol if there is no possibility of confusion, i.e. $dx^1 \wedge dx^2$ will be abbreviated as $dx^1 dx^2$. Clearly the degree k runs from 0 to $\dim M = n$, and $\wedge^k T_M$ has rank $\binom{n}{k}$. The degree 0 forms are functions of M in $\mathscr{A}^0(M)$, as denoted before. Similarly, we will set

$$\mathscr{A}^k(M) = \text{the space of smooth sections of } \wedge^k T_M^*;$$

$$\mathscr{A}^k(M, V) = \text{the space of smooth sections of } \wedge^k T_M^* \otimes V .$$

The top degree forms are densities that can be integrated over the manifold. We will work with orientable manifolds, where the notion of the integrable densities and top forms coincide.

The wedge product $\wedge : \mathscr{A}^{k_1}(M) \otimes \mathscr{A}^{k_2}(M) \to \mathscr{A}^{k_1+k_2}(M)$ gives the vector space $\bigoplus_k \mathscr{A}^k(M)$ the structure of a graded algebra. It is graded since

$$\omega_1 \wedge \omega_2 = (-1)^{k_1 k_2} \omega_2 \wedge \omega_1 , \tag{B.2.14}$$

where k_i is the degree of ω_i.

Contractions and Lie Derivatives

Tangent vectors are naturally linear functionals on $\mathscr{A}^1(M)$, i.e. they are dual to 1-forms. More generally, a vector $v \in \mathscr{X}(M)$ defines a linear map $v_{\llcorner}: \mathscr{A}^k(M) \to \mathscr{A}^{k-1}(M)$; this is the contraction map. In components we define the contraction as follows:

$$v_{\llcorner}\omega = \frac{1}{(k-1)!} v^{\mu_0} \omega_{\mu_0 \mu_1 \cdots \mu_{k-1}} dx^{\mu_1} \wedge \cdots \wedge dx^{\mu_{k-1}} . \tag{B.2.15}$$

Given a degree $m < k$ multi-vector $v \in \mathscr{A}^0(M, \wedge^m T_M)$, we set the contraction $v_{\llcorner}\mathscr{A}^k(M) \to \mathscr{A}^{k-m}(M)$ (also sometimes called the interior product) to be

$$v_{\llcorner}\omega = \frac{1}{m!} \frac{1}{(k-m)!} v^{\mu_1 \cdots \mu_m} \omega_{\mu_1 \cdots \mu_m \nu_1 \cdots \nu_{k-m}} dx^{\nu_1} \cdots dx^{\nu_{k-m}} . \tag{B.2.16}$$

A vector field $v \in \mathscr{X}(M)$ is a generator of an infinitesimal diffeomorphism on M. The action of this diffeomorphism on tensors is described by the Lie derivative $\mathcal{L}_v$; for any tensor T we have $\delta_v T = \mathcal{L}_v T$. In components, we have the following action on functions (f), vectors (u), and 1-forms (ω):

$$\mathcal{L}_v f = v(f) = v^\mu \partial_\mu f , \qquad (\mathcal{L}_v u)^\mu = ([v, u])^\mu = v^\nu u^\mu_{,\nu} - v^\mu_{,\nu} u^\nu ,$$

$$(\mathcal{L}_v \omega)_\mu = v^\nu \omega_{\mu,\nu} + v^\nu_{,\mu} \omega_\nu . \tag{B.2.17}$$

$\mathcal{L}_v$ obeys the Leibniz rule, so that the action on any tensor can be obtained from those just given; moreover, it satisfies $\mathcal{L}_v\mathcal{L}_u - \mathcal{L}_u\mathcal{L}_v = \mathcal{L}_{[v,u]}$.

Exercise B.2 Show that for any differential form ω

$$\mathcal{L}_v\omega = d(v_\llcorner\omega) + v_\llcorner d\omega . \tag{B.2.18}$$

□

Note that the derivatives in the Lie derivative definition may also be replaced by a covariant derivative with a torsion–free connection.

De Rham Cohomology

Another key notion is the de Rham differential operator $d : \mathscr{A}^k(M) \to \mathscr{A}^{k+1}(M)$. This is a linear nilpotent operator, and in local coordinates its action is as follows:

$$d\omega = \frac{1}{k!}\partial_{\mu_0}\omega_{\mu_1\mu_2\cdots\mu_k}dx^{\mu_0} \wedge dx^{\mu_1} \cdots \wedge dx^{\mu_k} . \tag{B.2.19}$$

A differential form ω is closed if $d\omega = 0$; if we can find a form η such that $\omega = d\eta$, then we say ω is exact. Because $d^2 = 0$ every exact form is closed, and the de Rham cohomology groups

$$H_d^k(M, \mathbb{R}) = \frac{\ker d \cap \mathscr{A}^k(M)}{\operatorname{im} d \cap \mathscr{A}^k(M)} . \tag{B.2.20}$$

characterize the space of closed forms modulo exact forms; a closed form $\omega \in \mathscr{A}^k(M)$ defines an equivalence class $[\omega] \in H_d^k(M)$, and any two representatives of the same equivalence class differ by an exact form.

Exercise B.3 If the reader is not familiar with this, it is a good exercise to check the product rule

$$d(\omega_1 \wedge \omega_2) = (d\omega_1) \wedge \omega_2 + (-1)^{k_1}\omega_1 \wedge d\omega_2 . \tag{B.2.21}$$

This is much like the familiar Leibniz rule except for the extra degree-dependent sign; for this reason one sometimes sees the terminology that d is an "anti-derivation."

The reader should also verify the crucial property

$$d^2 = 0 \tag{B.2.22}$$

and check that the wedge product gives a graded algebra structure to the vector space $\bigoplus_k H_d^k(M)$. That is, if ω_1 is a representative of a class in $H_d^{k_1}(M)$ and ω_2 represents a class in $H_d^{k_2}(M)$, then $[\omega_1 \wedge \omega_2] \in H_d^{k_1+k_2}(M)$, and this class is independent of the choice of representatives. □

The Hodge Star

So far nothing we said depends on a choice of metric on M. A metric leads to extra structure on the differential forms. The most important ingredient in this structure is the Hodge star, an metric–dependent linear isomorphism

$$*_g : \mathscr{A}^k(M) \to \mathscr{A}^{n-k}(M) . \tag{B.2.23}$$

We will omit the subscript g when it is unlikely to cause confusion.

A metric g on M determines a volume form

$$\mathrm{dVol}_g = \sqrt{\det g}\, dx^1 \wedge dx^2 \wedge \cdots \wedge dx^n . \tag{B.2.24}$$

Its integral yields the volume of the manifold with respect to the metric g.[3] For any $\beta \in \mathscr{A}^k(M)$ define the Hodge star $*\beta \in \mathscr{A}^{n-k}(M)$ by the property that for any $\alpha \in \mathscr{A}^k(M)$ we have

$$\alpha \wedge *\beta = \mathrm{dVol}_g \frac{1}{k!}\alpha_{i_1 \cdots i_k}\beta_{j_1 \cdots j_k} g^{i_1 j_1} \cdots g^{i_k j_k} , \tag{B.2.25}$$

where g^{ij} denotes the inverse of the metric: $g^{ij}g_{jk} = \delta_k^i$.

Exercise B.4   Show that in components

$$*_g \beta = \frac{\sqrt{\det g}}{k!(n-k)!}\beta^{i_1 \cdots i_k} \epsilon_{i_1 \cdots i_k j_1 \cdots j_{n-k}} dx^{j_1} \wedge \cdots \wedge dx^{j_{n-k}} . \tag{B.2.26}$$

The indices on β are raised with the inverse metric, and $\epsilon_{i_1 \cdots i_n}$ is fully antisymmetric and normalized to $\epsilon_{12 \cdots n} = +1$. Show also that

$$*_g^2 \mathscr{A}^k(M) = (-1)^{k(n-k)} \mathscr{A}^k(M) . \tag{B.2.27}$$

The diligent reader may pause here to show that $*_g\beta$ is invariant under changes of coordinates on M; while performing that check one may as well check that $d\beta$ is also invariant. □

---

[3] M must be orientable in order for this construction to make sense, and in writing dVol_g we made a choice of the orientation.

Evidently, $*$ exists and is a symmetric operation: $\alpha \wedge *\beta = \beta \wedge *\alpha$; moreover, on a compact manifold we can integrate this to define

$$\langle \alpha, \beta \rangle = \int_M \alpha \wedge *\beta \,, \tag{B.2.28}$$

a symmetric bilinear positive-definite form: $\langle \alpha, \alpha \rangle = 0$ if and only if $\alpha = 0$ pointwise on the manifold.

Exercise B.5 The Hodge star also interacts nicely with the contraction operation. Given a k-form η and a $k + l$ form ω, we define the multi-vector $\tilde{\eta}$ by raising the indices on η with the inverse metric g^{ij}. We can then make a slight abuse of notation and write $\eta \lrcorner \omega = \tilde{\eta} \lrcorner \omega$. That is, we define a contraction of a k-form into a $k + l$ form by converting the former into a multi-vector via the inverse metric, and then performing the contraction operation. With that definition in hand, show that the following relations hold:

$$\eta \lrcorner \omega = (-1)^{k(n-k-l)} * (\eta \wedge *\omega) \quad \Longleftrightarrow \quad *(\eta \lrcorner \omega) = (-1)^{kl} \eta \wedge *\omega \,. \tag{B.2.29}$$

$\square$

The Hodge–de Rham Laplacian

The Hodge inner product on differential forms can be used to define the formal adjoint of d, $d^\dagger$.[4] This is the unique linear operator

$$d^\dagger : \mathscr{A}^k(M) \to \mathscr{A}^{k-1}(M) \tag{B.2.30}$$

such that for any $\alpha \in \mathscr{A}^k$ and any $\beta \in \mathscr{A}^{k-1}(M)$

$$\langle d^\dagger \alpha, \beta \rangle = \langle \alpha, d\beta \rangle \,. \tag{B.2.31}$$

We can write down its explicit form in terms of $*$ and d as follows. We have

$$\langle d^\dagger \alpha, \beta \rangle = \langle \beta, d^\dagger \alpha \rangle = \int_M \beta \wedge *d^\dagger \alpha \tag{B.2.32}$$

On the other hand,

$$\langle \alpha, d\beta \rangle = \langle d\beta, \alpha \rangle = \int_M (d\beta) \wedge *\alpha \,. \tag{B.2.33}$$

---

[4]The "formal" in the definition refers to the fact that we have yet to specify the functional domain on which the operators d and $d^\dagger$ are acting. In keeping with our physics presentation, we will steadfastly refuse to delve into domain issues.

But, since

$$d(\beta \wedge *\alpha) = d\beta \wedge *\alpha + (-1)^{k-1}\beta \wedge d * \alpha , \tag{B.2.34}$$

as long as we can drop the total derivative term on the left (this is justified when M is compact without boundary),

$$\langle \alpha, d\beta \rangle = (-1)^k \int_M \beta \wedge d * \alpha , \tag{B.2.35}$$

and comparing (B.2.32) and (B.2.35), we find that

$$*d^\dagger \alpha = (-1)^k d * \alpha . \tag{B.2.36}$$

Finally, using $*^2 \mathscr{A}^{k-1}(M) = (-1)^{(k-1)(n-k+1)}\mathscr{A}^{k-1}(M)$, we obtain that for $\alpha \in \mathscr{A}^k(M)$

$$d^\dagger \alpha = -(-1)^{n(k-1)} * d * . \tag{B.2.37}$$

On an even-dimensional M this simplifies to $d^\dagger \alpha = -*d*$. Note that while $d^\dagger d^\dagger = 0$, $d^\dagger$ is not an anti-derivation with respect to the wedge product!

Once we have the adjoint operator, we define the Hodge-de Rham Laplacian by

$$\Delta_d = d^\dagger d + dd^\dagger . \tag{B.2.38}$$

This second order differential operator depends on the differential structure (i.e. on d) and on the Riemannian metric g. This is a formally self-adjoint positive semi-definite operator that commutes with $*$, d, and $d^\dagger$. Its kernel consists of forms that are closed (i.e. annihilated by d) and co-closed (i.e. annihilated by $d^\dagger$). To see the latter statement, we observe

$$\langle \alpha, \Delta_d \alpha \rangle = \langle d\alpha, d\alpha \rangle + \langle d^\dagger \alpha, d^\dagger \alpha \rangle . \tag{B.2.39}$$

We say that a form is harmonic if and only if it is closed and co-closed. Since Δ_d is linear, the space of harmonic forms of degree k, denoted by $\mathcal{H}^k(M)$, is a vector subspace of $\mathscr{A}^k(M)$. Note that if α is harmonic, then so is $*\alpha$; however, it is in general not true that the wedge product of two harmonic forms is harmonic. Constant functions and their Hodge duals (constant multiples of the volume form) are simple examples of harmonic forms.

With some basic analysis tools for elliptic operators on compact manifolds[163, 210], it is possible to prove that the solutions ω to the Laplace equation with C^∞ source η

$$\Delta_d \omega = \eta \tag{B.2.40}$$

are C^∞ and form a finite-dimensional vector space. In particular, the kernel of Δ_d is finite dimensional, i.e. $\dim \mathcal{H}^k < \infty$, and there is a projector onto harmonic forms Π_d and a Green's operator Σ, such that on every $\mathscr{A}^k(M)$

$$\mathbb{1} = \Pi_d + \Delta_d \Sigma \tag{B.2.41}$$

and there is a unique decomposition of any form into a sum of three orthogonal (with respect to the $\langle \cdot, \cdot \rangle$ inner product) terms: the harmonic piece, the exact piece, and the co-exact piece:

$$\omega = (\Pi_d \omega) + d(d^\dagger \Sigma \omega) + d^\dagger(d \Sigma \omega) . \tag{B.2.42}$$

Consequently the Laplace equation with source has solution if and only if $\Pi_d \eta = 0$, and the solution is unique up to addition of a harmonic form.

The decomposition implies that the degree k harmonic forms are in 1:1 correspondence with elements of the k-th de Rham cohomology group $H_d^k(M)$. Let ω be a representative of a class $[\omega] \in H_d^k(M)$. Since ω is closed it has no co-closed term in its decomposition, and therefore $\omega = \Pi_d \omega + d\eta$. But, then $\Pi_d \omega \in \mathcal{H}^k$, and $[\Pi_d \omega] = [\omega]$. In particular this shows that $H_d^k(M)$ is finite-dimensional.

De Rham's theorem [94] establishes the isomorphism $H_d^k(M) = H^k(M, \mathbb{R})$, where the right-hand-side is the cohomology group of M with real coefficients, a familiar invariant of M from algebraic topology. $H^k(M, \mathbb{R})$ is a topological invariant of M, and its dimension $b_k = \dim H^k(M, \mathbb{R})$ is the k-th Betti number of M. From this we learn the somewhat surprising fact that $\dim \mathcal{H}^k$ is also metric-independent, even though it relied on Δ_d, an operator that depends on the choice of metric! Hodge duality also shows that $b_k = b_{n-k}$.

B.2.3   Riemannian Geometry and Connections on Manifolds

Having fixed a metric g on M, we can also define the Levi-Civita connection that will allow us to differentiate any tensor. We will not bother with the axiomatic definition, as this is probably familiar, and merely state our conventions.

We will occasionally be working with a connection that has torsion, so we will first present some general results before moving on to the torsion–free case. Quite generally, we take ∇ to be a covariant derivative that acts on functions as $\nabla f = df$ and on 1-forms and vector fields as

$$\nabla_\mu \omega_\nu = \partial_\mu \omega_\nu - \Gamma_{\mu\nu}^\rho \omega_\rho , \qquad \nabla_\mu V^\nu = \partial_\mu V^\nu + \Gamma_{\mu\rho}^\nu V_\rho \tag{B.2.43}$$

The action on all other tensors extends by Leibniz rule. The connection Γ may have torsion: the tensor

$$T^{\alpha}_{\mu\nu} = \Gamma^{\alpha}_{\mu\nu} - \Gamma^{\alpha}_{\nu\mu} \tag{B.2.44}$$

need not be zero.

Because partial derivatives commute, the commutator of two covariant derivatives is a first-order differential operator. On functions

$$[\nabla_{\mu}, \nabla_{\nu}]f = -T^{\alpha}_{\mu\nu}\nabla_{\alpha}f \ , \tag{B.2.45}$$

while on more general tensors the action involves the curvature tensor. The form of the latter can be fixed from the action of ∇ on 1-forms, since the action on other tensors will follow from Leibniz rule. On 1-forms we have

$$[\nabla_{\mu}, \nabla_{\nu}]\omega_{\alpha} = (R_{\mu\nu})_{\alpha}{}^{\beta}\omega_{\beta} - T^{\beta}_{\mu\nu}\nabla_{\beta}\omega_{\alpha} \ . \tag{B.2.46}$$

The curvature tensor R has components

$$(R_{\mu\nu})_{\alpha}^{\beta} = \partial_{\nu}\Gamma^{\beta}_{\mu\alpha} - \partial_{\mu}\Gamma^{\beta}_{\nu\alpha} + \Gamma^{\beta}_{\nu\gamma}\Gamma^{\gamma}_{\mu\alpha} - \Gamma^{\beta}_{\mu\gamma}\Gamma^{\gamma}_{\nu\alpha} \tag{B.2.47}$$

and satisfies the Bianchi identity

$$(R_{\mu\nu})_{\alpha}^{\beta} + (R_{\alpha\mu})_{\nu}^{\beta} + (R_{\nu\alpha})_{\mu}^{\beta} = -\nabla_{\mu}T^{\beta}_{\nu\alpha} - \nabla_{\alpha}T^{\beta}_{\mu\nu} - \nabla_{\nu}T^{\beta}_{\alpha\mu} \ . \tag{B.2.48}$$

We often write the components of the curvature tensor as $R_{\mu\nu\alpha}{}^{\beta}$, and we lower the last index with the metric to obtain $R_{\mu\nu\alpha\beta} = R_{\mu\nu\alpha}{}^{\gamma}g_{\gamma\beta}$.

∇ is a metric connection if the metric is covariantly constant: $\nabla g = 0$. The Levi-Civita connection is the unique torsion-free metric connection, and its components are given by

$$\Gamma^{\alpha}_{\mu\nu} = \Gamma^{\alpha}_{\nu\mu} = \frac{1}{2}g^{\alpha\beta}\left(g_{\mu\alpha,\nu} + g_{\nu\alpha,\mu} - g_{\mu\nu,\alpha}\right) \ . \tag{B.2.49}$$

For the Levi-Civita connection the commutator $[\nabla_{\mu}, \nabla_{\nu}]$ is an algebraic operator, and the curvature obeys the well-known identities

$$R_{\mu\nu\alpha\beta} = -R_{\nu\mu\alpha\beta} = R_{\alpha\beta\mu\nu} \ ,$$

$$0 = R_{\mu\nu\alpha\beta} + R_{\alpha\mu\nu\beta} + R_{\nu\alpha\mu\beta} \ ,$$

$$0 = \nabla_{\alpha}R_{\mu\nu\beta\gamma} + \nabla_{\nu}R_{\alpha\mu\beta\gamma} + \nabla_{\mu}R_{\nu\alpha\beta\gamma} \ . \tag{B.2.50}$$

Given any point p on the manifold we can find coordinates such that the Levi-Civita connection vanishes at p; more generally, the torsion tensor T measures the obstruction to finding coordinates that make the connection vanish at a point.

Exercise B.6 Verify the following relationships between the Levi-Civita connection and the Hodge-de Rham Laplacian, where f is a function η is a $p + 1$-form, and ω is a 1-form:

$$(d^\dagger \eta)_{\mu_1 \cdots \mu_p} = -\nabla^{\mu_0} \eta_{\mu_0 \mu_1 \cdots \mu_p} ,$$

$$\Delta_d f = -\nabla^2 f = -g^{\mu\nu} \nabla_\mu \nabla_\nu f ,$$

$$(\Delta_d \omega)_\mu = -\nabla^2 \omega_\mu + [\nabla^\nu, \nabla_\mu] \omega_\nu . \tag{B.2.51}$$

□

Connections on a Vector Bundle

There is no canonical way to differentiate sections of a vector bundle $V \to M$. To define a covariant derivative, we pick a connection: on each open neighborhood U_α we choose a 1-form A_α valued in the Lie algebra $\mathfrak{g}$ of the structure group G_V, such that on overlaps $U_{\alpha\beta}$

$$A_\alpha = \lambda_{\alpha\beta} A_\beta \lambda_{\alpha\beta}^{-1} + (\lambda_{\alpha\beta}) d\lambda_{\alpha\beta}^{-1} . \tag{B.2.52}$$

Concretely, the A_α are 1-forms valued in $k \times k$ matrices that act on the fibers of V, but it is often useful to think of the A_α more intrinsically as valued in the Lie algebra $\mathfrak{g}_V$ of G_V; that way whenever we consider any other associated bundle, e.g. $V \otimes V^*$, then the appropriate connection is constructed by taking the A_α to be 1-forms valued in the appropriate representation of $\mathfrak{g}_V$.

Once A is chosen, the covariant derivative is defined as

$$D : \mathscr{A}^0(M, V) \to \mathscr{A}^1(M, V) ,$$

$$s \mapsto Ds = ds + As . \tag{B.2.53}$$

If we choose a basis for the local sections, then we can write this explicitly as

$$(Ds)^i = ds^i + A^i_j s^j . \tag{B.2.54}$$

By Leibniz rule we can then extend the action of D to $\mathscr{A}^k(M, V) \to \mathscr{A}^{k+1}(M, V)$, and there is a natural action on various tensor bundles constructed from V and V^*; for instance, if in the dual basis we have t_i denoting the components of a section of V^*, then $(Dt)_i = dt_i - A^j_i t_j$; this is natural since it ensures

$$d(t_i s^i) = (Dt)_i s^i + t_i Ds^i . \tag{B.2.55}$$

The difference of two connections A and A' on the same vector bundle satisfies

$$(A - A')_\alpha = \lambda_{\alpha\beta}(A_\beta - A'_\beta)\lambda_{\alpha\beta}^{-1} . \tag{B.2.56}$$

That is, any two connections differ by a section of $T_M^* \otimes \mathfrak{g}_V$, where $\mathfrak{g}_V \to M$ is the adjoint representation vector bundle.

Exercise B.7 The reader may wonder whether connections exist in general. To show that they do, consider a partition of unity ρ_γ subordinate to the cover $\{U_\gamma\}_{\gamma \in A}$ and satisfying the usual properties: the ρ_γ are smooth non-negative functions on M; the support of ρ_γ, that is the closure of the set where $\rho_\gamma \neq 0$, is contained in U_γ; and, finally, $\sum_\gamma \rho_\gamma = 1$.[5]
Now for each U_α set

$$A_\alpha = -\sum_\gamma \rho_\gamma (d\lambda_{\alpha\gamma})\lambda_{\alpha\gamma}^{-1} . \tag{B.2.57}$$

Argue that this is well-defined on each U_α and then use the cocycle condition $\lambda_{\alpha\beta}\lambda_{\beta\gamma}\lambda_{\gamma\alpha} = \mathbb{1}$ to show that (B.2.52) is satisfied.   □

The curvature of the connection, defined by

$$F_\alpha = dA_\alpha + A_\alpha \wedge A_\alpha , \tag{B.2.58}$$

is a section of $\wedge^2 T_M^* \otimes \mathfrak{g}_V$ and is the algebraic operator that measures the failure of D to be nilpotent: $D^2 s = Fs$.

It is appropriate here to add a few words on the meaning of gauge transformations in this global context. As we already described above, a gauge transformation ξ that acts on a vector bundle V should be thought of as a section of the bundle $\text{Aut}(V)$, so that ξ is valued in the representation of the structure group G and satisfies $\xi_\alpha = \lambda_{\alpha\beta}\xi_\beta\lambda_{\alpha\beta}^{-1}$ on $U_{\alpha\beta}$. The gauge transformations act linearly on the sections σ of V, with $\sigma_\alpha^\xi = \xi_\alpha\sigma_\alpha$. We can also define a gauge-transformed connection

$$A^\xi = \xi A\xi^{-1} - d\xi\xi^{-1} , \tag{B.2.59}$$

so that $D^\xi \sigma^\xi = (d\sigma^\xi + A^\xi\sigma^\xi)$ satisfies $(D^\xi\sigma^\xi)_\alpha = \xi_\alpha(D\sigma)_\alpha$.

We can linearize the action of the gauge transformation around the identity by taking $\xi = e^\epsilon$, where ϵ is now valued in $\mathfrak{g}$. This leads to an infinitesimal gauge variation

$$\delta_\epsilon A = -D\epsilon = -(d\epsilon + [A, \epsilon]) , \tag{B.2.60}$$

---

[5]If the reader wonders whether partitions of unity exist, it would be useful to consult an introductory book on differentiable manifolds. The theorem is that a connected differentiable manifold M has a partition of unity if and only if it is Hausdorff and has a countable basis for the open cover.

and

$$\delta_\epsilon F = -[\epsilon, F] \,. \tag{B.2.61}$$

Exercise B.8 Verify the preceding two equations and also show that $DF = dF + [A, F] = 0$. □

In the physics literature the term "gauge transformation" is often used in two senses: not only is it applied to the transformation we just described, but it may also be used to speak of the change in the trivialization that leads to isomorphic bundles. That is, given a collection of maps $\xi_\alpha : U_\alpha \to G$ and a bundle $V \to M$ with transition functions $g_{\alpha\beta}$, there is an isomorphic bundle V' with transition functions $\lambda'_{\alpha\beta} = \xi_\alpha^{-1} \lambda_{\alpha\beta} \xi_\beta$. We will not use this second sense: for us a gauge transformation will always be associated to a global section of Aut(V) or Aut($\mathcal{P}$).

The Frame Bundle and the Spin Connection

It is frequently useful to recast the tensorial manipulations in Riemannian geometry in the language of a vielbein, or a frame. We construct it as follows. We suppose our manifold is orientable, and we choose a cover for our manifold such that in each open set we find a basis for $\mathscr{A}^0(U_\alpha, T_M^*)$—$\{e_1, e_2, \ldots, e_n\}$, such that

$$g = g_{\mu\nu} dx^\mu dx^\nu = \eta_{ab} e^a \otimes e^b = \eta_{ab} e^a_\mu e^b_\nu dx^\mu dx^\nu \,, \tag{B.2.62}$$

where η_{ab} is the standard Euclidean metric in the case of Riemannian manifold and a Minkowski metric for a pseudo-Riemannian manifold. Orientability of M implies that we can order the basis such that the volume form is $\mathrm{dVol}(M) = e^1 \wedge e^2 \wedge \cdots \wedge e^n = \sqrt{\det g} d^n x$.

While the e^a are globally defined as far as diffeomorphisms are concerned, they are not global objects unless the tangent bundle happens to be parallelizable (i.e. trivial). The ordered set $\{e^1, e^2, \ldots, e^n\}$ is a section of the frame bundle associated to T_M^*. More precisely, given a cover of M as above, on non-trivial overlaps $U_{\alpha\beta}$

$$(e^a)_\alpha = (\lambda_{\alpha\beta})^a_b e^b \,, \tag{B.2.63}$$

where $\lambda_{\alpha\beta}$ are valued in SO(n)—the structure group of the frame bundle on an oriented Riemannian manifold.

The spin connection 1-form provides an analogue of the Levi-Civita connection. We define a covariant derivative D for sections of the frame bundle (and its dual) via

$$Ds^a = ds^a + \omega^a{}_b s^b \,, \qquad\qquad Ds_a = ds_a - \omega^b{}_a s_b \,. \tag{B.2.64}$$

This action extends to k-forms valued in the frame bundle by Leibniz. There is a unique 1-form ω^a_b such that

1. the connection is torsion free such that $de^a + \omega^a_b e^b = 0$;
2. the connection is metric: $D\eta_{ab} = 0 \iff \omega^a_b + \omega^b_a = 0$.

The relation between ω and Γ can be established by demanding that the vielbein coefficients are covariantly constant. That is,

$$\nabla_\nu e^a_\mu = e^a_{\mu,\nu} + (\omega_\nu)^a_b e^b_\mu - \Gamma^\rho_{\nu\mu} e^a_\rho = 0 \, . \tag{B.2.65}$$

The curvature of the spin connection is then computed in the standard way

$$R = d\omega + \omega \wedge \omega \, , \tag{B.2.66}$$

and its components are related to the Riemann tensor:

$$R = \frac{1}{2}(R_{\mu\nu})^a_b dx^\mu \wedge dx^\nu = \frac{1}{2}\eta_{bc}e^a_\rho e^c_\sigma R_{\mu\nu\rho\sigma} dx^\mu \wedge dx^\nu \, . \tag{B.2.67}$$

This construction generalizes in a straightforward fashion to any real vector bundle. Namely, for any real vector bundle $E \to M$ of rank r we may choose an orthonormal frame, and therefore reduce the structure group to $O(r)$; the structure group reduces to $SO(r)$ if and only if E is orientable. Similarly, for any complex vector bundle, with fiber $\mathbb{C}^n$, we may choose a unitary frame, and therefore reduce the structure group to $U(n)$ [94].

B.2.4   Two Quantization Results

In this little section we describe two facts that everyone knows, but whose justification is perhaps not so apparent. The first fact is that the curvature of the connection for a rank 2 vector bundle has quantized periods. The second fact is that while the abelian Chern-Simons action is in general gauge-variant, its exponential, with an appropriate coefficient, is well-defined.

The First Chern Class and Sections of a Line Bundle

The presentation given here closely follows chapter 11 of [94].

Let $\pi_v : E \to M$ be a rank 2 vector bundle with structure group $SO(2) = U(1)$ over a compact Riemann surface without boundary. We can think of the fiber as a copy of $\mathbb{C}$, with transition functions on overlaps $U_{\alpha\beta}$ denoted by $\lambda_{\alpha\beta} = e^{i\sigma_{\alpha\beta}}$. We can always take $\sigma_{\alpha\beta} + \sigma_{\beta\alpha} = 0$, and the cocycle condition on $U_{\alpha\beta\gamma}$ is

$$\sigma_{\alpha\beta} + \sigma_{\beta\gamma} + \sigma_{\gamma\alpha} = 2\pi n_{\alpha\beta\gamma} \ , \tag{B.2.68}$$

where $n_{\alpha\beta\gamma} \in \mathbb{Z}$ is completely antisymmetric in the open set indices.

A connection A_α patches on $U_{\alpha\beta}$ according to

$$A_\alpha = A_\beta - id\sigma_{\alpha\beta} \ . \tag{B.2.69}$$

The curvature $F = dA_\alpha$ is then a closed global 2-form on M. Since $H^2(M, \mathbb{Z})$ is torsion-free, the first Chern class of E can be represented by the cohomology class of the curvature: $c_1(E) = [\frac{i}{2\pi}F]$. This class is independent of choice of connection, since a change in the connection shifts F by a global exact form.

Our first goal is to verify directly that $\frac{i}{2\pi}\int_M F \in \mathbb{Z}$. The first step is to introduce the concept of a global angular form associated to a circle bundle $\pi : X \to M$. The circle bundle is constructed with the same transition functions as the ones we used for E: namely, X is a manifold with local charts $U_\alpha \times S^1$, and the circle coordinates ϕ_α patch on overlaps $U_{\alpha\beta}$ according to

$$\phi_\alpha = \phi_\beta + \sigma_{\alpha\beta} \ . \tag{B.2.70}$$

For each patch we define the angular 1-form ψ_α via

$$\psi_\alpha = d\phi_\alpha - i\pi^*(A_\alpha) \ . \tag{B.2.71}$$

Since pull-back commutes with the differential, it follows that

$$d\psi = -i\pi^*(F) \ . \tag{B.2.72}$$

Exercise B.9 Show that $\psi_\alpha = \psi_\beta$ on $U_{\alpha\beta}$, i.e. ψ defines a global 1-form ψ on X.
                                                                                                            □

While a vector bundle $\pi_v : E \to M$ has many sections, it is in general impossible to find sections that are everywhere non-zero. Indeed, as we saw above, a rank r vector bundle is trivial if and only if it admits r linearly independent sections at every point in the base manifold. However, it is always possible to choose a section s that is non-zero on the complement of a finite number of points $\{x_1, \ldots, x_q\} \in M$. It is possible to refine the cover $\{U_\alpha\} \to \{V_\alpha\}$ so that every x_i is contained in an open neighborhood W_i, with $x_j \notin W_i$ unless $j = i$, and each V_α contains at most one neighborhood W_i [94].

With these tools in hand, for each i we choose an open disk D_i around x_i with closure $\overline{D_i} \subset W_i$, and we set $M' = M \setminus \cup_i D_i$. Choose a connection as in (B.2.57) for the cover $\{V_\alpha\}$ and observe that it is non-zero only on double overlaps $V_{\alpha\beta}$; in particular the connection and its curvature F vanish on every D_i, and therefore

$$\int_M F = \int_{M'} F \ . \tag{B.2.73}$$

The section $s : M \to E$ trivializes the bundle $\pi'_v : E \to M'$, and if we locally write $s_\alpha = r_\alpha(x)e^{i\theta_\alpha(x)}$, then the $\theta_\alpha(x)$ define a global section θ for the bundle $\pi : X \to M'$. By definition $\pi\theta : M' \to M'$ is an isomorphism, and therefore, using (B.2.72),

$$\int_{M'} F = \int_{M'} \theta^* \pi^*(F) = i \int_{M'} \theta^*(d\psi) \ . \tag{B.2.74}$$

Now we use Stokes theorem, being careful about orientation of the boundary of M' relative to that of the D_i, to write

$$\int_{M'} \theta^*(d\psi) = -\sum_i \int_{\partial \overline{D}_i} \theta^*(\psi) \ . \tag{B.2.75}$$

Each $\partial \overline{D}_i$ is topologically a circle parametrized by, say, an angle χ, and since the connection vanishes on every W_i, we know that on each patch $\psi_\alpha = d\phi_\alpha$, and therefore its pull-back by the section is simply

$$\theta^*(\psi) = \frac{d\theta}{d\chi} d\chi \ . \tag{B.2.76}$$

Because θ is a global section $\theta : M' \to X$, $e^{i\theta(\chi)}$ is a single-valued function of χ on $\partial \overline{D}_i$, i.e. $\theta = k_i \chi + f_i(\chi)$, where $f_i(\chi)$ is any smooth periodic function and k_i is an integer. So, putting it all together, we obtain

$$\int_M F = -i \sum_i 2\pi \times k_i \ . \tag{B.2.77}$$

The integer k_i defines the local degree of vanishing of the section s at point x_i, which is defined as the degree of the map $s|_{\partial \overline{D}_i} : S^1 \to S^1$. A simple case to keep in mind is if z is a complex coordinate centered on x_i, then setting $s = z^m \overline{z}^n$, we obtain a section with degree $k = m - n$. Precisely when $m = n$ we can make a small deformation to obtain a non-zero section.

This basic result has many important consequences. First, we now see that the first Chern class of any complex line bundle $\mathcal{L} \to Y$ over a compact manifold Y satisfies $[\frac{i}{2\pi} F] \in H^2(Y, \mathbb{Z})$. The reason is that we can pull back the bundle to any 2-cycle $M \subset Y$, and by the presentation we just gave, the integral of $\frac{i}{2\pi} F$ on M must yield an integer.

Second, there is a generalization to any complex vector bundle $E \to Y$ with fiber $\mathbb{C}^n$. There is an associated sphere bundle $X \to Y$ with fiber S^{2n-1} with its angular form. The bundle E then still admits a section that is non-zero away from a finite set of points y_i in Y, and the zeroes then contribute to the Euler class $\chi(E) \in H^{\text{top}}(Y, \mathbb{Z})$ such that

$$\int_Y \chi(E) = \sum_i \deg(s(y_i)) . \tag{B.2.78}$$

The Euler class is simply the top Chern class: $\chi(E) = c_{\text{top}}(E)$. The Euler class is an obstruction to finding a non-zero section, and this obstruction vanishes if Y is odd–dimensional.

One natural bundle we can always consider on a manifold Y is its tangent bundle T_Y. In this case, The Poincaré–Hopf theorem [94] states that $\int_Y \chi(T_Y) = \chi(Y)$, where Y is the topological Euler characteristic $\chi(Y) = \sum_k (-1)^k b_k$.

Abelian Chern-Simons Forms

We mentioned in the text the ambiguity in the integral of a Chern-Simons form over a manifold, and in this section we will make this more precise. We fix notation as in the previous section but now take $\pi : X \to Y$ to be a principal circle bundle over a 3-manifold Y. The Chern-Simons form over an open set U_α is $A_\alpha d A_\alpha$. Picking a partition of unity ρ_α subordinate to the cover, the Chern-Simons action on Y is

$$S[Y, A] = \sum_\alpha \int_{U_\alpha} \rho_\alpha A_\alpha d A_\alpha . \tag{B.2.79}$$

The trouble with this definition is that it is gauge-dependent. However, as we will now show, $\exp\{\frac{i}{2\pi} S[Y, A]\}$ is gauge-invariant because the gauge ambiguity shifts S by an integer multiple of $4\pi^2$.

The demonstration rests on two facts:

1. $\frac{i}{2\pi} F_\alpha = \frac{i}{2\pi} d A_\alpha$ is a global 2-form with integral periods;
2. an abelian gauge transformation, i.e. a map $g : Y \to S^1$ defines a global closed 1-form $\omega_g = g^{-1} dg$, and $\frac{i}{2\pi} \omega_g$ also has integral periods.

The first fact was demonstrated in the previous section. The following exercise establishes the second fact.

Exercise B.10 Parametrize the S^1 in the map $g : Y \to S^1$ by $e^{i\chi}$, so that $\omega_g = i g^*(d\chi)$. The canonical 1-form on S^1 then satisfies $\int_{S^1} d\chi = 2\pi$. Now let $\mathsf{c} : S^1 \to Y$ be any simple curve, so that the composition $g\mathsf{c}$ is a map $g\mathsf{c} : S^1 \to S^1$.

Use continuity to argue that $\int_{S^1} \mathbf{c}^* g^*(d\chi) = 2\pi N_g(\mathbf{c})$, where $N_g(\mathbf{c}) \in \mathbb{Z}$ only depends on the homotopy class $[\mathbf{c}] \in \pi_1(Y)$ and in fact defines a homomorphism $\pi_1(Y) \to \pi_1(S^1) \simeq \mathbb{Z}$. □

With this basic result in hand, we return to the abelian Chern-Simons action. Any two connections A' and A on $\pi : X \to Y$ differ by a global 1 form, i.e. $A'_\alpha - A_\alpha = \xi_\alpha$, with $\xi_\alpha = \xi_\beta$ on $U_{\alpha\beta}$. Thus, the difference in the Chern-Simons actions is

$$S[Y, A'] - S[Y, A] = \sum_\alpha \int_{U_\alpha} \rho_\alpha \left[A d\xi + \xi dA + \xi d\xi \right] . \tag{B.2.80}$$

Now suppose that g is a gauge transformation, so that $A^g = A - g^{-1}dg$. The difference of the connections $\xi = -g^{-1}dg$ is a closed global 1-form, and therefore

$$S[Y, A^g] - S[Y, A] = -\sum_\alpha \int_{U_\alpha} \rho_\alpha \xi dA = -\int_Y \xi F = 4\pi^2 \times \text{integer} .$$

$$\tag{B.2.81}$$

The first equality follows because $\xi \mathcal{F}$ is a global 3-form, and the last equality follows from the integral periods of ξ and $\mathcal{F}$.[6] Our claim now follows.

B.3 Complex Geometry

We begin our discussion of complex geometry with a coarser notion of an almost complex structure. An almost complex structure on M is a smooth map $J : \mathscr{A}^0(T_M) \to \mathscr{A}^0(T_M)$ that satisfies $J^2 = -\mathbb{1}$, or in components, $J_\nu^\mu J_\rho^\nu = -\delta_\rho^\mu$. There are topological obstructions to the existence of an almost complex structure [290]; for instance, the sphere S^4 does not admit an almost complex structure, while S^6 does.

The notion of an almost complex structure is equivalent to that of an almost symplectic structure. We say M is almost symplectic if and only if it admits a non-degenerate two-form $\omega \in \mathscr{A}^2(M)$. Since an odd-dimensional anti-symmetric matrix is necessarily degenerate, an almost symplectic structure can only exist on a manifold of even dimension.

---

[6]Our erudite reader might wonder: if two closed forms have integral periods, does it follow that their wedge product also has integral periods? The answer is yes, and the only subtlety in the story is associated to torsion in homology—a topic we have conscientiously and consciously avoided throughout in these notes. More details may be found in standard references on algebraic topology, such as [94, 223], but the basic point is that the integral period forms are dual to integral homology cycles, and the integrals of the wedge products of the forms are then related to intersection numbers of the homology cycles, which are, of course, integers.

Exercise B.11 Prove that an almost complex structure is equivalent to an almost symplectic structure [290]. First, assume M has an almost complex structure J and show that for any metric g the two-form

$$\omega_{\mu\nu} = \tfrac{1}{2}(J^\rho_\mu g_{\rho\nu} - J^\rho_\nu g_{\rho\mu})$$

is non-degenerate. [Hint: show that $\omega(v, J(v))$ is non-zero if v is non-zero.]

Next, start with an almost symplectic structure ω on M and construct an almost complex structure. This is a little trickier. To achieve the goal recall that if A is an invertible anti-symmetric matrix, then $-A^2 = AA^T$ is a positive definite symmetric matrix with a positive square root, a symmetric matrix B such that $-A^2 = AA^T = B^2$. Show that $[A, B] = 0$ and $[A, B^{-1}] = 0$ [Hint: compute $A[A, B]A$] and conclude that $J = AB^{-1}$ satisfies $J^2 = -\mathbb{1}$. Now apply these linear algebra results to $A^\rho_\mu = \omega_{\mu\lambda}g^{\lambda\rho}$.

The notion of an almost complex structure has a number of refinements. The most basic of these is an almost-Hermitian structure: this is a choice of J and metric g such that $\omega_{\mu\lambda} = J^\rho_\mu g_{\rho\lambda}$ is anti-symmetric. The remaining ones that we will consider are complex, symplectic, and Kähler structures.

Symplectic Manifolds

M admits a symplectic structure if there exists a closed non-degenerate $\omega \in \mathscr{A}^2(M)$. The local structure of symplectic geometry is trivial: the Darboux theorem asserts that for any sufficiently small open neighborhood there exist coordinates $\{q^1, p_1, q^2, p_2, \ldots, q^n, p_n\}$ such that

$$\omega = dq^1 \wedge dp_1 + dq^2 \wedge dp_2 + \cdots + dq^n \wedge dp_n . \tag{B.3.1}$$

More generally, if ω is an almost symplectic form, then the three-form $d\omega$ is the obstruction to having a compatible symplectic structure and the existence of corresponding Darboux coordinates.

A symplectic manifold comes with an orientation and a symplectic volume form

$$\mathrm{dVol}_\omega(M) = \frac{1}{n!}\omega^n . \tag{B.3.2}$$

If M is compact, then the volume form cannot be exact, and therefore ω cannot be exact either; in other words, $[\omega]$ defines a non-zero class in $H^2(M, \mathbb{R})$.

Complex Manifolds

There are several equivalent ways of describing a complex structure. Consider the cotangent bundle T^*_M. Its sections are the 1-forms in $\mathscr{A}^1(M)$, and we can consider these with complex coefficients. The space of all such forms is a complex vector space $\mathscr{A}^1(M, \mathbb{C})$. If M has an almost complex structure J, then we define the projector operators

$$\Pi^{1,0} = \frac{1}{2}(\mathbb{1} - iJ), \qquad \Pi^{0,1} = \frac{1}{2}(\mathbb{1} + iJ) \tag{B.3.3}$$

and therefore decompose the 1-forms as

$$\mathscr{A}^1(M, \mathbb{C}) = \mathscr{A}^{1,0}(M) \oplus \mathscr{A}^{0,1}(M). \tag{B.3.4}$$

In a similar fashion, we define the projectors $\Pi^{p,q}$ on $\mathscr{A}^{p+q}(M, \mathbb{C})$, such that

$$\mathscr{A}^{p,q}(M) = \Pi^{p,q}\mathscr{A}^{p+q}(M, \mathbb{C}) \tag{B.3.5}$$

and

$$\mathscr{A}^k(M, \mathbb{C}) = \bigoplus_{p+q=k} \mathscr{A}^{p,q}(M). \tag{B.3.6}$$

An almost complex structure J is said to be an integrable complex structure (or complex for short) if and only if for any $\omega \in \mathscr{A}^{p,q}(M)$

$$d\omega \in \mathscr{A}^{p+1,q}(M) \oplus \mathscr{A}^{p,q+1}(M). \tag{B.3.7}$$

In this case we can decompose the de Rham differential as $d = \partial + \bar{\partial}$, where

$$\partial : \mathscr{A}^{p,q}(M) \to \mathscr{A}^{p+1,q}(M), \tag{B.3.8}$$

and

$$\bar{\partial} : \mathscr{A}^{p,q}(M) \to \mathscr{A}^{p,q+1}(M). \tag{B.3.9}$$

The operator $\bar{\partial}$ is the Dolbeault operator on M with respect to the complex structure J.

Exercise B.12 If J is an almost complex structure on M, show that it is complex if and only if it has a vanishing Nijenhuis tensor $\mathcal{N}(J) \in \mathscr{A}^2(M, T_M)$ defined by

$$\mathcal{N}_{\nu\rho}^\mu = J_\nu^\alpha(\partial_\alpha J_\rho^\mu - \partial_\rho J_\alpha^\mu) - J_\rho^\alpha(\partial_\alpha J_\nu^\mu - \partial_\nu J_\alpha^\mu). \tag{B.3.10}$$

Show that if $\dim M = 2$ then $\mathcal{N}(J) = 0$ for any almost complex structure. □

An almost complex structure J with $\mathcal{N}(J) = 0$ is said to be an integrable almost complex structure , and a manifold admitting such a J is said to be complex.

Suppose M admits a holomorphic atlas. That is, M has a cover by open neighborhoods $U_\alpha \simeq \mathbb{C}^n$, and holomorphic transition functions. In other words, given any two such neighborhoods U and V with $U \cap V \neq \emptyset$ and coordinates z^a, $a = 1, \ldots, n$ on U and w^a, $a = 1, \ldots, n$ on V, the $z^a(w^1, \ldots, w^n)$ are holomorphic and invertible functions on $U \cap V$. The differential forms in $\mathscr{A}^k(M, \mathbb{C})$ can then be decomposed in terms of the holomorphic and anti-holomorphic differentials dz^a and $d\bar{z}^{\bar{a}}$, and, indeed, M is complex by the definitions given above, since

$$
J = i \frac{\partial}{\partial z^a} \otimes dz^a - i \frac{\partial}{\partial \bar{z}^{\bar{a}}} \otimes d\bar{z}^{\bar{a}} \tag{B.3.11}
$$

is a complex structure.

Exercise B.13 Write $z^a = x^a + iy^a$ and $\bar{z}^{\bar{a}} = x^a - iy^a$ where x^a, y^a are real coordinates. We then have $dz^a = dx^a + idy^a$, and $\frac{\partial}{\partial z^a} = \frac{1}{2}\left(\frac{\partial}{\partial x^a} - i\frac{\partial}{\partial y^a}\right)$, and similarly for $d\bar{z}^{\bar{a}}$ and $\frac{\partial}{\partial \bar{z}^{\bar{a}}}$. Work out the form of J in the real coordinates.   □

A key result—the Newlander–Nirenberg theorem—asserts that every complex manifold M admits a holomorphic atlas; $\mathcal{N}(J)$ is therefore the obstruction to finding a set of holomorphic coordinates in which J takes the form (B.3.11).

Compatibility of Complex and Symplectic Structure
In general it is a difficult problem to show that an almost complex manifold M admits either a complex or a symplectic structure. For instance, S^6 has an almost complex structure but no symplectic structure because $H^2(S^6, \mathbb{R})$ is empty; it is not known whether S^6 admits a complex structure. As another class of examples, it is known that every even-dimensional complex Lie group G, e.g. $G = S^3 \times S^3$, admits a complex structure but has no symplectic structure, again because $H^2(G, \mathbb{R}) = 0$.

An almost Hermitian geometry on M is a choice of an almost complex structure J and a Hermitian metric g. The latter is defined with respect to J by the requirement

$$
g(J\cdot, J\cdot) = g(\cdot, \cdot) . \tag{B.3.12}
$$

Such metrics are easy to construct: let h be any Riemannian metric on M; then

$$
g(\cdot, \cdot) = h(\cdot, \cdot) + h(J\cdot, J\cdot) \tag{B.3.13}
$$

is a Hermitian metric on M. Associated to the pair (J, g) is the Hermitian form $\omega \in \mathscr{A}^2(M, \mathbb{R})$ with components

$$
\omega_{\mu\nu} = J^\lambda_\mu g_{\lambda\nu} . \tag{B.3.14}
$$

Note that if our starting point is an almost complex structure J and an almost symplectic structure ω, then the two may not be compatible with a Hermitian metric:

$\omega_{\mu\nu} J^{\nu}_{\lambda}$ need not be symmetric or positive-definite. An almost Hermitian geometry with almost complex structure J and Hermitian form ω is said to be:

1. almost Kähler if $d\omega = 0$;
2. Hermitian if J is an integrable complex structure;
3. Kähler if J is an integrable complex structure and $d\omega = 0$.

The Hermitian and Kähler geometries are those of most relevance to (0,2) theories.

B.3.1 Differential Forms on Complex Manifolds

Let M be a complex manifold with $\dim_{\mathbb{C}} M = n$, and let z^a, $a = 1, \ldots, n$ denote local holomorphic coordinates. We denote their anti-holomorphic complex conjugates by $\overline{z^a} = \overline{z}^{\overline{a}}$, and we set

$$\partial_a = \frac{\partial}{\partial z^a}, \qquad\qquad \overline{\partial}_a = \frac{\partial}{\partial \overline{z}^a}. \qquad (B.3.15)$$

A (p,q) differential form $\psi \in \mathscr{A}^{p,q}(M, \mathbb{C})$ is then given by (the wedges are understood)

$$\psi = \frac{1}{p!q!} \psi_{a_1 \cdots a_p \overline{b}_1 \cdots \overline{b}_q} dz^{a_1} \cdots dz^{a_p} d\overline{z}^{\overline{b}_1} \cdots d\overline{z}^{\overline{b}_q}. \qquad (B.3.16)$$

The de Rham differential d splits as $d = \partial + \overline{\partial}$, where ∂ and $\overline{\partial}$ satisfy

$$\partial^2 = 0, \qquad\qquad \partial\overline{\partial} + \overline{\partial}\partial = 0, \qquad\qquad \overline{\partial}^2 = 0. \qquad (B.3.17)$$

In local coordinates

$$\partial\psi = \frac{1}{p!q!} \partial_{a_0} \psi_{a_1 \cdots a_p \overline{b}_1 \cdots \overline{b}_q} dz^{a_0} dz^{a_1} \cdots dz^{a_p} d\overline{z}^{\overline{b}_1} \cdots d\overline{z}^{\overline{b}_q},$$

$$\overline{\partial}\psi = \frac{(-1)^p}{p!q!} \overline{\partial}_{\overline{b}_0} \psi_{a_1 \cdots a_p \overline{b}_1 \cdots \overline{b}_q} dz^{a_0} dz^{a_1} \cdots dz^{a_p} d\overline{z}^{\overline{b}_0} \cdots d\overline{z}^{\overline{b}_1} d\overline{z}^{\overline{b}_q}. \qquad (B.3.18)$$

The Dolbeault operator $\overline{\partial}$ will play a particularly important role in the following.

We also define the complex conjugate (q,p) form $\overline{\psi}$ by

$$\overline{\psi} = \frac{1}{q!p!} \overline{\psi_{a_1 \cdots a_p \overline{b}_1 \cdots \overline{b}_q}} d\overline{z}^{\overline{a}_1} \cdots d\overline{z}^{\overline{a}_p} dz^{b_1} \cdots dz^{b_q}$$

$$= \frac{(-1)^{pq}}{q!p!} \overline{\psi_{a_1 \cdots a_p \overline{b}_1 \cdots \overline{b}_q}} dz^{b_1} \cdots dz^{b_q} d\overline{z}^{\overline{a}_1} \cdots d\overline{z}^{\overline{a}_p}. \qquad (B.3.19)$$

Note that complex conjugation defined this way commutes with differentials $\overline{\partial \psi} = \bar{\partial} \, \overline{\psi}$ and $\overline{\bar{\partial} \psi} = \partial \overline{\psi}$, as well as with the wedge product: $\overline{\alpha \wedge \beta} = \bar{\alpha} \wedge \bar{\beta}$. We are following here the convention of [271] that the "bar" can act on the indices of a tensor. This may at first be a bit confusing, but it is a very useful notation. To help clarify the point, the components of $\overline{\psi}$ are

$$\overline{\psi}_{b_1 \cdots b_q \bar{a}_1 \cdots \bar{a}_p} = (-1)^{pq} (\overline{\psi})_{\bar{a}_1 \cdots \bar{a}_p b_1 \cdots b_q} = (-1)^{pq} \overline{\psi_{a_1 \cdots a_p \bar{b}_1 \cdots \bar{b}_q}} \ . \tag{B.3.20}$$

The object in the last equality is just the usual complex conjugate of the particular component of the tensor.

Hermitian Metric and the Hodge Star

The complexified tangent bundle on a complex manifold decomposes into a direct sum of a holomorphic and an anti-holomorphic component:

$$T_M \otimes \mathbb{C} = T_M^{1,0} \oplus T_M^{0,1} \ . \tag{B.3.21}$$

This is an instance of reduction of structure on the manifold: while the tangent bundle of a real manifold of dimension $2n$ has in general transition functions in $GL(2n, \mathbb{R})$, a complex manifold has transition functions in $GL(n, \mathbb{C})$.

For instance, sections of $T_M^{1,0}$ have the local form

$$V = V^a(z, \bar{z}) \frac{\partial}{\partial z^a} \ , \tag{B.3.22}$$

and because M has holomorphic transition functions, this makes sense globally. We define the complex conjugate $\overline{V}$, as section of $T_M^{0,1}$ in the same way as we defined the complex conjugate differential forms above:

$$\overline{V} = \overline{V^a} \frac{\partial}{\partial \bar{z}^a} \ . \tag{B.3.23}$$

A Hermitian metric on M is then a map $g : T_M^{1,0} \otimes T_M^{0,1} \to \mathbb{C}$

$$g = g_{a\bar{b}} dz^a \otimes d\bar{z}^{\bar{b}} \tag{B.3.24}$$

such that $g_{a\bar{b}} = \overline{g_{b\bar{a}}}$ is a positive-definite Hermitian matrix. This means that g gives a Hermitian inner product on $T_M^{1,0}$. For any two sections $V, W \in \mathscr{A}^0(T_M^{1,0})$ we set

$$g(V, W) = V^a g_{a\bar{b}} \overline{W^b} \ , \tag{B.3.25}$$

and with our assumptions this satisfies $\overline{g(V, W)} = g(W, V)$ and $g(V, V) > 0$ for all $V \neq 0$. We will write $g^{\bar{b}a}$ for the components of the inverse metric. The associated Hermitian form is

$$\omega = ig_{a\bar{b}}dz^a \wedge d\bar{z}^{\bar{b}} . \tag{B.3.26}$$

If we write $z^a = x^a + iy^a$ and label the x^a, y^a collectively by X^μ, we obtain the Riemannian metric on T_M and the line element

$$ds^2 = G_{\mu\nu}dX^\mu dX^\nu = g_{a\bar{b}}dz^a d\bar{z}^{\bar{b}} + \text{c.c.} . \tag{B.3.27}$$

With this normalization $g_{a\bar{b}} = \frac{1}{2}\delta_{a\bar{b}}$ leads to the standard Euclidean metric on $\mathbb{C}^n$.

Exercise B.14 Compute the top power of the Hermitian form to obtain the Hermitian volume form

$$\frac{1}{n!}\omega^n = \det g \prod_{k=1}^{n} idz^a \wedge d\bar{z}^a = 2^n(\det g)dx^1 \wedge dy^1 \cdots dx^n \wedge dy^n . \tag{B.3.28}$$

This gives a canonical orientation to M. Check that on $\mathbb{C}^n$ with $g_{a\bar{b}} = \frac{1}{2}\delta_{a\bar{b}}$ we obtain the canonical form for g and ω:

$$ds^2 = (dx^1)^2 + (dy^1)^2 + \cdots + (dx^n)^2 + (dy^n)^2 ,$$
$$\omega = dx^1 \wedge dy^1 + \cdots + dx^n \wedge dy^n . \tag{B.3.29}$$

$\square$

With those preliminaries, we can now define a Hermitian inner product on (p,q) forms. Let $\alpha, \beta \in \mathscr{A}^{p,q}(M, \mathbb{C})$ and set

$$(\alpha, \beta) = \frac{1}{p!q!}\alpha_{a_1 \cdots a_p \bar{b}_1 \cdots \bar{b}_q}\overline{\beta_{c_1 \cdots c_p \bar{d}_1 \cdots \bar{d}_q}}g^{\bar{c}_1 a_1} \cdots g^{\bar{c}_p a_p}g^{\bar{b}_1 d_1} \cdots g^{\bar{b}_q d_q} . \tag{B.3.30}$$

Clearly $\overline{(\alpha, \beta)} = (\beta, \alpha)$, and $(\cdot, \cdot)$ is positive-definite point-wise on M.
The Hodge star is taken to be the unique map

$$* : \mathscr{A}^{p,q}(M) \to \mathscr{A}^{n-q,n-p}(M) \tag{B.3.31}$$

such that for any $\alpha, \beta \in \mathscr{A}^{p,q}(M)$

$$\alpha \wedge *\bar{\beta} = (\alpha, \beta)\frac{1}{n!}\omega^n . \tag{B.3.32}$$

Exercise B.15 Show that $*$ on $\mathscr{A}^{p,q}$ is given by

$$*\psi = \frac{i^n(-1)^{pn+n(n-1)/2}\det g}{p!q!(n-p)!(n-q)!}$$
$$\psi^{\overline{b}_1\cdots\overline{b}_p a_1\cdots a_q}\epsilon_{a_1\cdots a_n}\epsilon_{\overline{b}_1\cdots\overline{b}_n}dz^{a_{q+1}}\cdots dz^{a_n}d\overline{z}^{\overline{b}_{p+1}}\cdots d\overline{z}^{\overline{b}_n} \, , \quad\quad \text{(B.3.33)}$$

and verify

$$*^2\mathscr{A}^{p,q}(M) = (-1)^{p+q}\mathscr{A}^{p,q}(M) \, , \quad\quad \text{(B.3.34)}$$

as well as

$$*\overline{\psi} = \overline{*\psi} \, . \quad\quad \text{(B.3.35)}$$

$\square$

With this exercise in hand, we define the Hermitian inner product on (p,q) forms as

$$\langle \alpha, \beta \rangle = \int_M (\alpha, \beta)\frac{\omega^n}{n!} = \int_M \alpha \wedge *\overline{\beta} \, . \quad\quad \text{(B.3.36)}$$

We could now proceed to define a refinement of de Rham cohomology that follows from the decomposition $d = \partial + \overline{\partial}$, but we will first pause to discuss a few notions of holomorphic vector bundles. These will allow us to build a much more general framework for this refined structure, known as Dolbeault cohomology.

B.3.2 Holomorphic Vector Bundles

A holomorphic vector bundle over a complex manifold M is simply a vector bundle that respects the underlying complex structure of M: the fibers are isomorphic to $\mathbb{C}^k$, where k is the rank of the bundle, and the transition functions are holomorphic functions of the coordinates. The first condition means the bundle is complex and has well-defined Chern classes; the second is much stronger and implies that the total space of the bundle is itself a complex manifold, with the complex structure respecting that of the base manifold. A holomorphic bundle $\mathcal{E} \to M$ has a canonical holomorphic dual bundle $\mathcal{E}^* \to M$, as well as a complex conjugate bundle $\overline{\mathcal{E}} \to M$; the latter is of course anti-holomorphic.

The tangent bundle $T_M^{1,0}$, its dual $(T_M^{1,0})^*$, or their tensor powers are all examples of holomorphic bundles. We will pare down the notation a bit and abbreviate $T_M^{1,0}$ by T_M, and similarly, T_M^* will denote the dual bundle $(T_M^{1,0})^*$; there will be a similar story on the anti-holomorphic side: $\overline{T}_M = T_M^{0,1}$, and $\overline{T}_M^* = (T_M^{0,1})^*$.

A line bundle is a holomorphic vector bundle of rank 1. Every complex manifold is equipped with a canonical line bundle given by the top exterior power of T_M^*; it is frequently denoted by

$$K_M = \wedge^n T_M^* \, . \tag{B.3.37}$$

For any holomorphic bundle $\mathcal{E} \to M$ we can consider the bundle of $\mathcal{E}$-valued (p,q) differential forms, $\mathcal{E} \otimes \wedge^p T_M^* \otimes \wedge^q \overline{T}_M^*$, and we denote the smooth sections of such a bundle by $\mathscr{A}^{p,q}(M, \mathcal{E})$. Clearly

$$\mathscr{A}^{p,q}(M, \mathcal{E}) = \mathscr{A}^{0,q}(M, \wedge^p (T_M^{1,0})^* \otimes \mathcal{E}) \, . \tag{B.3.38}$$

The preceding discussion of (p,q) differential forms is then just a special case where we take $\mathcal{E}$ to be the trivial bundle over M. We will denote this by $\mathcal{O}$, and with a slight abuse of notation write $\mathscr{A}^{p,q}(M) = \mathscr{A}^{p,q}(M, \mathcal{O})$.

On a general real vector bundle there is no canonical choice of connection and therefore no canonical differential operator. This is not the case for holomorphic bundles, which come with a canonical differential operator $\bar{\partial}$. Let η_A, $A = 1, \ldots, k$ be a holomorphic frame for the bundle $\mathcal{E}$: that is, on each patch the η_A form a basis for $\mathbb{C}^k$, and the η_A have holomorphic transition functions in $\mathrm{GL}(k, \mathbb{C})$ on overlaps. We then define $\bar{\partial}$ as follows:

$$\bar{\partial} : \mathscr{A}^{0,0}(M, \mathcal{E}) \to \mathscr{A}^{0,1}(M, \mathcal{E})$$
$$\bar{\partial} : s = s^A \eta_A \mapsto \bar{\partial} s = d\bar{z}^{\bar{a}} (\bar{\partial}_{\bar{a}} s^A) \otimes \eta_A \, . \tag{B.3.39}$$

We can extend the action to $\mathscr{A}^{0,q}(M, \mathcal{E})$ by enforcing the Leibniz rule. If

$$s = \frac{1}{k!} s^A_{\bar{a}_1 \cdots \bar{a}_k} d\bar{z}^{\bar{a}_1} \wedge \cdots \wedge d\bar{z}^{\bar{a}_k} \otimes \eta_A \in \mathscr{A}^{0,k}(M, \mathcal{E}) \, , \tag{B.3.40}$$

then

$$\bar{\partial} s = \frac{1}{k!} \bar{\partial}_{\bar{a}_0} s^A_{\bar{a}_1 \cdots \bar{a}_k} d\bar{z}^{\bar{a}_0} \wedge d\bar{z}^{\bar{a}_1} \wedge \cdots \wedge d\bar{z}^{\bar{a}_k} \otimes \eta_A \in \mathscr{A}^{0,k+1}(M, \mathcal{E}) \, . \tag{B.3.41}$$

In the same fashion the holomorphic differential ∂ has a canonical action on sections of the anti-holomorphic bundle $\overline{\mathcal{E}}$.

Suppose that the holomorphic bundle $\mathcal{E}$ is equipped with a Hermitian metric. In other words, given $\mathcal{E}$, with its complex conjugate bundle denoted by $\overline{\mathcal{E}}$, we have h a section of $\mathcal{E}^* \otimes \overline{\mathcal{E}}^*$, such that for any two sections $s_1, s_2 \in \Gamma(\mathcal{E})$ we have

$$h(s_1, s_2) = \overline{h(s_2, s_1)} = h_{A\bar{B}} s_1^A \overline{s_2^B} \, , \tag{B.3.42}$$

and $h_{A\bar{B}}$ is a positive-definite Hermitian matrix. Much like Riemannian metrics on manifolds, Hermitian metrics always exist on holomorphic bundles. This means that we are always free to work with a unitary frame for the basis of sections, where in each patch the metric is simply identity, and therefore the transition functions are valued in $U(n)$. As the following exercise illustrates, a holomorphic vector bundle of rank n is, in particular, a real vector bundle of rank $2n$.

Exercise B.16 Start with a little linear algebra: prove that any unitary matrix U may be written as $U = A + iB$, where A and B are commuting real matrices that satisfy $AA^T + BB^T = \mathbb{1}$, $A^T A + B^T B = \mathbb{1}$, $A^T B = B^T A$, and $AB^T = BA^T$. Use that to show that the block matrix

$$M = \begin{pmatrix} A & B \\ -B & A \end{pmatrix} \in SO(2n, \mathbb{R}) .$$
(B.3.43)

Now apply this to the transition functions of $\mathcal{E}$ to demonstrate that $\mathcal{E}$ is isomorphic to E, a rank $2n$ vector bundle with structure group $U(n) \subset SO(2n)$ given by the explicit embedding in the previous equation.   □

We extend the definition of Hodge star and the inner product defined in (B.3.36) to sections in $\mathscr{A}^{p,q}(\mathcal{E})$:

$$\langle s_1, s_2 \rangle = \int_M s_1 \wedge *_{\mathcal{E}} \overline{s_2} = \int_M (s_1, s_2) \frac{\omega^n}{n!} ,$$
(B.3.44)

where

$$(s_1, s_2) = \frac{1}{p!q!} (s_1)^A_{a_1 \cdots a_p \bar{b}_1 \cdots \bar{b}_q} \overline{(s_2)^B_{c_1 \cdots c_p \bar{d}_1 \cdots \bar{d}_q}} g^{\bar{c}_1 a_1} \cdots g^{\bar{c}_p a_p} g^{\bar{b}_1 d_1} \cdots g^{\bar{b}_q d_q} h_{A\bar{B}} .$$
(B.3.45)

Just as in our discussion of (p,q) forms, it is useful to think of the map $s \rightarrow *_{\mathcal{E}} \bar{s}$ as a composition of two isomorphisms:

$$\mathscr{A}^{p,q}(M, \mathcal{E}) \simeq \mathscr{A}^{q,p}(M, \overline{\mathcal{E}}) \simeq \mathscr{A}^{n-p,n-q}(M, \mathcal{E}^*) ,$$
(B.3.46)

where the first isomorphism is obtained by complex conjugation, while the second one is the Hodge star $*_{\mathcal{E}}$:

$$*_{\mathcal{E}} : \mathscr{A}^{p,q}(M, \mathcal{E}) \rightarrow \mathscr{A}^{n-q,n-p}(M, \overline{\mathcal{E}}^*) .$$
(B.3.47)

The wedge operator in (B.3.44) should be understood to include the dual pairing between $\mathcal{E}$ and $\mathcal{E}^*$, i.e if $s_1 \in \mathscr{A}^{p_1,q_1}(M, \mathcal{E})$ and $s_2 \in \mathscr{A}^{p_2,q_2}(M, \mathcal{E}^*)$, then we set

$$s_1 \wedge s_2 = s_1^A \wedge s_{2A} .$$
(B.3.48)

We can now construct the formal adjoint of the Dolbeault operator. Demanding that for all $s \in \mathscr{A}^{p,q-1}(M, \mathcal{E})$ and $t \in \mathscr{A}^{p,q}(M, \mathcal{E})$

$$\langle \bar{\partial} s, t \rangle = \langle s, \bar{\partial}^\dagger t \rangle , \tag{B.3.49}$$

we find

$$\bar{\partial}^\dagger = - *_{\mathcal{E}} \, \partial \, *_{\mathcal{E}} . \tag{B.3.50}$$

Exercise B.17 Check the form of $\bar{\partial}^\dagger$. □

In a completely analogous fashion to the Hodge-de Rham discussion, the $\bar{\partial}$-Laplace operator

$$\Delta_{\bar{\partial}} = \bar{\partial} \bar{\partial}^\dagger + \bar{\partial}^\dagger \bar{\partial} , \tag{B.3.51}$$

is formally self-adjoint on $\mathscr{A}^{p,q}(M, \mathcal{E})$ and commutes with $\bar{\partial}$ and $\bar{\partial}^\dagger$. Moreover, we have the relation

$$\Delta_{\bar{\partial}}(\overline{*_{\mathcal{E}}s}) = \overline{*_{\mathcal{E}} \Delta_{\bar{\partial}} s} . \tag{B.3.52}$$

On a compact manifold $\Delta_{\bar{\partial}}$ is elliptic, and has a finite-dimensional kernel [210]. Letting $\Pi^{p,q}_{\bar{\partial}}(\mathcal{E})$ denote the projector onto the space of zero modes of $\Delta_{\bar{\partial}}$ on $\mathscr{A}^{p,q}(M, \mathcal{E})$—we denote this subspace by $\mathcal{H}^{p,q}_{\bar{\partial}}(\mathcal{E})$—there is a Green's operator $\Sigma_{\bar{\partial}}$ such that

$$\mathbb{1} = \Pi^{p,q}_{\bar{\partial}} + \Delta_{\bar{\partial}} \Sigma_{\bar{\partial}} , \tag{B.3.53}$$

and therefore there is a decomposition of $\mathscr{A}^{p,q}(M, \mathcal{E})$ into three orthogonal terms: the harmonic forms, the exact forms, and the co-exact forms:

$$s = \Pi^{p,q}_{\bar{\partial}} s + \bar{\partial}(\bar{\partial}^\dagger \Sigma_{\bar{\partial}} s) + \bar{\partial}^\dagger(\bar{\partial} \Sigma_{\bar{\partial}} s) . \tag{B.3.54}$$

All of this is entirely analogous to the Hodge decomposition on a real manifold discussed above, and there is also a cohomological parallel given by Dolbeault cohomology.

The Dolbeault cohomology groups are defined by

$$H^{p,q}_{\bar{\partial}}(M, \mathcal{E}) = \frac{\ker \bar{\partial} \cap \mathscr{A}^{p,q}(M, \mathcal{E})}{\operatorname{im} \bar{\partial} \cap \mathscr{A}^{p,q}(M, \mathcal{E})} , \tag{B.3.55}$$

and by the previous results we have the isomorphism $H^{p,q}_{\bar{\partial}}(M, \mathcal{E}) = \mathcal{H}^{p,q}_{\bar{\partial}}(\mathcal{E})$. The relation (B.3.52) implies the Kodaira-Serre duality:

$$H^{p,q}_{\bar{\partial}}(M, \mathcal{E}) \simeq \overline{H^{n-p,n-q}_{\bar{\partial}}(M, \mathcal{E}^*)} . \tag{B.3.56}$$

There is an important difference between de Rham cohomology and Dolbeault cohomology. The former only depends on the differential structure on M and is a topological invariant of the manifold; the latter requires a choice of complex structure to define $\bar{\partial}$, and in general the Dolbeault cohomology groups change as the complex structure is varied. Thus, unlike the Betti numbers b_k, the dimensions

$$h^{p,q}(M, \mathcal{E}) = \dim_{\mathbb{C}} H^{p,q}_{\bar{\partial}}(M, \mathcal{E}) \tag{B.3.57}$$

vary with complex structure of M, but they do so in a way consistent with Kodaira-Serre:

$$h^{p,q}(M, \mathcal{E}) = h^{n-p, n-q}(M, \mathcal{E}^*) . \tag{B.3.58}$$

Čech Cohomology and Vector Bundles

A holomorphic vector bundle $\mathcal{E} \to M$ is an example of a more general notion—a sheaf on M. In this section we briefly quote from [210] some standard ideas that turn out to be of use in (0,2) explorations. For a relatively gentle introduction to some of the more advanced ideas, such as the precise characterization of which sheaves are equivalent to vector bundles and various cohomological notions, the physicist reader can consult [30]; a thorough readable treatment is in [373].

A sheaf $\mathcal{F}$ on M is an assignment to each open set $U \subset M$ of a group (usually additive) of sections $\mathcal{F}(U)$ and for reach pair $U \subset V$ a restriction map $\mathcal{F}(V) \to \mathcal{F}(U)$ that can be composed in the obvious fashion. In addition, the sections satisfy two key properties: (1) given sections $\sigma \subset \mathcal{F}(U)$ and $\tau \in \mathcal{F}(V)$ that agree on $U \cap V$, there exists a section $\rho \in \mathcal{F}(U \cup V)$ that restricts to σ on U and to τ on V; (2) a section $\sigma \in \mathcal{F}(U \cup V)$ that restricts to zero on U and on V is necessarily zero.

There are several cohomology theories that can be defined for a sheaf $\mathcal{F}$ on M. One of these is Čech cohomology. To describe it, we work with our good cover $U = \{U_\alpha\}_{\alpha \in A}$ and define

$$C^0(U, \mathcal{F}) = \prod_\alpha \mathcal{F}(U_\alpha) , \qquad\qquad C^1(U, \mathcal{F}) = \prod_{\alpha\beta} \mathcal{F}(U_{\alpha\beta}) ,$$

$$C^2(U, \mathcal{F}) = \prod_{\alpha\beta\gamma} \mathcal{F}(U_{\alpha\beta\gamma}) , \qquad \text{etc.} \tag{B.3.59}$$

A p-cochain $\sigma \in C^p(U, \mathcal{F})$ is then a collection of sections $\sigma_I \in \mathcal{F}(V_I)$, where I runs over the open sets that appear in C^p. The coboundary operator $\delta : C^p(U, \mathcal{F}) \to C^{p+1}(U, \mathcal{F})$ is a signed linear combination of restrictions, defined so that $\delta^2 = 0$. To get an idea of this, we quote δ for $\sigma \in C^0$ and $\tau \in C^1$:

$$\sigma = \{\sigma_\alpha\}_{\alpha \in A} , \qquad\qquad\qquad \tau = \{\tau_{\alpha\beta}\}_{\alpha, \beta \in A} ,$$

$$(\delta\sigma)_{\alpha\beta} = -\sigma_\alpha|_{U_{\alpha\beta}} + \sigma_\beta|_{U_{\alpha\beta}} ,$$

$$(\delta\tau)_{\alpha\beta\gamma} = \tau_{\alpha\beta}|_{U_{\alpha\beta\gamma}} + \tau_{\beta\gamma}|_{U_{\alpha\beta\gamma}} - \tau_{\alpha\gamma}|_{U_{\alpha\beta\gamma}} . \tag{B.3.60}$$

Since $\delta^2 = 0$, it is sensible to take cohomology of the complex

$$0 \longrightarrow C^0 \xrightarrow{\ \delta\ } C^1 \xrightarrow{\ \delta\ } C^2 \xrightarrow{\ \delta\ } \cdots \tag{B.3.61}$$

$$H^p(U, \mathcal{F}) = \frac{\ker \delta \cap C^p(U, \mathcal{F})}{\operatorname{im} \delta \cap C^p(U, \mathcal{F})} . \tag{B.3.62}$$

In particular, we see an immediate interpretation of $H^0(U, \mathcal{F})$—this is the set of global sections of the sheaf $\mathcal{F}$ on M. The higher cohomology groups do not have such a simple interpretation and may depend on the choice of the over U. It is possible to sufficiently refine the cover, i.e. to take the open sets to be smaller and smaller, so that the groups become independent of the cover. The limit defines the Čech cohomology groups $\check{H}^p(M, \mathcal{F})$. An important fact is that any good cover is "good enough," in the sense that its Čech cohomology is isomorphic to $\check{H}^p(M, \mathcal{F})$ [94]. When M is compact, so that the good cover may be taken to be finite, this immediately implies that the cohomology groups $\check{H}^p(M, \mathcal{F})$ are finite.

Exercise B.18 The transition functions of a line bundle define a Čech cohomology class. Taking a look back to (B.2.68), the transition functions satisfy

$$\sigma_{\alpha\beta} + \sigma_{\beta\gamma} + \sigma_{\gamma\alpha} = 2\pi n_{\alpha\beta\gamma} \tag{B.3.63}$$

on every $U_{\alpha\beta\gamma}$, where $n_{\alpha\beta\gamma} \in \mathbb{Z}$. Argue that $n = \{n_{\alpha\beta\gamma}\} \in C^2(U, \mathbb{Z})$, and $\delta n = 0$. Thus, $[n] \in \check{H}^2(M, \mathbb{Z})$. This is precisely the first Chern class of the bundle encoded via Čech cohomology.   □

Let us now return to the case of a holomorphic bundle $\mathcal{E} \to M$. There are two sheaves that are naturally associated to this bundle. First, we might take $\mathscr{A}^{p,q}(\mathcal{E})$, i.e. the sheaf of C^∞ (p,q)-forms valued in $\mathcal{E}$. Quite generally, this sheaf has a very boring Čech cohomology theory: $\check{H}^k(M, \mathscr{A}^{p,q}(\mathcal{E})) = 0$ for $k > 0$.[7] On the other hand, since $\mathcal{E}$ is holomorphic, we can also consider the sheaf of holomorphic p-forms valued in $\mathcal{E}$ and often denoted by $\Omega^p(\mathcal{E})$. The sheaf of holomorphic sections of $\mathcal{E}$ is often denoted not by $\Omega^0(\mathcal{E})$ but rather by $\mathcal{O}(\mathcal{E})$.

---

[7]This is due to the existence of partitions of unity for a C^∞ sheaf [210].

The Čech cohomology groups for sheaves of holomorphic sections are isomorphic to the Dolbeault cohomology of the corresponding bundles:

$$H_{\bar\partial}^{p,q}(M,\mathcal{E}) = H_{\bar\partial}^{0,q}(M,\mathcal{E}\otimes\wedge^p T_M^*) = \check{H}^q(M,\Omega^p(\mathcal{E}))$$

$$= \check{H}^q(M,\mathcal{O}(\wedge^p T_M^*\otimes\mathcal{E}))\,. \tag{B.3.64}$$

This isomorphism leads to some natural and useful abuse of notation. For instance, it is not unusual to write $H^q(M,\wedge^p T_M^*\otimes\mathcal{E})$ for $\check{H}^q(M,\Omega^p(\mathcal{E}))$.

We will denote the dimensions of these cohomology groups by

$$h^q(M,\mathcal{E}) = \dim_{\mathbb{C}}\check{H}^q(M,\mathcal{O}(\mathcal{E}))\,, \qquad h^{p,q}(M,\mathcal{E}) = \dim_{\mathbb{C}} H_{\bar\partial}^{p,q}(M,\mathcal{E})\,. \tag{B.3.65}$$

Exercise B.19 To gain some practice with the notation, show that the Kodaira-Serre isomorphism implies

$$h^q(M,\mathcal{E}) = h^{n-q}(M,\mathcal{E}^*\otimes K_M)\,. \tag{B.3.66}$$

□

Local Geometry and the Chern Connection

As on any vector bundle, we would like to define a connection on a holomorphic bundle $\mathcal{E}\to M$ and a corresponding differential operator

$$D:\mathscr{A}^p(\mathcal{E})\to\mathscr{A}^{p+1}(\mathcal{E})\,, \tag{B.3.67}$$

which we can write as $D = D' + D''$, where

$$D':\mathscr{A}^{p,q}(\mathcal{E})\to\mathscr{A}^{p+1,q}(\mathcal{E})\,, \qquad D'':\mathscr{A}^{p,q}(\mathcal{E})\to\mathscr{A}^{p,q+1}(\mathcal{E})\,. \tag{B.3.68}$$

While we have the natural choice $D'' = \bar\partial$, there is no correspondingly canonical choice for D'. However, if we equip the bundle with a Hermitian metric, then there is a unique connection that is compatible with the complex structure ($D'' = \bar\partial$) and preserves the metric. This is the Chern connection, which we now describe in detail.

Let $\mathcal{E}\to M$ be a rank k holomorphic vector bundle equipped with a hermitian metric with explicit components $h_{A\bar{B}}$. The Chern connection has components $\mathcal{A}_B^A = h^{\bar{A}A}\partial h_{B\bar{A}}$, so that the covariant derivative of a section is

$$(D's)^A = \partial s^A + \mathcal{A}_B^A s^B\,, \qquad (D''s)^A = \bar\partial s^A\,. \tag{B.3.69}$$

As in our general vector bundle discussion, these equations should be understood as holding over the patches U_α and related by transition functions on $U_{\alpha\beta}$.

Exercise B.20 Show that the $(2,0)$ component of $D^2 s$ vanishes for the Chern connection. That is, $D'^2 s = (\partial \mathcal{A} + \mathcal{A} \wedge \mathcal{A})s = 0$. Furthermore, $D^2 s = \mathcal{F} s$, where $\mathcal{F} = \bar{\partial} \mathcal{A} \in \mathscr{A}^{1,1}(M, \mathcal{E} \otimes \mathcal{E}^*)$.

Next, show that the connection is metric compatible: if s, t are two sections of $\mathcal{E}$, then

$$d(h_{A\bar{B}} s^A \bar{t}^B) = h_{A\bar{B}}(Ds)^A \bar{t}^B + h_{A\bar{B}} s^A \overline{(Dt)^B} . \tag{B.3.70}$$

Finally, show that metric compatibility and $D'' = \bar{\partial}$ uniquely determine the connection coefficients in terms of the metric h. □

An important special case of this construction is provided by $\mathcal{E} = T_M$ equipped with Hermitian metric $g = g_{a\bar{b}} dz^a d\bar{z}^b$ and Chern connection $\mathcal{A}_b^a = g^{\bar{b}a} \partial g_{a\bar{b}}$. While this connection is compatible with the metric and complex structure, it is not in general torsion-free. The reader can verify that the torsion of the connection vanishes if and only if $d\omega = 0$.

A Few Remarks on Variations of Complex Structure

Suppose M is a complex manifold. In physical applications we are often interested in the dependence of various quantities on the complex structure of M. What does it mean to vary the complex structure of a manifold? In general this is a fairly complicated idea, so it is good to keep in mind some basic examples. Perhaps the simplest non-trivial example is an elliptic curve M_τ defined as the quotient of the plane:

$$M_\tau = \mathbb{C}/\Lambda_\tau , \tag{B.3.71}$$

where $\Lambda_\tau = \langle 1, \tau \rangle$ is a lattice generated by 1 and τ, i.e. consists of all points $m + n\tau$, where m, n are integers. The complex parameter τ is defined to be in the upper-half-plane, and locally it parametrizes deformations of the torus complex structure. Globally there are additional identifications: $M_\tau \simeq M_{\tau'}$ as complex manifolds if and only if τ and τ' are related by an SL$(2, \mathbb{Z})$ transformation:

$$\tau' = \frac{a + b\tau}{c + d\tau} , \qquad \begin{pmatrix} a & b \\ c & d \end{pmatrix} \in \text{SL}(2, \mathbb{Z}) . \tag{B.3.72}$$

This example is presumably very familiar to any string theorist.

Another basic example is a hypersurface in a projective space. Let $\mathbb{P}^n$ be a projective space with projective coordinates $[z_0, \ldots, z_n]$. The vanishing locus of a generic homogeneous polynomial of degree d will define a compact complex (in fact Kähler) hypersurface M in $\mathbb{P}^n$:

$$f(z_0, \ldots, z_n) = z_0^d + z_1^2 + \cdots z_n^d + \text{other degree } d \text{ monomials} . \tag{B.3.73}$$

The coefficients of the monomials then parameterize some of the deformations of complex structure of M.

Fundamental work by Kodaira, Spenser, and others showed how to define these notions intrinsically in terms of the geometry of M [271]. Infinitesimally, the deformations of complex structure are characterized by the cohomology group $H^1(M, T_M)$. These may be obstructed, and the obstruction theory is governed by maps $H^1(M, T_M) \rightarrow H^2(M, T_M)$.[8] In particular, when $H^2(M, T_M) = 0$, there are no obstructions, but there are also situations, say for Calabi-Yau manifolds, where although $H^2(M, T_M) \neq 0$, the first order deformations are nevertheless unobstructed, as we discuss a bit further in the text.

The key idea to describe the deformations is to construct a holomorphic fibration $\pi_1 : \mathcal{M} \rightarrow B$, where $\mathcal{M}$ is a complex manifold, and for a point $0 \in B$ the fiber $\pi_1^{-1}(0)$ is the manifold M. Motion on B then describes deformations of complex structure of M, i.e. at least for some open neighborhood $\mathcal{U}$ of $0 \in B$, for every point $b \in \mathcal{U}$ we have $\pi^{-1}(b) = M_b$, where M_b is a complex manifold diffeomorphic to M_0. This intrinsic point of view is conceptually very useful, but it is computationally challenging.

Given a holomorphic vector bundle $\pi_0 : \mathcal{E}_0 \rightarrow M_0$, not every deformation of complex structure of M lifts to a deformation of complex structure of $\mathcal{E}$; the obstruction is measured by the Atiyah class discussed in further detail in the text. Suppose, on the other hand, that we can build a holomorphic vector bundle $\pi_2 : \mathcal{F} \rightarrow \mathcal{M}$, such that $\pi_3 = \pi_1 \pi_2$ satisfies the property that $\pi_3^{-1} = \mathcal{E}_b$, where $\pi_b : \mathcal{E}_b \rightarrow M_b$ is a holomorphic vector bundle, i.e. $\pi_b \pi_3^{-1}(b) = \pi_1^{-1}(b) = M_b$. This describes a family of holomorphic vector bundles that change with complex structure on M.

We now have a very important result that follows from the Čech-Dolbeault isomorphism: because the Dolbeault Laplace operator is semi-positive-definite, the number of its zero modes can only decrease when we move a small distance from the original point. Therefore, the Hodge numbers obey an upper-semi-continuity relation: for any b in some open neighborhood of $b = 0$ we have

$$h^{p,q}(M_0, \mathcal{E}_0) \geq h^{p,q}(M_b, \mathcal{E}_b) . \tag{B.3.74}$$

This is very reminiscent of the possible jumping in the chiral rings of a (0,2) SCFT discussed in the text; in each case the underlying reasoning is no different from that in perturbation theory in basic quantum mechanics: the degeneracy of the ground state cannot increase under an arbitrarily small positive-semi-definite deformation of the Hamiltonian.

---

[8]Compare this to the discussion of the Atiyah class in the text.

B.3.3   Hodge and Betti Numbers and the $\partial\bar{\partial}$-Lemma

We now return to the simple choice where we set $\mathcal{E} = \mathcal{O}$, the trivial bundle. In that case we speak of the Hodge numbers of M as $h^{p,q}(M) = h^{p,q}(M, \mathcal{O})$. In general there is no simple relation between these dimensions and the Betti numbers b_k. On an arbitrary compact complex manifold M the best one has is the Frölicher inequality

$$b_k \le \sum_{p+q=k} h^{p,q}(M) \,. \tag{B.3.75}$$

The inequality is saturated if M satisfies the $\partial\bar{\partial}$-lemma, i.e. that on M

$$\ker \partial \cap \ker \bar{\partial} \cap \operatorname{im} d = \ker \partial \cap \ker \bar{\partial} \cap \operatorname{im} \partial \oplus \ker \partial \cap \ker \bar{\partial} \cap \operatorname{im} \bar{\partial} = \operatorname{im} \partial\bar{\partial} \,. \tag{B.3.76}$$

A very readable discussion of the $\partial\bar{\partial}$-lemma, including examples, equivalent characterizations, and related cohomology theories can be found in [20]. Here we will just quote some of the results. First, it may perhaps be helpful to restate the lemma as follows. Let $\gamma \in \mathscr{A}^{p,q}(M)$ be a differential form that satisfies $\partial\gamma = 0$ or $\bar{\partial}\gamma = 0$, and suppose γ is d-exact, i.e. $\gamma = d\eta$ for some $\eta \in \mathscr{A}^{p+q-1}(M)$. If the $\partial\bar{\partial}$-lemma holds, then there exists $\lambda \in \mathscr{A}^{p-1,q-1}(M)$ such that $\gamma = \partial\bar{\partial}\lambda$.

For our applications it is useful to keep in mind three basic facts.

1. The $\partial\bar{\partial}$-lemma is stable under deformations of complex structure. In the language introduced above, if $b \in B$ parameterizes a family of complex manifolds M_b, and M_0 satisfies the $\partial\bar{\partial}$-lemma, then so does M_b for any b in a sufficiently small open neighborhood around $b = 0$ [373].
2. The $\partial\bar{\partial}$-lemma implies that $h^{p,q}(M)$ are locally constant because their sum satisfies $b_k = \sum_{p+q=k} h^{p,q}(M)$, and each summand obeys (B.3.74).
3. A compact Kähler manifold is a prototypical example of a manifold satisfying the $\partial\bar{\partial}$-lemma.

B.3.4   Kähler Geometry

There are several equivalent definitions of a Kähler manifold. Let M be a complex manifold. We say that M is Kähler if it admits a Hermitian metric with closed Hermitian form, i.e. $d\omega = 0$. Equivalently, in each open neighborhood we can choose complex coordinates z^a, $\bar{z}^{\bar{a}}$ so that the metric takes the form $g_{a\bar{a}} = \frac{1}{2}\delta_{a\bar{a}} + O(z^2, z\bar{z}, \bar{z}^2)$. A Kähler manifold admits compatible complex and symplectic structures.

Local Hermitian Geometry

Let M be a complex manifold with Hermitian metric $g_{a\bar{a}}$ (we do not yet assume the Kähler property); denote by $g^{\bar{b}b}$ the inverse metric. In that case, the Hermitian Levi-Civita connection of (B.2.49) on $T_M \otimes \mathbb{C} = T_M^{1,0} \oplus T_M^{0,1}$ has the following components:

$$\Gamma_{bc}^a = \tfrac{1}{2}g^{\bar{a}a}(g_{b\bar{a},c} + g_{c\bar{a},b})\,, \qquad \Gamma_{bc}^{\bar{a}} = 0\,, \qquad \Gamma_{\bar{a}c}^b = \tfrac{1}{2}g^{\bar{b}b}(g_{c\bar{b},\bar{a}} - g_{c\bar{a},\bar{b}})\,,$$

$$\Gamma_{\bar{a}c}^{\bar{b}} = \tfrac{1}{2}g^{\bar{b}b}(g_{b\bar{a},c} - g_{c\bar{a},b})\,, \tag{B.3.77}$$

as well as complex conjugate expressions, e.g. $\Gamma_{\bar{b}\bar{c}}^{\bar{a}} = \tfrac{1}{2}g^{\bar{a}a}(g_{a\bar{b},\bar{c}} + g_{a\bar{c},\bar{b}})$.

Exercise B.21 Check the Hermitian Levi-Civita connection by verifying $\nabla G = 0$, where G is the Riemannian metric, as well as $[\nabla_\mu, \nabla_\nu]f = 0$ for any function f, where μ, ν take on the possibilities (a, b), $(a, \bar{b})$, $(\bar{a}, \bar{b})$, and $G_{ab} = G_{\bar{a}\bar{b}} = 0$, while $G_{a\bar{b}} = g_{a\bar{b}}$ and $G_{\bar{b}a} = g_{a\bar{b}}$.

Also observe that given a vector $V \in \mathcal{A}^0(T_M^{1,0})$, its covariant derivative with respect to the Levi-Civita connection will in general have components in

$$\nabla V \in \mathcal{A}^{1,0}(T_M^{1,0}) \oplus \mathcal{A}^{0,1}(T_M^{1,0}) \oplus \mathcal{A}^{0,1}(T_M^{0,1})\,, \tag{B.3.78}$$

i.e., for a generic vector $V = V^a \frac{\partial}{\partial z^a}$, $\nabla_a V^b$, $\nabla_{\bar{a}} V^b$ and $\nabla_{\bar{a}} V^{\bar{b}}$ are all non-zero; the only component that vanishes is $\nabla_{\bar{a}} V^{\bar{b}}$. This shows that we cannot think of the Hermitian Levi-Civita connection as a connection on $T_M^{1,0}$; rather it is a connection on $T_M \otimes \mathbb{C}$.   □

While the "pure" component of the connection, Γ_{bc}^a is non-tensorial, the remaining non-zero terms are tensors that encode the non-closure of the Hermitian form:

$$d\omega = \tfrac{i}{2}(g_{a\bar{b},c} - g_{c\bar{b},a})dz^c \wedge dz^a \wedge d\bar{z}^{\bar{b}} + \tfrac{i}{2}(g_{a\bar{b},\bar{c}} - g_{a\bar{c},\bar{b}})dz^a \wedge d\bar{z}^{\bar{b}} \wedge d\bar{z}^{\bar{c}}\,. \tag{B.3.79}$$

Exercise B.22 The complex structure takes the form $\mathcal{J}_b^a = i\delta_b^a$ and $\mathcal{J}_{\bar{b}}^{\bar{a}} = -i\delta_{\bar{b}}^{\bar{a}}$. While it is constant, it may not be covariantly constant. Show that the non-vanishing covariant derivatives of $\mathcal{J}$ are

$$\nabla_{\bar{c}}\mathcal{J}_b^{\bar{a}} = g^{\bar{a}a}(\partial\omega)_{ba\bar{c}}\,, \qquad \nabla_c \mathcal{J}_{\bar{b}}^a = -g^{\bar{a}a}(\bar{\partial}\omega)_{c\bar{a}\bar{b}}\,. \tag{B.3.80}$$

Thus, $d\omega = 0$ if and only if $\mathcal{J}$ is covariantly constant with respect to the Levi-Civita connection.   □

Local Kähler Geometry

If M is Kähler, then we can choose a Hermitian metric so that $d\omega = 0$ and therefore $g_{a\bar{b},\bar{c}} = g_{a\bar{c},\bar{b}}$, as well as $g_{a\bar{b},c} = g_{c\bar{b},a}$. Comparing this to the Levi-Civita connection, we find another equivalent characterization of a Kähler metric: a Hermitian metric is Kähler if and only if the Hermitian Levi-Civita connection on $T_M \otimes \mathbb{C}$ decomposes into a direct sum of Chern connections on $T_M^{1,0}$ and its complex conjugate on $T_M^{0,1}$:

$$\Gamma^a_{bc} = g^{\bar{a}a} g_{b\bar{a},c}, \qquad\qquad \Gamma^{\bar{a}}_{\bar{b}\bar{c}} = g^{\bar{a}a} g_{a\bar{b},\bar{c}}, \qquad\qquad (B.3.81)$$

and all the other "impure" components of Γ are zero.

These local geometric statements are equivalent to the existence of a Kähler potential, real function $K(z, \bar{z})$, such that

$$g_{a\bar{a}} = \frac{\partial^2 K}{\partial z^a \partial \bar{z}^{\bar{a}}}. \qquad\qquad (B.3.82)$$

For example, on flat space we may take

$$K = \tfrac{1}{2} \sum_a |z^a|^2. \qquad\qquad (B.3.83)$$

Although in this example K is a well-defined function on all of $\mathbb{C}^n$, that is typically not the case. More generally, we can only hope to define K patch by patch on an open cover, with a relation on the overlaps $U_{\alpha\beta}$

$$K_\alpha = K_\beta + \left[f_{\alpha\beta}(z) + \text{c.c.} \right]. \qquad\qquad (B.3.84)$$

M does not need to be topologically trivial to admit a global Kähler potential. For instance, suppose that M obeys the $\partial\bar{\partial}$ lemma and has an exact Hermitian form ω. In that case there exists a well-defined function K such that $\omega = i\partial\bar{\partial}K$.[9] There are obstructions to this scenario. For instance, since the volume form on a Hermitian manifold is $\frac{1}{n!}\omega^n$, it follows that ω cannot be exact on a compact manifold. More generally, if M contains a compact holomorphic submanifold $\phi : X \to M$, then $\phi^*(\omega)$ defines a Kähler form on X, and since d commutes with pull-back, ω cannot be exact on M. Some amusing applications of these observations to moduli space geometry of superconformal theories are given in [151].

---

[9]As we stated above, the $\partial\bar{\partial}$-lemma holds automatically for compact Kähler manifolds; on a non-compact manifold, however, it is an independent condition.

Let M be compact and Kähler, with Kähler metric g. The purity of the Levi-Civita connection allows us to express the differentials ∂ and $\bar{\partial}$ in terms of the Levi-Civita connection. If $\eta \in \mathscr{A}^{p,q}(M)$, then

$$\partial\eta = \frac{1}{p!q!}\nabla_{i_0}\eta_{i_1\cdots i_p\bar{J}_1\cdots\bar{J}_q}dz^{i_0\cdots i_p}d\bar{z}^{\bar{J}_1\cdots\bar{J}_q},$$

$$\bar{\partial}\eta = \frac{(-)^p}{p!q!}\nabla_{\bar{J}_0}\eta_{i_1\cdots i_p\bar{J}_1\cdots\bar{J}_q}dz^{i_1\cdots i_p}d\bar{z}^{\bar{J}_0\cdots\bar{J}_q}. \tag{B.3.85}$$

We introduced a useful shorthand $dz^{i_0\cdots i_p} = dz^{i_0} \wedge dz^{i_1} \wedge \cdots \wedge dz^{i_p}$.

This leads to a simple form for the adjoint operators and simplifies the Lefschetz decomposition we are about to introduce. Let η and ψ be a $(p-1, q)$ forms. In components

$$(\partial\eta)_{i_1\cdots i_p\bar{J}_1\cdots\bar{J}_q} = p\nabla_{[i_1}\eta_{i_2\cdots i_p]\bar{J}_1\cdots\bar{J}_q},$$

$$(\partial\eta, \psi) = \frac{1}{(p-1)!q!}\eta_{i_2\cdots i_p\bar{J}_1\cdots\bar{J}_q}\overline{(-\nabla_{\llcorner}\psi)}^{i_2\cdots i_p\bar{J}_1\cdots\bar{J}_q}\frac{\omega^n}{n!} + \nabla(\cdots), \tag{B.3.86}$$

where

$$(\nabla_{\llcorner}\psi)_{i_2\cdots i_p\bar{J}_1\cdots\bar{J}_q} = \nabla^{i_1}\psi_{i_1\cdots i_p\bar{J}_1\cdots\bar{J}_q}. \tag{B.3.87}$$

Thus, we conclude

$$\partial^{\dagger}\psi = -\frac{1}{(p-1)!q!}(\nabla_{\llcorner}\psi)_{i_2\cdots i_p\bar{J}_1\cdots\bar{J}_q}dz^{i_2\cdots i_p}d\bar{z}^{\bar{J}_1\cdots\bar{J}_q}. \tag{B.3.88}$$

Starting with η a $(p, q-1)$ form, a very similar computation yields

$$(\bar{\partial}\eta, \psi) = -\frac{1}{p!(q-1)!}\eta_{i_1\cdots i_p\bar{J}_2\cdots\bar{J}_q}\overline{(\bar{\nabla}_{\llcorner}\psi)}^{i_1\cdots i_p\bar{J}_2\cdots\bar{J}_q}\frac{\omega^n}{n!} + \nabla(\cdots), \tag{B.3.89}$$

where

$$\bar{\nabla}_{\llcorner}\psi = \frac{(-)^p}{p!(q-1)!}\bar{\nabla}^{\bar{J}_1}\psi_{i_1\cdots i_p\bar{J}_1\cdots\bar{J}_q}. \tag{B.3.90}$$

Thus,

$$\bar{\partial}^{\dagger}\psi = -\frac{1}{p!(q-1)!}(\bar{\nabla}_{\llcorner}\psi)_{i_1\cdots i_p\bar{J}_2\cdots\bar{J}_q}dz^{i_1\cdots i_p}d\bar{z}^{\bar{J}_2\cdots\bar{J}_q}. \tag{B.3.91}$$

B.3.5 An Extended Example

Exercise B.23 A simple example of $\mathbb{P}^1$ illustrates many of the concepts we just encountered. We label points by projective coordinates $[Z_0 : Z_1]$, where Z_0 and Z_1 do not vanish simultaneously, and for any non-zero $\lambda \in \mathbb{C}$ $[Z_0 : Z_1] \sim [\lambda Z_0 : \lambda Z_1]$. The affine chart defined by $Z_0 \neq 0$ covers everything but the south pole of the sphere, and we take the coordinate on it to be $z = Z_1 / Z_0$. It is related to the familiar spherical coordinates θ, ϕ via

$$z = \frac{\cos\theta - 1}{\sin\theta} e^{i\phi} . \tag{B.3.92}$$

i. Show that for a round sphere of radius r we have

$$ds^2 = \frac{4r^2}{(1 + z\bar{z})^2} dz d\bar{z} = r^2 (d\theta^2 + \sin^2\theta d\phi^2) . \tag{B.3.93}$$

ii. Show that the round metric can be derived from a Kähler potential:

$$ds^2 = g_{z\bar{z}} dz d\bar{z} + \text{c.c.}, \qquad g_{z\bar{z}} = \partial_z \partial_{\bar{z}} K , \qquad K = 2r^2 \log(1 + z\bar{z}) . \tag{B.3.94}$$

Note that we can write K in affine coordinates as

$$K = 2r^2 \log(Z_0 \bar{Z}_0 + Z_1 \bar{Z}_1) , \tag{B.3.95}$$

and show that if we consider the two affine charts defined, respectively, by $z = Z_1 / Z_0$ and $w = Z_0 / Z_1$, then on their overlap (B.3.84) is obeyed.

iii. The obvious generalization of K to $n + 1$ projective coordinates leads to the Fubini-Study metric on $\mathbb{P}^n$. Argue that $\mathbb{P}^n$ can be covered with $n + 1$ affine charts and compute the metric $g_{i\bar{j}}$ in one of them.

iv. Returning to $\mathbb{P}^1$, we can have a bit of practice with our definitions of conjugation and Hodge star. For instance, let $\alpha, \beta \in \mathscr{A}^{0,1}(\mathbb{P}^1)$, and show that

$$\alpha \wedge *\bar{\beta} = (\alpha_{\bar{z}} \bar{\beta}_z g^{\bar{z}z}) \times i g_{z\bar{z}} dz \wedge d\bar{z} . \tag{B.3.96}$$

Note that the integral is in fact independent of the choice of metric! That is a special feature of the product of $(1, 0)$ or $(0, 1)$ forms on a Riemann surface.

v. Next, verify the volume normalization:

$$\int_{\mathbb{P}^1} *_g 1 = \int_{\mathbb{P}^1} g_{z\bar{z}} i dz \wedge d\bar{z} = 2r^2 \int \frac{d^2 z}{(1 + z\bar{z})^2} = 4\pi r^2 . \tag{B.3.97}$$

vi. Holomorphic line bundles on $\mathbb{P}^1$ are classified by the first Chern class, a single integer k, and denoted by $\mathcal{O}(k)$. To describe these explicitly, work with the two affine charts as above and define the transition function on $U \cap V$, so that if $s_U \in \mathscr{A}^0(U, \mathcal{O}(k))$ and $s_V \in \mathscr{A}^0(V, \mathcal{O}(k))$ are sections on the two charts, then on $U \cap V = \mathbb{C}^*$, we have

$$s_U(z) = z^k s_V(w) = z^k s_V(z^{-1}) . \tag{B.3.98}$$

Show that $h = (1 + z\bar{z})^{-k}$ is a Hermitian metric on $\mathcal{O}(k)$, with corresponding Chern connection $\mathcal{A} = -k\bar{z}dz/(1+z\bar{z})$. Verify that $c_1(\mathcal{O}(k)) = k$ as promised. Argue that $\mathcal{O}(k_1) \otimes \mathcal{O}(k_2) = \mathcal{O}(k_1 + k_2)$.

vii. Show that for $k \geq 0$ $H^0(\mathbb{P}^1, \mathcal{O}(k)) = \mathbb{C}^{k+1}$, and these global sections can be represented by homogeneous degree k polynomials in the Z_0, Z_1 coordinates. Show that the tangent bundle is isomorphic to $T_{\mathbb{P}^1} = \mathcal{O}(2)$, while the canonical bundle is $K_{\mathbb{P}^1} = \mathcal{O}(-2)$. The global sections of $T_{\mathbb{P}^1}$ describe the conformal Killing vectors on the sphere. Also argue that $H^0(\mathbb{P}^1, \mathcal{O}(k)) = 0$ for $k < 0$.

viii. Use Kodaira-Serre duality and preceding discoveries to show that $H^1(\mathbb{P}^1, \mathcal{O}(k)) = 0$ if $k > -2$ and $H^1(\mathbb{P}^1, \mathcal{O}(k)) = \mathbb{C}^{-1-k}$ for $k \leq -2$.

$\square$

Presumably many of the results of the preceding exercise are familiar in one form or another, but perhaps the statements about line bundles are a little less familiar. At any rate, it is also good to keep in mind a few associated facts:

1. A theorem of Grothendieck shows that every holomorphic vector bundle over $\mathbb{P}^1$ is holomorphically isomorphic to a direct sum $\oplus_{k \in I} \mathcal{O}(k)$.
2. Holomorphic line bundles on a Riemann surface Σ are also topologically classified by degree, and the canonical bundle is $\mathcal{O}(2g - 2)$, where g is the genus of the Riemann surface. The bundle cohomology groups for the general $\mathcal{O}(k)$ are then determined by three observations. First, for $k \geq 0$

$$h^0(\Sigma, \mathcal{O}(k)) = \begin{cases} 1, k \leq g \\ k + 1 - g, k > g . \end{cases} \tag{B.3.99}$$

Second, there is the Riemann-Roch formula (a special case of the Hirzebruch-Riemann-Roch formula quoted in the text),

$$h^0(\Sigma, \mathcal{O}(k)) - h^1(\Sigma, \mathcal{O}(k)) = k + 1 - g . \tag{B.3.100}$$

Finally, by Kodaira-Serre duality we have $h^1(\Sigma, \mathcal{O}(k)) = h^0(\Sigma, \mathcal{O}(2g-2-k))$.

The Lefschetz Decomposition

We will now introduce the Lefschetz decomposition on a compact Kähler manifold M, following the treatment in [210]. Let $L : \mathscr{A}^{p,q}(M) \to \mathscr{A}^{p+1,q+1}(M)$ denote the operation of wedging with the Kähler form. In components,

$$L(\eta)_{i_0 \cdots i_p \bar{J}_0 \cdots \bar{J}_q} = (-)^p (p+1)(q+1) \omega_{[i_0 \bar{J}_0} \eta_{i_1 \cdots i_p \bar{J}_1 \cdots \bar{J}_q]}. \tag{B.3.101}$$

Its adjoint $L^\dagger$, defined by $(L(\eta), \psi) = (\eta, L^\dagger(\psi))$, is a map $L^\dagger : \mathscr{A}^{p+1,q+1} \to \mathscr{A}^{p,q}$ given by

$$L^\dagger(\psi)_{i_1 \cdots i_p \bar{J}_1 \cdots \bar{J}_q} = i(-)^{p+1} g^{\bar{J}_0 i_0} \psi_{i_0 \cdots i_p \bar{J}_0 \cdots \bar{J}_q}. \tag{B.3.102}$$

Because the operators are algebraic, we can determine their commutation relations generally by studying the problem on $\mathbb{C}^n$. Let $E_k, \overline{E}_{\bar{k}}$ denote a basis for T_M dual to the $dz^k, d\bar{z}^{\bar{k}}$, with induced anti-commutative product $E_k E_m + E_m E_k = 0$, etc. Duality implies the relations

$$E_k dz^m + dz^m E_k = 2\delta_k^m, \qquad E_k d\bar{z}^m + d\bar{z}^m E_k = 0, \tag{B.3.103}$$

as well as their complex conjugates. In terms of these operators,

$$L = \frac{i}{2} \sum_k dz^k d\bar{z}^k, \qquad\qquad L^\dagger = -\frac{i}{2} \sum_k \overline{E}_k E_k, \tag{B.3.104}$$

so, $LL^\dagger = \frac{1}{4} \sum_{k,m} dz^m d\bar{z}^m \overline{E}_k E_k$, and

$$L^\dagger L = \frac{1}{4} \sum_{k,m} \overline{E}_k E_k dz^m d\bar{z}^m = n - \frac{1}{2} \sum_m (d\bar{z}^m \overline{E}_m + dz^m E_m) + LL^\dagger. \tag{B.3.105}$$

But, $\frac{1}{2} \sum_m dz^m E_m$ and $\frac{1}{2} \sum_m d\bar{z}^m \overline{E}_m$ act as number operators on (p, q) monomials, returning, respectively p and q. Hence, we conclude

$$[L, L^\dagger]\mathscr{A}^{p,q} = (p + q - n)\mathscr{A}^{p,q}. \tag{B.3.106}$$

We can write this more abstractly by introducing a third generator

$$E = \sum_{k=0}^{2n} (n - k)\Pi^k, \tag{B.3.107}$$

where Π^k is the projector onto k-forms. Then we conclude that $E, L, L^\dagger$ are generators of an $\mathfrak{sl}_2 \mathbb{C}$ algebra acting on the differential forms:

$$[L^\dagger, L] = E, \qquad [E, L] = -2L, \qquad [E, L^\dagger] = 2L^\dagger. \tag{B.3.108}$$

Since these operations are algebraic, we can understand their structure one point $p \in M$ at a time, and therefore standard properties of finite dimensional representations of $\mathfrak{sl}_2 \, \mathbb{C}$ guarantee the isomorphisms

$$L^k : \mathscr{A}^{n-k}(M, \mathbb{C}) \to \mathscr{A}^{n+k}(M, \mathbb{C}),$$

$$\ker L^{k+1} \cap \mathscr{A}^{n-k}(M, \mathbb{C}) \simeq \ker L^{\dagger} \cap \mathscr{A}^{n-k}(M, \mathbb{C}). \tag{B.3.109}$$

For $m \leq n$ we say that a form $\eta \in \mathscr{A}^m(M, \mathbb{C})$ is primitive if and only if $\omega^{n+1-m} \wedge \eta = 0$; by the last isomorphism this is equivalent to $\omega \llcorner \eta = 0$.

As we will now see, these isomorphisms descend to the cohomology on M.

Exercise B.24 Using the identities for $\bar{\partial}$ and ∂ in terms of the Levi-Civita connection, as well as the covariant constancy of ω, prove the following relations:

$$[L, \bar{\partial}] = 0, \quad [L, \partial] = 0, \quad [L^{\dagger}, \bar{\partial}] = -i\partial^{\dagger}, \quad [L^{\dagger}, \partial] = i\bar{\partial}^{\dagger}. \tag{B.3.110}$$

Taking adjoints, we also see

$$[L^{\dagger}, \partial] = i\bar{\partial}^{\dagger}, \qquad\qquad [L, \bar{\partial}^{\dagger}] = -i\partial. \tag{B.3.111}$$

$\square$

The Hodge Decomposition and Global Considerations

As we already observed, on a complex manifold there are three different Laplacians: Δ_d, Δ_{∂}, and $\Delta_{\bar{\partial}}$. In general there is no simple relation between these operators or their eigenspaces, but on a Kähler manifold (B.3.110) implies $\Delta_d = 2\Delta_{\partial} = 2\Delta_{\bar{\partial}}$. To prove this use $d = \partial + \bar{\partial}$ to write

$$\Delta_d = \Delta_{\partial} + \Delta_{\bar{\partial}} + \{\partial, \bar{\partial}^{\dagger}\} + \{\bar{\partial}, \partial^{\dagger}\}. \tag{B.3.112}$$

Now observe that

$$\{\partial, \bar{\partial}^{\dagger}\} = i\{[L, \bar{\partial}^{\dagger}], \bar{\partial}^{\dagger}\} = \frac{i}{2}[L, \{\bar{\partial}^{\dagger}, \bar{\partial}^{\dagger}\}] = 0. \tag{B.3.113}$$

The other term, $\{\bar{\partial}, \partial^{\dagger}\} = 0$ in a similar way. Moreover,

$$\partial\partial^{\dagger} = i\partial[L^{\dagger}, \bar{\partial}] = \bar{\partial}^{\dagger}\bar{\partial} + i[L^{\dagger}, \partial\bar{\partial}], \quad \text{and}$$

$$\partial^{\dagger}\partial = \bar{\partial}\bar{\partial}^{\dagger} + i[L^{\dagger}, \partial\bar{\partial}] \implies \Delta_{\partial} = \Delta_{\bar{\partial}}. \tag{B.3.114}$$

$\Delta_d = 2\Delta_{\bar{\partial}}$ follows and in particular implies that Δ_d preserves the type of form.

Exercise B.25 Show that Δ_d also commutes with $\mathfrak{sl}_2\,\mathbb{C}$ by proving and utilizing

$$L\partial\partial^\dagger = \partial\partial^\dagger L + i\partial\bar\partial \tag{B.3.115}$$

and similar relations.   □

The exercise implies that the $\mathfrak{sl}_2\,\mathbb{C}$ commutes with Δ_d and therefore yields an action on the complexified space of Δ_d-harmonic forms. By the de Rham isomorphism this implies an action on the cohomology $H^k(M,\mathbb{C})$ that leads to the decomposition

$$H^m(M,\mathbb{C}) = \bigoplus_{k=0}^{n} L^k P^{m-2k}(M)\,, \tag{B.3.116}$$

where $\dim_\mathbb{C} M = n$, and the primitive cohomology $P^{n-k}(M) \subset H^{n-k}(M,\mathbb{C})$ is defined by

$$P^{n-k}(M) \equiv \ker(L^{k+1} : H^{n-k}(M,\mathbb{C}) \to H^{n+k+2}(M,\mathbb{C}))$$
$$\simeq \ker\left(L^\dagger : H^{n-k}(M,\mathbb{C}) \to H^{n-k-2}(M,\mathbb{C})\right)\,. \tag{B.3.117}$$

This is a crucial topological restriction: a complex manifold without this structure on the cohomology cannot admit a Kähler metric.

Returning to the Laplacians, because $\Delta_d = 2\Delta_{\bar\partial}$ preserves the (p,q) type of form, we obtain a remarkable refinement of (B.2.41): the equation

$$\mathbb{1} = \Pi_d + \Delta_d \Sigma \tag{B.3.118}$$

holds separately on every $\mathscr{A}^{p,q}(M)$, and we obtain the Hodge decomposition for a Kähler manifold. The complexified space of Δ_d—harmonic forms $\mathcal{H}_d^k(M,\mathbb{C})$ decomposes as

$$\mathcal{H}_d^k(M,\mathbb{C}) = \bigoplus_{p+q=k} \mathcal{H}_d^{p,q}(M,\mathbb{C})\,, \tag{B.3.119}$$

and therefore the cohomology groups decompose as well:

$$H_d^k(M,\mathbb{C}) = \bigoplus_{p+q=k} H_d^{p,q}(M,\mathbb{C}) = \bigoplus_{p+q=k} H_{\bar\partial}^{p,q}(M,\mathbb{C})\,. \tag{B.3.120}$$

The last equality indicates that the decomposition only depends on the complex structure of M and not the metric. Since Δ_d is real, it also follows that $\mathcal{H}_d^{p,q} = \overline{\mathcal{H}_d^{q,p}}$ and therefore $h^{p,q}(M) = h^{q,p}(M)$. As a result, we find a topological restriction obeyed by Kähler manifolds:

$$b_{2k+1} = \sum_{p+q=2k+1} h^{p,q}(M) \in 2\mathbb{Z} . \qquad (B.3.121)$$

In particular, $b_1 = 2h^{1,0}$.

It also follows that any Kähler manifold satisfies the $\partial\bar{\partial}$-lemma. Suppose $\eta \in$ im d. The Hodge decomposition and $\Delta_d = 2\Delta_{\bar{\partial}}$ then imply that for some form σ

$$\eta = d(\bar{\partial}\bar{\partial}^\dagger + \bar{\partial}^\dagger\bar{\partial})\sigma = \partial\bar{\partial}(\bar{\partial}^\dagger - \partial^\dagger)\sigma + \bar{\partial}^\dagger\bar{\partial}\partial\sigma + \partial^\dagger\partial\bar{\partial}\sigma . \qquad (B.3.122)$$

To derive this we used the anti-commutation of ∂ with $\bar{\partial}$ and with $\bar{\partial}^\dagger$. In this form $\partial\eta = 0$ implies

$$0 = \partial\eta = \partial\partial^\dagger\partial\bar{\partial}\sigma , \qquad (B.3.123)$$

but that is possible only if $\partial\bar{\partial}\sigma = 0$, i.e.

$$\eta = \partial\bar{\partial}\tau , \qquad\qquad \tau = (\bar{\partial}^\dagger - \partial^\dagger)\sigma . \qquad (B.3.124)$$

Exercise B.26 Prove the following two useful results for M a compact Kähler 3-fold with Kähler form ω.

First, for any $\beta \in \mathscr{A}^{1,1}(M, \mathbb{R})$; show that there exists a real function f such that

$$\partial\bar{\partial}\beta \wedge \omega = \partial\bar{\partial}f \wedge \omega^2 . \qquad (B.3.125)$$

To do this, first observe that the equation is equivalent to $(L^\dagger)^3 \left[L\partial\bar{\partial}\beta - L^2\partial\bar{\partial}f \right] = 0$, and then use the commutation relations to reduce that equation to $\Delta_d L^\dagger \beta - i\bar{\partial}^\dagger\partial^\dagger\beta = 2\Delta_d f$. Since the left-hand-side is orthogonal to all harmonic functions, there exists a requisite f.

Next, prove that a real $(1,1)$ form σ on M can be decomposed as

$$\sigma = \psi\omega + \kappa , \qquad (B.3.126)$$

where ψ is a real function and κ is a real $(1,1)$ form that satisfies $\omega \wedge \partial\bar{\partial}\kappa = 0$. [Hint: use the previous result!]   □

B.3.6   Some Global Considerations

Kähler manifolds constitute a remarkable subset of complex manifolds. Clearly there are necessary conditions that a complex manifold M must satisfy in order to admit a Kähler metric. We already discussed the most famous of these: the odd

Betti numbers of M must be even. In the case of complex surfaces, i.e. compact manifolds of complex dimension 2, this is in fact sufficient. Additional topological restrictions follow from the Lefschetz decomposition.

While Kähler manifolds are rather special, it is easy to construct many examples. Smooth projective varieties constitute an important and well-studied class. These are smooth compact manifolds defined by the vanishing locus of a finite number of homogeneous polynomials in projective space. A simple example is a degree k Fermat hypersurface in $\mathbb{P}^n$. Such a manifold is automatically compact and complex, and the pull-back of the Kähler form on $\mathbb{P}^n$ then defines a Kähler metric on the variety.

Which Kähler manifolds are projective? The answer is provided by the Kodaira embedding theorem. Let X be Kähler. Since the Kähler form is closed, it automatically defines a cohomology class $[\omega] \in H^{1,1}(X, \mathbb{C})$. If X is compact (which must hold if X is to be projective), then this class is non-trivial. The Kodaira embedding theorem asserts that X is projective if and only if $[\omega] \in H^2(X, \mathbb{Q})$, i.e. $[\omega]$ defines a rational class in cohomology.

The condition depends on the complex structure of X, since this determines how the integral lattice $H^2(X, \mathbb{Z})$ is embedded in $H^2(X, \mathbb{C})$. The complex structure determines the Picard lattice $\mathrm{Pic}(X) = H^{1,1}(X, \mathbb{C}) \cap H^2(X, \mathbb{Z})$, and it may be empty. In this case X is not projective. The complex surfaces T^4 and K3 provide examples where the Picard lattice rank varies as a function of the complex structure and is generically zero. Of course there are many other examples where $H^2(X, \mathbb{C}) = H^{1,1}(X, \mathbb{C})$; in this case the Picard lattice is isomorphic to $H^2(X, \mathbb{Z})$, and X is projective. For instance, all Calabi-Yau 3-folds with $H^{0,2}(M, \mathbb{C}) = 0$ have this property.

These examples may give the impression that it is always possible to adjust the complex structure in a suitable way to produce a projective embedding. Remarkably, this is false: there exist Kähler manifolds X that are not homeomorphic to any projective manifold. We leave this fascinating topic to the references and the careful presentation given in [373].

Given a Kähler manifold with some fixed Kähler form ω_0, we observe that we can always slightly perturb $\omega_0 \to \omega = \omega_0 + \delta\omega$, where $[\delta\omega] \in H^{1,1}(X, \mathbb{R})$ is chosen sufficiently small such that the deformed metric remains non-degenerate. The tangent space of such deformations is isomorphic to $H^{1,1}(X, \mathbb{R})$. As we increase $\delta\omega$, it may be that the metric becomes singular; for instance, the volume of some holomorphic curve $\iota : C \to X$ vanishes: $\int_C \iota^*(\omega) = 0$. It can be shown that the classes of non-degenerate Kähler metrics lie in a cone in $H^{1,1}(X, \mathbb{R})$—this is the Kähler cone. Its properties are discussed in [282].

Bibliography

1. Adam, I.: On the marginal deformations of general (0,2) non-linear sigma-models. Proc. Symp. Pure Math. **90**, 171–179 (2015). http://arxiv.org/abs/1710.07431

2. Adams, A., Basu, A., Sethi, S.: (0,2) duality. Adv. Theor. Math. Phys. **7**, 865–950 (2004). http://arxiv.org/abs/hep-th/0309226

3. Adams, A., Distler, J., Ernebjerg, M.: Topological heterotic rings. Adv. Theor. Math. Phys. **10**, 657–682 (2006). http://arxiv.org/abs/hep-th/0506263

4. Adams, A., Dyer, E., Lee, J.: GLSMs for non-Kahler geometries. J. High Energy Phys. **1301**, 044 (2013). http://arxiv.org/abs/1206.5815

5. Adams, A., Ernebjerg, M., Lapan, J.M.: Linear models for flux vacua. http://arxiv.org/abs/hep-th/0611084

6. Affleck, I., Haldane, F.D.M.: Critical theory of quantum spin chains. Phys. Rev. **B36**, 5291–5300 (1987). http://dx.doi.org/10.1103/PhysRevB.36.5291

7. Affleck, I., Dine, M., Seiberg, N.: Dynamical supersymmetry breaking in supersymmetric QCD. Nucl. Phys. **B241**, 493–534 (1984)

8. Aldazabal, G., Ibanez, L.E.: A note on 4D heterotic string vacua, FI-terms and the swampland. Phys. Lett. **B782**, 375–379 (2018). http://dx.doi.org/10.1016/j.physletb.2018.05.061; http://arxiv.org/abs/1804.07322

9. Alexandrov, S., Louis, J., Pioline, B., Valandro, R.: $\mathcal{N} = 2$ heterotic-type II duality and bundle moduli. J. High Energy Phys. **08**, 092 (2014). http://dx.doi.org/10.1007/JHEP08(2014)092; http://arxiv.org/abs/1405.4792

10. Alvarez-Gaume, L., Ginsparg, P.H.: The structure of gauge and gravitational anomalies. Ann. Phys. **161**, 423 (1985)

11. Alvarez-Gaume, L., Witten, E.: Gravitational anomalies. Nucl. Phys. **B234**, 269 (1984). http://dx.doi.org/10.1016/0550-3213(84)90066-X

12. Alvarez-Gaume, L., Freedman, D.Z., Mukhi, S.: The background field method and the ultraviolet structure of the supersymmetric nonlinear sigma model. Ann. Phys. **134**, 85 (1981). http://dx.doi.org/10.1016/0003-4916(81)90006-3

13. Anderson, L.B.: Heterotic and M-theory compactifications for string phenomenology. PhD thesis, Oxford University (2008). http://arxiv.org/abs/0808.3621; https://inspirehep.net/record/793857/files/arXiv:0808.3621.pdf

14. Anderson, L.B., Feng, H.: New evidence for (0,2) target space duality. http://arxiv.org/abs/1607.04628

15. Anderson, L.B., Gray, J., Lukas, A., Palti, E.: Heterotic line bundle standard models. J. High Energy Phys. **06**, 113 (2012). http://dx.doi.org/10.1007/JHEP06(2012)113; http://arxiv.org/abs/1202.1757

© Springer Nature Switzerland AG 2019
I. V. Melnikov, *An Introduction to Two-Dimensional Quantum Field Theory with (0,2) Supersymmetry*, Lecture Notes in Physics 951,
https://doi.org/10.1007/978-3-030-05085-6

16. Anderson, L.B., Apruzzi, F., Gao, X., Gray, J., Lee, S.-J.: A new construction of Calabi–Yau manifolds: generalized CICYs. Nucl. Phys. **B906**, 441–496 (2016). http://dx.doi.org/10.1016/j.nuclphysb.2016.03.016; http://arxiv.org/abs/1507.03235

17. Anderson, L.B., Gray, J., Sharpe, E.: Algebroids, heterotic moduli spaces and the Strominger system. http://arxiv.org/abs/1402.1532

18. Anderson, L.B., Gray, J., Lukas, A., Ovrut, B.: The Atiyah class and complex structure stabilization in heterotic Calabi-Yau compactifications. http://arxiv.org/abs/1107.5076

19. Angelantonj, C., Israel, D., Sarkis, M.: Threshold corrections in heterotic flux compactifications. J. High Energy Phys. **08**, 032 (2017). http://dx.doi.org/10.1007/JHEP08(2017)032; http://arxiv.org/abs/1611.09442

20. Angella, D.: Cohomological Aspects in Complex Non-Kähler Geometry. Lecture Notes in Mathematics, vol. 2095. Springer, Cham (2014). https://doi.org/10.1007/978-3-319-02441-7

21. Angella, D., Ugarte, L.: On small deformations of balanced manifolds. Differ. Geom. Appl. **54**(part B), 464–474 (2017)

22. Apruzzi, F., Hassler, F., Heckman, J.J., Melnikov, I.V.: UV completions for non-critical strings. J. High Energy Phys. **07**, 045 (2016). http://dx.doi.org/10.1007/JHEP07(2016)045; http://arxiv.org/abs/1602.04221

23. Apruzzi, F., Hassler, F., Heckman, J.J., Melnikov, I.V.: From 6D SCFTs to dynamic GLSMs. Phys. Rev. **D96**(6), 066015 (2017). http://dx.doi.org/10.1103/PhysRevD.96.066015; http://arxiv.org/abs/1610.00718

24. Argyres, P.C.: An introduction to global supersymmetry. DIY (2000)

25. Argyres, P.C., Plesser, M.R., Seiberg, N.: The moduli space of vacua of N=2 SUSY QCD and duality in N=1 SUSY QCD. Nucl. Phys. **B471**, 159–194 (1996). http://arxiv.org/abs/hep-th/9603042

26. Argyres, P.C., Plesser, M.R., Shapere, A.D.: N=2 moduli spaces and N=1 dualities for SO(n(c)) and USp(2n(c)) superQCD. Nucl. Phys. **B483**, 172–186 (1997). http://dx.doi.org/10.1016/S0550-3213(96)00583-4; http://arxiv.org/abs/hep-th/9608129

27. Arnold, V.: Mathematical Methods of Classical Mechanics. Springer, Berlin (1989)

28. Ashmore, A., De La Ossa, X., Minasian, R., Strickland-Constable, C., Svanes, E.E.: Finite deformations from a heterotic superpotential: holomorphic Chern–Simons and an L_∞ algebra. http://arxiv.org/abs/1806.08367

29. Aspinwall, P.S.: A McKay-like correspondence for (0,2)-deformations. http://arxiv.org/abs/1110.2524

30. Aspinwall, P.S.: D-branes on Calabi-Yau manifolds. http://arxiv.org/abs/hep-th/0403166

31. Aspinwall, P.S.: K3 surfaces and string duality. http://arxiv.org/abs/hep-th/9611137

32. Aspinwall, P.S.: The Moduli space of N=2 superconformal field theories. http://arxiv.org/abs/hep-th/9412115

33. Aspinwall, P.S., Gaines, B.: Rational curves and (0,2)-deformations. J. Geom. Phys. **88**, 1–15 (2014). http://dx.doi.org/10.1016/j.geomphys.2014.09.012; http://arxiv.org/abs/1404.7802

34. Aspinwall, P.S., Morrison, D.R.: Chiral rings do not suffice: N=(2,2) theories with nonzero fundamental group. Phys. Lett. **B334**, 79–86 (1994). http://dx.doi.org/10.1016/0370-2693(94)90594-0; http://arxiv.org/abs/hep-th/9406032

35. Aspinwall, P.S., Plesser, M.R.: General mirror pairs for gauged linear sigma models. J. High Energy Phys. **11**, 029 (2015). http://dx.doi.org/10.1007/JHEP11(2015)029; http://arxiv.org/abs/1507.00301

36. Aspinwall, P.S., Plesser, M.R.: Decompactifications and massless D-branes in hybrid models. http://arxiv.org/abs/0909.0252

37. Aspinwall, P.S., Plesser, M.R.: Elusive worldsheet instantons in heterotic string compactifications. http://arxiv.org/abs/1106.2998

38. Aspinwall, P.S., Greene, B.R., Morrison, D.R.: The monomial divisor mirror map. Int. Math. Res. Not. **1993**(12), 319–337 (1993). http://arxiv.org/abs/alg-geom/9309007

39. Aspinwall, P.S., Greene, B.R., Morrison, D.R.: Calabi-Yau moduli space, mirror manifolds and spacetime topology change in string theory. Nucl. Phys. **B416**, 414–480 (1994). http://arxiv.org/abs/hep-th/9309097

40. Aspinwall, P.S., Bridgeland, T., Craw, A., Douglas, M.R., Kapustin, A., Moore, G.W., Gross, M., Segal, G., Szendroi, B., Wilson, P.M.H.: Dirichlet Branes and Mirror Symmetry. Clay Mathematics Monographs, vol. 4. AMS, Providence (2009). http://people.maths.ox.ac.uk/cmi/library/monographs/cmim04c.pdf

41. Aspinwall, P.S., Melnikov, I.V., Plesser, M.R.: (0,2) elephants. J. High Energy Phys. **1201**, 060 (2012). http://arxiv.org/abs/1008.2156

42. Aspinwall, P.S., Plesser, M.R., Wang, K.: Mirror symmetry and discriminants. http://arxiv.org/abs/1702.04661

43. Atiyah, M.F.: Complex analytic connections in fibre bundles. Trans. Am. Math. Soc. **85**(1), 181–207 (1957)

44. Atiyah, M., Hitchin, N.J., Singer, I.: Selfduality in four-dimensional Riemannian geometry. Proc. R. Soc. Lond. **A362**, 425–461 (1978)

45. Ballmann, W.: Lectures on Kähler Manifolds. ESI Lectures in Mathematics and Physics. European Mathematical Society (EMS), Zürich (2006). http://dx.doi.org/10.4171/025

46. Banks, T., Dixon, L.J.: Constraints on string vacua with space-time supersymmetry. Nucl. Phys. **B307**, 93–108 (1988)

47. Banks, T., Seiberg, N.: Nonperturbative infinities. Nucl. Phys. **B273**, 157 (1986)

48. Banks, T., Dixon, L.J., Friedan, D., Martinec, E.J.: Phenomenology and conformal field theory or can string theory predict the weak mixing angle? Nucl. Phys. **B299**, 613–626 (1988)

49. Barth, W.P., Hulek, K., Peters, C.A.M., Van de Ven, A.: Compact Complex Surfaces, vol. 4, 2nd edn. Springer, Berlin (2004)

50. Basu, A., Sethi, S.: World-sheet stability of (0,2) linear sigma models. Phys. Rev. **D68**, 025003 (2003). http://arxiv.org/abs/hep-th/0303066

51. Batyrev, V.V.: Quantum cohomology rings of toric manifolds. Astérisque **218**, 9–34 (1993). http://arxiv.org/abs/alg-geom/9310004

52. Batyrev, V.V.: Dual polyhedra and mirror symmetry for Calabi-Yau hypersurfaces in toric varieties. J. Algebraic Geom. **3**, 493–545 (1994). http://arxiv.org/abs/arXiv:alg-geom/9310003

53. Batyrev, V.: The stringy Euler number of Calabi-Yau hypersurfaces in toric varieties and the Mavlyutov duality. http://arxiv.org/abs/1707.02602

54. Batyrev, V.V., Materov, E.N.: Toric residues and mirror symmetry. Mosc. Math. J. **2**(3), 435–475 (2002)

55. Batyrev, V., Nill, B.: Combinatorial aspects of mirror symmetry. http://arxiv.org/abs/math/0703456

56. Baume, F., Keren-Zur, B., Rattazzi, R., Vitale, L.: The local Callan-Symanzik equation: structure and applications. J. High Energy Phys. **08**, 152 (2014). http://dx.doi.org/10.1007/JHEP08(2014)152; http://arxiv.org/abs/1401.5983

57. Beasley, C., Witten, E.: Residues and world-sheet instantons. J. High Energy Phys. **10**, 065 (2003). http://arxiv.org/abs/hep-th/0304115

58. Beasley, C., Witten, E.: New instanton effects in supersymmetric QCD. J. High Energy Phys. **0501**, 056 (2005). http://dx.doi.org/10.1088/1126-6708/2005/01/056; http://arxiv.org/abs/hep-th/0409149

59. Becker, K., Dasgupta, K.: Heterotic strings with torsion. J. High Energy Phys. **11**, 006 (2002). http://dx.doi.org/10.1088/1126-6708/2002/11/006; http://arxiv.org/abs/hep-th/0209077

60. Becker, K., Becker, M., Dasgupta, K., Green, P.S.: Compactifications of heterotic theory on non-Kaehler complex manifolds. I. J. High Energy Phys. **04**, 007 (2003). http://arxiv.org/abs/hep-th/0301161

61. Becker, K., Becker, M., Green, P.S., Dasgupta, K., Sharpe, E.: Compactifications of heterotic strings on non-Kaehler complex manifolds. II. Nucl. Phys. **B678**, 19–100 (2004). http://arxiv.org/abs/hep-th/0310058

62. Becker, K., Becker, M., Fu, J.-X., Tseng, L.-S., Yau, S.-T.: Anomaly cancellation and smooth non-Kaehler solutions in heterotic string theory. Nucl. Phys. **B751**, 108–128 (2006). http://arxiv.org/abs/hep-th/0604137

63. Behr, N., Konechny, A.: Renormalization and redundancy in 2d quantum field theories. J. High Energy Phys. **1402**, 001 (2014). http://dx.doi.org/10.1007/JHEP02(2014)001; http://arxiv.org/abs/1310.4185

64. Behtash, A., Dunne, G.V., Schaefer, T., Sulejmanpasic, T., Unsal, M.: Complexified path integrals, exact saddles and supersymmetry. Phys. Rev. Lett. **116**(1), 011601 (2016). http://dx.doi.org/10.1103/PhysRevLett.116.011601; http://arxiv.org/abs/1510.00978

65. Belavin, A., Polyakov, A.M., Zamolodchikov, A.: Infinite conformal symmetry in two-dimensional quantum field theory. Nucl. Phys. **B241**, 333–380 (1984). http://dx.doi.org/10.1016/0550-3213(84)90052-X

66. Benini, F., Bobev, N.: Exact two-dimensional superconformal R-symmetry and c-extremization. Phys. Rev. Lett. **110**, 061601 (2013). http://dx.doi.org/10.1103/PhysRevLett.110.061601; http://arxiv.org/abs/1211.4030

67. Benini, F., Bobev, N.: Two-dimensional SCFTs from wrapped branes and c-extremization. J. High Energy Phys. **06**, 005 (2013). http://dx.doi.org/10.1007/JHEP06(2013)005; http://arxiv.org/abs/1302.4451

68. Benini, F., Cremonesi, S.: Partition functions of N=(2,2) gauge theories on S2 and vortices. http://arxiv.org/abs/1206.2356

69. Benini, F., Eager, R., Hori, K., Tachikawa, Y.: Elliptic genera of 2d N=2 gauge theories. http://arxiv.org/abs/1308.4896

70. Benini, F., Eager, R., Hori, K., Tachikawa, Y.: Elliptic genera of two-dimensional N=2 gauge theories with rank-one gauge groups. http://arxiv.org/abs/1305.0533

71. Berglund, P., Hubsch, T.: A generalized construction of mirror manifolds. Nucl. Phys. **B393**, 377–391 (1993). [AMS/IP Stud. Adv. Math.9,327(1998)]. http://dx.doi.org/10.1016/0550-3213(93)90250-S; http://arxiv.org/abs/hep-th/9201014

72. Berglund, P., Hubsch, T.: A generalized construction of Calabi-Yau models and mirror symmetry. SciPost Phys. **4**, 009 (2018). http://dx.doi.org/10.21468/SciPostPhys.4.2.009; http://arxiv.org/abs/1611.10300

73. Berglund, P., Candelas, P., de la Ossa, X., Derrick, E., Distler, J., et al.: On the instanton contributions to the masses and couplings of E(6) singlets. Nucl. Phys. **B454**, 127–163 (1995). http://arxiv.org/abs/hep-th/9505164

74. Berglund, P., Johnson, C.V., Kachru, S., Zaugg, P.: Heterotic coset models and (0,2) string vacua. Nucl. Phys. **B460**, 252–298 (1996). http://dx.doi.org/10.1016/0550-3213(95)00641-9; http://arxiv.org/abs/hep-th/9509170

75. Bergshoeff, E., de Roo, M.: The Quartic effective action of the heterotic string and supersymmetry. Nucl. Phys. **B328**, 439 (1989). http://dx.doi.org/10.1016/0550-3213(89)90336-2

76. Berkovits, N.: Explaining the pure spinor formalism for the superstring. J. High Energy Phys. **01**, 065 (2008). http://dx.doi.org/10.1088/1126-6708/2008/01/065; http://arxiv.org/abs/0712.0324

77. Bershadsky, M., Cecotti, S., Ooguri. H., Vafa, C.: Kodaira-Spencer theory of gravity and exact results for quantum string amplitudes. Commun. Math. Phys. **165**, 311–428 (1994). http://arxiv.org/abs/hep-th/9309140

78. Bershadsky, M., Intriligator, K.A., Kachru, S., Morrison, D.R., Sadov, V., et al.: Geometric singularities and enhanced gauge symmetries. Nucl. Phys. **B481**, 215–252 (1996). http://arxiv.org/abs/hep-th/9605200

79. Bertolini, M.: Testing the (0,2) mirror map. http://arxiv.org/abs/1806.05850

80. Bertolini, M., Plesser, M.R.: Worldsheet instantons and (0,2) linear models. J. High Energy Phys. **8**, 081 (2015) . http://dx.doi.org/10.1007/JHEP08(2015)081; http://arxiv.org/abs/1410.4541

81. Bertolini, M., Plesser, M.R.: (0,2) hybrid models. http://arxiv.org/abs/1712.04976

82. Bertolini, M., Romo, M.: Aspects of (2,2) and (0,2) hybrid models. http://arxiv.org/abs/1801.04100

83. Bertolini, M., Melnikov, I.V., Plesser, M.R.: Accidents in (0,2) Landau-Ginzburg theories. J. High Energy Phys. **12**, 157 (2014). http://dx.doi.org/10.1007/JHEP12(2014)157; http://arxiv.org/abs/1405.4266

84. Bertolini, M., Melnikov, I.V., Plesser, M.R.: Hybrid conformal field theories. http://arxiv.org/abs/1307.7063

85. Billera, L.J., Filliman, P., Sturmfels, B.: Constructions and complexity of secondary polytopes. Adv. Math. **83**(2), 155–179 (1990). https://doi.org/10.1016/0001-8708(90)90077-Z

86. Blumenhagen, R., Rahn, T.: Landscape study of target space duality of (0,2) heterotic string models. J. High Energy Phys. **1109**, 098 (2011). http://arxiv.org/abs/1106.4998

87. Blumenhagen, R., Schimmrigk, R., Wisskirchen, A. The (0,2) exactly solvable structure of chiral rings, landau-ginzburg theories, and Calabi-Yau manifolds. Nucl. Phys. **B461**, 460–492 (1996). http://arxiv.org/abs/hep-th/9510055

88. Blumenhagen, R., Schimmrigk, R., Wisskirchen, A.: (0,2) mirror symmetry. Nucl. Phys. **B486**, 598–628 (1997) . http://arxiv.org/abs/hep-th/9609167

89. Blumenhagen, R., Jurke, B., Rahn, T.: Computational tools for cohomology of toric varieties. Adv. High Energy Phys. **2011**, 152749 (2011). http://dx.doi.org/10.1155/2011/152749; http://arxiv.org/abs/1104.1187

90. Blumenhagen, R., Lüst, D., Theisen, S.: Basic Concepts of String Theory. Theoretical and Mathematical Physics. Springer, Heidelberg (2013). http://dx.doi.org/10.1007/978-3-642-29497-6

91. Bogomolov, F.A.: Hamiltonian Kählerian manifolds. Dokl. Akad. Nauk SSSR **243**(5), 1101–1104 (1978)

92. Bohr, C., Hanke, B., Kotschick, D.: Cycles, submanifolds, and structures on normal bundles. Manuscr. Math. **108**(4), 483–494 (2002). https://doi.org/10.1007/s002290200279

93. Borisov, L.A., Kaufmann, R.M.: On CY-LG correspondence for (0,2) toric models. http://arxiv.org/abs/1102.5444

94. Bott, R., Tu, L.W.: Differential Forms in Algebraic Topology. Graduate Texts in Mathematics, vol. 82. Springer, New York (1982)

95. Boucher, W., Friedan, D., Kent, A.: Determinant formulae and unitarity for the N=2 superconformal algebras in two-dimensions or exact results on string compactification. Phys. Lett. **B172**, 316 (1986). http://dx.doi.org/10.1016/0370-2693(86)90260-1

96. Bouwknegt, P., Schoutens, K.: W symmetry in conformal field theory. Phys. Rep. **223**, 183–276 (1993). http://dx.doi.org/10.1016/0370-1573(93)90111-P; http://arxiv.org/abs/hep-th/9210010

97. Bradlow, S.B., Daskalopoulos, G.D.: Moduli of stable pairs for holomorphic bundles over Riemann surfaces. Int. J. Math. **2**(5), 477–513 (1991)

98. Braun, V., Kreuzer, M., Ovrut, B.A., Scheidegger, E.: Worldsheet instantons and torsion curves. Part A: direct computation. J. High Energy Phys. **10**, 022 (2007). http://arxiv.org/abs/hep-th/0703182

99. Braun, A.P., Knapp, J., Scheidegger, E., Skarke, H., Walliser, N.-O.: PALP - a user manual. In: Rebhan, A., Katzarkov, L., Knapp, J., Rashkov, R., Scheidegger, E. (eds.) Strings, Gauge Fields, and the Geometry Behind: The Legacy of Maximilian Kreuzer. World Scientific, Singapore (2012). http://arxiv.org/abs/1205.4147

100. Braun, V., Kreuzer, M., Ovrut, B.A., Scheidegger, E.: Worldsheet instantons and torsion curves. http://arxiv.org/abs/0801.4154

101. Bruns, W., Herzog, J.: Cohen-Macaulay Rings. Cambridge Studies in Advanced Mathematics, vol. 39. Cambridge University Press, Cambridge (1993)

102. Bryant, R.L.: An introduction to Lie groups and symplectic geometry. In: Geometry and Quantum Field Theory. Proceedings, Graduate Summer School on the Geometry and Topology of Manifolds and Quantum Field Theory, Park City, 22 June–20 July 1991, pp. 7–181. American Mathematical Society, Providence (1991)

103. Buchbinder, E., Lukas, A., Ovrut, B., Ruehle, F.: Heterotic instanton superpotentials from complete intersection Calabi-Yau manifolds. J. High Energy Phys. **10**, 032 (2017). http://dx.doi.org/10.1007/JHEP10(2017)032; http://arxiv.org/abs/1707.07214

104. Buchbinder, E.I., Lin, L., Ovrut, B.A.: Non-vanishing heterotic superpotentials on elliptic fibrations. http://arxiv.org/abs/1806.04669

105. Buchdahl, N.P.: Hermitian-Einstein connections and stable vector bundles over compact complex surfaces. Math. Ann. **280**(4), 625–648 (1988). http://dx.doi.org/10.1007/BF01450081

106. Buividovich, P.V., Dunne, G.V., Valgushev, S.N.: Complex path integrals and saddles in two-dimensional gauge theory. Phys. Rev. Lett. **116**(13), 132001 (2016). http://dx.doi.org/10.1103/PhysRevLett.116.132001; http://arxiv.org/abs/1512.09021

107. Callan, J., Curtis, G., Martinec, E., Perry, M., Friedan, D.: Strings in background fields. Nucl. Phys. **B262**, 593 (1985). http://dx.doi.org/10.1016/0550-3213(85)90506-1

108. Candelas, P., de la Ossa, X.: Moduli space of Calabi-Yau manifolds. Nucl. Phys. **B355**, 455–481 (1991)

109. Candelas, P., Horowitz, G.T., Strominger, A., Witten, E.: Vacuum configurations for superstrings. Nucl. Phys. **B258**, 46–74 (1985)

110. Candelas, P., De La Ossa, X.C., Green, P.S., Parkes, L.: A Pair of Calabi-Yau manifolds as an exactly soluble superconformal theory. Nucl. Phys. **B359**, 21–74 (1991). http://dx.doi.org/10.1016/0550-3213(91)90292-6

111. Candelas, P., De La Ossa, X., Font, A., Katz, S.H., Morrison, D.R.: Mirror symmetry for two parameter models. I. Nucl. Phys. **B416**, 481–538 (1994). http://arxiv.org/abs/hep-th/9308083

112. Candelas, P., Constantin, A., Mishra, C.: Calabi-Yau threefolds with small hodge numbers. Fortschr. Phys. **66**, 1800029 (2018). http://dx.doi.org/10.1002/prop.201800029; http://arxiv.org/abs/1602.06303

113. Cecotti, S., Vafa, C.: Topological antitopological fusion. Nucl. Phys. **B367**, 359–461 (1991). http://dx.doi.org/10.1016/0550-3213(91)90021-O

114. Chen, Y.-H., Wilczek, F., Witten, E., Halperin, B.I.: On anyon superconductivity. Int. J. Mod. Phys. **B3**, 1001 (1989). http://dx.doi.org/10.1142/S0217979289000725

115. Closset, C., Cremonesi, S., Park, D.S.: The equivariant A-twist and gauged linear sigma models on the two-sphere. J. High Energy Phys. **06**, 076 (2015). http://dx.doi.org/10.1007/JHEP06(2015)076; http://arxiv.org/abs/1504.06308

116. Closset, C., Gu, W., Jia, B., Sharpe, E.: Localization of twisted $\mathcal{N} = (0, 2)$ gauged linear sigma models in two dimensions. J. High Energy Phys. **03**, 070 (2016). http://dx.doi.org/10.1007/JHEP03(2016)070; http://arxiv.org/abs/1512.08058

117. Coleman, S.R.: There are no Goldstone bosons in two-dimensions. Commun. Math. Phys. **31**, 259–264 (1973). http://dx.doi.org/10.1007/BF01646487

118. Coleman, S.: Aspects of Symmetry. Cambridge University Press, Cambridge (1985). http://dx.doi.org/10.1017/CBO9780511565045

119. Collins, T.C., Xie, D., Yau, S.-T.: K stability and stability of chiral ring. http://arxiv.org/abs/1606.09260

120. Cox, D.A., Katz, S.: Mirror Symmetry and Algebraic Geometry, 469pp. AMS, Providence (2000)

121. Cox, D., Little, J., O'Shea, D.: Using Algebraic Geometry. Graduate Texts in Mathematics. Springer, New York (1998)

122. Cox, D., Little, J., Schenck, H.: Toric Varieties. Graduate Studies in Mathematics, vol. 124. AMS, Providence (2011)

123. Creutzig, T., Ridout, D.: Logarithmic conformal field theory: beyond an introduction. J. Phys. **A46**, 4006 (2013). http://dx.doi.org/10.1088/1751-8113/46/49/494006; http://arxiv.org/abs/1303.0847

124. Dai, X.-z., Freed, D. S.: Eta invariants and determinant lines. J. Math. Phys. **35**, 5155–5194 (1994). [Erratum: J. Math. Phys. **42**, 2343 (2001)]. http://dx.doi.org/10.1063/1.530747; http://arxiv.org/abs/hep-th/9405012

125. Dasgupta, K., Rajesh, G., Sethi, S.: M theory, orientifolds and G-flux. J. High Energy Phys. **08**, 023 (1999). http://arxiv.org/abs/hep-th/9908088

126. de Boer, J., Manschot, J., Papadodimas, K., Verlinde, E.: The Chiral ring of AdS(3)/CFT(2) and the attractor mechanism. J. High Energy Phys. **03**, 030 (2009). http://dx.doi.org/10.1088/1126-6708/2009/03/030; http://arxiv.org/abs/0809.0507

127. de la Ossa, X., Svanes, E.E.: Holomorphic Bundles and the Moduli Space of N=1 Supersymmetric Heterotic Compactifications. J. High Energy Phys. **10**, 123 (2014). http://dx.doi.org/10.1007/JHEP10(2014)123; http://arxiv.org/abs/1402.1725

128. de la Ossa, X., Svanes, E.E.: Connections, field redefinitions and heterotic supergravity. J. High Energy Phys. **12**, 008 (2014). http://dx.doi.org/10.1007/JHEP12(2014)008; http://arxiv.org/abs/1409.3347

129. Dedushenko, M.: Chiral algebras in Landau-Ginzburg models. http://arxiv.org/abs/1511.04372

130. Deser, S., Schwimmer, A.: Geometric classification of conformal anomalies in arbitrary dimensions. Phys. Lett. **B309**, 279–284 (1993). http://arxiv.org/abs/hep-th/9302047

131. Di Francesco, P., Mathieu, P., Senechal, D.: Conformal Field Theory. Graduate Texts in Contemporary Physics. Springer, New York (1997). http://dx.doi.org/10.1007/978-1-4612-2256-9

132. Di Vecchia, P., Petersen, J., Zheng, H.: N=2 extended superconformal theories in two-dimensions. Phys. Lett. **B162**, 327 (1985). http://dx.doi.org/10.1016/0370-2693(85)90932-3

133. Di Vecchia, P., Knizhnik, V.G., Petersen, J.L., Rossi, P.: A supersymmetric Wess-Zumino lagrangian in two-dimensions. Nucl. Phys. **B253**, 701–726 (1985). http://dx.doi.org/10.1016/0550-3213(85)90554-1

134. Dijkgraaf, R., Verlinde, H.L., Verlinde, E.P.: Notes on topological string theory and 2-D quantum gravity. Based on lectures given at Spring School on Strings and Quantum Gravity, Trieste, 24 Apr–2 May 1990 and at Cargese Workshop on Random Surfaces, Quantum Gravity and Strings, Cargese, 28 May–1 Jun 1990

135. Dine, M., Lee, C.: Remarks on (0,2) models and intermediate scale scenarios in string theory. Phys. Lett. **B203**, 371–377 (1988)

136. Dine, M., Seiberg, N.: (2,0) superspace. Phys. Lett. **B180**, 364 (1986)

137. Dine, M., Seiberg, N.: Are (0,2) models string miracles? Nucl. Phys. **B306**, 137 (1988)

138. Dine, M., Seiberg, N.: Microscopic knowledge from macroscopic physics in string theory. Nucl. Phys. **B301**, 357 (1988)

139. Dine, M., Seiberg, N., Wen, X.G., Witten, E.: Nonperturbative effects on the string world sheet. Nucl. Phys. **B278**, 769 (1986)

140. Dine, M., Seiberg, N., Wen, X.G., Witten, E.: Nonperturbative effects on the string world sheet. 2. Nucl. Phys. **B289**, 319 (1987)

141. Dine, M., Seiberg, N., Witten, E.: Fayet-Iliopoulos terms in string theory. Nucl. Phys. **B289**, 589 (1987)

142. Distler, J.: Resurrecting (2,0) compactifications. Phys. Lett. **B188**, 431–436 (1987)

143. Distler, J.: Notes on (0,2) superconformal field theories. http://arxiv.org/abs/hep-th/9502012

144. Distler, J., Greene, B.R.: Aspects of (2,0) string compactifications. Nucl. Phys. **B304**, 1 (1988)

145. Distler, J., Kachru, S.: (0,2) Landau-Ginzburg theory. Nucl. Phys. **B413**, 213–243 (1994). http://arxiv.org/abs/hep-th/9309110

146. Distler, J., Kachru, S.: Duality of (0,2) string vacua. Nucl. Phys. **B442**, 64–74 (1995). http://arxiv.org/abs/hep-th/9501111

147. Distler, J., Sharpe, E.: Heterotic compactifications with principal bundles for general groups and general levels. Adv. Theor. Math. Phys. **14**, 335–398 (2010). http://arxiv.org/abs/hep-th/0701244

148. Dixon, L.J.: Some world sheet properties of superstring compactifications, on orbifolds and otherwise. Lectures given at the 1987 ICTP Summer Workshop in High Energy Phsyics and Cosmology, Trieste, 29 June–7 Aug 1987

149. Dixon, L.J., Kaplunovsky, V., Vafa, C.: On four-dimensional gauge theories from type II superstrings. Nucl. Phys. **B294**, 43–82 (1987)

150. Dixon, L.J., Kaplunovsky, V., Louis, J.: On effective field theories describing (2,2) vacua of the heterotic string. Nucl. Phys. **B329**, 27–82 (1990)

151. Donagi, R., Morrison, D.R.: Conformal field theories and compact curves in moduli spaces. J. High Energy Phys. **05**, 021 (2018). http://dx.doi.org/10.1007/JHEP05(2018)021; http://arxiv.org/abs/1709.05355

152. Donagi, R., Lu, Z., Melnikov, I.V.: Global aspects of (0,2) moduli space: toric varieties and tangent bundles.http://arxiv.org/abs/1409.4353

153. Donagi, R., Guffin, J., Katz, S., Sharpe, E.: Physical aspects of quantum sheaf cohomology for deformations of tangent bundles of toric varieties. http://arxiv.org/abs/1110.3752

154. Donagi, R., Guffin, J., Katz, S., Sharpe, E.: A mathematical theory of quantum sheaf cohomology. http://arxiv.org/abs/1110.3751

155. Donaldson, S.K., Kronheimer, P.B.: The Geometry Of Four-Manifolds. Oxford Mathematical Monographs. The Clarendon Press/Oxford University Press, New York (1990)

156. Doroud, N., Gomis, J., Le Floch, B., Lee, S.: Exact results in D=2 supersymmetric gauge theories. http://arxiv.org/abs/1206.2606

157. Dryden, J.: Fables Ancient and Modern. HardPress Publishing, Los Angeles (2012)

158. Dumitrescu, T.T., Seiberg, N.: Supercurrents and brane currents in diverse dimensions. J. High Energy Phys. **1107**, 095 (2011). http://dx.doi.org/10.1007/JHEP07(2011)095; http://arxiv.org/abs/1106.0031

159. Eguchi, T., Taormina, A.: On the unitary representations of N=2 and N=4 superconformal algebras. Phys. Lett. **B210**, 125 (1988). http://dx.doi.org/10.1016/0370-2693(88)90360-7

160. Eguchi, T., Taormina, A.: Extended superconformal algebras and string compactifications. In: Trieste Spring School and Workshop on Superstrings (SUPERSTRINGS '88), Trieste, 11–22 Apr 1988. http://alice.cern.ch/format/showfull?sysnb=0102114

161. Eguchi, T., Gilkey, P.B., Hanson, A.J.: Gravitation, gauge theories and differential geometry. Phys. Rep. **66**, 213 (1980). http://dx.doi.org/10.1016/0370-1573(80)90130-1

162. Eisenbud, D.: Commutative Algebra. Graduate Texts in Mathematics, vol. 150. Springer, New York (1995)

163. Evans, L.C.: Partial Differential Equations. Graduate Studies in Mathematics, vol. 19, 2nd edn. American Mathematical Society, Providence (2010) https://doi.org/10.1090/gsm/019

164. Fei, T., Huang, Z., Picard, S.: A construction of infinitely many solutions to the Strominger system. http://arxiv.org/abs/1703.10067

165. Ferrara, S., Lust, D., Theisen, S.: World sheet versus spectrum symmetries in heterotic and type II superstrings. Nucl. Phys. **B325**, 501 (1989)

166. Florakis, I., Garcia-Etxebarria, I., Lust, D., Regalado, D.: 2d orbifolds with exotic supersymmetry. http://arxiv.org/abs/1712.04318

167. Fre, P., Girardello, L., Lerda, A., Soriani, P.: Topological first order systems with Landau-Ginzburg interactions. Nucl. Phys. **B387**, 333–372 (1992). http://arxiv.org/abs/hep-th/9204041

168. Freed, D.: Determinants, torsion, and strings. Commun. Math. Phys. **107**, 483–513 (1986). http://dx.doi.org/10.1007/BF01221001

169. Freed, D.S.: Special Kaehler manifolds. Commun. Math. Phys. **203**, 31–52 (1999). http://arxiv.org/abs/hep-th/9712042

170. Freed, D., Harvey, J.A.: Instantons and the spectrum of Bloch electrons in a magnetic field. Phys. Rev. **B41**, 11328 (1990). http://dx.doi.org/10.1103/PhysRevB.41.11328

171. Freed, D.S., Uhlenbeck, K.K.: Instantons and Four-Manifolds. Mathematical Sciences Research Institute Publications, 2nd edn. Springer, New York (1991)

172. Friedan, D.H.: Nonlinear models in two + epsilon dimensions. Ann. Phys. **163**, 318 (1985). Ph.D. Thesis. http://dx.doi.org/10.1016/0003-4916(85)90384-7

173. Friedan, D., Konechny, A.: Gradient formula for the beta-function of 2d quantum field theory. J. Phys. **A43**, 215401 (2010). http://dx.doi.org/10.1088/1751-8113/43/21/215401; http://arxiv.org/abs/0910.3109

174. Friedan, D., Konechny, A.: Curvature formula for the space of 2-d conformal field theories. J. High Energy Phys. **09**, 113 (2012). http://dx.doi.org/10.1007/JHEP09(2012)113; http://arxiv.org/abs/1206.1749

175. Friedan, D., Martinec, E.J., Shenker, S.H.: Conformal invariance, supersymmetry and string theory. Nucl. Phys. **B271**, 93 (1986)
176. Fu, J.-X., Yau, S.-T.: The theory of superstring with flux on non-Kaehler manifolds and the complex Monge-Ampere equation. J. Differ. Geom. **78**, 369–428 (2009). http://arxiv.org/abs/hep-th/0604063
177. Fu, J., Yau, S.-T.: A note on small deformations of balanced manifolds. C. R. Math. Acad. Sci. Paris **349**(13–14), 793–796 (2011)
178. Fulton, W.: Introduction to Toric Varieties. Princeton University Press, Princeton (1993)
179. Gaberdiel, M.R., Konechny, A., Schmidt-Colinet, C.: Conformal perturbation theory beyond the leading order. J. Phys. **A42**, 105402 (2009). http://dx.doi.org/10.1088/1751-8113/42/10/105402; http://arxiv.org/abs/0811.3149
180. Gadde, A., Gukov, S.: 2d index and surface operators. http://arxiv.org/abs/1305.0266
181. Gadde, A., Putrov, P.: Exact solutions of (0,2) Landau-Ginzburg models. http://arxiv.org/abs/1608.07753
182. Gadde, A., Gukov, S., Putrov, P.: Fivebranes and 4-manifolds. http://arxiv.org/abs/1306.4320
183. Gadde, A., Gukov, S., Putrov, P.: Exact solutions of 2d supersymmetric gauge theories. http://arxiv.org/abs/1404.5314
184. Gadde, A., Gukov, S., Putrov, P.: (0,2) trialities. http://arxiv.org/abs/1310.0818
185. Gaiotto, D., Moore, G.W., Witten, E.: Algebra of the infrared: string field theoretic structures in massive $\mathcal{N} = (2, 2)$ field theory in two dimensions. http://arxiv.org/abs/1506.04087
186. Garavuso, R.S., Sharpe, E.: Analogues of Mathai–Quillen forms in sheaf cohomology and applications to topological field theory. J. Geom. Phys. **92**, 1–29 (2015). http://arxiv.org/abs/1310.5754
187. Garcia-Etxebarria, I., Hayashi, H., Ohmori, K., Tachikawa, Y., Yonekura, K.: 8d gauge anomalies and the topological Green-Schwarz mechanism. J. High Energy Phys. **11**, 177 (2017). http://dx.doi.org/10.1007/JHEP11(2017)177; http://arxiv.org/abs/1710.04218
188. Garcia-Fernandez, M.: Lectures on the Strominger system. http://arxiv.org/abs/1609.02615
189. Gates, S.J., Hull, C., Rocek, M.: Twisted multiplets and new supersymmetric nonlinear sigma models. Nucl. Phys. **B248**, 157 (1984). http://dx.doi.org/10.1016/0550-3213(84)90592-3
190. Gauntlett, J.P., Martelli, D., Waldram, D.: Superstrings with intrinsic torsion. Phys. Rev. **D69**, 086002 (2004). http://arxiv.org/abs/hep-th/0302158
191. Gelfand, I.M., Kapranov, M.M., Zelevinsky, A.V.: Discriminants, Resultants, and Multidimensional Determinants. Mathematics: Theory & Applications. Birkhauser, Boston (1994). http://dx.doi.org/10.1007/978-0-8176-4771-1
192. Gepner, D.: Exactly solvable string compactifications on manifolds of SU(N) holonomy. Phys. Lett. **B199**, 380–388 (1987)
193. Gepner, D.: Space-time supersymmetry in compactified string theory and superconformal models. Nucl. Phys. **B296**, 757 (1988)
194. Gepner, D.: Lectures on N=2 string theory. Lectures at Spring School on Superstrings, Trieste, 3–14 Apr 1989
195. Gerchkovitz, E., Gomis, J., Komargodski, Z.: Sphere partition functions and the Zamolodchikov metric. J. High Energy Phys. **11**, 001 (2014). http://dx.doi.org/10.1007/JHEP11(2014)001; http://arxiv.org/abs/1405.7271
196. Gerhardus, A., Jockers, H., Ninad, U.: The geometry of gauged linear sigma model correlation functions. http://arxiv.org/abs/1803.10253
197. Ginsparg, P.H.: Applied conformal field theory. http://arxiv.org/abs/hep-th/9108028
198. Goddard, P., Kent, A., Olive, D.I.: Virasoro algebras and coset space models. Phys. Lett. **B152**, 88 (1985). http://dx.doi.org/10.1016/0370-2693(85)91145-1
199. Goldstein, E., Prokushkin, S.: Geometric model for complex non-Kaehler manifolds with SU(3) structure. Commun. Math. Phys. **251**, 65–78 (2004). http://arxiv.org/abs/hep-th/0212307
200. Gomis, J., Lee, S.: Exact kahler potential from gauge theory and mirror symmetry. J. High Energy Phys. **1304**, 019 (2013). http://dx.doi.org/10.1007/JHEP04(2013)019; http://arxiv.org/abs/1210.6022

201. Gomis, J., Hsin, P.-S., Komargodski, Z., Schwimmer, A., Seiberg, N., Theisen, S.: Anomalies, conformal manifolds, and spheres. J. High Energy Phys. **03**, 022 (2016). http://dx.doi.org/10.1007/JHEP03(2016)022; http://arxiv.org/abs/1509.08511

202. Gomis, J., Komargodski, Z., Ooguri, H., Seiberg, N., Wang, Y.: Shortening anomalies in supersymmetric theories. J. High Energy Phys. **01**, 067 (2017). http://dx.doi.org/10.1007/JHEP01(2017)067; http://arxiv.org/abs/1611.03101

203. Grana, M., Minasian, R., Petrini, M., Waldram, D.: T-duality, generalized geometry and non-geometric backgrounds. J. High Energy Phys. **04**, 075 (2009). http://dx.doi.org/10.1088/1126-6708/2009/04/075; http://arxiv.org/abs/0807.4527

204. Green, M.B., Seiberg, N.: Contact interactions in superstring theory. Nucl. Phys. **B299**, 559 (1988). http://dx.doi.org/10.1016/0550-3213(88)90549-4

205. Green, M.B., Schwarz, J.H., West, P.C.: Anomaly free chiral theories in six-dimensions. Nucl. Phys. **B254**, 327–348 (1985). http://dx.doi.org/10.1016/0550-3213(85)90222-6

206. Green, M., Schwarz, J., Witten, E.: Superstring Theory, Volume 1. Cambridge University Press, Cambridge (1987)

207. Green, M., Schwarz, J., Witten, E.: Superstring Theory, Volume 2. Cambridge University Press, Cambridge (1987)

208. Green, D., Komargodski, Z., Seiberg, N., Tachikawa, Y., Wecht, B.: Exactly marginal deformations and global symmetries. J. High Energy Phys. **1006**, 106 (2010). http://dx.doi.org/10.1007/JHEP06(2010)106; http://arxiv.org/abs/1005.3546

209. Greene, B.R., Plesser, M.R.: Duality in Calabi-Yau moduli space. Nucl. Phys. **B338**, 15–37 (1990)

210. Griffiths, P., Harris, J.: Principles of Algebraic Geometry. Wiley, New York (1978)

211. Grisaru, M.T., van de Ven, A., Zanon, D.: Two-dimensional supersymmetric sigma models on Ricci flat Kahler manifolds are not finite. Nucl. Phys. **B277**, 388 (1986). http://dx.doi.org/10.1016/0550-3213(86)90448-7

212. Gromov, M.: Pseudoholomorphic curves in symplectic manifolds. Invent. Math. **82**(2), 307–347 (1985)

213. Gross, D.J., Harvey, J.A., Martinec, E.J., Rohm, R.: Heterotic string theory. 1. The free heterotic string. Nucl. Phys. **B256**, 253 (1985). http://dx.doi.org/10.1016/0550-3213(85)90394-3

214. Grünbaum, B.: Convex Polytopes. Graduate Texts in Mathematics, vol. 221, 2nd edn. Springer, New York (2003). Prepared and with a preface by Volker Kaibel, Victor Klee and Günter M. Ziegler. http://dx.doi.org/10.1007/978-1-4613-0019-9

215. Gu, W., Sharpe, E.: A proposal for nonabelian mirrors. http://arxiv.org/abs/1806.04678

216. Guffin, J., Katz, S.: Deformed quantum cohomology and (0,2) mirror symmetry. http://arxiv.org/abs/arXiv:0710.2354

217. Guffin, J., Sharpe, E.: A-twisted heterotic Landau-Ginzburg models. http://arxiv.org/abs/0801.3955

218. Guillemin, V.: Moment Maps and Combinatorial Invariants of Hamiltonian T^n-Spaces. Progress in Mathematics, vol. 122. Birkhäuser, Boston (1994). http://dx.doi.org/10.1007/978-1-4612-0269-1

219. Haase, C., Melnikov, I.V.: The reflexive dimension of a lattice polytope. Ann. Comb. **10**(2), 211–217 (2006)

220. Harlow, D., Maltz, J., Witten, E.: Analytic continuation of Liouville theory. J. High Energy Phys. **12**, 071 (2011). http://dx.doi.org/10.1007/JHEP12(2011)071; http://arxiv.org/abs/1108.4417

221. Harris, J.: Algebraic Geometry: A First Course. Graduate Texts in Mathematics, vol. 133. Springer, New York (1992)

222. Hartshorne, R.: Algebraic Geometry, 8th edn. Springer, Berlin (1997)

223. Hatcher, A.: Algebraic Topology. Cambridge University Press, Cambridge (2002)

224. Hatcher, A.: Vector Bundles and K-Theory. Online, 2.1 edn. (2009)

225. Herbst, M., Hori, K., Page, D.: Phases of N=2 theories in 1+1 dimensions with boundary. http://arxiv.org/abs/0803.2045

226. Hirzebruch, F.: Ueber eine Klasse von einfach-zusammenhaengenden komplexen Mannigfaltigkeiten. Math. Ann. **124**, 77–86 (1951)

227. Hirzebruch, F., Höfer, T.: On the Euler number of an orbifold. Math. Ann. **286**(1–3), 255–260 (1990). https://doi.org/10.1007/BF01453575

228. Hitchin, N.J.: Lectures on special Lagrangian submanifolds. In: Proceedings, Winter School on Mirror Symmetry and Vector Bundles, Cambridge, MA, 4–15 Jan 1999, pp. 151–182. http://arxiv.org/abs/math/9907034

229. Hohenberg, P.C.: Existence of long-range order in one and two dimensions. Phys. Rev. **158**, 383–386 (1967). http://dx.doi.org/10.1103/PhysRev.158.383

230. Hollands, S.: Action principle for OPE. Nucl. Phys. **B926**, 614–638 (2018). http://dx.doi.org/10.1016/j.nuclphysb.2017.11.013; http://arxiv.org/abs/1710.05601

231. Honecker, G.: Massive U(1)s and heterotic five-branes on K3. Nucl. Phys. **B748**, 126–148 (2006). http://arxiv.org/abs/hep-th/0602101

232. Hori, K.: Duality in two-dimensional (2,2) supersymmetric non-abelian gauge theories. J. High Energy Phys. **10**, 121 (2013). http://dx.doi.org/10.1007/JHEP10(2013)121; http://arxiv.org/abs/1104.2853

233. Hori, K., Kapustin, A.: Duality of the fermionic 2-D black hole and N=2 liouville theory as mirror symmetry. J. High Energy Phys. **0108**, 045 (2001). http://arxiv.org/abs/hep-th/0104202

234. Hori, K., Knapp, J.: Linear sigma models with strongly coupled phases - one parameter models. J. High Energy Phys. **11**, 070 (2013). http://dx.doi.org/10.1007/JHEP11(2013)070; http://arxiv.org/abs/1308.6265

235. Hori, K., Tong, D.: Aspects of non-abelian gauge dynamics in two-dimensional N=(2,2) theories. J. High Energy Phys. **05**, 079 (2007). http://dx.doi.org/10.1088/1126-6708/2007/05/079; http://arxiv.org/abs/hep-th/0609032

236. Hori, K., Vafa, C.: Mirror symmetry. http://arxiv.org/abs/hep-th/0002222

237. Hori, K., Katz, S., Klemm, A., Pandharipande, R., Thomas, R., Vafa, C., Vakil, R., Zaslow, E.: Mirror Symmetry. Clay Mathematics Monographs, vol. 1. American Mathematical Society, Providence (2003). With a preface by Vafa

238. Hosono, S., Klemm, A., Theisen, S., Yau, S.-T.: Mirror symmetry, mirror map and applications to Calabi-Yau hypersurfaces. Commun. Math. Phys. **167**, 301–350 (1995). http://arxiv.org/abs/hep-th/9308122

239. Howe, P.S., Papadopoulos, G.: Anomalies in two-dimensional supersymmetric nonlinear sigma models. Class. Quantum Gravity **4**, 1749–1766 (1987)

240. Howe, P.S., Papadopoulos, G.: Further remarks on the geometry of two-dimensional nonlinear sigma models. Class. Quantum Gravity **5**, 1647–1661 (1988)

241. Hubsch, T.: Calabi-Yau Manifolds: A Bestiary for Physicists. World Scientific, Singapore (1992)

242. Hughes, J., Polchinski, J.: Partially broken global supersymmetry and the superstring. Nucl. Phys. **B278**, 147 (1986). http://dx.doi.org/10.1016/0550-3213(86)90111-2

243. Hull, C.: Compactifications of the Heterotic Superstring. Phys. Lett. **B178**, 357 (1986). http://dx.doi.org/10.1016/0370-2693(86)91393-6

244. Hull, C.M., Townsend, P.K.: Finiteness and conformal invariance in nonlinear σ models. Nucl. Phys. **B274**, 349–362 (1986). http://dx.doi.org/10.1016/0550-3213(86)90289-0

245. Hull, C.M., Townsend, P.K.: World sheet supersymmetry and anomaly cancellation in the heterotic string. Phys. Lett. **B178**, 187 (1986)

246. Hull, C.M., Witten, E.: Supersymmetric sigma models and the heterotic string. Phys. Lett. **B160**, 398–402 (1985)

247. Huybrechts, D.: Lectures on k3 Surfaces. Cambridge University Press, Cambridge (2010). Online

248. Intriligator, K.A., Seiberg, N.: Lectures on supersymmetric gauge theories and electric-magnetic duality. Nucl. Phys. Proc. Suppl. **45BC**, 1–28 (1996). http://arxiv.org/abs/hep-th/9509066

249. Israel, D., Sarkis, M.: New supersymmetric index of heterotic compactifications with torsion. J. High Energy Phys. **12**, 069 (2015). http://dx.doi.org/10.1007/JHEP12(2015)069; http://arxiv.org/abs/1509.05704

250. Israel, D., Sarkis, M.: Dressed elliptic genus of heterotic compactifications with torsion and general bundles. J. High Energy Phys. **08**, 176 (2016). http://dx.doi.org/10.1007/JHEP08(2016)176; http://arxiv.org/abs/1606.08982

251. Ivanov, S., Ugarte, L.: On the Strominger system and holomorphic deformations. http://arxiv.org/abs/1705.02792

252. Jardine, I.T., Quigley, C.: Conformal invariance of (0, 2) sigma models on Calabi-Yau manifolds. J. High Energy Phys. **03**, 090 (2018). http://dx.doi.org/10.1007/JHEP03(2018)090; http://arxiv.org/abs/1801.04336

253. Jia, B., Sharpe, E., Wu, R.: Notes on nonabelian (0,2) theories and dualities. http://arxiv.org/abs/1401.1511

254. Jockers, H., Kumar, V., Lapan, J.M., Morrison, D.R., Romo, M.: Two-sphere partition functions and Gromov-Witten invariants. http://arxiv.org/abs/1208.6244

255. Joyce, D.D.: Compact Manifolds with Special Holonomy. Oxford Mathematical Monographs. Oxford University Press, Oxford (2000)

256. Joyce, D.D.: Riemannian Holonomy Groups and Calibrated Geometry. Oxford Graduate Texts in Mathematics, vol. 12. Oxford University Press, Oxford (2007)

257. Jow, S.-Y.: Cohomology of toric line bundles via simplicial Alexander duality. J. Math. Phys. **52**, 033506 (2011). http://arxiv.org/abs/1006.0780v1

258. Kachru, S., Vafa, C.: Exact results for N=2 compactifications of heterotic strings. Nucl. Phys. **B450**, 69–89 (1995). http://dx.doi.org/10.1016/0550-3213(95)00307-E; http://arxiv.org/abs/hep-th/9505105

259. Kachru, S., Witten, E.: Computing the complete massless spectrum of a Landau- Ginzburg orbifold. Nucl. Phys. **B407**, 637–666 (1993). http://arxiv.org/abs/hep-th/9307038

260. Kapranov, M.M.: A characterization of A-discriminantal hypersurfaces in terms of the logarithmic Gauss map. Math. Ann. **290**(2), 277–285 (1991)

261. Kapustin, A.: Chiral de Rham complex and the half-twisted sigma-model. http://arxiv.org/abs/hep-th/0504074

262. Karu, K.: Toric residue mirror conjecture for Calabi-Yau complete intersections. J. Algebraic Geom. **14**(4), 741–760 (2005)

263. Kastor, D.A., Martinec, E.J., Shenker, S.H.: RG Flow in N=1 discrete series. Nucl. Phys. **B316**, 590–608 (1989)

264. Katz, S.H., Sharpe, E.: Notes on certain (0,2) correlation functions. Commun. Math. Phys. **262**, 611–644 (2006). http://arxiv.org/abs/hep-th/0406226

265. Katzarkov, L., Kontsevich, M., Pantev, T.: Bogomolov-Tian-Todorov theorems for Landau-Ginzburg models. J. Differ. Geom. **105**(1), 55–117 (2017). http://arxiv.org/abs/1409.5996

266. Kawai, T., Mohri, K.: Geometry of (0,2) Landau-Ginzburg orbifolds. Nucl. Phys. **B425**, 191–216 (1994). http://arxiv.org/abs/hep-th/9402148

267. Ketov, S.: Quantum Non-linear Sigma Models. Springer, Berlin (2000)

268. Klyachko, A.A.: Equivariant bundles over toric varieties. Izv. Akad. Nauk SSSR Ser. Mat. **53**(5), 1001–1039, 1135 (1989)

269. Knizhnik, V.G., Zamolodchikov, A.B.: Current algebra and Wess-Zumino model in two-dimensions. Nucl. Phys. **B247**, 83–103 (1984). http://dx.doi.org/10.1016/0550-3213(84)90374-2

270. Knutson, A., Sharpe, E.R.: Sheaves on toric varieties for physics. Adv. Theor. Math. Phys. **2**, 865–948 (1998). http://arxiv.org/abs/hep-th/9711036

271. Kodaira, K.: Complex Manifolds and Deformation of Complex Structures. Classics in Mathematics. Springer, Berlin (2005)

272. Krawitz, M., Priddis, N., Acosta, P., Bergin, N., Rathnakumara, H.: FJRW-rings and mirror symmetry. Commun. Math. Phys. **296**(1), 145–174 (2010). https://doi.org/10.1007/s00220-009-0929-7

273. Kreuzer, M., Nill, B.: Classification of toric Fano 5-folds. Adv. Geom. **9**(1), 85–97 (2009). https://doi.org/10.1515/ADVGEOM.2009.005

274. Kreuzer, M., Skarke, H.: On the classification of quasihomogeneous functions. Commun. Math. Phys. **150**, 137 (1992). http://dx.doi.org/10.1007/BF02096569; http://arxiv.org/abs/hep-th/9202039

275. Kreuzer, M., Skarke, H.: PALP: a package for analyzing lattice polytopes with applications to toric geometry. Comput. Phys. Commun. **157**, 87–106 (2004). http://arxiv.org/abs/math/0204356

276. Kreuzer, M., McOrist, J., Melnikov, I.V., Plesser, M.: (0,2) deformations of linear sigma models. J. High Energy Phys. **1107**, 044 (2011). http://dx.doi.org/10.1007/JHEP07(2011)044; http://arxiv.org/abs/1001.2104

277. Kronheimer, P.B.: The construction of ALE spaces as hyper-Kähler quotients. J. Differ. Geom. **29**(3), 665–683 (1989)

278. Kumar, V., Taylor, W.: Freedom and constraints in the K3 landscape. J. High Energy Phys. **0905**, 066 (2009). http://dx.doi.org/10.1088/1126-6708/2009/05/066; http://arxiv.org/abs/0903.0386

279. Kutasov, D.: Geometry on the space of conformal field theories and contact terms. Phys. Lett. **B220**, 153 (1989)

280. Kutasov, D., Lin, J.: (0,2) dynamics from four dimensions. Phys. Rev. **D89**(8), 085025 (2014). http://dx.doi.org/10.1103/PhysRevD.89.085025; http://arxiv.org/abs/1310.6032

281. Lawson, H.B., Jr., Michelsohn, M.-L.: Spin Geometry. Princeton Mathematical Series, vol. 38. Princeton University Press, Princeton (1989)

282. Lazarsfeld, R.: Positivity in Algebraic Geometry. I. A Series of Modern Surveys in Mathematics, vol. 48. Springer, Berlin (2004)

283. Lerche, W., Vafa, C., Warner, N.P.: Chiral rings in N=2 superconformal theories. Nucl. Phys. **B324**, 427 (1989)

284. Li, J.: Hermitian-Yang-Mills connections and beyond. In: Surveys in Differential Geometry 2014. Regularity and Evolution of Nonlinear Equations. Surveys in Differential Geometry, vol. 19, pp. 139–149. International Press, Somerville (2015). http://dx.doi.org/10.4310/SDG.2014.v19.n1.a6

285. Li, J., Yau, S.-T.: Hermitian-Yang-Mills connection on non-Kähler manifolds. In: Mathematical Aspects of String Theory (San Diego, California, 1986). Advanced Series in Mathematical Physics, vol. 1, pp. 560–573. World Scientific Publishing, Singapore (1987)

286. Lindstrom, U., Rocek, M., von Unge, R., Zabzine, M.: Generalized Kahler manifolds and off-shell supersymmetry. Commun. Math. Phys. **269**, 833–849 (2007). http://dx.doi.org/10.1007/s00220-006-0149-3; http://arxiv.org/abs/hep-th/0512164

287. Losev, A., Nekrasov, N., Shatashvili, S.L.: The Freckled instantons. http://arxiv.org/abs/hep-th/9908204

288. Lutken, C., Ross, G.G.: Taxonomy of heterotic superconformal field theories. Phys. Lett. **B213**, 152 (1988). http://dx.doi.org/10.1016/0370-2693(88)91016-7

289. Martinec, E.J.: Algebraic geometry and effective lagrangians. Phys. Lett. **B217**, 431 (1989)

290. McDuff, D., Salamon, D.: Introduction to Symplectic Topology. Oxford Graduate Texts in Mathematics, 3rd edn. Oxford University Press, Oxford (2017). https://doi.org/10.1093/oso/9780198794899.001.0001

291. McOrist, J.: On the effective field theory of heterotic vacua. Lett. Math. Phys. **108**(4), 1031–1081 (2018). http://dx.doi.org/10.1007/s11005-017-1025-0; http://arxiv.org/abs/1606.05221

292. McOrist, J., Melnikov, I.V.: Summing the instantons in half-twisted linear sigma models. J. High Energy Phys. **02**, 026 (2009). http://arxiv.org/abs/0810.0012

293. McOrist, J., Melnikov, I.V.: Old issues and linear sigma models. Adv. Theor. Math. Phys. **16**, 251–288 (2012). http://arxiv.org/abs/1103.1322

294. Melnikov, I.V.: (0,2) Landau-Ginzburg models and residues. J. High Energy Phys. **09**, 118 (2009). http://arxiv.org/abs/0902.3908

295. Melnikov, I.V., Minasian, R.: Heterotic sigma models with N=2 space-time supersymmetry. J. High Energy Phys. **1109**, 065 (2011). http://dx.doi.org/10.1007/JHEP09(2011)065; http://arxiv.org/abs/1010.5365

296. Melnikov, I.V., Plesser, M.R.: The Coulomb branch in gauged linear sigma models. J. High Energy Phys. **0506**, 013 (2005). http://dx.doi.org/10.1088/1126-6708/2005/06/013; http://arxiv.org/abs/hep-th/0501238

297. Melnikov, I.V., Plesser, M.R.: A-model correlators from the Coulomb branch. J. High Energy Phys. **02**, 044 (2006). http://arxiv.org/abs/hep-th/0507187

298. Melnikov, I.V., Plesser, M.R.: A (0,2) mirror map. J. High Energy Phys. **1102**, 001 (2011). http://dx.doi.org/10.1007/JHEP02(2011)001; http://arxiv.org/abs/1003.1303

299. Melnikov, I.V., Sharpe, E.: On marginal deformations of (0,2) non-linear sigma models. Phys. Lett. **B705**, 529–534 (2011). http://dx.doi.org/10.1016/j.physletb.2011.10.055; http://arxiv.org/abs/1110.1886

300. Melnikov, I.V., Quigley, C., Sethi, S., Stern, M.: Target spaces from chiral gauge theories. J. High Energy Phys. **1302**, 111 (2013). http://dx.doi.org/10.1007/JHEP02(2013)111; http://arxiv.org/abs/1212.1212

301. Melnikov, I.V., Minasian, R., Theisen, S.: Heterotic flux backgrounds and their IIA duals. J. High Energy Phys. **07**, 023 (2014). http://dx.doi.org/10.1007/JHEP07(2014)023; http://arxiv.org/abs/1206.1417

302. Melnikov, I.V., Minasian, R., Sethi, S.: Heterotic fluxes and supersymmetry. J. High Energy Phys. **06**, 174 (2014). http://dx.doi.org/10.1007/JHEP06(2014)174; http://arxiv.org/abs/1403.4298

303. Melnikov, I.V., Minasian, R., Sethi, S.: Spacetime supersymmetry in low-dimensional perturbative heterotic compactifications. http://arxiv.org/abs/1707.04613

304. Mermin, N.D., Wagner, H.: Absence of ferromagnetism or antiferromagnetism in one-dimensional or two-dimensional isotropic Heisenberg models. Phys. Rev. Lett. **17**, 1133–1136 (1966). http://dx.doi.org/10.1103/PhysRevLett.17.1133

305. Michelsohn, M.L.: On the existence of special metrics in complex geometry. Acta Math. **149**(3–4), 261–295 (1982)

306. Mikhalkin, G.: Amoebas of algebraic varieties and tropical geometry. In: Different Faces of Geometry. International Mathematical Series (New York), vol. 3. Kluwer/Plenum, New York (2004). http://dx.doi.org/10.1007/0-306-48658-X_6; http://arxiv.org/abs/math/0403015

307. Milnor, J.W., Stasheff, J.D.: Characteristic Classes. Annals of Mathematics Studies, vol. 76. Princeton University Press, Princeton (1974)

308. Moore, G.W., Nelson, P.C.: The etiology of sigma model anomalies. Commun. Math. Phys. **100**, 83 (1985). http://dx.doi.org/10.1007/BF01212688

309. Moore, G.W., Seiberg, N.: Classical and quantum conformal field theory. Commun. Math. Phys. **123**, 177 (1989). http://dx.doi.org/10.1007/BF01238857

310. Moroianu, A.: Lectures on Kähler Geometry, vol. 69. Cambridge University Press, Cambridge (2007)

311. Morrison, D.R., Plesser, M.R.: Summing the instantons: quantum cohomology and mirror symmetry in toric varieties. Nucl. Phys. **B440**, 279–354 (1995). http://arxiv.org/abs/hep-th/9412236

312. Morrison, D.R., Plesser, M.R.: Towards mirror symmetry as duality for two dimensional abelian gauge theories. Nucl. Phys. Proc. Suppl. **46**, 177–186 (1996). http://arxiv.org/abs/hep-th/9508107

313. Mumford, D., Fogarty, J., Kirwan, F.: Geometric Invariant Theory. Ergebnisse der Mathematik und ihrer Grenzgebiete (2) [Results in Mathematics and Related Areas (2)], vol. 34, 3rd edn. Springer, Berlin (1994)

314. Nakayama, Y.: Scale invariance vs conformal invariance. Phys. Rep. **569**, 1–93 (2015). http://dx.doi.org/10.1016/j.physrep.2014.12.003; http://arxiv.org/abs/1302.0884

315. Nekrasov, N.A.: Lectures on curved beta-gamma systems, pure spinors, and anomalies. http://arxiv.org/abs/hep-th/0511008

316. Nemeschansky, D., Sen, A.: Conformal invariance of supersymmetric sigma models on Calabi-Yau manifolds. Phys. Lett. **B178**, 365 (1986). http://dx.doi.org/10.1016/0370-2693(86)91394-8

317. Nibbelink, S.G.: Heterotic orbifold resolutions as (2,0) gauged linear sigma models. Fortschr. Phys. **59**, 454–493 (2011). http://dx.doi.org/10.1002/prop.201100002; http://arxiv.org/abs/1012.3350

318. Nibbelink, S.G., Horstmeyer, L.: Super Weyl invariance: BPS equations from heterotic worldsheets. http://arxiv.org/abs/1203.6827

319. Orlov, D.: Triangulated categories of singularities and D-branes in Landau-Ginzburg models. Trudy Steklov Mat. Inst. **246**, 240–262 (2004). http://arxiv.org/abs/math/0302304

320. Osborn, H.: Local renormalization group equations in quantum field theory. In: 2nd JINR Conference on Renormalization Group Dubna, USSR, 3–6 Sept 1991, pp. 128–138

321. Osborn, H.: Weyl consistency conditions and a local renormalization group equation for general renormalizable field theories. Nucl. Phys. **B363**, 486–526 (1991). http://dx.doi.org/10.1016/0550-3213(91)80030-P

322. Pappadopulo, D., Rychkov, S., Espin, J., Rattazzi, R.: OPE convergence in conformal field theory. Phys. Rev. **D86**, 105043 (2012). http://dx.doi.org/10.1103/PhysRevD.86.105043; http://arxiv.org/abs/1208.6449

323. Payne, S.: Moduli of toric vector bundles. Compos. Math. **144**(5), 1199–1213 (2008)

324. Pestun, V., et al.: Localization techniques in quantum field theories. J. Phys. **A50**(44), 440301 (2017). http://dx.doi.org/10.1088/1751-8121/aa63c1; http://arxiv.org/abs/1608.02952

325. Poland, D., Rychkov, S., Vichi, A.: The conformal bootstrap: theory, numerical techniques, and applications. http://arxiv.org/abs/1805.04405

326. Polchinski, J.: Scale and conformal invariance in quantum field theory. Nucl. Phys. **B303**, 226 (1988). http://dx.doi.org/10.1016/0550-3213(88)90179-4

327. Polchinski, J.: String Theory, Volume 2. Cambridge University Press, Cambridge (1998)

328. Polchinski, J.: String Theory. Volume 1: An Introduction to the Bosonic String. Cambridge University Press, Cambridge (2007)

329. Quigley, C., Sethi, S.: Linear sigma models with torsion. J. High Energy Phys. **1111**, 034 (2011). http://arxiv.org/abs/1107.0714

330. Quigley, C., Sethi, S., Stern, M.: Novel branches of (0,2) theories. J. High Energy Phys. **1209**, 064 (2012). http://arxiv.org/abs/1206.3228

331. Ramond, P., Schwarz, J.H.: Classification of dual model gauge algebras. Phys. Lett. **B64**, 75 (1976). http://dx.doi.org/10.1016/0370-2693(76)90361-0

332. Reid, M.: Young person's guide to canonical singularities. In: Algebraic Geometry, Bowdoin, 1985 (Brunswick, Maine, 1985). Proceedings of Symposia in Pure Mathematics, vol. 46, pp. 345–414. American Mathematical Society, Providence (1987)

333. Rocek, M., Verlinde, E.P.: Duality, quotients, and currents. Nucl. Phys. **B373**, 630–646 (1992). http://dx.doi.org/10.1016/0550-3213(92)90269-H; http://arxiv.org/abs/hep-th/9110053

334. Rohm, R., Witten, E.: The antisymmetric tensor field in superstring theory. Ann. Phys. **170**, 454 (1986). http://dx.doi.org/10.1016/0003-4916(86)90099-0

335. Sagnotti, A.: A note on the Green-Schwarz mechanism in open string theories. Phys. Lett. **B294**, 196–203 (1992). http://dx.doi.org/10.1016/0370-2693(92)90682-T; http://arxiv.org/abs/hep-th/9210127

336. Salamon, S.M.: Hermitian geometry. In: Invitations to Geometry and Topology. Oxford Graduate Texts in Mathematics, vol. 7, pp. 233–291. Oxford University Press, Oxford (2002)

337. Samelson, H.: A class of complex-analytic manifolds. Port. Math. **12**, 129–132 (1953)

338. Schafer-Nameki, S., Weigand, T.: F-theory and 2d (0, 2) theories. J. High Energy Phys. **05**, 059 (2016). http://dx.doi.org/10.1007/JHEP05(2016)059; http://arxiv.org/abs/1601.02015

339. Schottenloher, M.: A Mathematical Introduction to Conformal Field Theory. Lect. Notes Phys. 759, 1–237 (2008). http://dx.doi.org/10.1007/978-3-540-68628-6

340. Schwimmer, A., Seiberg, N.: Comments on the N=2, N=3, N=4 superconformal algebras in two-dimensions. Phys. Lett. **B184**, 191 (1987). http://dx.doi.org/10.1016/0370-2693(87)90566-1

341. Seiberg, N.: Electric - magnetic duality in supersymmetric nonAbelian gauge theories. Nucl. Phys. **B435**, 129–146 (1995). http://arxiv.org/abs/hep-th/9411149

342. Seiberg, N., Tachikawa, Y., Yonekura, K.: Anomalies of duality groups and extended conformal manifolds. http://arxiv.org/abs/1803.07366

343. Sen, A.: (2, 0) supersymmetry and space-time supersymmetry in the heterotic string theory. Nucl. Phys. **B278**, 289 (1986)

344. Sen, A.: Supersymmetry restoration in superstring perturbation theory. J. High Energy Phys. **12**, 075 (2015). http://dx.doi.org/10.1007/JHEP12(2015)075; http://arxiv.org/abs/1508.02481

345. Sen, K., Tachikawa, Y.: First-order conformal perturbation theory by marginal operators. http://arxiv.org/abs/1711.05947

346. Sevrin, A., Troost, W., Van Proeyen, A., Spindel, P.: Extended supersymmetric sigma models on group manifolds. 2. Current algebras. Nucl. Phys. **B311**, 465 (1988). http://dx.doi.org/10.1016/0550-3213(88)90070-3

347. Sharpe, E.: Notes on certain other (0,2) correlation functions. http://arxiv.org/abs/hep-th/0605005

348. Shatashvili, S.L., Vafa, C.: Superstrings and manifold of exceptional holonomy. Sel. Math. **1**, 347 (1995). http://dx.doi.org/10.1007/BF01671569; http://arxiv.org/abs/hep-th/9407025

349. Silverstein, E., Witten, E.: Global U(1) R symmetry and conformal invariance of (0,2) models. Phys. Lett. **B328**, 307–311 (1994). http://arxiv.org/abs/hep-th/9403054

350. Silverstein, E., Witten, E.: Criteria for conformal invariance of (0,2) models. Nucl. Phys. **B444**, 161–190 (1995). http://arxiv.org/abs/hep-th/9503212

351. Spindel, P., Sevrin, A., Troost, W., Van Proeyen, A.: Extended supersymmetric sigma models on group manifolds. 1. The complex structures. Nucl. Phys. **B308**, 662 (1988)

352. Strominger, A.: Superstrings with torsion. Nucl. Phys. **B274**, 253 (1986)

353. Strominger, A.: Special geometry. Commun. Math. Phys. **133**, 163–180 (1990)

354. Sturmfels, B.: Solving systems of polynomial equations. In: Regional Conference Series in Mathematics, vol. 97. American Mathematical Society, Providence (2002)

355. Szenes, A., Vergne, M.: Toric reduction and a conjecture of Batyrev and Materov. Invent. Math. **158**(3), 453–495 (2004)

356. Tachikawa, Y.: N=2 Supersymmetric Dynamics for Pedestrians, vol. 890. Springer, Berlin (2014). http://dx.doi.org/10.1007/978-3-319-08822-8; http://arxiv.org/abs/1312.2684

357. Tan, M.-C.: Two-dimensional twisted sigma models and the theory of chiral differential operators. Adv. Theor. Math. Phys. **10**, 759–851 (2006). http://arxiv.org/abs/hep-th/0604179

358. Tan, M.-C., Yagi, J.: Chiral algebras of (0,2) sigma models: beyond perturbation theory. Lett. Math. Phys. **84**, 257–273 (2008). http://arxiv.org/abs/0801.4782

359. Taubes, C.H.: Self-dual Yang-Mills connections on non-self-dual 4-manifolds. J. Differ. Geom. **17**(1), 139–170 (1982)

360. Taylor, W.: TASI lectures on supergravity and string vacua in various dimensions. http://arxiv.org/abs/1104.2051

361. Teschner, J.: Liouville theory revisited. Class. Quantum Gravity **18**, R153–R222 (2001). http://dx.doi.org/10.1088/0264-9381/18/23/201; http://arxiv.org/abs/hep-th/0104158

362. The Sage Developers: SageMath, the Sage Mathematics Software System (Version 8.2) (2018). http://www.sagemath.org

363. The Stacks Project Authors: Stacks project. http://stacks.math.columbia.edu (2018)

364. Tian, G.: Smoothness of the universal deformation space of compact Calabi-Yau manifolds and its Petersson-Weil metric. In: Mathematical Aspects of String Theory (San Diego, California, 1986). Advanced Series in Mathematical Physics, vol. 1, pp. 629–646. World Scientific Publishing, Singapore (1987)

365. Todorov, A.: Weil-Petersson volumes of the moduli spaces of CY manifolds. Commun. Anal. Geom. **15**(2), 407–434 (2007). http://dx.doi.org/10.4310/CAG.2007.v15.n2.a8

366. Tong, D.: Quantum vortex strings: a review. Ann. Phys. **324**, 30–52 (2009). http://dx.doi.org/10.1016/j.aop.2008.10.005; http://arxiv.org/abs/0809.5060

367. Tseytlin, A.A.: σ model Weyl invariance conditions and string equations of motion. Nucl. Phys. **B294**, 383–411 (1987). http://dx.doi.org/10.1016/0550-3213(87)90588-8

368. Tsikh, A., Yger, A.: Residue currents. Complex analysis. J. Math. Sci. (N. Y.) **120**(6), 1916–1971 (2004)

369. Uhlenbeck, K., Yau, S.: On the existence of Hermitian-Yang-Mills connections in stable vector bundles. Commun. Pure Appl. Math. **39**, S257–S293 (1986)

370. Vafa, C.: String vacua and orbifoldized L-G models. Mod. Phys. Lett. **A4**, 1169 (1989)

371. Vafa, C.: Topological Landau-Ginzburg models. Mod. Phys. Lett. **A6**, 337–346 (1991)

372. Vafa, C., Warner, N.P.: Catastrophes and the classification of conformal theories. Phys. Lett. **B218**, 51 (1989)

373. Voisin, C.: Hodge Theory and Complex Algebraic Geometry. I and II. Cambridge Studies in Advanced Mathematics. Cambridge University Press, Cambridge (2007)

374. Wang, H.-C.: Closed manifolds with homogeneous complex structure. Am. J. Math. **76**, 1–32 (1954)

375. Weinberg, S.: The Quantum Theory of Fields, vol. 2. Cambridge University Press, Cambridge (1996)

376. West, P.: Introduction to Supersymmetry and Supergravity. World Scientific, Singapore (1990)

377. Wilson, P.M.H.: Erratum: "The Kähler cone on Calabi-Yau threefolds" [Invent. Math. **107** (1992), no. 3, 561–583; MR1150602 (93a:14037)]. Invent. Math. **114**(1), 231–233 (1993)

378. Witten, E.: θ vacua in two-dimensional quantum chromodynamics. Nuovo Cim. **A51**, 325 (1979). http://dx.doi.org/10.1007/BF02776593

379. Witten, E.: Constraints on supersymmetry breaking. Nucl. Phys. **B202**, 253 (1982)

380. Witten, E.: Supersymmetry and Morse theory. J. Differ. Geom. **17**, 661–692 (1982)

381. Witten, E.: Nonabelian bosonization in two dimensions. Commun. Math. Phys. **92**, 455–472 (1984)

382. Witten, E.: Global gravitational anomalies. Commun. Math. Phys. **100**, 197 (1985). http://dx.doi.org/10.1007/BF01212448

383. Witten, E.: Global anomalies in string theory. In: Bardeen, W.A. (ed.) Argonne Symposium on Geometry, Anomalies and Topology. Argonne, Lemont (1985)

384. Witten, E.: New issues in manifolds of SU(3) holonomy. Nucl. Phys. **B268**, 79 (1986)

385. Witten, E.: Topological quantum field theory. Commun. Math. Phys. **117**, 353 (1988). http://dx.doi.org/10.1007/BF01223371

386. Witten, E.: Topological sigma models. Commun. Math. Phys. **118**, 411 (1988)

387. Witten, E.: Introduction to cohomological field theories. Int. J. Mod. Phys. **A6**, 2775–2792 (1991)

388. Witten, E.: Phases of N = 2 theories in two dimensions. Nucl. Phys. **B403**, 159–222 (1993). http://arxiv.org/abs/hep-th/9301042

389. Witten, E.: On the Landau-Ginzburg description of N=2 minimal models. Int. J. Mod. Phys. **A9**, 4783–4800 (1994). http://arxiv.org/abs/hep-th/9304026

390. Witten, E.: Mirror manifolds and topological field theory. http://arxiv.org/abs/hep-th/9112056

391. Witten, E.: Superstring perturbation theory revisited. http://arxiv.org/abs/1209.5461

392. Witten, E.: The Verlinde algebra and the cohomology of the Grassmannian. http://arxiv.org/abs/hep-th/9312104

393. Witten, E.: Two-dimensional models with (0,2) supersymmetry: perturbative aspects. http://arxiv.org/abs/hep-th/0504078

394. Yagi, J.: Chiral algebras of (0,2) models. http://arxiv.org/abs/1001.0118

395. Yau, S.T.: On the Ricci curvature of a compact Kähler manifold and the complex Monge-Ampère equation. I. Commun. Pure Appl. Math. **31**(3), 339–411 (1978). https://doi.org/10.1002/cpa.3160310304

396. Yau, S.-T.: A survey of Calabi-Yau manifolds. In: Surveys in Differential Geometry. Vol.
     XIII. Geometry, Analysis, and Algebraic Geometry: Forty Years of the Journal of Differential
     Geometry. Surveys in Differential Geometry, vol. 13, pp. 277–318. Internatinal Press,
     Somerville (2009). http://dx.doi.org/10.4310/SDG.2008.v13.n1.a9
397. Zamolodchikov, A.: Conformal symmetry and multicritical points in two-dimensional quan-
     tum field theory (in Russian). Sov. J. Nucl. Phys. **44**. 529–533 (1986)
398. Zamolodchikov, A.B.: Renormalization group and perturbation theory near fixed points in
     two-dimensional field theory. Sov. J. Nucl. Phys. **46**, 1090 (1987) [Yad. Fiz.46,1819(1987)]
399. Zumino, B.: Supersymmetry and Kahler manifolds. Phys. Lett. **87B**, 203 (1979). http://dx.
     doi.org/10.1016/0370-2693(79)90964-X

Printed in the United States
By Bookmasters